Green Petroleum

Scrivener Publishing
100 Cummings Center, Suite 41J
Beverly, MA 01915-6106

Publishers at Scrivener
Martin Scrivener (martin@scrivenerpublishing.com)
Phillip Carmical (pcarmical@scrivenerpublishing.com)

Green Petroleum

How Oil and Gas Can Be Environmentally Sustainable

Edited by
M.R. Islam, M.M. Khan, and A.B. Chhetri

Co-published by John Wiley & Sons, Inc. Hoboken, New Jersey, and Scrivener Publishing LLC, Salem, Massachusetts.
Published simultaneously in Canada.

For general information on our other products and services or for technical support, please contact our Customer Care Department within the United States at (800) 762-2974, outside the United States at (317) 572-3993 or fax (317) 572-4002.

Wiley also publishes its books in a variety of electronic formats. Some content that appears in print may not be available in electronic formats. For more information about Wiley products, visit our web site at www.wiley.com.

For more information about Scrivener products please visit www.scrivenerpublishing.com.

Cover design by Russell Richardson

Library of Congress Cataloging-in-Publication Data:

Islam, Rafiqul, 1959-
Green petroleum: how oil and gas can be environmentally sustainable / M.R. Islam. – 1st ed.
p. cm.
Includes bibliographical references and index.
ISBN 978-1-118-07216-5 (hardback)
1. Petroleum engineering. 2. Gas engineering. 3. Petroleum engineering–Environmental aspects. 4. Sustainable engineering. 5. Environmental protection. I. Title.
TN870.I834 2012
665.5028′6–dc23

2012007304

ISBN 978-1-118-07216-5

Printed in the United States of America

10 9 8 7 6 5 4 3 2 1

Contents

Preface

An earlier volume entitled *The Greening of Petroleum Operations* (Wiley-Scrivener, 2011), by the same group of authors as the present work, has been well-received in various sectors of the contemporary engineering profession. At the same time, its appeal is inherently limited by the required principal focus on the concerns and practical preoccupations of working petroleum engineers and the postgraduate-level research community. In the current volume, the authors recap the main lines of argument of that earlier work, paying greater attention to its general line and method. With a view to further enriching the formation and developing outlook of undergraduate engineering students, the authors have accordingly adjusted the focus away from the detail about the actual engineering and its associated analytical methods, tests, and criteria, and more onto the broad principles of contemporary scientific investigation and data analysis.

We believe this to be a worthy development, and welcome this opportunity to say why we think this is so, and so necessary at this particular time. To illuminate the point, a Google search was conducted on the phrase "medical science versus old-wives'-tales and home cures." The overwhelming majority of "hits" argued thus: old-wives'-tales and home remedies have long been a hit-or-miss affair because they largely lack the backing of "sound" modern science, yet somehow, despite everything, some of these seem actually to work... FULL STOP! That is, while the vast majority of hits grudgingly acknowledge some truth lurking in "old-wives'-tales," this is coupled with a *pro forma* ritual obeisance to the "authority" of "modern science" and its "methods." At the same time, however, nothing, absolutely nothing, is said or even hinted about redirecting deployment of any of these hallowed "modern scientific methods" to establish why some of these home remedies and old-wives'-tales in particular could possibly work, and establishing exactly why "modern" science could have missed seeing this before. Thus, what we actually have here is *not* the joining of a common search for knowledge of the truth between forces armed with very different kinds of "authority," but something else actually quite disturbing. What we have here on display are voices echoing the highest levels of the medical-scientific establishment desperate to maintain not only their grip but especially their aura of superiority and "authority" over all other approaches to knowledge. Instead of deploying scientific and consensus-building methods to resolve the contradictions and move society forward, these differences are quasi-criminalized as some kind of threat to established authority, hyped and distorted to the point of anarchy, and this anarchy is then raised to the level of authority by those elements from the establishment that feel the threat most palpably: "you are either with us, or with the forces of Darkness and Ignorance..."

In the current Information Age, it has become more transparent and more undeniable than ever before that the truth-content of knowledge has nothing to do with so-called authority. The essence of any deed, including the deed of the kinds of research brought together in these and companion volumes of this series, is intention. Hewing closely to that principle, the authors of these volumes explicitly repudiate the idea that the products of honestly-intended scientific investigation must ever be the monopoly of some aristocracy, "natural" or otherwise, of intellectual or other talent. Let the most rabid defenders and perpetrators of "New Science" prejudice continue their hollow rejections and dismissals of everything scientific learned before Newton as "fanciful," "speculative," and "inexact". In the sheer breadth of sources and insights drawn upon, ranging from the social and natural sciences, history and even religion, this volume and its predecessor serve to demonstrate why the merited insights of all honestly-intended investigations into the natural order, without reference to the social or political or intellectual standing of their source, must always be welcomed by all those engaged in the quest for meaningful knowledge.

G.V. Chilingar, University of Southern California, USA
H. Vaziri, BP America, USA
G. Zatzman, EEC Research Organisation, Canada

1

Introduction

1.1 'Greening': What does it Entail?

The devastating cultural impacts of extremely fundamental ideas start to register very loudly and sharply at moments of massive crisis such as the present. This is true with environmental concerns, and it is even truer in economics, which is the focus of the post-Renaissance modern world. However, few recognize those impacts, and even fewer understand from where the impacts originate. In this process, the entire world either continues to ignore the impacts, or does not see or hear these impacts coming.

In today's world, focused on tangible and externals, greening is often synonymous with coloring an object green. If this were the case, greening of petroleum operations would simply mean painting gas stations green (this has actually been done, as easily evidenced in the rush to paint green). In order that "sustainability" or "green" status be deemed real and provided with a scientific basis, there are criteria to be set forth and met. Nowhere is this need clearer or more crucial than in the characterization of energy sources. Petroleum fuels being the principal driver of today's energy needs, one must establish the role of petroleum fluids in the overall energy picture. One must not forget that petroleum products (crude oil and gas) are natural resources and they cannot be inherently unsustainable, unless there is an inherent problem with nature. Here, it is crucial to understand nature and how nature operates, without any preconceived bias. This book is based on the premise that true green exists in nature and we must emulate nature in order to develop technologies that are green. In this, we cannot rely on what has happened in the modern age in terms of technology development. After all, none other than Nobel Laureate chemists have designated our time as a 'technological disaster'. We do not have to wait for the 'Occupy Wall street' folks to tell us there is a need for change.

1.2 The Science of Change: How will our Epoch be Remembered?

Energy policies have defined our modern civilization. Politicizing energy policies is nothing new, but bipartisan bickering is new for the Information Age. The overwhelming theme is "change" (similar to the term "paradigm shift"), and both sides of the "change" debate remain convinced that the other party is promoting a flat-earth theory. One side supports petroleum production and usage, and the other side supports the injection of various "other" energy sources, including nuclear, wind, and solar. This creates consequences for scientific study. The petroleum industry faces the temptation of siding with the group that promotes petroleum production and continued usage with only cosmetic change to the energy consumption side, namely in the form of "energy saving" utilities. The other side, of course, has a vested interest in opposing this move and spending heavily on infrastructure development using renewable energy sources. Both sides seem to agree on one thing: there is no sustainable solution to the energy crisis, and the best we can do is to minimize the economic and environmental downfall. This book shows, with scientific arguments, that both sides are wrong, and there is indeed a sustainable solution to petroleum production and operations. With the proposed schemes, not only would the decline of the economic and environmental conditions be arrested, but one could improve both these conditions, launching our civilization onto an entirely new path.

This book is about scientific change that we can believe in. This is not about repeating the same doctrinal lines that got us in this modern-day "technological disaster" mode (in the word of Nobel Laureate chemist Robert Curl). The science of true change is equated with the science of sustainability. This change is invoked by introducing both a natural source and a natural pathway. This book summarizes an essential, critical distinction between the outcomes of natural processes and the outcomes of engineered processes that conventional science discourse and work have either missed or dismissed. In contrast to what defines a change in a natural process, the outcomes of engineered processes can change if there is a change in only the source or only along the pathway, and there may be no net change in outcome if changes at the source cancel out changes along the pathway, or vice-versa.

Today, the entire focus has been on the source (crude oil in petroleum engineering), and the role of the pathway has been completely misunderstood or deliberately ignored. Numerous schemes are being presented as sustainable alternatives (sustainable because the source has been replaced with another source while keeping the process intact). This mode of cognition has been a very typical philosophy for approximately the last 900 years and has many applications in other disciplines, including mathematics (i.e., theory of chaos). This book deconstructs this philosophy and presents scientific analysis that involves both the source and the pathway. As a result, all the analyses are consistent with the first premise, and no question remains unanswered.

1.3 Are Natural Resources Finite and Human Needs Infinite?

Over a decade ago, Lawrence Lerner, Professor Emeritus in Physics and Astronomy at the University of Chicago, was asked to evaluate how Darwin's theory of evolution was being taught in each state of the United States (Lerner 2000). In addition to his attempt to find a standard in K-12 teaching, he made some startling revelations. His recommendations created controversy, and many suggested that he was promoting "bad science" in name of "good science." However, no one singled out another aspect of his findings. He observed that "some Native American tribes consider that their ancestors have lived in the traditional tribal territories forever." He then equated "forever" with "infinity" and continued his comment stating, "Just as the fundamentalist creationists underestimate the age of the earth by a factor of a million or so, the Black Muslims overestimate by a thousand-fold and the Indians are off by a factor of infinity." (Lerner 2001). This confusion between "forever" and "infinity" is not new in modern European culture. In the words of Albert Einstein, "There are two things that are infinite, human stupidity and the Universe, and I am not so sure about the Universe." Even though the word "infinity" emerges from a Latin word, *infinitas*, meaning "unboundedness," for centuries this word has been applied in situations in which it promotes absurd concepts. In Arabic, the equivalent word means "never-ending." In Sanskrit, similar words exist, and those words are never used in mathematical terms as a number. This use of infinity to enumerate something (e.g., infinite number of solutions) is considered to be absurd in other cultures.

Nature is infinite – in the sense of being all-encompassing – within a closed system. Somewhat paradoxically, nature as a system is closed in the sense of being self-closing. This self-closure property has two aspects. First, everything in a natural environment is used. Absent anthropogenic interventions, conditions of net waste or net surplus would not persist for any meaningful period of time. Secondly, nature's closure system operates without the benefit of, or dependence upon, any internal or external boundaries. Because of this infinite dimension, we may deem nature, considered in net terms as a system overall, to be perfectly balanced. Of course, within any arbitrarily selected finite time period, any part of a natural system may appear out of balance. However, to look at nature's system without acknowledging all of the subtle dependencies that operate at any given moment introduces a bias that distorts any conclusion that is asserted on the basis of such a narrow approach.

From where do the imbalance and unsustainability that seem so ubiquitously manifest in the atmosphere, the soil, and the oceans originate? As the "most intelligent creation of nature," men were expected to at least stay out of the natural ecosystem. Einstein might have had doubts about human intelligence or the infinite nature of the Universe, but human history tells us that human beings have always managed to rely on the infinite nature of nature. From Central American Mayans to Egyptian Pharaohs, from Chinese Hans to the Mannaeans of Persia, and from the Edomites of the Petra

Valley to the Indus Valley civilization of the Asian subcontinent, all managed to remain in harmony with nature. They were not necessarily free from practices that we no longer consider (Pharaohs sacrificed humans to accompany the dead royal for the resurrection day), but they did not produce a single gram of an inherently anti-nature product, such as DDT. In modern times, we have managed to give a Nobel Prize (in medicine) for that invention.

The first few chapters of this book examine how our ancestors dealt with energy needs and the knowledge they possessed that is absent in today's world. Regardless of the technology these ancient civilizations lacked that many might look for today, our ancestors were concerned with not developing technologies that might undo or otherwise threaten the perceived balance of nature that, today, seems desirable and worth emulating. Nature remains and will remain truly sustainable.

1.4 The Standard of Sustainable Engineering

Early in the twentieth century, alcohol was placed under Prohibition in the United States, even for medicinal purposes. Today, the most toxic and addictive form of alcohol is not only permitted, but is promoted as a part of a reputedly "refined" life style. Only about four to six generations ago, in the mid- to late-19th century, interracial marriages and marriages between cousins were forbidden (some still are, e.g., Paul and Spencer 2008), women and African-Americans did not have the right to vote in elections, and women (after marriage) and slaves (after sale) were required to change their surname and identity. In many parts of rural Quebec, well into the 20th century, women were required to replace all their teeth with a denture as a gift to the groom. Today, as part of the reaction to the extreme backwardness of these reactionary social practices, same-sex marriage is allowed in Canada, much of the United States, and Europe. Marriage among siblings is even allowed in some "enlightened" parts of Europe, and changing one's surname has become a sign of backwardness. Although the religious establishment's various sanctions surrounding these relationships – not to mention the status of these various relations themselves – have actually changed very little, a vast propaganda was loosed nonetheless upon the world, proclaiming the alleged modernization of all human and social relations represented by such "reversals." However, all that has "changed" is the standard as to what is acceptable. Similarly, about one to two generations ago, organic food was still the most abundant and most affordable food. Then, along came the notorious "Green Revolution," fostered mainly in developing countries by U.S.-based agribusiness interests often acting through governments. "Productivity" reportedly doubled and tripled in less than a decade. Today, organic food in general costs three times more (200% increase) than non-organic. In this process, the actual quality of the food declined. Yet, the standard had been shifted again, rendering possible an extensive widening of profit margins in the most powerfully positioned sectors of food production and distribution. When, where, and how does such a reversal in the trend of quality and pricing start, like the reversal in the trend of the quality of certain social relations and the value placed on them? In either case, investigating and establishing true sustainability entails a deep analysis of the entire matter of what constitutes a standard, what social forces are in a position to shift standards, how the process of rewriting standards operates, and where and when may people intervene to empower themselves and put an end to being victimized in such processes.

Chapter 3 discusses and discloses the problem inherent in the standards that we use today – standards or ideals that are not natural. Chapter 4 explains that, by forcing a non-natural standard or ideal in all engineering calculations, all subsequent conclusions are falsified. This chapter also makes it clear that with New Science we are simply incapable of getting of the current 'environmental disaster'.

Nature exists in a state of dynamic balance, both in space and time. In its attempts to comprehend fundamental changes of state within a natural environment, the conventional science of tangibles hits a wall, especially when it comes to the treatment of time's role at such bifurcation points. Why? Such a situation follows from the fact that the actual rate at which time unfolds at such points within a natural process is itself part of that process. This means that time cannot be treated as varying independently of that process. Very much akin to the problem of standards falling under the dictate of special, usually private, interests, the mathematics used by the science of tangibles becomes hijacked. For centuries, mathematics warned its users about the falsehoods that will arise when differentiating a discontinuous function as though it were continuous, or integrating over a region of space or time that is discontinuous as though it were continuous. In practice, meanwhile, pragmatism has often prevailed, and the reality of a natural system's output often bearing little or no relationship to what the theoretical mathematical model predicted is treated as an allowable source of error. However, what could be expected to eventuate if the standpoint of the science of intangibles, based on a conception of nature's system as one that is perfect (in the sense of complete and self-contained), were adopted instead? Many of these howling contradictions that emerge from retaining the conventional science of tangibles in areas where its modeling assumptions no longer apply would turn out to be removable paradoxes.

The conundrum arises in the first place mainly (and/or only) because the tangible aspects of any phenomenon do not go beyond the very small element in space, i.e., $\Delta s \to 0$, and even a smaller element in time, i.e., $\Delta t = 0$ (meaning, time t = "right now"). Within the space-time of a purely mathematical universe, Newton's calculus gives reliable answers concerning the derivative of a function based on taking the limit of the difference quotient of the function as change in any selected variable of the said function approaches zero. However, regardless of that fact, is the underlying first assumption correct? That is to say, what fit may be expected between the continuity of processes in a natural system and the continuity of mathematical space time that undergirds whether we can even speak of a function's derivative? In general, the results in the mathematical reality and the natural reality don't match, at least not without "fudging" of some kind. Is it reasonable to consign such a mismatch to "error," or is something else at work here? The authors believe the incompatibility has a deeper source, namely in an insurmountable incompatibility between the continuum of mathematical space-time and the essentially dynamic balance of natural systems.

Since the dawn of the Industrial Revolution, the only models used and developed continually have been based on what we characterize in this work as the science of tangibles. This book reviews the outcome of these models as manifested in the area of energy management. The prejudicial components of "steady state" based analysis and assumptions have begun to emerge in their true light mostly as an unintended byproduct of the rise of the Information Age. From this perspective, it becomes possible to clarify how and why modeling anything in nature in terms of a steady state has become unsustainable.

The unexpected fallout that we are ascribing to the emergence of the Information Age is simply this: in light of the undreamt-of expansion in the capacity to gather, store, and manipulate unprecedented quantities of data on anything, the science of tangibles calling itself "New Science," that developed out of the European Renaissance has turned out to be a double-edged sword. All its models are based on the short term, so short that they practically eliminate the time dimension (equivalent to assigning $\Delta t = 0$). However, these models are promoted as "steady state" models with the assertion that, as Δt approaches ∞, a steady state is reached. This syllogism is based on two false premises: (1) that there is such a state as steady state and (2) that nature is never in balance.

By proceeding according to a perspective that accepts and embraces the inherent overall dynamic balance of natural systems as given, it soon emerges that all these models are inherently flawed and are primarily responsible for transforming the truth into falsehood. That is because their continued promotion obscures key differences between real (natural) and artificial (created by violating natural process). Models based on steady-state have been developed and promoted by all the great names of natural and social science over the last 400 years, from Sir Isaac Newton and Lord Kelvin to the economist John Maynard Lord Keynes. However, although presented as the only acceptable bridging transition from natural science to engineering, all such models are in fact freighted with the enormous baggage of a Eurocentric cultural bias. A most glaring feature of technological development derived on the basis of this "steady state" bridging transition from theory to practice has been its denaturing of how time actually operates, reducing the meaningful sense of time to whatever exists "right now."

Thus, for example, in medical science this has strengthened the tendency to treat symptoms first and worry about understanding how disease actually works later. In economic development, it amounts to increasing wasteful habits in order to increase GDP. In business, it amounts to maximizing quarterly income even if it means resorting to corruption. In psychology, it means maximizing pleasure and minimizing pain (both in the short-term). In politics, it amounts to obliterating the history of a nation or a society. In mathematics, it means obsessions with numbers and exact (and unique) solutions. In technology, it means promoting comfort at the expense of long-term damage. In philosophy, it means positivism, behaviorism, and materialism. In religion, it means obsession with ritual and short-term gains. This steady state doesn't exist anywhere and contradicts fundamental traits of nature, which is inherently dynamic. When it was recognized that steady states were non-existent, the time function was introduced in practically all analysis, this time function being such that as $t \rightarrow \infty$, and the aphenomenal steady state emerged. That should have triggered investigation into the validity of the time function. Instead, it was taken as a proof that the universe is progressively moving toward a state of heat death, an aphenomenal concept promoted by Kelvin.

Chapter 5 presents a comprehensive theory that can answer all the questions that remain unanswered with the current engineering tools. This theory combines mass and energy balance to demonstrate that mass and energy cannot be treated in isolation if one has to develop sustainable energy management schemes. This theory exposes the shortcomings of New Science on this score and is a powerful tool for deconstructing key spurious concepts, such as the following. The concept that "if you cannot see it, it does not exist" denies all aspects of intangibles, yet it forms the basis of environmental and medical science. The concept that "chemicals are chemicals," originally promoted by Linus Pauling, a two-time Nobel Laureate, assumes that the pathway of a chemical does not matter and is used in the entire pharmaceutical, chemical, and agricultural industry. Numerous physicists, including Einstein, Rutherford, and Fermi, believed the notion that "heat is heat," which was originally inspired by Planck.

With the above formulation, two important fallacies at the core of currently unsustainable engineering practice are removed: (1) the fallacy that human need is infinite and (2) the fallacy that natural resources are finite.

These notions were not only accepted, but they were presented as the only knowledge. Yet, they clearly violate the fundamental trait of nature. If nature is perfect, it is balanced. It is inherently sustainable. However, it does not follow that it can be artificially sustained. If human beings are the best creation of nature, it cannot be that the sustainability of nature is being threatened by human activities, unless these activities are based on flawed science. In Chapter 4, the core reasons behind this apparent imperfection of nature are analyzed, and the science that prompted the fallacious conclusions is deconstructed. It is shown that within the core of current engineering design practices lies a fundamentally flawed notion of ideal and standard. This ideal is mentioned as the first premise of Newton's work, followed by the first premise of Lord Kelvin. Both used the first premise that is aphenomenal. In philosophical terms, this first premise is equivalent to saying that nature is imperfect and is degrading to a lower quality as time progresses, and, in order to remove this degradation, we must "engineer" nature. Chapter 3 shows that, by making a paradigm shift (starting from the first premise), i.e. "nature is perfect" is used as the first premise, the resulting model can answer all questions that are posed in the wake of the environmental consciousness in the Information Age. This approach has been long sought-after, but has not been implemented until now (Yen 2007).

1.5 Can Nature be Treated as if it were Static?

The most significant contribution of the previous section has been the recognition that the time function is the most important dimension in engineering design. If this is the case, one must then ask what the duration of this time function is in which one should observe the effect of a certain engineering design. It is of utmost importance to make sure that this duration is long enough to preserve the direction of the changes invoked by an engineering design. Philosophically, this is equivalent to saying that at any time the short-term decisions cannot be based on an observation that is not absolutely true or something that would be proven to be false as a matter of time. This notion is linked to the concept of sustainability in Chapter 6.

The term "sustainable" has become a buzzword in today's technology development. Commonly, the use of this term creates an inference that the process is acceptable for a certain duration of time. True sustainability cannot be a matter of arbitrary definition, nor can it be a matter of a policy objective lacking any prospect of physical achievement in reality. In this book, a scientific criterion for determining sustainability is presented. The foundation of this criterion is "time-testedness."

In Chapter 6, a detailed analysis of different features of sustainability is presented in order to understand the importance of using the concept of sustainability in every technology development model. A truly sustainable process conforms to natural phenomena both in its source, or its root, and in its process, or pathway.

However, as applied to resource engineering nominally intent on preserving a healthy natural environment, the science of tangibles has frequently given rise to one-sided notions of sustainability. For example, a phenomenal root, such as natural gas supplies, is addressed with principal attention focused on whether there will be a sufficient supply over some finite projected duration of time. Such a bias does not consider which uses of the said resource should be expected to continue into the future and for how long into the future. For example, should natural gas production in the Canadian province of Alberta be sustainable to the end of being used for (1) feedstock for heavy-oil upgraders, (2) export to residential and commercial markets in the U.S., (3) a principal supply for Canadian residential and commercial markets, or (4) some combination of two or more of the above?

Of course, absent any other considerations, sufficient supply to meet future demand can only be calculated by assuming that current demand continues into the future, including its current rate of growth, and, hence, is utterly aphenomenal. Both the origin and destiny of this source of natural gas, and the mechanics by which it will enter the market, are matters of speculation and not science. However, consciousness of this problem is obscured by retrofitting a pathway that might attain all potential projected demands targeted for this resource. Inconvenient facts, such as the likelihood that nuclear power is being strongly pushed to replace Alberta's natural gas as the fuel for future upgraders *in situ* at the tar sands, are not taken into account. Whatever the pathway, it is computed in accordance to a speculated prediction and, hence, is utterly aphenomenal. Meanwhile, whether that pathway is achievable or even desirable, given current engineering practices, is neither asked nor answered. The initial big "plus" supposedly in natural gas' favor was that it is "cleaner" than other petroleum-based sources. Given the quantity of highly toxic amines and glycols that must be added in order to make it commercially competitive for supplying residential markets, however, this aspect of its supposed sustainability would seem to raise more uncomfortable questions.

In addition, the natural criterion alluded to above means that true long-term considerations of humans should include the entire ecosystem. Some have called this inclusion "humanization of the environment" and have put this phenomenon as a pre-condition to true sustainability (Zatzman and Islam, 2007). The inclusion of the entire ecosystem is only meaningful when the natural pathway for every component of the technology is followed. Only such design can assure both short-term (tangible) and long-term (intangible) benefits.

1.6 Can Human Intervention Affect Long-Term Sustainability of Nature?

It is commonly said that any set of data doesn't reach statistical significance unless several cycles are considered. For instance, if the lifecycle of a forest is 100 years, a statistical study should cover at least several centuries. Then, one must question what the duration of the life cycle of humans is since they started to live in a community. Past experience has shown that putting 10,000 years on the age of Adam (the alleged first human) was as dangerous as calling the earth flat. Recent findings show that it is not unreasonable that the humans lived as a society for over a million years. What, then, should be the statistically significant time period for studying the impact of human activities? In Chapter 6, the focus lies on developing a sustainability criterion that is valid for time tending toward infinity. This criterion would be valid for both tangible (very large number) and intangible (no-end) meanings of the word infinity. If this criterion is the true scientific criterion, then there should be no discontinuity between so-called renewable and non-renewable natural resources. In fact, the characterization should be re-cast on the basis of sustainable and non-sustainable. After this initial characterization is done, only then comparisons among various energy sources can be made. Chapter 7 offers this characterization.

It is shown in the following chapters that sustainable energy sources can be rendered fully sustainable (including the refining and emission capture), making them even more environmentally appealing in the short-term (long-term being already covered by the fact they meet the sustainability criterion). In order to characterize various energy sources, the concept of global efficiency is introduced. This efficiency automatically shows that the unsustainable technologies are also the least efficient ones. It is shown that crude oil and natural gas are compatible with organic processes that are known to produce no harmful oxidation products.

1.7 Can an Energy Source be Isolated from Matter?

The most important feature of human beings is their ability to think (*homo sapiens* literally meaning the thinking man). If one simply thinks about this feature, one realizes that thinking is necessary to decide between two processes. In that sense, our brain is just like a computer, the option always being 0 or 1. The history of philosophy supports this cognition process, as evidenced by Aristotle's theory of exclusions of the middles. Even though dogmatic application of this theory to define enemies with tangible features (the with us or against us syndrome) drew severe criticism from people of conscience, this theory indeed was instrumental throughout history in discerning between right and wrong, true and false, and real and artificial. This is not just a powerful tool. It is also an essential and sufficient decision-making tool because it provides one with the fundamental basis of "go" or "no go." This has been known as *Al-furqan* (the criterion) in Arabic and has been used throughout Qu'ranic history as the most important tool for decision-makers and revolutionaries (Zatzman and Islam 2007b).

Aristotle considered the speed of light to be infinity. Father of modern optics, Ibn Al-Haytham (also known as Al-Hazen) realized that this theory does not meet the fundamental logical requirement that a light source be an integral part of the light particles. Using this logic, he concluded that the speed of light must be finite. Many centuries later, Satyendra Nath Bose (1894–1974) supported Al-Hazen's theory. In addition, he added that the speed of light must be a function of the media density. Even though he didn't mention anything about the light source being isolated from the light particles, he did not oppose Al-Hazen's postulate. Until today, Bose's theory has been considered to be the hallmark of material research (see recent Nobel prizes in physics that refer to the Bose-Einstein theory), but somehow the source has been isolated from the light particles. This was convenient for promoting artificial light as being equivalent to real light (sunlight).

Chapter 8 characterizes lights from various sources and based on their sources. It shows that with such scientific characterization it becomes evident why sunlight is the essence of life and artificial light (dubbed as "white light") is the essence of death. The problems associated with artificial lights, ranging from depression and breast cancer to myopia (Chhetri and Islam 2008), are explained in terms of the characterization done in this chapter. It is shown that a natural light source is a necessary condition of sustainability. However, it is not sufficient, as the process of converting the energy source into light must not be unsustainable.

1.8 Is it Possible that Air, Water, and Earth became our Enemies?

For thousands of years of known history, air, water, fire, and earth matter were considered to be the greatest assets available for the sustainability of human civilization. On the same token, humans (earthlings, in English inhabitants of the "earth," which comes from old English, *earthe*, old German *erda*, and is very close to Arabic, الأرْض (*al-ardha*) that means "the natural habitat" of humans) were the greatest assets that a family, a tribe, or a community could have. Land was the measure of worldly success, whereas organic earth contents, produces, and derivatives (e.g., crops, vegetation, and domestic animals) were the symbol of productivity. Ever since the reign of King David (CNN 2008a), non-organic minerals also joined the value-added possessions of humans. Never in history were these natural resources considered to be liabilities. The natural value addition was as follows:

Air → land → water → organic matter earth matter → non-organic earth matter

In the above path, air represents the most essential (for human survival) and abundant natural resource available. Without air, a human cannot survive beyond a few minutes. Land comes next because humans, who have the same composition as earth matter, need to connect with the earth for survival. This is followed by water, without which humans cannot survive beyond a few days. Water is also needed for all the organic matter that is needed for human survival as well. Organic matter is needed for food, and without it humans cannot survive beyond weeks. Non-organic earth matter is least abundant, although readily available, but its usefulness is also non-essential for human survival.

In the quest of survival and betterment of our community, the human race discovered fire. Even though the sun is the root energy source available to the earth, fire was not first discovered from solar energy. Instead, it was from natural wood, which itself was a naturally processed source of energy and matter. Nature did not act on energy and matter separately, and both energy and mass do conserve with inherent interdependency. No civilizations had the illusion that energy and matter could be created. They all acted on the premise that energy and matter are interrelated. The discovery of coal as an energy source was another progression in human civilization. There was no question, however, that the direct burning of coal was detrimental to the environment. Coal was just like green wood, except more energy was packed in a certain volume of coal than in the same volume of green wood. Next came the petroleum era. Once again, a naturally processed source of energy was found that was much more efficient than coal. Even though petroleum fluids have been in use for millennia, the use of these fluids for burning and producing heat is relatively new, and is a product of modern times. There is no logical or scientific reasoning behind the notion that the emissions of petroleum products are harmful and the same from wood is not. Yet, that has been the biggest source of controversy in the scientific community over the last four decades.

Ironically, the scientists who promoted the notion that "chemicals are chemicals," meaning carbon dioxide is independent of the source or the pathway, are the same ones who became the most ardent proponent of the "carbon dioxide from petroleum is evil" mantra. How could this be? If carbon dioxide is the essence of photosynthesis, which is essential for the survival of plants, and those plants are needed for sustaining the entire ecosystem, how could the same carbon dioxide be held responsible for "destroying the planet?" The same group promotes that nuclear energy is "clean" energy, considers that genetically modified, chemical fertilizer- and pesticide-infested crop derivatives processed through toxic means are "renewable," and proclaims that electricity collected with toxic silicon photovoltaics and stored with even more toxic batteries (all to be utilized through most toxic "white light") is sustainable. In the past, the same logic has been used in the "I can't believe it's not butter" culture, which saw the dominance of artificial fat (transfat) over real fat (saturated fat).

Chapter 9 demystifies the above doctrinal philosophy that has perplexed the entire world, led by scientists who have shown little appetite for solving the puzzle, resorting instead to be stuck in the

Einstein box. This chapter discusses how the build-up of carbon dioxide in the atmosphere results in irreversible climate change, and presents theories that show why such build-up has to do with the type of CO_2 that is emitted. For the first time, carbon dioxide is characterized based on various criteria, such as the origin, the pathway it travels, and the isotope number. In this chapter, the current status of greenhouse gas emissions from various anthropogenic activities is summarized. The role of water in global warming is also discussed. Various energy sources are classified based on their global efficiencies. The assumptions and implementation mechanisms of the Kyoto Protocol are critically reviewed. Also, a series of sustainable technologies that produce natural CO_2, which does not contribute to global warming, are presented.

1.9 Can we Compare Diamonds with Enriched Uranium?

Engineers are always charged with the task of comparing one scheme with another in order to help decide on a scheme that would be ideal for an application. It is commonly understood that a single-criterion analysis (e.g., fuel efficiency for energy management) will be inherently skewed because other factors (e.g., environmental concerns, economic factors, and social issues) are not considered. The same engineers are also told that they must be linear thinkers (the line promoted even in the engineering classroom being "engineers love straight lines"). This inherent contradiction is very common in post-Renaissance science and engineering.

Chapter 10 demystifies the characterization principle involved in energy management. Various energy systems are characterized based on 1) their sustainability and 2) their efficiency. This characterization removes the paradox of attempting to characterize diamonds (source being carbon) and enriched uranium (source being uranium ore), inherently showing that diamonds are a less efficient energy source than enriched uranium. This paradox is removed by including the time function (essence of intangibles), which clearly shows that a sustainable energy source is also the most efficient one. This characterization is an application of the single-criterion sustainability criterion, discussed in earlier chapters.

After the energy systems are classified under sustainable and unsustainable, energy sources are ranked under different categories.

Among sustainable ones, the classification led to improvements of design in order to achieve even better performance in terms of immediate benefits (long-term benefits being embedded in the sustainability criterion). For the unsustainable energy systems, it is shown how the long-term environmental impacts snowball into truly disastrous outcomes. This relates to the "tipping point" that many environmental pundits have talked about but that until now has not been introduced with a scientific basis.

1.10 Is Zero-Waste an Absurd Concept?

Lord Kelvin's theory leads to this engineering cognition: you move from point A to point B, then from point B to C, then back to point A. Because you came back to point A, you have not done any work. However, what if a person has actually worked (W) and spent energy (Q) to make the travel? Modern thermodynamics asserts that the claim of work is absurd and no *useful* work has been done. This is the engineering equivalent of stripping conscience participation of a worker. Rather than finding the cause of this confusion, which is as easy as saying that the time function to the movement should be included and that you have actually traveled from Point A (time 1) to Point A (time 2), what has been introduced is this: because you didn't do any *useful* work, W, the energy that you have spent, Q, is actually 0.

The above example is not allegorical, it is real and anyone who attempted to design an engineering system using conventional thermodynamics principles would understand this. That's why any attempt to include real work (as opposed to *useful* work) or real heat (as opposed to heat to produce *useful* work) would blow up the engineering calculations with divisions by zero all over the place. This is equivalent to how economic calculations blow up if the interest rate is written equal to zero[1].

Truly sustainable engineering systems require the use of zero-waste. This, however, would make it impossible to move further in engineering design using the conventional tool that has no tolerance to zero-waste (similar to zero-interest rate in economic

---

[1] The standard practice of financial organizations that are using software is to put a small number, e.g., 0.1% interest rate, to simulate 0% interest rate.

models). This is why an entire chapter (Chapter 11) is dedicated to showing how petroleum engineering design can be done with the zero-waste mode. The scientific definition of a zero-waste scheme is followed by an example of zero-waste with detailed calculations showing how this scheme can be formulated. Following this, various stages of petroleum engineering are discussed in light of the zero-waste scheme.

1.11 How can we Determine Whether Natural Energy Sources Last Forever?

This would not be a valid question in the previous epochs of human civilization. Everyone knew and believed that nature was infinite and would continue to sustain itself. In the modern age, we have been told that human needs are infinite, humans are liabilities, and nature is finite. We are also told that carbon is the enemy and enriched uranium is the friend. We are told that carbon dioxide is the biggest reason our atmosphere is in crisis (yet carbon dioxide is the essence of photosynthesis), and organic matter is the biggest reason our water is polluted (yet organic matter is the essence of life). If we seriously address the questions that are asked in previous questions, the above question posed in this section becomes a matter of engineering details.

Chapter 12 proposes techniques for greening of refining and gas processing schemes. It is shown that refining, as used in modern times, remains the most important reason behind the toxic outcome of petroleum product utilization for natural gas, liquid petroleum, and solid final products (e.g., plastics). Alternatives are proposed so that refining is done in such a way that the refined products retain real value, yet do not lose their original environmental sustainability (as in crude oil).

Chapter 13 identifies root causes of toxicity in fluid transport processes. It is shown that the main causes of such unsustainability arise from the use of artificial chemicals in combating corrosion, hydrate formation, wax deposits, and aspheltene deposition. The pathway analysis of these chemicals shows clearly how detrimental they are to the environment. For each of these chemicals, alternate chemicals are proposed that do not suffer from the shortcomings of the artificial chemicals and prevent the flow assurance problems that prompted the usage of artificial chemicals.

Chapter 14 proposes a host of green chemicals for enhanced oil recovery (EOR) applications. Historically, EOR techniques have been considered the wave of the future, and are believed to be a major source of increasing petroleum production in the upcoming decades. However, to date, all EOR techniques adopt unsustainable practices, and if these practices are not rendered sustainable, an otherwise sustainable recovery scheme will become unsustainable.

1.12 Can Doing Good be Bad Business?

Chapter 15 removes all the above paradoxes and establishes the economics of sustainable engineering. This chapter shows that doing good is actually good business and deconstructs all the models that violate this time-honored first premise. This chapter brings back the pricing system that honors real value, replacing the artificial pricing system that has dogged the petroleum industry for the longest time. Despite the length of this chapter, it does not cover all of the details because that would be beyond the scope of this book. However, it provides enough details so that decision-makers can be comforted with enough economic backing. After all, engineering is all about practical applications of science. No practical application can take place without financial details. This is true even for sustainable engineering.

1.13 Greening of Petroleum Operations: A Fiction?

This book is all about true paradigm shift, from ignorance to knowledge. A true paradigm shift amounts to revolution because it challenges every concept, every first premise, and every process. No revolution can take place if false perceptions and misconceptions persist. Chapter 2 begins with highlighting the misconceptions that have been synonymous with the modern age. In Chapter 16, the outcomes of those misconceptions are deconstructed. It shows how Enron should have never been promoted as the most creative energy management company of our time, not unlike how DDT should not have been called the "miracle powder," and we did not have to wait for decades to find out what false claims were made.

Most importantly, this chapter shows that we must not fall prey to the same scheme.

Chapter 16 discusses how if the misconceptions of Chapter 2 were addressed, all of the contradictions of modern times would not come to deprive us of a sustainable lifestyle that has become the hallmark of our current civilization. In particular, this chapter presents and deconstructs a series of engineering myths that have been deeply rooted in the energy sector with an overwhelming impact on modern civilization.

Once the mythical drivers of our energy sector are removed, it becomes self evident that we have achieved complete reversal of slogan, a true paradigm shift. Chapter 17 summarizes all the changes that would take place if the sustainable schemes were implemented. Reversing global warming would just be icing on the cake. The true accomplishment would be the reversal of the pathway to knowledge, like with the scientific meaning of *homo sapiens* in Latin or *insan* in Arabic[2]. If we believe that "humans are the best creation of nature," then the expectations of this book are neither unreal nor unrealistic.

---

[2] The Arabic word, *insan*, is derived from root that means sight, senses, knowledge, science, discovery by senses, and responsible behavior. It also means the one that can comprehend the truth and the one that can comprehend the creation or nature.

2

From the Pharaonic Age to the Information Age: Have we Progressed in Technology Development Skills?

2.1 Introduction

In 2003, Nobel Laureate chemist Robert Curl called our time a 'technological disaster.' Of course, back then it was not yet fashionable to be an environmental activist. The point to be made here is that our time is full of paradoxes and we cannot rely on new science scientists to solve these paradoxes. What we need is a fresh outlook, free from paradoxical narration and dogmatic thinking. This chapter reviews the history of modern technology development in order to answer the principal question asked in the title of the chapter.

Lord Kelvin, whose 'laws' are a must for modern day engineering design, believed that the earth is progressively moving to worse status, which would eventually lead to the 'heat death' of the habitat of the 'best creation of God.' So, if Kelvin were to be correct, we are progressively moving to greater energy crisis, and indeed we need to worry about how to fight this 'natural' death of our planet. Kelvin also believed flying an airplane was an absurd idea, so absurd that he did not care to be a member of the aeronautical club. Anyone would agree that it is not unreasonable to question this assertion of Lord Kelvin, but the moment one talks about nature progressively improving, if left alone (by humans, of course), many scientists break out in utter contempt and invoke all kinds of arguments of doctrinal fervor. How do these scientists explain then, if the earth is progressively dying, how it happened that life evolved from the non-biological materials and eventually very sophisticated creatures, called *homo sapiens* (thinking group) came to exist? Their only argument becomes the one that has worked for all religions, 'you have to believe.' All of a sudden, it becomes a matter of faith, and all the contradictions that arise from that assertion of Lord Kelvin become paradoxes and we mere humans are not supposed to understand them. Today, the internet is filled with claims that Kelvin is actually a god, and there is even a society that worships him. This line of argument cannot be scientific.

In 2007, when the United States Secretary of State Condoleezza Rice hailed nuclear energy as 'clean' on the occasion of the US-India nuclear accord, many scientists cringed, alongside even the most enthusiastic supporters of the Bush Doctrine. It is not as though we were not already used to such practices as 'purifying' with toxic catalysts, dry 'cleaning' with CCl_4, or 'disinfecting' with toxic chemicals. Rather, the timing of Madame Secretary's statement was so acutely ... embarrassing. It came within days of North Korea, one of the three points forming President George W. Bush's 'axis [*sic*] of evil', declaring that they indeed possessed a nuclear weapon – a declaration that 'justified' the original assertion of the existence of an 'axis of evil'. The obvious palpable question just hung there in the air, unspoken:

> *If nuclear weapons are evil, how is it possible that nuclear energy is clean?*

No one seems to acknowledge this 1000-pound gorilla. In fact, during the 2008 Presidential campaign, both candidates touted nuclear energy, with petroleum products being the culprit designate. In the following years, Al Gore visited numerous Middle Eastern capitals, promoting nuclear energy with dogmatic fervor, until the Fukushima nuclear plant disaster cracked open the peril of nuclear energy once again in early 2011.

The latest talk is that of hydrogen fuel. The world that has been obsessed with getting rid of wrong enemies is now obsessed with getting rid of ... carbon. Hydrogen fuel is attractive because it is ... not carbon. The slogan is so overpowering that a number of universities have opened up 'hydrogen chairs' in order to advance humanity out of the grip of carbon. In 2005, the President of Shell Canada talked about hydrogen being the wave of the future. All quite puzzling, really, even by post-9/11 standards: *what hydrogen are we talking about*? Could it be the hydrogen that is present in hydrocarbons (of course, after we get rid of the carbon)? Of course, this question would be discarded as 'silly.' Everyone should know that he meant hydrogen as in fuel cells, hydrogen as in after dissociation of ultra-pure water, and so on. This is the hydrogen, one would argue, that produces clean water as a byproduct, and nothing but water. As petroleum engineers, we are supposed to marvel: what a refreshing change from nature that always produces water infested with countless chemicals, many of which are not even identifiable. No one dares question if this is possible, let alone beneficial. Until such a question is raised, and actually investigated, however, any claims can be made: aye, there's the rub! After all, we have taken for granted that "if it cannot be seen, it does not exist." An even more convincing statement would be, "If the Establishment says so, it exists." What progress has been made, on what pathways, in rolling hydrogen out as the energy source of the future? Apart from the fact that produced water will invariably contain toxic residue of the catalysts that are being used (especially more so because high-temperature systems are being used), one must also worry about where the hydrogen will come from. Shell proudly invested billions of dollars in answering this question and the answer is... "water." Thankfully, water doesn't have any carbon, so nothing will be 'dirty,' if we can just separate hydrogen from its ever-so-hard-to-break bonds with oxygen. Visualize this, then: we are breaking water so we can produce hydrogen, so it can combine with oxygen, to produce water. In the meantime, we produce a lot of energy. Is this a miracle, or what?

It is less a miracle than a magic trick. It is a miracle as long as the science that is used is the same as that which called DDT a miracle powder. How could such a preposterous process be sold with a straight face? What is possibly the efficiency of such a process? It turns out the efficiency is quite high, but only so long as the focus is restricted to the hydrogen and oxygen reaction, in a cocoon completely isolated from the rest of the world, a status that does not exist in nature. If one analyzes the efficiency using a new definition that includes more components than just those on which current discussion has so tightly focused, the efficiency indeed becomes very low. How low? As much as 100 times less than what is promoted as a marketing gimmick. However, low efficiency is not the only price to pay for such an anti-nature technology. Recent findings of Cal Tech researchers indicate that the focus on creating hydrogen fuel will be disastrous, and far worse than the one created by the 'oil addiction.' Can we wait for another few decades to 'verify' this claim?

Then come the energy panaceas: wind energy, photovoltaics, and electric cars. No one can argue that they are anything but renewable and perfectly clean, right? Wrong. If one considers what is involved in producing and utilizing these 'clean' energy sources, it becomes clear they are neither clean nor productive. Consider the analysis depicted in Figure 2.1.

This figure shows how the solar panel itself is both toxic (emitting silicon dioxide continuously) and inefficient (efficiency around 15%). As the photovoltaic charges the battery, efficiency goes down, and toxicity is further increased. When the electricity reaches the utilization stage (light source), triple deterioration occurs. 1) Fluorescent bulbs contain mercury vapor; 2) efficiency drops further; and 3) the light that is emitted is utterly toxic to the brain (for details, see Chhetri and Islam, 2008).

The latest buzz is renewable energy. The worldwide production of biodiesel in 2005 was reported to be 2.2 billions gallons (Martinot, 2005). Even though this is less than 7% of that year's total diesel need of the United States, it is considered to be significant because it is a move in the right direction. Even though a few

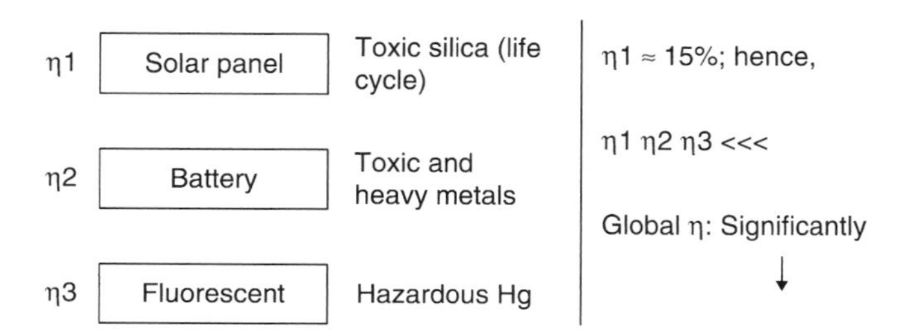

Figure 2.1 Environmental unsustainability of photovoltaics.

anti-globalization activists raised the question of what wisdom could be behind converting food into fuel, particularly in view of the global food shortage in 2007 and global financial crisis in 2008, few scientists considered biofuel as unsustainable. Chhetri *et al.* (2008) analyzed the biofuel production process, using biodiesel as an example. They found biodiesel to be anything but 'green' for the following reasons: 1) the source is either chemical fertilizer and pesticide-infested or genetically modified; 2) the process of making biodiesel is not the one Mr. Diesel (who died or was murdered before his dream process of using vegetable oil to produce combustible fuel came true) proposed to do; and 3) by using a catalytic converter, the emitted CO_2 would be further contaminated and the quality of CO_2 would not be much different from the one emitted from non-biodiesel. This would nullify practically all of the allegedly pro-environmental reasons behind investing in biodiesel. The following figure (Fig. 2.2) shows a schematic of how various functional groups of a source (Item 1 above) are linked to one another, accounting for overall balance of the ecosystem. This figure shows clearly that in nature it is impossible to 'engineer' nature and expect to keep track of its impacts. This is particularly true because of the special features of nature that do not allow any finite boundary to be drawn. Even though mass balance, the most fundamental engineering equation, is based on a finite volume enclosed inside a boundary, this first premise of a mass balance equation is clearly aphenomenal. This is an important issue, because if one ignores the fact that nature is continuous (meaning a particular segment can never be isolated), then no engineering design could ever

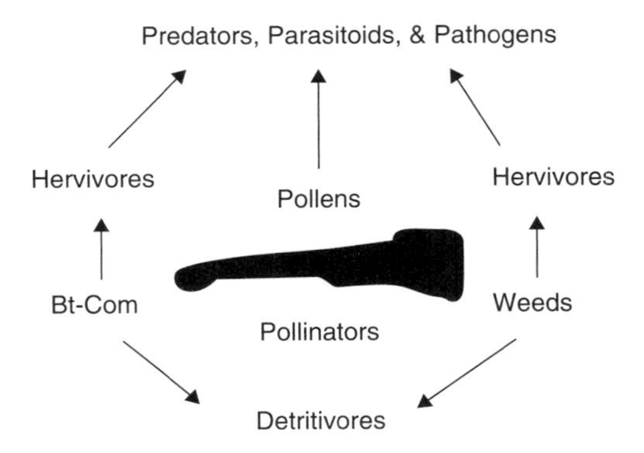

Figure 2.2 The role of genetically-engineered insecticidal Bt-corn in altering the ecosystem (*redrawn from* Losey *et al.*, 2004).

deem any of these processes unsustainable, even if the definition of sustainability is a strictly scientific one.

Zatzman and Islam (2007a) recognized that an 'artificial' object, even though it comes to reality by its mere presence, behaves totally differently than the object it was supposedly emulating. This would explain why different forms of vitamin C act differently depending on their origin (*e.g.*, organic or synthetic), and so does every other artificial product, including antibiotics (Chhetri *et al.*, 2007; Chhetri and Islam, 2007). However, not uncharacteristically, a number of papers have appeared that deny any connection between any cause and effect (for instance, Dively, 2007 in the context of genetic engineering and colony collapse disorder). This is entirely expected. The nature of natural phenomena is such that there cannot be a single cause behind a phenomenon. As a consequence, each cause is a suspect, but with the new science the suspicion can be buried within statistical uncertainties, depending on which agency is paying for the study. This is beyond the conventionally accepted conflict between corporate profit and public access to information (Makeig, 2002), it is rather about creating disinformation in order to increase profit margin, as noted earlier (Lähateenmäkia *et al.*, 2002).

Because economics is the driver of modern engineering, short-term is the guiding principle behind all engineering calculations. This focus on short-term poses a serious problem in terms of scientific investigation. The new science says: there is no need, or room, for intangibles unless one can verify their presence and role with some experimental program (experimental meaning controlled conditions, probably in a laboratory, with experiments that are designed through the same science that one is set out to 'prove'). In contrast, Khan and Islam (2007a; 2007b) argued that the science of tangibles so far has not been able to account for disastrous outcomes of numerous modern technologies. In the same way that scientists cannot determine the cause of global warming with the science that assumes all molecules are identical, thereby making it impossible to distinguish between organic CO_2 and industrial CO_2, scientists cannot determine the cause of diabetes unless there is a paradigm shift that distinguishes between sucrose in honey and sucrose in Aspartame (Chhetri and Islam, 2007).

Have we progressed in terms of knowledge in last few centuries? Why is it, then, that we still do not know how a dead body can be preserved without using toxic chemicals as was done for mummies from the days of ancient Egypt? Of course, it is said, after Newton, everything has changed and we have evolved into our modern-day civilized status. Which of Newton's laws was necessary to design the structural marvels of the pyramids? Which part of Ohm's Law was used to run the fountains of the Taj Mahal? Whence came the Freon that must have been used to run the air conditioners of Persian palaces? Which engineering design criteria were used to supply running water in Cordova? Which ISO standard was used to design the gardens of Babylon? One of the worst weapons of disinformation has been that things can be compared on the basis of a single dimension. The most progress that this disinformation machine is willing to accept is the extension of three spatial dimensions. If the time dimension is not included as a continuous function, the extension to 3D or even 4D (*e.g.*, using either a discrete or a continuous temporal function) cannot distinguish between a man caressing his child or hitting her. It also cannot distinguish between a dancing person from a sleeping person, depending on the frequency of the temporal function. If that is preposterous in a social setting, imagine not being able to distinguish between the process of mummifying and modern-day 'preservation,' marble and cement, coral and concrete, mother's milk and baby formula, a beeswax candle and a paraffin wax candle, wood and plastic, silk and polyester, or honey and sugar? How can we call ourselves civilized if our science made it more difficult to discern between DDT and snake venom, fluorescent light and sunlight, and lightning energy and nuclear energy?

2.2 Fundamental Misconceptions of the Modern Age

2.2.1 Chemicals are Chemicals and Energy is Energy

Paul Hermann Müller was credited with inventing dichlorodiphenyltrichloroethane (DDT) and was awarded a Nobel Prize in medicine and physiology. This marks the beginning of a chemical addiction that has paralyzed modern society. In Müller's invention, there was no distinction between natural DDT or synthetic DDT; the thrilling news was that synthetic DDT could be mass-produced, whereas natural DDT could not. (Even the most devoutly religious scientist would not question why this mantra would make God a far less efficient creator than humans.) This very old misconception of 'chemicals are chemicals,' premised on the useless and actually harmful fact that a natural substance and its synthesized non-natural 'equivalent' have similar tangible features at $\Delta t = 0$ [an equilibrium state that exists nowhere in nature at any time], got new life. And just as international diplomacy today has redefined nuclear energy to be 'clean' when 'we' do it [and 'dirty' if Iranians or Koreans do it], it was the diplomacy unfolded at the dawn of the Cold War initiated by the United States against the Soviet Union that was ultimately responsible. What happened with DDT and many other products of high-tech chemical engineering was that they were developed as part of so-called 'peaceful competition' between differing social systems. U.S.-based industry patented all the artificial shortcuts; Soviet industry started copying and replicating so as not to lose place in the competition, even though the peoples of the USSR had developed natural alternative pesticides that were perfectly safe. For example, malaria was eliminated in the southern parts of the former Soviet Union in the 1920s by spreading unrefined crude oil among the swamps in which the malaria-carrying mosquitoes bred (Brecht, 1947).

Noting the accumulation of DDT in the environment by the early 1960s, the American writer Rachel Carson famously condemned its potential to kill off bird life and bring about a 'silent spring' (the title of her book), but she missed the even greater long-term damage inflicted on scientific thinking and the environment by this 'peaceful' competition, a disaster that was likely not less than anything that could follow the 'thermonuclear holocaust' which the 'peaceful' competition was supposed to avoid. Synthetic chemicals were acceptable because no matter which individual chemical constituent you chose, it existed in nature. The fact that almost none of the *synthetic combinations* of these otherwise natural-occurring individual elements had ever existed in nature was ignored. A top scientist with DuPont claimed dioxin (the most toxic product emitted from polyvinyl chloride, PVC) existed in nature; therefore, synthetic PVC should not be harmful. The mere fact something exists in nature tells us nothing about the *mode* of its existence. This is quite crucial, when it is remembered that synthetic products are used up and dumped as waste in the environment without any consideration being given (at the time their introduction is being planned) to the consequences of their possible persistence or accumulation in the environment. All synthetic products "exist" (as long as the focus is on the most tangible aspect) in nature in a timeframe in which $\Delta t = 0$. That is the mode of their existence, and that is precisely where the problem lies.

With this mode, one can justify the use of formaldehyde in beauty products, anti-oxidants in health products, dioxins in baby bottles, bleaches in toothpaste, all the way up to every pharmaceutical product promoted today. Of course, the same focus, based on the new science, is applied to *processes*, as well as *products*.

This very old misconception (dating back to Aristotle's time, as will be discussed later in this chapter) got a new life, inspired by the work of Linus Pauling (the two-time Nobel Prize winner in chemistry and peace, see Fig. 2.3) who justified using artificial products because artificial and natural both have similar tangible (at $\Delta t = 0$) features. He considered the organic source of vitamin C a 'waste of money' because artificial vitamin C has the same ascorbic acid content and it is cheaper. By 2003, it became evident that Pauling was wrong, as organic vitamin C was found to prevent cancer, as opposed to the artificial one that would actually induce cancer (it is worth noting that Pauling, who promoted vitamin C therapy, and he and his wife both died of cancer). However, the

Figure 2.3 Linus Pauling, the two-time Nobel Prize winner, did not see the difference between natural and artificial vitamin C.

actual reason behind why natural vitamin C acts opposite to artificial vitamin C remains a mystery (Baillie-Hamilton, 2004). This very finding unravels the need of identifying the shortcomings of new science. Even the Linus Pauling Institute considers Pauling's stance on artificial vitamin C as erroneous. However, the old justification of introducing synthetic materials continues. Synthetic chemicals were okay because no matter which individual chemical constituent you chose, it existed in nature; the fact that almost none of the *synthetic combinations* of these otherwise natural-occurring individual elements had ever existed in nature was ignored.

Figure 2.4 shows artificial material on the left and natural material on the right. With conventional analysis, they are both high in carbon and hydrogen component. In fact, their bulk composition is approximately the same. New science discards non-bulk materials and ignores the dynamic nature of atomic or subatomic particles. The result of this analysis automatically makes further analysis inherently biased. With that analysis, if one burns the plastic material on the left, it would produce the same waste as the one on the

Figure 2.4 Artificial material may appear both jazzy and cleaner than natural material, but that should not be a criterion for determining sustainability.

right. Yet, the one on the left (including the plastic container that holds holy water) will continuously emit dioxin to nature, while the one on the right will emit materials necessary for renewing and sustaining life. Of course, it is common knowledge that a wood stove is good for barbecue or blackened Cajun chicken, or that after every forest fire, renewal of ecosystem occurs, but what is not known in new science is that we have no mechanism to distinguish the long-term fate of carbon or carbon dioxide from a wood stove and the one emitted from artificial heat sources.

As long as atoms are all the same and only bulk composition is considered, there will be no distinction between organic food and chemical-infested food or genetically-modified food, or between sugar and honey, or between coke and water, or snake venom and pesticide, or extra-virgin olive oil and ultra-refined olive oil, or stone-ground flour and steel-ground flour, or free-range eggs and farm eggs, or white bread and whole-wheat bread, or white rice and brown rice, or real leather and artificial leather, or between farm fish and ocean fish, and the list truly goes on. While this is contrary to any common sense, no one seems to question the logic behind selecting the atom as a unit of mass and assigning uniform characteristic to every atom. Yet, this is the most important reason behind the misconception that chemicals are chemicals.

Just like assuming that atoms are rigid spheres that can be treated as the fundamental building block has created confusion regarding the quality of matter, assuming photons are the fundamental unit of light has created confusion regarding how nature uses light to produce mass (see Fig. 2.5). Because new science considers energy as energy, irrespective of the source (violating the fundamental continuity in nature), there is no way to distinguish between light from the sun and the light from the fluorescent light bulb. We do not even have the ability to distinguish between microwave oven heat and wood stove heat. It is known, however, that microwaves can destroy 97% of the flavonoids of certain vegetables (Chhetri and Islam, 2008). The current techniques do not allow us to include the role of flavonoids in any meaningful way (similar to the role of catalysts), and certainly do not explain why microwaves would do so much damage while the heating level is the same as, say, boiling on an electric stove. Similarly, nuclear fission in an atomic weapon is considered to be the same as what is going on inside the sun (the common saying is that "there are trillions of nuclear bombs going

Figure 2.5 Artificial light can only insult the environment, while natural light can sustain the environment.

off every second inside the sun"). In other words: 'energy is energy.' This is the misconception shared by all nuclear physicists, including Nobel physics laureates Enrico Fermi and Ernest Rutherford, that served to rationalize nuclear energy as clean and efficient in contrast to 'dirty,' 'toxic,' and even 'expensive' fossil fuel. The theory developed by Albert Einstein, who is credited with discovering the theory that led to the invention of the atomic bomb, spoke of the dynamic nature of mass and the continuous transition of mass into energy, which could have been used clearly to show that 'chemicals are not chemicals' and 'energy is not energy.' Its abuse, as a prop for aphenomenal notions about the artificial, synthesized processes and output of industry being an improvement upon and even a repairing of nature, violates two deeply fundamental principles, that 1) everything in nature is dynamic; and 2) nature harbors no system or sub-system that can be considered 'closed' and/or otherwise isolated from the rest of the environment. This is only conceivable where $\Delta t = 0$, an assumption that violates the principle that everything in nature is dynamic.

Based on this misconception, Dr. Müller was awarded a Nobel Prize in 1948. Sixty years later, even though we now know Dr. Muller's first premise was false, we continue to award Nobel Prizes to those who base their study on the same false premise. In 2007, the Nobel Prize in medicine was offered to three researchers for their 'discovery of the principle for introducing specific gene modifications in mice by the use of embryonic stem cells.' What is the first premise of this discovery? Professor Stephen O'Rahilly of the University of Cambridge said, "The development of the gene targeting technology in mice has had a profound influence on medical research ... Thanks to this technology we have a much better understanding of the function of specific genes in pathways of the whole organism and a greater ability to predict whether drugs acting on those pathways are likely to have beneficial effects in disease." No one seems to know why only beneficial effects are to be anticipated from the introduction of "drugs acting on those pathways." When did intervention in nature (meaning at this level of very real and even profound ignorance about the pathway) yield any beneficial result? Can one example be cited from the history of the world since Renaissance? We have learned nothing from Dr. Müller's infamous 'miracle powder,' also called DDT.

In 2008, the Chemistry Nobel Prize was awarded to three scientists who discovered the so-called green fluorescent protein (GFP). While discovering something that occurs naturally is a lofty goal, New science does not allow the use of this knowledge for anything other than making money, without regard to long-term impact or the validity of the assumptions behind the application. This Nobel Prize-winning technology is being put to work by implanting these proteins in other animals, humans included. Two immediate applications are: 1) the monitoring of brain cells of Alzheimer's patients; and 2) the use as a signal to monitor others (including crops infested with a disease) that need to be interfered with. Both of these are money-making ventures, based on scientifically false premises. For instance, the first application assumes that the implantation (or mutation) of these 'foreign' proteins will not alter the natural course of brain cells (affected by Alzheimer's or not). So, what will be monitored is not what would have taken place; it is rather what is going to take place after the implant is in place. The two pathways are not and cannot be identical. More in-depth research (something not allowed to grow out of new science) would show this line of application is similar to the use of a CT scan (at least 50 times more damaging than an X-ray) for detecting cancer, whereas the CT-scanning process itself is prone to causing cancer (Brenner and Hall, 2007). The CT-scan study was financed by tobacco companies, the worst perpetrators of cancer in the modern age. In sum: going back in time, the 1979 Nobel Prize was awarded to Hounsfield and Cormack for CT scan technology, and thirty years later this Nobel Prize-winning technology is found responsible for causing many of the cancers that it purports to detect.

2.2.2 If you Cannot See, it Does not Exist

It is well known that the mantra "dilution is the solution to pollution" has governed the environmental policies of the modern age. Rather than addressing the cause and removing the source of pollution, this policy has enabled operators to 'minimize' impact by simply diluting the concentration. Instead of refraining from artificial fluids, more refrigeration fluids were engineered that had only one good quality: they were not Freon. Because they are not Freon, they must be acceptable to the ecosystem. This line of thinking is very similar to accepting nuclear energy because it is not carbon.

On the regulation side, this misconception plays through the lowering of the 'acceptable level' of a contaminant, playing the same mantra: dilution is the solution to pollution.

The defenders of PVC often state that just because there is chlorine in PVC does not mean PVC is bad. After all, they argue, chlorine is bad as an element, but PVC is not an element, it is a compound! Here two implicit spurious assumptions are invoked: 1) chlorine can and does exist as an element; and 2) the toxicity of chlorine arises from it being able to exist as an element. This misconception, combined with 'chemicals are chemicals,' makes up an entire aphenomenal process of 'purifying' through concentration. H_2S is essential for human brain activities, yet concentrated H_2S can kill. Water is essential to life, yet 'purified' water can be very toxic and leach out minerals rather than nourishing living bodies.

Many chemical reactions are thought to be valid or operative only in some particular range of temperature and pressure. It is thought so because we consider it so: we cannot detect the products in the range outside of the one we claim the reaction to be operational. Temperature and pressure are themselves neither matter nor energy, and are therefore not deemed to be participating in the reaction. We simply exclude them on the grounds of this intangibility.

It is the same story in mathematics: data output from some process are mapped to some function-like curve that, truth be told, includes discontinuities where data could not be obtained. Calculus teaches that one cannot differentiate across the discontinuous bits because "the function does not exist there." So, instead of figuring out how to treat (as part of one and the same phenomenon) both the continuous areas and the discontinuities that have been detected, the derivative of the function obtained from actual observation according only to the intervals in which the function "exists" is treated as definitive – clearly a fundamentally dishonest procedure.

2.2.3 Simulation Equals Emulation

We proclaim that we emulate nature. It is stated that engineering is based on natural science. Invention of toxic chemicals and anti-nature processes are called chemical engineering. People seriously thought sugar is concentrated sweet, just like honey, so the process must be the same: "we are emulating nature!" This was the first thought, but then, because honey cannot be mass-produced and sugar can be, we proceeded to "improve" nature.

There are countless *simulators* and *simulations* "out there." Routinely, they disregard, or try to linearize/smooth over, the non-linear changes-of-state. The interest and focus is entirely on what can be taken *from* nature. This immediately renders such simulators and simulations useless for planning the long-term.

Emulations and emulators, on the other hand, start with what is available *in* nature, and how to sustain that. Emulation requires establishing a 1:1 correspondence between operation "X_c" of the computer and operation "X_b" of the human brain. Obviously, there exists as yet no such correspondence.

Could it be that this 'science' is actually a disinformation machine, carefully doctored to obscure the difference between real and artificial? Our research shows why the 'chemicals are chemicals' mantra is not promoted out of ignorance, it is rather promoted out of necessity to uphold the aphenomenal model that is incapable of existing or coexisting with knowledge or the truth. Quantity and 'choices' (for example: honey, or sugar, or saccharine, or aspartame) proliferate in such profusion that the degradation of quality, and the consequences thereof, are overlooked or ignored. Yet asking obvious questions, like that posed by the child in the fairy tale who noted the Emperor's state of *deshabille*, immediately discloses the problem: how anti-bacterial is aspartame? How many cancers will consumption of any amount of bees' honey engender? This is no instructional fairy tale, but a nightmare unfolding in broad daylight. The truth, it is said, has consequences. Yet, the consequences of the untruths underpinning aphenomenal models, hardly spoken about at all, seem far worse.

2.2.4 Whatever Works is True

The essence of the pragmatic approach is, "whatever works is true." This fundamentally aphenomenal principle is the driver of the short-term approach. Any short-term approach perverts the process, even if once in a while a conclusion appears to be correct. The problem in this approach is in the definition of 'work.' If one can define and portray anything as working, even for a temporary period, anything becomes true during that period of time.

Table 2.1 Analysis of "breakthrough" technologies.

Product	**Promise (knowledge at t = 'right now'**	**Current knowledge (closer to reality)**
Microwave oven	Instant cooking (bursting with nutrition)	97% of the nutrients destroyed; produces dioxin from baby bottles
Fluorescent light (white light)	Simulates the sunlight and can eliminate 'cabin fever'	Used for torturing people, causes severe depression
Prozac (the wonder drug)	80% effective in reducing depression	Increases suicidal behavior

(Continued)

Table 2.1 (Cont.) Analysis of "breakthrough" technologies.

Product	Promise (Knowledge at t = 'right now'	Current Knowledge (Closer to Reality)
Anti-oxidants	Reduces aging symptoms	Gives lung cancer
Vioxx	Best drug for arthritis pain, no side effect	Increases the chance of heart attack
Coke	Refreshing, revitalizing	Dehydrates; used as a pesticide in India
Transfat	Should replace saturated fats, incl. high-fiber diets	Primary source of obesity and asthma
Simulated wood, plastic gloss	Improve the appearance of wood	Contains formaldehyde that causes Alzheimer's
Cell phone	Empowers, keep connected	Gives brain cancer, decreases sperm count among men
Chemical hair colors	Keeps young, gives appeal	Gives skin cancer
Chemical fertilizer	Increases crop yield, makes soil fertile	Harmful crop; soil damaged
Chocolate and 'refined' sweets	Increases human body volume, increasing appeal	Increases obesity epidemic and related diseases
Pesticides, MTBE	Improves performance	Damages the ecosystem
Desalination	Purifies water	Necessary minerals removed
Wood paint/ varnish	Improves durability	Numerous toxic chemicals released
Leather technology	Won't wrinkle, more durable	Toxic chemicals
Freon, aerosol, etc.	Replaced ammonia that was 'corrosive'	Global harms immeasurable and should be discarded

Source: Chhetri and Islam, 2008.

Figure 2.6 Olive oil press (left, millennia-old technology that produces no toxin) and modern refinery that produces only toxins.

Hitler's lies took 12 years to unfold. For 12 years, Hitler's agenda 'worked.'

Shapiro *et al.* (2007) examined how the pragmatic approach has been used to systematically disinform. They explained how polling is routinely launched as a purportedly scientific approach to analyzing public opinion and trends within it. It brings out the manipulation of public opinion carried on under the guise of sampling how many believe something whose truth or falsehood has not first been established. Routinely, theories are formulated that have no phenomenal basis and experiments are conducted to support the theory. If there is any discrepancy, it is said to have been caused by 'experimental error,' which implicitly suggests that the theory is the truth and experiment is the illusion.

With the above misconceptions, we have come to a point that every technology that was developed based on those misconceptions has resulted in breaking the promise that was given when the technology was being introduced. This is evident from Table 2.1. Figure 2.6 sums up the dilemma.

3

How Long Has This 'Technological Disaster' Been in the Making? Delinearized History of Civilization and Technology Development

3.1 Introduction

It is becoming increasingly clear that the current mode of technology development is not sustainable (Mittelstaedt, 2007). Most agree with Nobel chemistry laureate Dr. Robert Curl that the current technological marvels are 'technological disasters.' Few, however, understand the root cause of this failure. This failure now endangers the future of entire species, including our own. Many theories have been put forward, including those branded as 'conspiracy theories,' 'pseudoscience,' and 'creationism,' but none provide answers to the questions that face the current perilous state of the world.

This lack of fundamentally correct theories is felt in every sector. The outcome is seen in the prevailing popular mood throughout Western countries that "*t*here *i*s *n*o *a*lternative." This so-called TINA syndrome was identified once again by Zatzman and Islam

(2007a), who credited former British Prime Minister (now Baroness) Margaret Thatcher with its enunciation during the 1980s, as the secular equivalent of Holy Write among Anglo-American policymakers. The essence of this approach upholds and promotes the status quo as the only way to move forward.

While some prominent individuals openly despair humanity's acquisition of technology in the first place as a form of societal 'original sin,' others have pinpointed the severe pollution burdens, mounting especially in the energy resource exploitation sector, to single out petroleum operations and the petroleum industry for special blame. In the context of increasing oil prices, the petroleum sector has become an almost trivially easy target, and this situation has been compounded by pronouncements from former United States President George W. Bush about "oil addiction." Even his most ardent detractors welcomed this particular comment as a signal insight (Khan and Islam, 2007a). Numerous alternate fuel projects have been launched, but they propose the same inefficient and contaminated processes that got us into trouble with fossil fuel.

Albert Einstein famously stated, "The thinking that got you into the problem, is not going to get you out." Meanwhile, there is no evidence yet that modern Eurocentric civilization is ready to propose a way out of this technological conundrum. The symptoms are ubiquitous, from addiction to toxic chemicals (Mittelstaedt, 2006) to global warming (Chhetri and Zatzman, 2008). How is the crying need to treat these symptoms being addressed? Soon after it was revealed that the farmers of India have been committing suicide in record numbers, and that the much-vaunted 'green revolution' was actually a fraud keeping humanity in a chokehold (Saunders, 2007), the Congressional Medal of Honour was bestowed on Dr. Norman Borlaug. This individual orchestrated the key experimental research undertaken with Mexican maize in the 1940s that underpinned the eventual introduction of costly chemical pesticides, herbicides and other 'aids' for enhancing the productivity of individual tillers throughout the developing world, in one crop after another (Editorial, 2007).

In the chemical sector (food, drug, and lifestyle), similar absurdities abound. Chemical companies have 'celebrated' 100 years of PVC, as though PVC has done humanity good. Yet, China is being accused of using PVC and other toxic chemicals in children's toys, leading to the recall (only in the United States and Europe) of millions of toys (CNN, 2007). Weeks before the recall, the head of China's Food and Drug Administration was executed for allowing unwanted chemicals in drugs. That followed weeks of other major scandals (or accusations) regarding Chinese chemical treatment of food and health products (e.g., toothpaste) earmarked for export to the United States. In the same year, debates raged around the possible connections to chemical fertilizer, chemical pesticides, and genetically modified crops to honey bee colony collapse disorder (CCD) among European honey bees.

As time progresses, we have become more addicted to low efficiency and high consumption, all being justified by aesthetic value, rather than real quality.

At one level, the official discourse is rigidly maintained that hormones are no longer just hormones, and that stem cell research is absolutely necessary because there is no substitute to real (natural) cells. Yet, even Prince Charles "gets it." He joined the choir of pro-nature scientists by asking the Sheikhs of Abu Dhabi (site of the second-highest rate of child diabetes in the world), "Have you considered banning McDonald's?" The Prince might not be willing to admit the connection of British-invented sugar culture to diabetes, but he is delighted to report that he owns an organic farm.

Somehow, when it comes to scientists, trained with the science of tangibles, organic products are 'not necessarily' better than other products, such as crops chemically-grown using toxic pesticide. One food scientist wrote to one of the authors, "I have yet to be convinced that the history of a molecule affects its function and that natural is necessarily better." Apart from the all-pervasive culture of fear that prevents one's taking a stand in case a lawsuit looms on the horizon (in geology, the identical syndrome hides behind the frequently-encountered claim that 'the possibility of finding oil cannot be ruled out entirely'), this comment from a food scientist contains a wry reminder of what has corrupted modern science in the most insidious manner. This has been the dilemma of modern age. Scientists today do not, or find that they cannot, simply uphold the truth. Those that speak, when they speak, uphold self-interest and the status quo. If natural is not necessarily better than artificial, and if chemicals are just chemicals, why should we have a different branch called food science? For that matter, why differentiate chemical engineering, nuclear engineering, pharmacy, military science, or anything else? With such an attitude, why not declare defeat and place them all hereafter under a single umbrella: the Science of Disinformation. Comparing opinion polling methodologies in general with the results of actually surveying in detail a defined population from and in a developing country, Shapiro *et al.* (2007) were able to elaborate in some detail the *modus operandi* of this science of disinformation. Has the overall state of the scientific world become so grim that the general public has to resort to trusting social-political activists (e.g., Al Gore's *An Inconvenient Truth*, Michael Moore's *Sicko!* and even a relative novice like Morgan Spurlock and his documentary, *Supersize Me!*) ahead of professionally knowledgeable scientists? This chapter attempts to find the cause of such helplessness in a science whose modernity seems to consist entirely of promoting only the culture of tangibles under the slogan "More is better — because there is more of it."

3.2 Delinearized History of Time, Science, and Truth

Thousands of years ago, Indian philosophers commented about the role of time, as a space (or dimension), in unraveling the truth, the essential component of knowledge (Zatzman and Islam, 2007a).

The phrase used by Ancient Indian philosophers was that *the world reveals itself*. Scientifically, it would mean that time is the dimension in which all the other dimensions completely unfold, so that truth becomes continuously known to humans, who use science (as in critical thinking). Another very well-known principle from ancient India is the connection among *Chetna* (inspiration), *dharma* (inherent property), *karma* (deeds arising from *Chetna*), and *chakra* (wheel, symbolizing closed loop of a sustainable life style). Each of these concepts scientifically bears the intangible meanings, which cannot be expressed with conventional European mathematical approach (Joseph, 2000). Ketata *et al.* (2006a; 2006b; 2006c; 2006d) recognized this fact and introduced a series of mathematical tools that can utilize the concept of meaningful zero and infinity in computational methods.

These ancient principles contain some of the most useful hints, extending far back into the oldest known human civilizations, of true sustainability as a state of affairs requiring the involvement of infinite time as a condition of maintaining a correct analysis as well as ensuring positive pro-social conclusions (Khan and Islam, 2007b). Moving from ancient India to ancient China, the Chinese philosophers provide one with some useful insight into very similar principles of sustainability and knowledge. The well-known statement, although rarely connected to science, of Confucius (551–479 B.C.) relates unraveling of the truth to creating *balance* (the statement is: *Strive for balance that remains quiet within*). For Confucius, balance had the essential condition of 'quiet within.' This idea is of the essence of intangibles in the "knowledge" sense (Zatzman and Islam, 2007b).

In the Qur'an (first and only version compiled in mid-7th century), humans' time on earth and time in nature are all part of one vast expanse of time. Although it arrives by quite a different route, this position is also entirely consistent with the notion that the world reveals itself.[1] Life in this world is the 'first life' in contrast to

---

[1] In terms of the role of intention, the most famous saying of The Prophet, the very first cited in the Bukhari's collection of the *hadiths*, is that any deed is based on the intention (*Hadiths of The Prophet*, 2007). A review of human history reveals that what is commonly cast or understood as "the perpetual conflict between Good and Evil" has in fact always been about opposing intentions. The Good has always been characterized by an intention to serve a larger community, while Evil has been characterized as the intention to serve the self-interest. What was known in ancient India as the purpose of life (serving humanity) is promoted in the Qur'an

the life after.[2] This Qur'anic description of human life, man's role in this world, and his eternal life is quite consistent with other Eastern philosophies that considerably predate the period in which the Qur'an was first compiled in its tangible form.

There is a notion widespread in the Western world that the monotheistic premises of each of the three Abrahamic religions — Judaism, Christianity and Islam — point to broad but unstated other cultural common ground. Reinforcing the philological analysis of Qur'anic Arabic terms in footnotes 1 and 2 *supra*, the historical record itself further suggests such has not been the case when it comes to certain fundamental premises of the outlook on and approaches taken to science and the scientific method.

The position of mainstream Greek, i.e., Eurocentric, philosophy on the key question of the nature of the existence of the world

---

as serving self-interest in the long-term (as in infinity, see discussion later). Because nature itself is such that any act of serving others leads to serving the self in the long term, it is conceivable that all acts of serving others in fact amount to self-interest in the long-term (Islam, 2005b). In terms of *balance*, the Qur'an promoted the notion of *qadar* (as in *Faqaddarahu*, فقدّرهٗ, meaning 'thereby proportioned him', 80:19, the Qur'an), meaning proportionate or balanced in space as well as time. The Qur'an is also specific about the beginning and end of hum?Author? Personal communication with a PhD from MIT (as reported in Abou-Kassem *et al.*, 2007) resulted in the following discussion. Thus for example the word "*aakherat*" (as in *Al-aakherat*, الآخِرَةِ, 2:4 the Qur'an) is the feminine form of the adjective "*aakher*" (آخر), meaning "last", as in last one to arrive, or generally last one to take place. This is only the simple or immediate meaning, however. According to a well-understood but wider connotation, the word *aakher*, meaning the last when it describes a period of time, also connotes the infinite in that period.

[2] Thus, *Aakherat*, referring to the life after, means it is an infinite life. In other words: while man is considered mortal, man is in fact eternal. Man's body is mortal, but his *self* is eternal. This is the only kind of creation with this quality, being eternal. Hence, man approaches divine nature with thenal,?Author? that is the existence in $+\infty$ of time while it differs with the attribute of having a beginning, meaning man lacks existence in time = 0. The same after-life is sometimes referred to as *ukhraa* (أخرى), the feminine form of *aakhar*. Note the slight difference from the previous *aakher*, meaning last. The word *aakhar* means 'second', but used only when there is no 'third.' Arabic and Sanskrit distinguish grammatically between one, two and more than two, unlike European languages that distinguish between one and more-than-one (as reported by Islam and Zatzman, 2006). This *ukhraa* is not used in the Qur'an. In contrast to *aakherat*, two words are used in the Qur'an. One is *ulaa* (as in Al-Ulaa, ولى ألأ, meaning The First, 92:13, the Qur'an). The other word used in the Qur'an is *dunya* (as in Al-Dunya, ألدُّنيَا, 79:38, the Qur'an), which means the 'nearer' or 'lower.' So, this life is nearer to us, time-wise, and also of lower quality.

external to any human observer is: *everything is either A or not-A.* That is Aristotle's law of the excluded middle which assumes time t = "right now" (Zatzman and Islam, 2007b). Scientifically, this assumption is the beginning of what would be termed as the steady-state models, for which Δt approaches 0. This model is devoid of the time component, a spurious state even if a time-dependent term is added in order to render the model 'dynamic' (Abou-Kassem *et al.*, 2007). Aristotle's model finds its own root in ancient Greek philosophy (or mythology) that assumes that 'time begins when Chaos of the Void ended' (Islam, 2005b). Quite similar to Aristotle's law of excluded middle, the original philosophy also disconnected both time function and human intention by invoking the assumption, "the gods can interrupt human intention at any time or place." This assertion essentially eliminates any relationship between individual human acts with a sense of responsibility. This particular aspect was discussed in details by Zatzman and Islam (2007a), who identified the time function and the intention as the most important factors in conducting scientific research. Their argument will be presented later in this section.

The minority position of Greek philosophy, put forward by Heraclitus, was that matter is essentially atomic and that, at such a level, everything is in endless flux. Mainstream Greek philosophy of Heraclitus' own time buried his views because of their subversive implication that nature is essentially chaotic. Such an inference threatened the Greek mainstream view that Chaos was the Void that had preceded the coming into existence of the world, and that a natural order came into existence, putting an end to chaos.

What Heraclitus had produced was in fact a most precise description of what the human observer actually perceives of the world. However, he did not account for time at all, so changes in nature at this atomic level incorporated no particular direction or intention. In the last half of the 18th century, John Dalton reasserted the atomic view of matter, albeit now stripped of Heraclitus' metaphysical discussion and explanations. Newton's laws of motion dominated the scientific discourse of his day, so Dalton rationalized this modernized atomic view with Newton's object masses, and we ended up with matter composed of atoms rendered as spherical balls in three-dimensional space, continuously in motion throughout three-dimensional space, with time considered as an independent variable. This line of research seals the doom of any hope for incorporating time as a continuous function, which would effectively make the process infinite-dimensional.

Zatzman and Islam (2007b) have offered an extensive review of Aristotle's philosophy and provided one with scientific explanation of why that philosophy is equivalent to launching the science of tangibles. In economic life, tangible goods and services and their circulation provide the vehicles whereby intentions become, and define, actions. Locked inside those tangible goods and services, inaccessible to direct observation or measurement, are intangible relations, among the producers of the goods and services and between the producer and Nature, whose extent, cooperativeness, antagonism and other characteristic features are also framed and bounded by intentions at another level, in which the differing interests of producers and their employers are mutually engaged. In economic terms, Zatzman and Islam (2007b) identified two sources of distortion in this process. They are: 1) linearization of complex societal non-linear dependencies (functions and relationships) through the introduction of the theories of marginal utility (MU); and 2) lines in the plane intersect as long as they are not parallel, i.e., as long as the equation-relationships they are supposed to represent are not redundant. The first source removes important information pertaining to social interactions, and the second source enables the use of the "=" sign, where everything to its left is equated to everything to its right. Equated quantities can not only be manipulated, but interchanged, according to the impeccable logic, as sound as Aristotle (who first propounded it), which says: two quantities each equal to a third quantity must themselves be equal to one another or, symbolically, that 'A = C' and 'B = C' implies that 'A = B.' Scientific implications of this logic will be discussed in the latter part of this section. Here, in philosophical sense, the introduction of this logic led to the development of a 'solution.' As further arguments will be built on this 'solution,' soon this 'solution' will become 'the solution' as all relevant information is removed during the introduction of the aphenomenal process. This would lead to the emergence of 'equilibrium,' 'steady state,' and various other phenomena in all branches of new science. It will not be noticeable to ordinary folk that these are not natural systems. If anyone questions the non-existence of such a process, s/he will be marginalized as a 'conspiracy theorist', a 'pseudo-scientist', and numerous other derogatory designations. This line of thinking would explain why practically all scientists up until Newton had tremendous difficulty with the Establishment in Europe. In the post-Renaissance world, the collision between scientists and the Establishment was erased, *not* because the Establishment became pro-science, but more likely because the new scientists became equally obsessed with tangibles, devoid of the time function as well as intention (Zatzman and Islam, 2007a). Theoretically, both of these groups subscribed to the same set of misconceptions or aphenomenal bases that launched the technology development in the post-Renaissance era. Khan and Islam (2007a) identified these misconceptions as:

1. Chemicals are chemicals or energy is energy (meaning they are not a function of the pathway and are measured as a function of a single dimension);
2. If you cannot see, it does not exist (only tangible expressions, as 'measurable by certain standard' are counted); and
3. Simulation equals emulation (if there is agreement between reality and prediction at a given point in time, the entire process is being emulated). Zatzman and Islam (2007a) and Khan and Islam (2007a) attributed these misconceptions to the pragmatic approach (whatever works must be true), which can be traced back original Greek philosophy.

The immediate consequence of the science of tangibles is that every decision that emerges is scientifically false. The removal of a dimension is inherent to this consequence (Mustafiz, 2007). Because time is the most important dimension, the omission of this dimension has the severest consequences. Time also forms the pathway of the science of intangibles. Intention, on the other hand, forms the root or foundation of the science of intangibles. Ignoring any of these would render the process aphenomenal. Zatzman and Islam (2007b) cited a number of examples to establish this assertion. One of them is as follows:

Compare the circle to the sphere, in Figure 3.1 below. The shortest distance between two points A and B on the circle's circumference is a straight (secant) line joining them, whereas the curved arc of the circumference between the two points A and B is always longer than the secant. The shortest distance between two points on the surface of a sphere, on the other hand, is always a curve and can never be a straight line. Furthermore, between two points

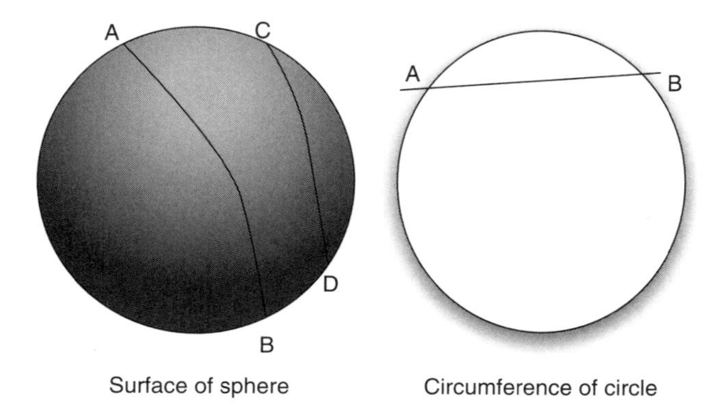

Figure 3.1 Sphere and circle compared (redrawn from Zatzman and Islam, 2007b).

A and B lying on any great circle of the sphere (i.e., lying along any of an infinite number of circles that may be drawn around the surface of the sphere whose diameter is identical to that of the sphere itself) and between any other two points, say C and D, a curved line of the same length joining each pair of points will subtend a different amount of arc. For the points not on the great circle, the same length-distance will subtend a greater arc than points on a great circle; curvature *k* in each case is uniform but different. On the other hand, what makes a circle circular rather than, say, elliptical, is precisely the condition whereby equal distances along its circumference subtend equal arcs, because curvature *k* anywhere along the circumference of a circle is the same.

Is anyone whose comprehension remains confined to cases of the circle in flat planar space likely to infer or extrapolate from such knowledge whatever they would need in order to grasp and use such differences found on the curved surface of the sphere? Of course not: indeed, any solution or solutions obtained for a problem formulated in two-dimensional space can often appear utterly aphenomenal when transferred or translated to three-dimensional space. In terms of the originating example of the fundamental metrics of the two kinds of surfaces: to those working in the environment of spherical surface-space, it becomes quickly obvious the shortest distance between two points that stand really close to one another on the surface of a sphere will approximate a straight line. In fact, of course, it is actually a curve and the idea that "the shortest distance between two points is a straight line" remains something that only approximates the situation in an extremely restricted subspace on the surface of a sphere. In a pragmatic sense, it is quite acceptable to propose that the shortest distance between two points is a straight line. However, in a real world which is anything but linear, this statement cannot form the basis for subsequent logical development, as it would falsify all subsequent logical trains. Zatzman and Islam (2007b) characterized this attempt to simplify a model used to account for complex phenomena by chopping dimensions away at the outset (with the idea of adding them back in at the end) as "about as practicable as trying to make water run uphill."

The above point is further expounded by giving a simple, yet thought-provoking example (Zatzman and Islam, 2007). This example involves an experiment carried out without the knowledge of the findings of Galileo and Newton. Someone could observe and write up everything about an experiment involving dropping a weighty lead ball from the top of the Tower of Pisa. On another occasion, the same could be done for an experiment entailing dropping a feather from the top of the Tower. Comparing the two, the key observation might well be that the lead ball fell to the ground faster and, from that fact, it might even be speculated that the difference in the time taken to reach the ground was a function of the difference in the weight of the feather and the lead ball. They argued that no matter how much more precise the measuring of the two experiments became, in the absence of any other knowledge or discovery, it would be difficult to overthrow or reject this line of reasoning. Now, let us say the measuring devices available for subsequent repetition of this pair of experiments over the next 10, 20 or 30 years develop digital readouts, to three or four decimal places of accuracy. Quite apart from what engineering marvels these 'precision clocks' have led to, the fact would remain that this advancement in engineering does not help to correct the conclusion. In fact, the engineering advancement without regard to the basis of fundamental science would actually increase confidence toward the wrong conclusion. This is the situation for which engineering (making of precision clocks) only serves to strengthen the prejudice of the science of tangibles.

Another example was cited by Abou-Kassem *et al.* (2007). This one involves the discussion of the speed of light. It is 'well recognized' in the new science world that the speed of light is the maximum achievable speed. The argument is based on Einstein's famous equation of $E=mc^2$. However, scientific scrutiny of this 'well recognized' statement in physics shows that this statement is aphenomenal. When the speed of light is stated, it is asserted that this is the speed of light (or radiation) within a vacuum. Then, when one asks what is a vacuum, it emerges that the vacuum in the true sense[3] does not exist and one must resort to a definition, rather than a phenomenal state.

Historically, there were always debates about the speed of light, but only in the post-Renaissance culture did that speed become a 'constant.'[4]

In modern science, the justification of a constant *c* is given through yet another definition, that of vacuum itself. The official NIST site (on Fundamental Physics constant) lists this constant as having a value of 299,792,458 m/s, with a standard uncertainty of zero (stated as 'exact'). Now, all of a sudden, a definition, rather than a phenomenon (measurable quantity) becomes the basis for future measurements. It is so because the fundamental SI unit of length,

---

[3] The Merriam-Webster Online Dictionary lists four meanings of "vacuum," none of them phenomenal:
1 : emptiness of space; **2 a :** a space absolutely devoid of matter; **b :** a space partially exhausted (as to the highest degree possible) by artificial means (as an air pump); **c :** a degree of rarefaction below atmospheric pressure; **3 a** : a state or condition resembling a vacuum : VOID <the power *vacuum* in Indochina after the departure of the French – Norman Cousins>; **b** : a state of isolation from outside influences <people who live in a *vacuum* ... so that the world outside them is of no moment – W. S. Maugham>; and **4 :** a device creating or utilizing a partial vacuum.

[4] For example, note the following quote from Wikipedia (Website 1, 2007):
"Many early Muslim philosophers initially agreed with Aristotle's view that light has an infinite speed. In the 1000s, however, the Iraqi Muslim scientist, Ibn al-Haytham (Alhacen), the 'father of optics,' using an early experimental scientific method in his *Book of Optics*, discovered that light has a finite speed. Some of his contemporaries, notably the Persian Muslim philosopher and physicist Avicenna, also agreed with Alhacen that light has a finite speed. Avicenna 'observed that if the perception of light is due to the emission of some sort of particles by a luminous source, the speed of light must be finite.'

The 14th century scholar Sayana wrote in a comment on verse Rigveda 1.50.4 (1700–1100 BCE — the early Vedic period): "Thus it is remembered: [O Sun] you who traverse 2202 yojanas [ca. 14,000 to 30,000 km] in half a nimesa [ca. 0.1 to 0.2 s]," corresponding to between 65,000 and 300,000 km/s, for high values of *yojana* and low values of *nimesa* consistent with the actual speed of light."

the meter, has been defined since October 21, 1983,[5] one meter being the distance light travels in a vacuum in 1/299,792,458 of a second. This implies that any further increase in the precision of the measurement of the speed of light will actually change the length of the meter, the speed of light being maintained at 299,792,458 m/s. This is equivalent to changing the base of a logical train from reality to aphenomenality.[6] This creates out of nothing a process that penalizes any improvement in discovering the truth (in this case, the true speed of light). Abou-Kassem *et al.* (2007) argued that this definition of meter is scientifically no more precise than the original definition that was instituted by the French in 1770s in the following form: one meter is defined as 1/10,000,000 of the distance from the North Pole to the Equator (going through Paris). They also discussed the spurious arrangement of introducing the unit of time as a second. It was not until 1832 that the concept of second was attached to the SI arrangement. The original definition was 1 second = 1 mean solar day/864,000. As late as 1960, the ephemeris second, defined as a fraction of the tropical year, officially became part of the new SI system. It was soon recognized that both mean solar day and mean tropical year both vary, albeit slightly, and the more 'precise' (the implicit assumption being that "more precise" means "closer to the truth") unit was introduced in 1967. It was defined as 9,192,631,770 cycles of the vibration of the cesium 133 atom. The assumption here is that vibration of cesium 133 atom is exact, this assumption being the basis of Atomic clock. It has been revealed that this assumption is not correct, creating an added source of error in the entire evaluation of the speed of light. On the other hand, if a purely scientific approach is taken, one would realize that the true speed of light is neither constant nor the highest achievable speed. Clayton and Moffat (1999) have discussed the phenomenon of variable light speed. Also, Schewe and Stein (1999) discuss the

---

[5] Decided at the Seventh Conférence générale des poids et mesures (CGPM; the same acronym is used in English, standing for "General Conference on Weights and Measures." It is one of the three organizations established to maintain the International System of Units (SI) under the terms of the Convention du Mètre (Meter Convention) of 1875. It meets in Paris every four to six years.

[6] The following non-serious riddle illuminates what is going on here perfectly: "QUESTION — how many Microsoft engineers does it take to change a light-bulb? ANSWER — None; Bill Gates [founder of Microsoft] just changes the standard of normal illumination from Light to Dark."

possibility of a very low speed of light. In 1998, the research group of Lene Hau showed that the speed of light can be brought down to as low as 61 km/hour (17 m/s) by manipulating the energy level of the medium (Hau *et al.*, 1999). Two years later, the same research group reported near halting of light (Liu *et al.*, 2001). The work of Bajcsy *et al.* (2003) falls under the same category, except that they identified the tiny mirror-like behavior of the media, rather than simply low energy level. Other work on the subject deals with controlling light, rather than observing its natural behavior (Ginsberg *et al.*, 2007). Abou-Kassem *et al.* (2007) used the arguments provided by previous physicists and constructed the following graph. It is clear from the graph (Figure 3.2) that the assumption that 'speed of light,' 'vacuum,' 'unit of time,' and 'unit of distance' are some arbitrarily set constants does not change the essential truth of change, development and motion within the natural order, which remain continuously dynamic. Note that media density can be converted into media energy, only if continuous transition between energy and mass is considered. Such transition, as will be clearer in the following section, is rarely talked about in the context of engineering (Khan *et al.*, 2007). This graph also reveals that that once definitions and assertions have been accepted in face values and are not subject to further scrutiny, the possibility of increasing knowledge (as in being closer to discover the truth about nature) is diminished.

A chemical engineer has argued that the above graph is only a trivial manifestation of what Bose-Einstein theory would have predicted some 100 years ago. According to this engineer, Bose-Einstein

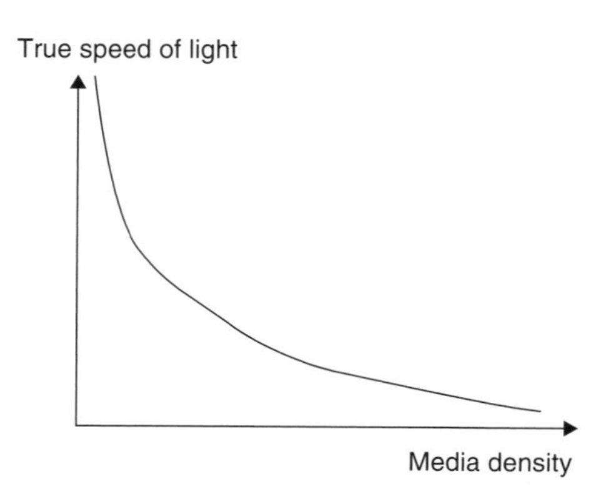

Figure 3.2 Speed of light as a function of media density (redrawn from Abou-Kassem *et al.*, 2007).

theory predicts that the speed of light within an infinitely dense medium must be zero. Similar to the attitude expressed by the food scientist (in the Introduction of the present chapter), this line of thinking unfortunately characterizes modern-day education, in which scientists become so focused on obvious tangible expressions that they are incapable of thinking beyond the most tangible aspect of the research. In 2001, Eric Cornell, Wolfgang Ketterle, and Carl Wieman were awarded the Nobel Prize in Physics "for the achievement of Bose-Einstein condensation in dilute gases of alkali atoms, and for early fundamental studies of the properties of the condensates." Their work, however, far from vindicating science of tangibles, has in fact made it more necessary to take an alternate approach. Eric Cornell's most popular invited lecture is titled "Stone Cold Science: Things Get Weird Around Absolute Zero." This 'weird'-ness cannot be predicted with the science of tangibles. One further citation of Eric Cornell is: "What was God thinking? Science can't tell" (Cornell, 2005). Far from being content that all discoveries have been made and, therefore, there is nothing more to discover, a true researcher in pursuit of knowledge readily picks up on these shortcomings of conventional thinking and, as history frequently shows, it is within the inquisitive mind that answers reside. Ironically, this would also mean that discoveries are made because one is not afraid to make mistakes and, more importantly, theorize without fear of the Establishment. Satyendranath Bose, who did not have a PhD himself, did his share of making mistakes. In fact, as the following quote from Wikipedia shows (Website 2, 2007), his findings were initially discarded by mainstream journals and he had to resort to sending Einstein his original work. It was only then that Einstein introduced the idea from light to mass (photon to atom) and the paper was finally published. If Bose had not had the tenacity, and if Einstein had not had the decency to consider thinking "outside the box," one could not begin to think what would be the state of lasers and all of the gadgets that we take for granted today.

> *"Physics journals refused to publish Bose's paper. It was their contention that he had presented to them a simple mistake, and Bose's findings were ignored. Discouraged, he wrote to Albert Einstein, who immediately agreed with him. His theory finally achieved respect when Einstein sent his own paper in support of Bose's to Zeitschrift für Physik, asking that they be published together. This was done in 1924."* (from Website 2, 2007)

In the core of any scientific research, the attitude must be such that there is no blind faith or automatic acceptance of an existing principle. In this particular case of Nobel Prize-winning work of Eric Cornell, he would not have been able to observe anomalies if he took Bose-Einstein theory as absolute true (Tung *et al.*, 2006). In fact, all experimental observations indicate there have to be major adjustments, if not re-formulations, made to the Bose-Einstein theory, as evidenced by follow-up research that won the Nobel Prize in 2005 (discussion below).

In 2005, the Nobel Prize in Physics was awarded jointly to Roy Glauber 'for his contribution to the quantum theory of optical coherence' and to John Hall and Theodor Hänsch 'for their contributions to the development of laser-based precision spectroscopy, including the optical frequency comb technique.' Indeed, these discoveries have something to do with Bose-Einstein theory, but far from being a regurgitation or redemption of the theory, they open up opportunities for new discoveries.

Even after the above series of Nobel Prize-quality research, the following questions remain unanswered. They are:

1. Bose envisioned dissimilar characteristics of any two photons, and Einstein extended that work to study with what degree of freedom two dissimilar atoms would behave within a balanced system. The work of the Nobel Physics laureate discovered the separation of certain atoms under very cold conditions, yet they concluded "all atoms are absolutely identical." What was the reason behind this conclusion? The difference between atoms could not be measured. Fleischhauer (2007) began to question this assumption by comparing millimeter distance in a lab in quantum scale with the distance between the moon and the earth in mega scale and calls 'indistinguishability' as applicable to 'far apart.' However, soon (in the same article) he yields to the pragmatic approach and concludes, "But it also shows that we are entering a state of unprecedented experimental control of coherent light and matter waves. That could bring very real technological benefits: applications that spring to mind include quantum information interfaces that allow the transfer of a quantum

bit encoded in a photon to a single atom, as well as ultra-sensitive rotation sensors and gravity detectors." One must wonder, what would happen if this pragmatic reasoning was eliminated?
2. With the above conclusion, how can we observe asymmetry in experimental observations (as reported by Tung *et al.*, 2006)? It must be noted that perfect symmetry is aphenomenal: Nature does not have any example of perfect symmetry. Such asymmetry is rarely taken in account in conventional theory (Dorogovtsev *et al.*, 2000).
3. Later work of 2005 comes up with similar conclusion about photons, with tremendous implications on light source development and monitoring (e.g., GPS). How would this explain the difference in various light sources, particularly between natural light and artificial light?
4. For both sets of discoveries, the focus is on 'engineering' phenomena rather than observing. What are the possible implications of this 'engineering' on quality of energy source or matter, e.g., nano materials that are being engineered?
5. What do these theories say about how transition between mass and energy takes place? Satyendra Nath Bose wrote (Website 2, 2007) the real gas equation of state some 100 years ago. What can be said today about real fluids that are still modeled with aphenomenal (unreal) first premises?
6. What does variable speed of light mean in terms of the definition of the unit of time? What would a paradigm shift in standards and technology development entail if one were to consider a truly natural unit of time (Zatzman, 2007)? Should the shift be to more 'engineering' or more natural?

Science news reported by (Moskowitz, 2011) has discussed the existence of a particle that can travel faster than light. Here is an excerpt:

> "Nothing goes faster than the speed of light. At least, we didn't think so.

> New results from the CERN laboratory in Switzerland seem to break this *cardinal rule* of physics, calling into question one of the most trusted *laws* discovered by Albert Einstein.
>
> Physicists have found that tiny particles called neutrinos are making a 454-mile (730-kilometer) underground trip faster than they should — more quickly, in fact, than light could do. If the results are confirmed, they could throw much of modern physics into upheaval.
>
> "The consequences would be absolutely revolutionary and very profound," said physicist Robert Plunkett of the Fermilab laboratory in Batavia, Ill., who was not involved in the new study. "That's why such a claim should be treated very carefully and validated as many ways as you can."
>
> **Rewriting the rules**
>
> The results come from the OPERA experiment, which sends sprays of neutrinos from CERN in Geneva to the INFN Gran Sasso Laboratory in Italy. Neutrinos do not interact with normal atoms, so they simply pass through the Earth as if it were a vacuum.
>
> After analyzing the results from 15,000 particles, it seems the neutrinos are crossing the distance at a velocity 20 parts per million faster than the speed of light. By making use of advanced GPS systems and atomic clocks, the researchers were able to determine this speed to an accuracy of less than *10 nanoseconds* (.00000001 seconds).
>
> "According to relativity, it takes an infinite amount of energy to make anything go faster than light," Plunkett told LiveScience. "If these things are going faster than light, then these rules would have to be rewritten." [emphasis added]

This entire hysteria about ground-breaking discovery can be rendered obvious if one simply asks for the definition of 'vacuum' (scientifically there should be no particle) and criterion used to define neutrinos as 'small particles.' Are they smaller than photons? How else would one explain a greater speed? In that case, what is the new 'mass,' considering that the mass of photons have already been declared 'zero'?

Following the above announcement, another story quickly broke in Bangladesh. Several University professors declared that they had made the same prediction in 2001 and attempted to publish in *Journal of Physics* that declined to publish because of the implication that Einstein's theory is false (Website 33; Website 44). What seems to be an encore of what happened to Bose some 100 years ago was put in the same vein. The excerpt follows.

> ***Faster Than Speed of Light***
>
> ***Bangladeshis claim formulating equations***
> Anwar Ali, Rajshahi
>
> Two Rajshahi physicists yesterday claimed to have formulated three equations and several fundamental constants suggesting that subatomic particles move faster than the speed of light.
>
> Speaking at a press conference at Juberi Bhaban of Rajshahi University (RU), one of the physicists Dr. M Osman Gani Talukder said they analysed the existing theories to reach their findings.
>
> Dr. Mushfiq Ahmad, another physicist who retired from RU Physics department in 2010, was present.
>
> Much of modern physics depends on the idea that nothing can exceed the speed of light, which was laid out in part by Albert Einstein in his theory of special relativity.
>
> On September 23, Cern, the world's largest physics lab, published its experiment results showing that neutrino particles move 60 billionths of a second faster than what they would have done if travelling at the speed of light, says a BBC report.
>
> However, Dr. Osman, a former teacher of RU applied physics and electronic engineering department, claimed their findings would not contradict Einstein, rather, help complete the theory.

Clearly, these authors are being careful not to contradict the first premise that Einstein can be wrong.

3.2.1 Role of First Premise

In our research group, we repeatedly emphasized that a false first premise will make the conclusion inherently false, even if at times it appears to be consistent with knowledge at a particular time (Zatzman and Islam, 2007; Islam and Khan, 2007; Islam *et al.*, 2010a; Islam *et al.*, 2010b; Islam and Khan, 2011). Historically, such a limit on cognition was emphasized by Averröes.

Nearly a millennium ago, long before the Renaissance reached Europe, Averröes (1126–1198 AD, known as Ibn Rushid outside of the Western world) pointed out that Aristotelian logic of the

New science has been all about eliminating history (or time function). With that model, chemicals are chemicals, heat is just heat, light is just light, truth is the perception of today, and the list goes on. The same model would put Lenin and the Tsar in the same picture and remove any information that would make it possible to cover up the past information.

excluded middle cannot lead to increasing knowledge unless the first premise is true. In other words, the logic can be used only to differentiate between true and false, as long as there is a criterion that discerns the truth from falsehood. The difficulty, all the way to the present modern age, has been the inability to propose a criterion that is time-honored (Zatzman and Islam, 2007a). For Averröes, the first premise was the existence of the (only) creator. It was not a theological sermon or a philosophical discourse, it was purely scientific. Inspired by the Qur'an that cites the root word *ilm* (علم, meaning 'science,' the verb, *yalamu*, يلم, standing for 'to understand,' something that is totally subjective and is the first step toward acquiring knowledge), second most frequently (only second to the word, *Allah*), he considered the Qur'an as the only available, untainted communication with the creator and linked the first premise to the existence of such a communication. Averröes based his logic on post-Islamic Arabia (630 AD onward). As stated earlier, similar to ancient India and China, Islamic philosophy outlined in the Qur'an did include intangibles, such as intention and the time function. The modern-day view holds that knowledge and solutions developed from and within nature might be either good, or neutral [zero net impact] in their effects, or bad, all depending on how developed and correct our initial information and assumptions are.

Averröes, known as the father of secular philosophy in Europe, promoted true source (the Creator) and continuous logic (science) as the requirement for increasing knowledge. Thomas Aquinas, known the father of doctrinal philosophy in Europe, promoted papal authority as the source, and holy writ as the means of all logic. Few realize that doctrinal approach is the governing principle of all aspects of European cognition, ranging from political science to hard core science, from social science to medical science, and that only the name of the 'Authority' has changed in the process.

A gold-plated bible (the Book) in a museum in Moscow. Fascination for external and tangible in matters of knowledge in the modern era dates back to Thomas Aquinas, who replaced the Qur'an, that has only one version (in original Arabic) since compilation some 1400 years ago with the Bible, that has over 40 versions in English alone.

A typical building in Paris. In modern engineering, primary focus is given to aesthetics and quarterly revenue for the industry.

The view of science in the period of Islam's rise was rather different. It was that since nature is an integrated whole in which humanity also has its roles, any knowledge and solutions developed according to how nature actually works will be *ipso facto* positive for humanity. Nature possesses an inbuilt positive intention of which people have to become conscious in order to develop knowledge and solutions that enhance nature. On the other hand, any knowledge or solutions developed by taking away from nature or going away from nature would be unsustainable. This unsustainability would mark such knowledge and solutions as inherently anti-nature.

Based on those intangibles, great strides in science, medicine, and all aspects of engineering was made in the Islamic era. Much of this was preserved, but by methods that precluded or did not include general or widespread publication. Thus, there could well have been almost as much total reliable knowledge 1400 years ago as today, but creative people's access and availability to that mass of reliable knowledge would have been far narrower. It has been established that Islamic scholars were doing mathematics some 1000 years ago of the same order that are thought to be discovered in the 1970s (Lu and Steinhardt, 2007), with the difference being that our mathematics can only track symmetry, something that does not exist in nature (Zatzman and Islam, 2007a). Knowledge definitely is not within the modern age. A three-dimensional PET-scan of a relic known as the 'Antikythera Mechanism' has demonstrated that it was actually a universal navigational computing device, with the difference being that our current-day versions rely on GPS, tracked and maintained by satellite (Freeth *et al.*, 2006). Ketata *et al.* (2007a) recognized that computational techniques that are based on ancient, but nonlinear counting techniques, such as the abacus, are far more superior to linear computing. Even in the field of medicine, one would be shocked to find out that what Ibn Sina ('Avicenna') said regarding nature being the source of all cure still holds true (Crugg and Newman, 2001), with the proviso that not a single quality given by nature in the originating source material of, for example, some of the most advanced pharmaceuticals used to "treat" cancer remains intact after being subject to mass production, and is accordingly stripped of its powers actually to cure and not merely "treat," i.e., delay, the onset or progress of symptoms. Therefore, there are examples from history that show that knowledge is directly linked with intangibles, and in fact, only when intangibles are included that science leads to knowledge (Vaziri *et al.*, 2007).

Thomas Aquinas (1225-1274 AD) took the logic of Averröes and introduced it to Europe with a simple, yet highly consequential modification: he would color the (only) creator as God and define the collection of Catholic church documentation on what eventuated in the neighborhood of Jerusalem some millennium ago as the only communication of God to mankind (hence the title, Bible, meaning the (only) Book). If Aristotle was the one who introduced the notion of removing intention and time function from all philosophical discourse, Thomas Aquinas is the one who legitimized the concept and introduced this as the only science (as in the process to gaining knowledge). Even though Thomas Aquinas is known to have adapted the logic of Averröes, his pathway as well as prescribed origin of acquiring knowledge was diametrically opposite to the science introduced by Averröes. This is because the intrinsic features of both God and Bible were dissimilar to the (only) creator and the Qu'ran, respectively (Armstrong, 1994). For old Europe and the rest of the world that it would eventually dominate, this act of Thomas Aquinas indeed became the bifurcation point between two pathways, with origin, consequent logic, and the end being starkly opposite. With Aristotle's logic, something either is or is not: if one is 'true,' the other must be false. Because Averröes' 'the creator' and Thomas Aquinas's 'God' both are used to denominate monotheist faith, the concept of science and religion became a matter of conflicting paradox (Pickover, 2004). Averröes called the (only) creator as 'The Truth' (in Qur'anic Arabic, the word 'the Truth' and 'the Creator' refer to the same entity). His first premise pertained to the book (the Qu'ran) that said, "Verily unto Us is the first and the last (of everything)"(89.13). Contrast this to a "modern" view of a creator. In Carl Sagan's words (Hawking, 1988), "This is also a book about God…or perhaps about the absence of God. The word God fills these pages. Hawking embarks on a quest to answer Einstein's famous question about whether God had any choice in creating the universe. Hawking is attempting, as he explicitly states, to understand the mind of God. And this makes all the more unexpected the conclusion of the effort, at least so far: a universe with no edge in space, no beginning or end in time, and nothing for a Creator to do."

This divergence in pathways was noted by Zatzman and Islam (2007a). Historically, challenging the first premise, where the divergence is set, has become such a taboo that there is no documented case of anyone challenging it and surviving the wrath of the Establishment (Church alone in the past, Church and Imperialism after the Renaissance). Even challenging some of the cursory premises have been hazardous, as demonstrated by Galileo. Today, we continue to avoid challenging the first premise, and even in the information age it continues to be hazardous, if not fatal, to challenge the first premise or secondary premises. It has been possible to keep this modus operandi because new "laws" have been passed to protect 'freedom of religion' and, of late, 'freedom of speech.' For special interest groups, this opens a Pandora's box for creating 'us *vs* them,' 'clash of civilizations,' and every aphenomenal model now in evidence (Zatzman and Islam, 2007b).

Avoiding discussion of any theological nature, Zatzman and Islam (2007a) nevertheless managed to challenge the first premise. Rather than basing the first premise on the truth *à la* Averröes, they mentioned the importance of individual acts. Each action would have three components: 1) origin (intention); 2) pathway; and 3) consequence (end). Averröes talked about origin being the truth; they talked about intention that is real. How can an intention be real or false? They equate real with natural. Their work outlines fundamental features of nature and shows there can be only two options: natural (true) or artificial (false). The paper shows Aristotle's logic of anything being 'either A or not-A' is useful only to discern between true (real) and false (artificial). In order to ensure the end being real, the criterion developed by Khan (2006) and Khan and Islam (2007b) is introduced. If something is convergent when time is extended to infinity, the end is assured to be real. In fact, if this criterion is used, one can be spared of questioning the 'intention' of an action. If any doubt, one should simply investigate where the activity will end up if time, t goes to infinity.

This absence of discussion of whatever happened to the tangible-intangible nexus involved at each stage of any of these developments is no merely accidental or random fact in the world. It flows directly from a Eurocentric bias that pervades, well beyond Europe and North America, the gathering and summation of scientific knowledge everywhere. Certainly, it is by no means a property inherent, either in technology as such, or in the norms and demands of the scientific method *per se*, or even within historical development, that time is considered so intangible as to merit being either ignored as a fourth dimension, or conflated with tangible space as something varying independently of any process underway within any or all dimensions of three-dimensional space. Mustafiz *et al.* (2007) have identified the need of including a continuous time function as a starting point of acquiring knowledge. According to them, the knowledge dimension does not get launched unless time as a continuous function is introduced. They further show that the knowledge dimension is not only possible, it is necessary. The knowledge is conditioned not only by the quantity of information gathered in the process of conducting research, but also by the depth of that research, i.e., the intensity of one's participation in finding things out. In and of themselves, the facts of nature's existence and of our existence within it neither guarantee nor demonstrate our consciousness of either, or the extent of that consciousness. Our perceptual apparatus enables us to record a large number of discrete items of data about the surrounding environment. Much of this information we organize naturally, and indeed unconsciously. The rest we organize according to the level to which we have trained, and/or come to use, our own brains. Hence, neither can it be affirmed that we arrive at knowledge directly or merely through perception, nor can we affirm being in possession at any point in time of a reliable proof or guarantee that our knowledge of anything in nature is complete.

Historically, what Thomas Aquinas' model did to European philosophy is the same as what Newton's model did to the new science (Figure 3.3). The next section examines Newton's models. Here, it would suffice to say that Newton's approach was not any different from the approach of Thomas Aquinas, or even Aristotle. One exception among scientists in Europe was Albert Einstein, who introduced the notion of time as the fourth dimension. However,

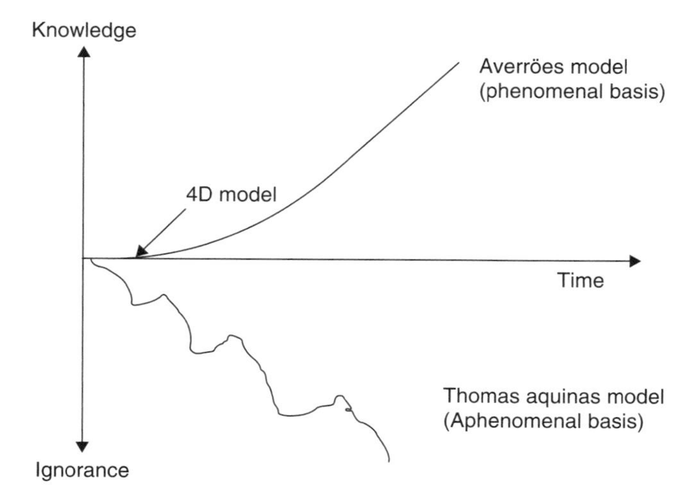

Figure 3.3 Logically, a phenomenal basis is required as the first condition to sustainable technology development. This foundation can be the truth as the original of any inspiration or it can be 'true intention,' which is the essence of intangibles (modified from Zatzman and Islam, 2007a and Mustafiz *et al.*, 2007). By using aphenomenal basis, all new science cognition has followed the doctrinal philosophy.

Newton, who confessed to have plotted to murder his mother and stepfather, wrote more about Christianity than on science, and held from the public most of his Alchemy work that was dedicated to turning lead into gold. Newton was readily accepted by and was the darling of the British monarch. Prior to Newton, most scientists were persecuted by the Establishment, such as Galileo, Copernicus, Mendeleev, Boltzman (committed suicide), Pythagoras (committed suicide), Lavoisier (executed), and many others.

his mathematical derivation used Maxwell's theory as the basis. Maxwell himself used the Newtonian mechanics model. In addition, no one followed up on the continuous time function aspect of Einstein's work, and it was considered that the addition of a time term in the Newton's so-called steady state models would suffice. Mustafiz (2007) recognized the need of including the time dimension as a continuous function and set the stage for modeling science of intangibles (Abou-Kassem *et al.*, 2007).

3.3 Other Considerations in Mathematics and Science

In both Sanskrit and Arabic, grammatically 'number' is singular, two-fold, or 'plural' (meaning more than two). This is important because adjectives and verbs have to agree with the number of the noun serving as the subject of the sentence. In all European languages, there is simply singular, meaning one only or fraction-of-one, and plural, meaning more-than-one.

In the tribal languages that we know about or that still exist and that go back to the period from 40000 to 12000 years ago (before the last Ice Age), there is no way to enumerate quantities past about 20 or 30. This might have something to do with how many lived together or in close contact in communities BEFORE versus AFTER the emergence of settled locales based on agriculture and animal husbandry.

The ancient Greeks have only very general ideas of large numbers, thus there is the myriad, meaning 10000, but it is for counting much larger numbers in "units" of 10000.

Ancient India houses several mathematical traditions that tackle very large numbers, most interestingly the Jaina mathematicians. It seems instead of fractions and rational numbers, the Jaina mathematicians redefine a whole so it has a very, very large number of "units" which exist in nature in some meaningful form. So, for example, the circuit of the sun around the earth can be measured in minutes of arc; that becomes the unit. Then the sun's circuit during the year is defined by the hundreds of millions of minutes in the total arc. The next threshold number is the number of such years that "the world" has existed according to the Jain priests' calculations, which is then crores of lakhs of minutes-of-arc.

Although schemas of this kind neatly skirt the apparent contradictions of fractions (what logic does a fraction have within a system that enumerates units by a decimal positioning system?), decimal representation was entirely normal in Indian maths of the late classical period, because this was or could be made consistent with the positioning system of representing number-values. However, these were numbers that existed as the result of subdividing by 10s, 100s, 1000s etc.

What the Indian mathematicians did not come up with was anything that would "reduce" 25 over 100, which could be represented as "0.25," to "1/4". And going the other way, Egyptian and Babylonian mathematics had fractions, but not positional representation of numerical value, so they, too, could not make the link.

Well into the 19th century, a value such as "66-2/3" was meaningful mainly only as the aliquot third part of 200. As "66.666..." on a number line, just ahead of 66.5 and just less than 66.75, it would have been meaningful mainly to people like the research mathematician Riemann, who was trying in the 1860s to define the real line as an infinitely divisible and ultimately non-denumerable continuum of numbers. There was little, if any, notion of physically measuring such a quantity from any natural phenomenon.

3.3.1 Numbers in the Qur'an

As mentioned earlier, the Qur'an is the only document that has survived in its original form for over 1400 years. This is also the only document that has used the word *ilm* (science, as in the process) a great many times. With this science, great strides were made for many centuries toward increasing knowledge and making tangible progress in human society. However, focus on tangibles was not the reason for this tangible progress. It was instead the focus on intangibles. Just after the introductory chapter, Chapter 2 (verses 3-4) outlines the conditions for true guidance. The first condition is set out to be 'believing in the intangibles' (the Arabic word being, "Al-Ghaib", الغيب).

ٱلَّذِينَ يُؤْمِنُونَ بِٱلْغَيْبِ وَيُقِيمُونَ ٱلصَّلَوٰةَ وَمِمَّا رَزَقْنَـٰهُمْ يُنفِقُونَ (٣[3])

(approximate translation of the meaning: those who believe in the intangible, establishes communication, and spends in charity from their lawful belongings)

It is no surprise that this document has the mention of very few numbers, all of which nevertheless have very significant meaning that is the subject of future research. In this chapter, we will simply state the numbers that appear in the Qur'an. To begin with, there is no word in Arabic that would bear similar meaning as in European mathematics. The word most commonly used as an equivalent to number is 'rukm,' رقم, as in Chapter 83, verse 9, كِتَٰبٌ مَّرْقُومٌ, *kitabum markum*, which means 'written record.' The word rukm has a number of other meanings, such as ornament, engrave, mark or sign, all indicating 'something to be known.' It seems this word in Arabic stands for tangible expressions of an intangible, and not something that is an image or aphenomenal, valid only for a time, t = 'right now.' Note that there is no mention of zero in the Qur'an, even though the Arabic equivalent word, *cipher*, existed even before the Qur'an came into tangible existence. There are *hadiths* of the prophet mentioning the word *cipher*, which stood for 'nothing' or 'empty.' This word, in fact, has been adapted into the English language (as used in 'decipher'), once again standing for 'meaningless' or empty zeros. At the same time, it is a place-holder: it can serve a function during computation without itself being a number. This raises an interesting point. According to the science of tangibles, in the matter of chemical reactions that can only take place in the presence of certain catalysts, it is repeated everywhere with almost the force of religious fervor: a catalyst does not take part in the reaction! Why don't the proponents of this "logic" not insist just as fervently that 0 as a placeholder takes no part in the computation? The introduction of 0 in computation was introduced by the Arabs, who adapted the concept and the symbol from ancient India. The influence of India in post-prophet Muhammad Arab mathematics is so dominant that the Indian numerals even replaced original Arab numerals (now adopted in European mathematics) that were actually more scientific than the Indian counterpart (e.g., every numeral represents the numeric value by the number of angle that it forms).

The Qur'an did not use any numerals, but used the following numbers in various concepts (some examples are given).

Number One (in the context of 'there is only one *ilah*[7]'):
(قل إنما هو اله **واحد** وإنني بريء مما تشركون)
(سورة الأنعام, الآية 19) Chapter 6, verse 19

---

[7] The word *ilah* is erroneously translated as 'god,' while the actual meaning is closer to shelter-giver, as in husbandry. *Al-Ilah* stands for 'the only shelter giver,' which might explain why it is commonly translated as the 'God.'

Number Two (in the context of 'do not choose two *ilahs*'):
(قال تعالى:(وقال الله لا تتخذوا إلهين ***اثنين*** إنما هو إله واحد
(سورة النحل, الآية 51) Chapter 17, verse 51

Number Three (in the context of 'not uttering Three (Trinity), stop because Allah is the only One *ilah*'):
(ولا تقولوا ***ثلاثة*** انتهوا خيرا لكم)
(سورة النساء, الآية 171) Chapter 4, verse 171

Number Four (in the context of allowing aphenomenal model worshippers for months for free movement):
(فسيحوا فالأرض ***أربعة*** أشهر)Chapter 9, verse 2
(سورة التوبة, الآية 2)

Number Five (in the context of the actual number of men in the story of the people of the Cave):
(ويقولون ***خمسة*** سادسهم كلبهم رجما بالغيب)
(سورة الكهف، الآية 22) Chapter 18, verse 22

Number Six (in the context of creating the Universe in six *yaums*'[8]):
(إن ربكم الله الذي خلق السماوات و الأرض في ***ستة*** أيام)
(سورة الأعراف,الآية 54) Chapter 7, verse 54

Number Seven (in the context of seven gates in hell):
(ها ***سبعة*** أبواب لكل باب منهم جزء مقسوم) Chapter
(سورة الحجر, الآية 44) Chapter 15, verse 44

Number Eight (in the context of eight angels upholding the Throne on the day of judgment, when there would be no secret):
(ويحمل عرش ربك فوقهم يومئذ ***ثمانية***)
(سورة الحاقة, الآية 17) Chapter 69, verse 17

Number Nine (in the context of nine persons making mischief and not reforming, during the period of *Thamud*, the ones who built the crystal valley in Jordan):
(وكان في المدينة ***تسعة*** رهط يفسدون في الأرض)
(سورة النمل, الآية 48) Chapter 27, verse 48

---

[8] The word *yaum* is errorneously translated as 'day.' It is closer to phase. This is supported by other statements that are indicative of phase and not 'day,' for instance, 'the day of judgment' is called '*yaum addin*.' For 'day,' as in daylight, there is a different word in Arabic, *nahar*. Also, in Arabic, Earth day begins when the sun sets.

Number 10 (in the context of fasting for 10 days if a pilgrim cannot afford to sacrifice an animal):
((تلك **عشرة** كامل))
(سورة البقرة, لآية 196) Chapter 2, verse 196

Number 11 (in the context of Prophet Joseph's description of his dream of 11 planets[9], the sun, and the moon prostrating toward him):
أحد عشر كوكبا والشمس و القمر رأيتهم لي ساجدين
(سورة يوسف, لآية 4) Chapter 12, verse 4

Number 12 (in the context of the number of months in a year as per Divine order):
إن عدة الشهور عند الله ***اثنا عشر*** شهرا في كتاب الله
(سورة التوبة, الآية 36) Chapter 9, verse 36

Number 19 (in the context of
(عليها ***تسعة عشر***
(سورة المدثر, الآية 30) Chapter 74, verse 30

Number 20 (in the context of 20 steadfast righteous people overcoming attack from 200 unrighteous ones):
إن يكن منكم ***عشرون*** صابرون يغلبوا مائتين
(سورة الأنفال,الآية 65) Chaper 8, verse 65

Number 30 (in the context of ordained bearing and weaning of a child for 30 months and maturity for man by 40 years):
(وحمله وفصاله ***ثلاثون*** شهرا
(سورة الأحقاف, الآية 15) Chapter 46, verse 15

Number 40 (in the context of assigning 40 nights of solitude):
(وإذ واعدنا موسى ***أربعين*** ليلة ثم اتخذتم العجل من بعده وأنتم ظالمون
(سورة البقرة, الآية 15) Chapter 2, verse 15

Number 50 (in the context of Noah spending 950 (1000 minus 50) years before the flood struck):
(ولقد أرسلنا نوحا إلى قومه فلبث فيهم ألف سنة إلا ***خمسين*** عاما
(سورة العنكبوت, الآية 14) Chapter 29, verse 14

[9] The Arabic word is كَوكَبًا, *kawakeb*. There are other words in Arabic to describe planets as well. The distinction of these terminologies is beyond the scope of this paper.

Number 60 (in the context of feeding 60 needy people in case someone is unable to fast two consecutive months as penalty to violating family code of conduct):

(فمن لم يستطع فإطعام**ستين** مسكينا

(سورة المجادلة, الآية 4) Chapter 58, verse 4

Number 70 (in the context of the length of the chain holding captives in hell):

(ثم في سلسة ذرعها**سبعون** ذراعا فسلكوه

(سورة الحاقة, الآية 32) Chapter 69, verse 32

Number 80 (in the context of punishment in the form of number of lashes for unfounded allegation against a chaste woman):

فاجلدوهم**ثمانون** جلدة ولا تقبلوا منهم شهادة أبدا

(سورة النور, الآية 4) Chapter 24, verse 4

Number 90 (in the context of two disputants (brothers) coming to David and asking to settle the dispute, one of them having nine and 90 ewes still wanted the other one's only one):

وهذا أخي له تسع و**تسعون** نعجة ولي نعجة واحدة

(سورة ص, الآية 22) Chapter 38, verse 22

Number 100 (in the context of someone being dead for 100 years yet answering "a day or part of a day" when asked "How long did you tarry (thus)?"):

قل بل لبثت**مائة** عام

(سورة البقرة, الآية 259) Chapter 2, verse 259

Number 200 (In the context of 20 righteous with steadfastness vanquishing 200 of the attackers):

إن يكن منكم عشرون صابرون يغلبوا **مائتين**

(سورة الأنفال,الآية 65) Chapter 8, verse 65

Number 300 (In the context of habitants of the cave sleeping for 300 years and nine days):

ولبثوا في كهفهم**ثلاث مائة** سنين و ازدادوا تسعا

(سورة الكهف, الآية 25) Chapter 18, verse 25

Number 1000 and 2000 (in the context of 1000 righteous people that are steadfast vanquishing 2000 attackers)[10]:

وإن يكن منكم**ألف** يغلبوا**ألفين** بإذن الله

(سورة الأنفال, الآية 66) Chapter 8, verse 66

---

[10] Note how the same verse stated earlier 20 righteous people vanquishing 200 attackers, as compared to 1000 righteous vanquishing 2000 attackers – the relationship is definitely not linear

Number 3000 (in the context of 3000 angels sent down to help the righteous who were under attack by wrongdoers):
إذ تقول للمؤمنين ألن يكفيكم أن يمدكم ربكم **بثلاثة آلاف** من الملائكة منزلين
(سورة آل عمران, الآية 124) Chapter 3, verse 124

Number 5000 (in the context of increasing the number of angels from 3000 to 5000 in order to help the righteous (see above)):
هذا يمددكم ربكم **بخمسة آلاف** من الملائكة مسومين
(سورة آل عمران, 125) Chapter 3, verse 125

Number 100,000 (in the context of Jonah being dispatched to a locality of 100,000 or more who were initially misguided):
وأرسلناه إلى **مائة ألف** أو يزيدون
(سورة الصافات, الآية 147), Chapter 37, verse 147

3.3.1.1 *Fractions*

Fractions in the Qur'an are principally used to denote natural division, as applied in sharing or just division of properties. Consequently, most divisions deal with family heritage laws. Following are some examples.

Number One Over Two (in the context of family heritage law):
ولكم **نصف** ما ترك أزواجكم إن لم يكن لهن ولد
(سورة النساء, الآية 12) Chapter 4, verse 12

Number One over Three (in the context of family heritage law):
إن لم يكن له ولد وورثه أبواه فلأمه **الثلث**
(سورة النساء, الآية 11) Chapter 4, verse 11

Number One over Four (in the context of family heritage law):
(قال تعالى:(فإن كان لهن ولد فلكم **الربع** مما تركن
(سورة النساء, الآية 12) Chapter 4, verse 12

Number One over Five (in the context of distribution of war booty[11]):
واعلموا إنما غنمتم من شيء فإن لله **خمسه**
(سورة الأنفال, الآية 41) Chapter 8, verse 41

[11] It is important to note the war booty in Qur'anic context applies strictly to defensive war. There was only one offensive war during the entire period of the revelation of the Qur'an. This war (the last one) had no casualty, generated no war booty, and was followed by general amnesty for the very people who attacked the Prophet's group in the form of four wars in less than 10 years.

Number One over Six (in the context of family heritage law):
فإن كان له إخوة فلأمه **السدس**
(سورة النساء,الآية 11) Chapter 4, verse 11

Number One over Eight (in the context of family heritage law):
(قال تعالى:(فإن كان لكم ولد فلهن **الثمن** مما تركتم
(سورة النساء, الآية 12) Chapter 4, verse 12

3.3.1.2 Natural Ranking

Various numeral rankings are used in the Qur'an. Some of the examples follow:

First (in the context of first one to be obedient to Allah):
قول إني أمرت أن أكون **أول** من أسلم
(سورة الأنعام, الآية 14) Chapter 6, verse 14

Second (in the context of the second one as a companion of the prophet when he was migrating from Mecca to Medina, being driven out by Arab pagans)[12]:
إلا تنصروه فقد نصره الله إذ أخرجه الذين كفروا **ثاني ثنين**
(سورة التوبة, الآية 40) Chapter 9, verse 40

Third (in the context of adding a third messenger because first two were not heeded):
إذ أرسلنا إليهم اثنين فكذبوهما فعززنا **بثالث** فقالوا إنا إليكم مرسلون
(سورة يس, الآية 14) Chapter 36, verse 14

Fourth (in the context of no secret can be kept from Allah as he makes the fourth when three human beings consult in secret):
ما يكون من نجوى ثلاثة إلا هو **رابعهم**
(سورة المجادلة, الآية 7) Chapter 58, verse 7

Fifth (in the context of taking a fifth oath invoking one's innocence in absence of material witness):
والخامسة أن لعنة الله عليه إن كان من الكاذبين
(سورة النور, الآية 7) Chapter 24, verse 7

Sixth (in the context of the sixth companion (a dog) of the inhabitants of the cave):

---

[12] Note that second, ثَانِى (*thani*), in Arabic only applies to ranking. The unit of time, 'second' is incorrectly translated in Arabic as *thani*. Such artificial time unit does not exist in Qur'anic Arabic (Zatzman, 2007).

ويقولون خمسة ***سادسهم*** كلبهم رجما بالغيب
(سورة الكهف, الآية 22) Chapter 18, verse 22

Eighth (in the context of the eighth companion (a dog) of the inhabitants of the cave, as hearsay):
ويقولون سبعة ***وثامنهم*** كلبهم
(سورة الكهف, الآية 22) Chapter 18, verse 22

3.3.1.3 Mathematical Operations

Mathematical operations in the Qur'an are all natural, i.e., addition to increase, subtract to decrease, multiply to strike or counter, and divide to breakdown in among natural recipients or compartments. No operation that is spurious would yield any result. These operations were mainly used by 7th century-onward Muslims to calculate compulsory charity as well as just shares of heritage properties. Examples are given below:

Addition (in the context of 300 and 9 years spent by the inhabitants of the cave):
قال تعالى:(ولبثوا في كهفهم ***ثلاث مائة سنين و ازدادوا تسعا***
(سورة الكهف, الآية 25) Chapter 8, verse 25
The operation is: 300 + 9

Subtraction (in the context of Noah spending 1000 minus 50 years with his tribe):
ولقد أرسلنا نوحا إلى قومه فلبث فيهم ***ألف سنة إلا خمسين عاما***
(سورة العنكبوت, الآية 14) Chapter 29, verse 14
The operation: 1000 – 50

Multiplication (in the context of righteous people fighting attackers and vanquishing them even when outnumbered)[13]:
إن يكن منكم ***عشرون*** صابرون يغلبوا ***مائتين***

[13] The root word used for 'multiplication' in Arabic is ضرب, which stands for numerous meanings. For example (Chapter and verse numbers from the Qur'an are indicated):

To travel, to get out: 3:156; 4:101; 38:44; 73:20; 2:273
To strike: 2:60,73; 7:160; 8:12; 20:77; 24:31; 26:63; 37:93; 47:4
To beat: 8:50; 47:27
To set up: 43:58; 57:13
To give (examples): 14:24,45; 16:75,76,112; 18:32,45; 24:35; 30:28,58; 36:78; 39:27,29; 43:17; 59:21; 66:10,11

وإن يكن منكم **ألف** يغلبوا **ألفين** بإذن الله

(سورة الأنفال, الآية 65, 66) Chapter 8, verses 65, 66

The operation: 20 striking against 200; 1000 striking against 2000

Division (in the context of fractions allocated to various recipients of a heritage) :

إن لم يكن له ولد وورثه أبواه فلأمه **الثلث** فإن كان له إخوة فلأمه **السدس**

(سورة النساء, الآية 11) Chapter 4, verse 11

3.3.1.4 The Number Seven and its Multiple in the Qur'an

In the last year, a book called إشراقات الرقم سبعة في القران الكريم, which won the Dubai International Holy Qur'an Award, talked about the number seven in the Qur'an. The book mentioned many things about the number seven and its multiple and here are a few examples:

1-The fact that there are seven heavens[14] were said seven times in the Qur'an:

(سورة البقرة, الآية 29) Chapter 2, verse 29
(سورة الإسراء, الآية 44)Chapter 17, verse 44
(سورة المؤمنون, الآية 86)Chapter 23, verse 86
(سورة فصلت, الآية 12) Chapter 41, verse 12
(سورة الطلاق, الآية 12) Chapter 65, verse 12
(سورة الملك, الآية 3) Chapter 67, verse 3
(سورة نوح, الآية 15) Chapter 71, verse 15

2- The fact of creation the earth and heavens in six *yaums* were also mentioned seven times:

(سورة الأعراف, الآية 54) Chapter 7, verse 54
(سورة يونس, الآية 3) Chapter 10, verse 3
(سورة هود,الآية 7) Chapter 11, verse 7
(سورة الفرقان, الآية 59) Chapter 25, verse 59
(سورة السجدة,الآية 32) Chapter 32, verse 32

To take away, to ignore: 43:5

- To condemn: 2:61
- To seal, to draw over: 18:11
- To cover: 24:31
- To explain: 13:17

[14] In the Qur'an, paradise (*jannah* or garden) is different from heaven (celestial). The سَمَٰوَٰتٍ :word used is which also stands for sky or a layer of stars.

(سورة ق, الآية 38) Chapter 50, verse 38
(سورة الحديد, الآية 4) Chapter 57, verse 4

3- The number of السور and the number of الآيات between the first time number seven was mention and the last time are a multiple of number seven:

First time (سورة البقرة, (2) الآية 29) Chapter 2, verse 29
Last time (سورة النبأ, (78) الآية 12) Chapter 78, verse 12
So, that:
78 – 2 = 77 = 7 * 11
The numbers of الآيات between them are 5649 = 7 * 807

4- The number of السور (Chapters) that begin with the praise of Allah are seven:

(سورة الإسراء, الآية 1)Chapter 17, verse 1
(سورة الحديد, الآية 1) Chapter 57, verse 1
(سورة الحشر, الآية 1) Chapter 59, verse 1
(سورة الصف, الآية 1) Chapter 61, verse 1
(سورة الجمعة, الآية 1) Chapter 62, verse 1
(سورة التغابن, الآية 1) Chapter 64, verse 1
(سورة الأعلى, الآية 1) Chapter 87, verse 1

3.4 Modeling Natural Phenomena in Multiple Dimensions

In relatively recent history, Einstein was the first scientist to recognize time as the fourth dimension. This recognition explained the transition from mass to energy and made it possible for scientists to consider combining mass- and energy-balance equations. The first use of the fourth dimension, however, was not by a physicist. Hermann Minkowski (1864–1909), a mathematician, used the fourth dimension to 'solve' the space-time continuum. In 1914, Gunnar Nordström (1881–1923), a theoretical physicist, included four dimensions in his gravitational theory, a theory that can be called the first general theory of relativity (Ravndal, 2003). His work on splitting of five-dimensional space into Einstein's and Maxwell's equations in four dimensions was the first recognition of more than four dimensions in modeling reality. Years later (1921), Kaluza (1885–1954) combined Maxwell's theory of electromagnetism and Einstein's theory of general relativity to develop a unified theory (Kaku and O'Keefe, 1994). This theory was useful for modeling galaxies, solar systems, and spacecraft for outer space travel. In 1926, Oskar Klein (1894–1977) hypothesized that the fourth spatial dimension is curled up in a circle of very small radius, so that a particle moving a short distance along that axis would return to where it began. This is perhaps the first non-linear, albeit circular, approximation of infinitely small distance in multi-dimensional analysis (as opposed to Newton's Δx approaching linearity). The distance a particle can travel before reaching its initial position is said to be the size of the dimension. This extra dimension is a compact set, and the phenomenon of having a space-time with compact dimensions is referred to as compactification. Following this initial development, many researchers worked on the development of string theory and its variations (e.g., superstring theory), a detailed description of which is available in a recent textbook (Becker *et al.*, 2007). It is generally recognized that string theory or its variations do not explain physical phenomena. However, only Abou-Kassem *et al.* (2007) attributed this shortcoming to spurious assumptions of various models and presented them in line with aphenomenal models that are incorrectly promoted as the correct models. One of these assumptions involves the cylindrical shape assumption. The research group of Wessen (e.g., Wessen *et al.*, 2000) eliminated this assumption and was 'successful' in solving space-time-continuum problems (Wessen, 2002). With time, many more dimensions have been recognized (as many as 11, as discussed by Caroll, 2004), but few have ventured into the possibility of modeling infinite dimension. Yet, this is precisely what is needed in solving natural phenomena (Mustafiz *et al.*, 2007).

In groundwater and hydrology applications, Bear (1972) is credited to highlighting the need for including greater dimensions in order for a natural value to be representative. He introduced the concept of Representative Elemental Volume (REV) that would essentially mean that unless a certain dimension is used, the value can fluctuate to such an extent that it is not meaningful. Islam (2002) used the same concept to introduce through numerical modeling that the volume of the sample is of extreme importance, both in physical and numerical modeling. Figure 3.4 shows his finding.

In the previous section, the shortest distance between two points has been discussed in the context of dimensionality. Such discussion is also of importance in determining the relationship between truth and dimensionality. Decades ago, in a discussion on

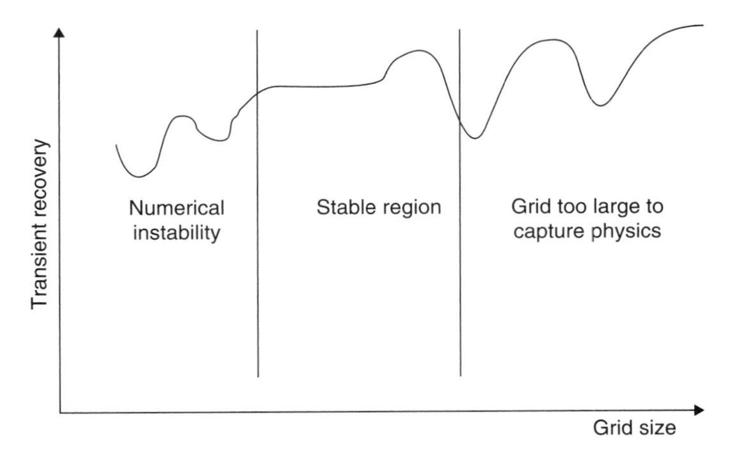

Figure 3.4 The effect of numerical grid sizes on the prediction of fluid flow behavior (From Islam, 2002).

fractals, a professor asked the allegorical question: How long is the Mississippi River? (Prof. M.R. Islam, personal communication). It was in line with the groundbreaking work of Benoît Mandelbrot (1967), who would later be known as the father of fractals. The intention behind this question was to demonstrate that the distance between two points depends on the scale used. It is easily understood by discussing the distance measured by an airplane, by a surveyor that uses a scale of 10s of meters, a surveyor that uses scale of 1 meter, a surveyor that uses his feet, all the way down to an ant that measure its feet to measure distances over pebbles. Figure 3.5 shows the effect of distance reported between two points for various dimensions used. To date, 12 dimensions have been identified. It is conceivable that this trend will continue and in the future many more dimensions will be revealed. The question one should ask is, what is the true distance between the two points that are being monitored in the following figure? As the dimension is increased, the distance is increased. Is there a limit to this trend? This was not the question that was answered by Benoît Mandelbrot. Instead, he introduced the concept of self-similarity and revolutionized 20th century mathematics. However, a close scrutiny of nature would reveal such self-similarity is nonexistent, making the fractal mathematics aphenomenal.

Mustafiz *et al.* (2007) discussed the effect of linearization (which means the dimensions are reduced to 1) on the predicted value.

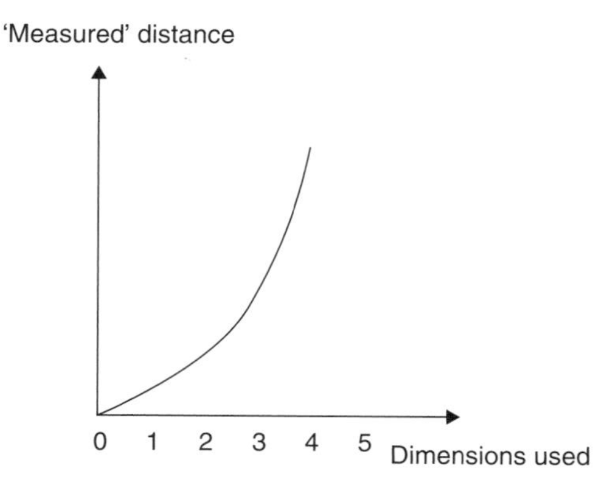

Figure 3.5 Relationship between dimensions used and measured distance (modified from Abou-Kassem *et al.*, 2007).

This work has been further extended to multiphase flow, for which more dimensions (philosophically) are added (Islam *et al.*, 2007). As expected, the effect of linearization of a higher order phenomenon gives even greater discrepancy between 1D and 3D results, quite similar to what was observed by Islam and Chilingar (1995) over a decade ago. The important point to be made in this graph is that with the inclusion of time as a continuous function, there is no need to include a finite number of dimensions, as has been the practice thus far in physics and cosmology. The use of time as a continuous function introduces infinite dimension. This dimension was characterized as 'Knowledge dimension' by Mustafiz (2007).

In engineering applications, the need for modeling with multiple dimensional approaches comes from practical need of explaining the existence of chaotic behavior (Ketata *et al.*, 2007b; 2007c). Even though it has been long been recognized that nature is chaotic, very little modeling work has been extended in order to simulate chaos deterministically (Ketata *et al.*, 2006c; 2006d). In the past, Islam and Nandakumar (1986) attempted to model multiple solutions as well as the bifurcation point for porous media problems with mixed convection. This work was later expanded by Islam and Nandakumar (1990) and others. Similar attempts have been made in other disciplines (e.g. Coriell *et al.*, 1998).

One of the first attempts to relate multiple solutions to dimensional analysis was reported by Islam and Chilingar (1995). In the context of microbial movement within a porous medium, a prime candidate for nonlinear dynamics, they showed that by simply

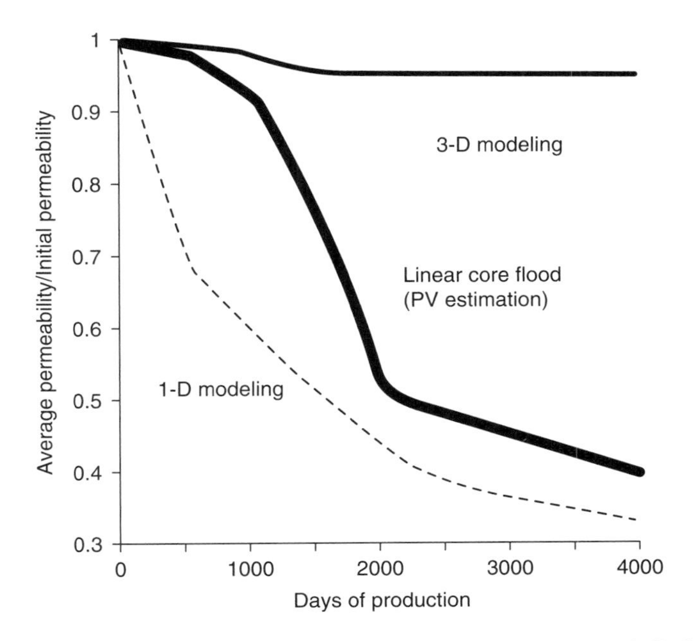

Figure 3.6 The effect of dimensionality on outcome (from Islam and Chilingar, 1995).

changing the dimension of the model, the outcome will change drastically (Figure 3.6) Such recognition was found to be useful in other analyses involving porous media characterization, which is prone to multiple solutions (Katz *et al.*, 1995).

3.4.1 Transition from Mathematics of Tangibles to Mathematics of Intangibles

The publication of the book *Principia Mathematica* by Sir Isaac Newton at the end of 17th century has been the most significant development in European-centered civilization. It is also evident that some of the most important assumptions of Newton were just as aphenomenal as the assertions of Thomas Aquinas, except that Newton did not talk about theology (Zatzman and Islam, 2007a). By examining the first assumptions involved, Zatzman and Islam (2007b) were able to characterize Newton's laws as aphenomenal, for three reasons that they 1) remove time-consciousness (nature is truly dynamic); 2) recognize the role of 'external force' (equivalent to 'gods playing with human intention' in pre-Aristotle Greek philosophy); and 3) do not include the role of intention. In brief, Newton's laws ignore, albeit implicitly, all intangibles from nature science.

Zatzman and Islam (2007b) identified the most significant contribution of Newton in mathematics as the famous definition of the derivative as the limit of a difference quotient involving changes in space or in time as small as anyone might like, but not zero, *viz.*

$$\frac{d}{dt} f(t) = \lim_{\Delta t \to 0} \frac{f(t + \Delta t) - f(t)}{\Delta t} \tag{3.1}$$

Without regard to further conditions being defined as to when and where differentiation would produce a meaningful result, it was entirely possible to arrive at "derivatives" that would generate values in the range of a function at points of the domain where the function was not defined or did not exist. Indeed, it took another century following Newton's death before mathematicians would work out the conditions, especially the requirements for continuity of the function to be differentiated within the domain of values, in which its derivative (the name given to the ratio-quotient generated by the limit formula) could be applied and yield reliable results. Kline (1972) detailed the problems involving this breakthrough formulation of Newton. However, no one in the past proposed an alternative to this differential formulation, at least not explicitly. The following figure (Figure 3.7) illustrates this difficulty.

In this figure, the economic index (it may be one of many indicators) is plotted as a function of time. In nature, all functions are very similar. They do have local trends as well as global trends (in time). One can imagine how the slope of this graph on a very small time frame would be quite arbitrary and how devastating it would be to take that slope to the long-term. One can easily show the trend emerging from Newton's differential quotient would be diametrically opposite to the real trend. In order to offer a substitute to taking the derivative at an infinitely small space, Zatzman and Islam (2007b) provided a new procedure.

Consider now the Nobel Prize-winning work of Saul Perlmutter and Brian Schmidt, who concluded after years of observation that the universe is expanding at an accelerated rate. They attribute such 'counter-intuitive' and anti-Big Bang theory concepts to 'dark energy.' Out of the previously estimated 14 billion years (this is not being disputed by the Nobel Laureates) of existence of the universe,

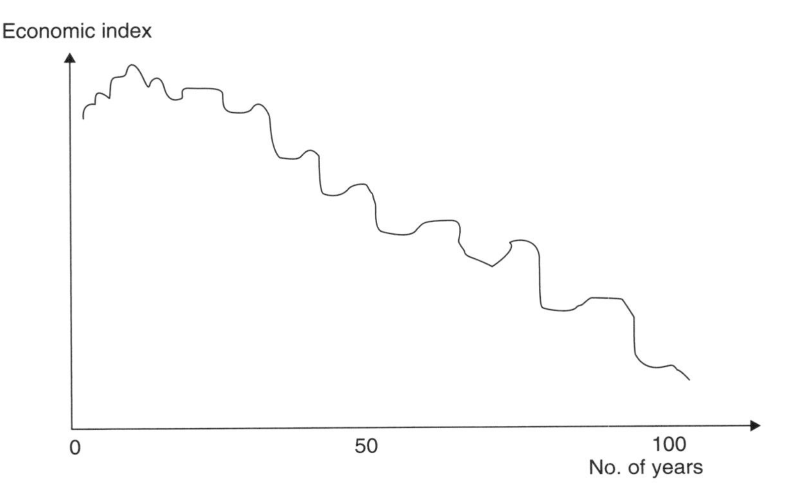

Figure 3.7 Economic wellbeing is known to fluctuate with time (adapted from Abou-Kassem *et al.*, 2007).

they make an observation of a few years and 'discover' that the universe system is expanding at an accelerated rate! Clearly, such a conclusion can only be valid if the acceleration rate is constant, which translates into yet another absurd concept.

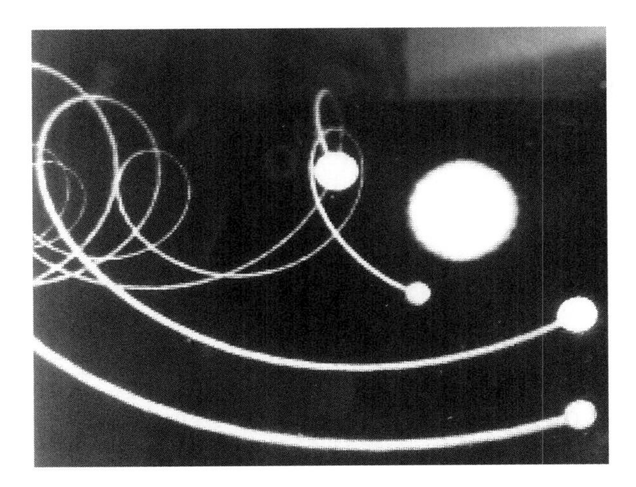

The reality of Eurocentric cognition is that even the simplest conclusion, e.g., the Earth orbits the sun, is inherently false. In this animation, it is shown that the earth's principal direction of motion is parallel to that of the sun and it never orbits the sun following the same path.

In a different discourse, Abou-Kassem *et al.* (2006) introduced the so-called engineering approach in order to solve reservoir engineering problems. Abou-Kassem *et al.* (2006) pointed out that there is no need to go through this process of expressing in differential equation form (avoiding Newton's ratio-quotient), followed by discretization. In fact, by setting up the algebraic equations directly, one can make the process simple and yet maintain accuracy (Mustafiz *et al.*, 2007).

Figure 3.8 shows how formulation with the engineering approach ends up with the same linear algebraic equations if the inside steps are avoided. Even though the engineering approach was known for decades (known as the control volume approach), no one in the past identified the advantage of removing in-between steps.

In analyzing further the role of mathematical manipulation in solving a natural problem, Abou-Kassem *et al.* (2007) cited the example of the manipulation of the simple function $y = 5$. The following steps show how this simple function can take the route of knowledge or ignorance, based on the information that is exposed or hidden, respectively.

Step 1: $y = 5$. This is an algebraic equation that means y is a constant with a value of 5. This statement is an expression of tangibles, which becomes clear if the assumptions are pointed out. The assumptions are (Islam and Zatzman, 2006): a) y has the same dimension as 5 (meaning dimensionless); and b) nothing else matters (this one actually is a clarification of the condition a). Therefore, the above function implies that y cannot be a function of anything (including space and time). The mere fact that there is nothing in nature that is constant makes the function aphenomenal. However, subsequent manipulations (as in Step 2) make the process even more convoluted.

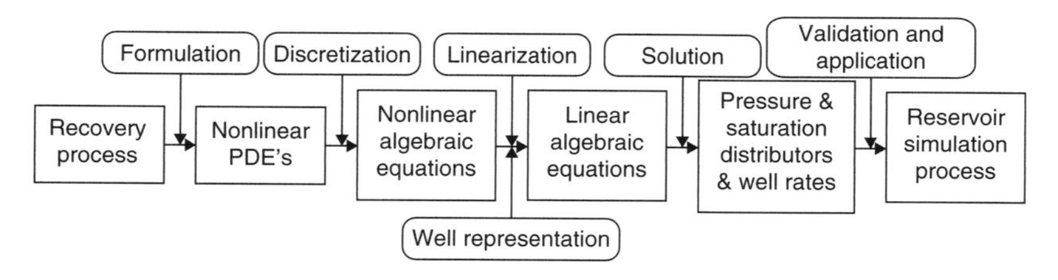

Figure 3.8 Major steps used to develop reservoir simulators (redrawn from Abou-Kassem *et al.*, 2006).

Step 2: $dy/dx = 0$. This simple derivation is legitimate in calculus that originates from Newton's ratio of quotient theory. In this, even a partial derivative would be allowed with equal legitimacy, as nothing states in conventional calculus that such an operation is illegitimate. By adding this derivative in x (as in x direction in Cartesian coordinate), a spurious operation is performed. In fact, if Step 1 is true, one can add any dimension to this and the differential would still be 0, a statement that is 'technically' true but hides background information that becomes evident in Step 3.

Step 3: If one integrates dy/dx, one obtains $y = C$, where C is a constant that can have infinite number of values, depending on which special case it is being solved for. All of a sudden, it is clear that the original function ($y = 5$) has disappeared.

Step 4: One special case of $y = C$ is $y = 5$. To get back the original and unique function (as in Step 1), one now is required to have boundary conditions that are no longer attached to the mathematical formulation. If a special case of $C = 5$ is created, similarly one can have $y =$ 1, 2, 3, 4, 5, 6, and so forth. How does one know which solution will give the 'true' solution? Based on pragmatism (whatever works is true, see Zatzman and Islam (2007a) for further details), one resorts to eliminating solutions that do not meet the immediate need. In this particular case, one by one, solutions are called 'spurious' solutions, because they failed to match the solution of interest, i.e., $y = 5$.

This simple example shows how imposing Newton's differential and integrating procedure convolutes the entire process, while losing information that would allow anyone to trace back to the original algebraic function. One the other hand, if one is looking at an actual phenomenon, then $dy/dx = 0$ could mean that we are at the very start of something, or at the very end of something. However, we actually don't know, because in physical nature, spatial transformations that do not incorporate a time element are like the two-dimensional person who could be sleeping or dancing. If we look at $\partial y/\partial x$, on the other hand, then we have to look also at $\Delta y/\Delta t$, and then we also have to consider the situation where $\Delta y/\Delta x = 0$ but $\Delta y/\Delta t$ is non-0. This might very well be a branch-point, a point of bifurcation or, generally speaking, something marking a change from an old state to a new state. Branch-points in physical natural reality clearly imply infinite solutions, since the process could go literally anywhere from that branch-point. This approach of locating bifurcation phenomena has eluded previous researchers engaged in modeling chaos (Gleick, 1987).

The engineering implication of the Newtonian approach was highlighted by Abou-Kassem *et al.* (2007). The following steps were highlighted. Note the similarity of these steps with the one shown above regarding a simpler function.

Step 1) Mass balance + Darcy's law → It is an algebraic equation.
Step 2a) Time variable is added to it through Newton's differential quotient → time increment is allowed to approach to 0.
Step 2b) Space variable is added through Newton's differential quotient → space increment is allowed to approach 0.
Step 2c) Differential equations emerge, burying assumptions added in 2a and 2b.
Step 3) Differential equations are integrated analytically, and original solutions will not be recovered. The integrated values will have the possibility of having infinite solutions, depending on the boundary conditions, initial conditions, and so forth.
Step 4) Three special cases (out of an infinite number of solutions possible) are identified as part of the integrated solution of the differential equation in Step 3. They are:
Step 4a) Steady state, compressible. This is Mass balance and Darcy's law, as in Step 1.
Step 4b) Unsteady state, slightly compressible fluid. This is the typical equation that gives rise to the diffusivity equation. This one would have liquid density as the only equation of state (density as a function of pressure).
Step 4c) Unsteady state, compressible fluid case. This is the typical gas flow equation. This would have gas compressibility (z factor that is a function of pressure and temperature). Here the term 'unsteady' means there is a time derivative. As time extends to infinity, this term drops off, making the model steady state. This alone shows that these 'unsteady state' models are not dynamic models and do not emulate nature that is dynamic at all times.

This development is seen as a great accomplishment of modern mathematics. We started off with one equation, but ended up with three equations that describe other fluids. The pragmatic approach says if the results are okay, the process must be okay.

Step 5) Because analytical methods are very limited in solving a PDE and require additional assumptions regarding boundary conditions, numerical techniques were introduced. This one essentially involves discretization through Taylor series approximation and subsequent elimination of higher order terms, arguing that at higher order they are too small (e.g. if Δx is less than 1, Δx^2 is $<< 1$; $\Delta x^3 <<< 1$).

Step 6) The removal of higher order terms, as in Step 5, is equivalent to eliminating the additions of space and time variables as in Steps 2a and 2b. We, therefore, recover original mass balance and Darcy's law (Momentum balance equation) as in Step 1. The engineering approach works with these algebraic equations (Step 1) rather than working with PDE and subsequent discretized forms. This explains why the engineering approach coefficients are the same as the 'mathematical approach' coefficients.

When we go from $y = 5$ to $dy/dx = 0$, we add the possibility of adding an infinite series. When we then go from dy/dx to $y =$ constant, we are left with infinite solutions, all because of the previous addition. On the other hand, if we do this integration numerically (by invoking Taylor series approximation), we end up having the original solution only if we ignore the leftover terms of the infinite series that may or may not be convergent. It is important to see the actual derivation of Taylor series expansion. This really *is* magical, as in **aphenomenal**!

3.5 Conclusions

The approach to knowledge in this chapter has challenged many previous assumptions, not only in their contents but especially in their "linearity" and the implicit assumption of a more or less smooth continuous path of knowledge from the past into the present and the future. That is what this chapter has set out to "de-linearize."

From the evidence presented in this chapter, does it seem reasonable to maintain that new knowledge accumulates on the basis of using earlier established findings, with the entire body of knowledge then being passed on to later generations? To begin to answer that question, we must first know: on what basis does an individual investigator cognize the existing state of knowledge? If the individual investigator cognizes the existing state of knowledge on the basis of his/her own re-investigation of the bigger picture surrounding his/her field of interest, that is a conscious approach that shows the investigator operating according to conscience.

If, on the other hand one accepts as given the so-called conclusions reached up to now by others, such a consideration could introduce a problem: what were the *pathways* by which those earlier conclusions were reached? An investigator who declines to investigate those pathways is negating conscience.

Enormous blackmail was exercised against scientists, aimed at compelling them to negate their conscience. This accounts for Galileo's resorting to defensive maneuvers (claiming he was not out to disprove Scripture), a tactic of conceding a small lie in order to be able to continue nailing down a larger more important truth. Why mix such hypocrisy into such matters? Because *it had worked for other investigators in the past*. What was new in Galileo's case was the decision of the Church of that time not to permit him that private space in which to maneuver, in order to make of him an example with which to threaten less-talented researchers coming after him. The worst we can say against Galileo after that point is that once an investigator (in order to get along in life) goes along with this, s/he destroys some part of her/his usefulness as an investigator.

Prior to the conflict with Church authority occasioned by the matter of Galileo's teachings, conflicts had already broken out about the relative weight of current knowledge discovered experimentally and what appeared to be the meaning and direction of earlier and most ancient knowledge. The Church's authority was vested in the unchanging character of the key conclusions of earlier investigators. This authority was never vested in the integrity and depth of probing by earlier investigators and investigations into all the various pathways and possibilities.

In medieval Europe, especially during the two centuries preceding Galileo, the resort to experimental methods did not arise on the basis of rejecting or breaking with Church authority. Rather, it was justified instead by a Christian-theological argument, along the following lines:

- knowledge of God is what makes humans right-thinking and good and capable of having their souls saved in Eternity;
- this knowledge should be accessible wherever humans live and work;
- the means should be at hand for any right-thinking individual to verify the truth or eliminate the error in their knowledge.

These "means" were then formulated as the starting point of what would become the "scientific method."

Combining the matter of the absence of any sovereign authority for the scientific investigator's conscience, and the Christian-theological justification for certain methods of investigation that might not appear to have been provided by any previously-existing authority: what was the result? Even with scientific methods, such as experiments, the conscience of an investigator who separated his/her responsibility for the truth from the claims of Church authority, but without opposing or rebelling against that authority, could not ensure that his/her investigation could or would increase knowledge of the truth.

As long as knowledge was being published and disseminated by monks copying scrolls of works compiled hundreds of years and even millennia earlier, this did not surface as a threat to Church authority. However, with the establishment in Europe of printing by means of movable type, a technology brought from China, all this would change. The basis on which publication of new findings and research would now take place became a burning issue. While every case has its own features, certain elements run commonly throughout. Publication served to advance knowledge in rapid and great strides **IF AND ONLY IF** authority was vested in the integrity and depth of probing by earlier investigators and investigations into all the various pathways and possibilities. Otherwise, the societal necessity and usefulness for publication would be readily and easily subverted. The best known form of that subversion is what is sometimes called the "the Culture of Patents."

Today, under the conditions of what is known as the "information age," this increasingly takes the form of what is known as "monopoly right." Under present conditions and for the foreseeable future, **IF AND ONLY IF** we put first the matter of the *actual conduct* of scientific investigations, as well as the 'politics' attached to that conduct (meaning the *ways and means* by which new results are enabled to build humanity's store of knowledge), **THEN AND ONLY THEN** can we hope to reconstruct the actual line of development. With the knowledge of this actual line of development, for any given case, we can then proceed to critique, isolate and eliminate the thinking and underlying ideological outlook that keeps scientific work and its contents traveling down the wrong path on some given problem or question.

4

Is Modern Science Capable of Discerning Between True and False?

4.1 Introduction

In 2006, the authors started with a simple question in order to determine our ability to discern between truth and falsehood. The question was: what is true? After six months of research, the following criteria for truth were developed. This logic was used to define natural cognition or natural material or natural energy.

1. Must have true base or source
2. Must be continuous
3. Any break-up of continuity or exception must be supported by true criterion or bifurcation point.

The item 3 above sets apart scientific cognition from doctrinal or dogmatic cognition. Even though it is long recognized that Thomas Aquinas is the father of doctrinal philosophy and Averroës the father of secular philosophy, our research shows all aspects of scientific developments in modern Europe has been based on doctrinal philosophy, while claiming secular status. If one sets apart the hypothesis that modern new science is based on non-dogmatic logic, it becomes clear that the modern science is full of paradoxes and contradictions precisely because original premises are unreal or unnatural or nonexistent. Because of the inherent flaws of our education system, it is impossible for our system to make the necessary paradigm shift that requires change in both source and process of scientific cognition. Table 4.1 summarizes what we teach in our education system and how more time spent in the education

Table 4.1 Long-term implications of what we teach.

What we Teach	What it Leads to
Short-term is important (quiz, exams, surprise test, etc. all are for making sure students can reproduce something that they have learned over months)	Overwhelming pressure to make quick decision; reliance on spontaneous surveys, quick money schemes, instant success schemes, etc.
The teacher knows the solution to every problem given in class (the implicit assumption being: Every problem has unique solution)	Looking for answers that would please the boss (trying to find out what the boss is thinking), the public (especially before election), and other authorities that can influence our life.
Read, do literature search, find out what others have done	Constantly limiting thoughts, trying to emulate others; developed world emulating the past behavior; developing world emulating the developed world.
$E = mc^2$	Einstein cannot be wrong; c is constant, m is not changing; time is NOT a dimension.
Ohm's law Darcy's law Fick's law Poiseuille law Navier Stokes equation Fourier's equation	Knowledge is finite; Establishment has the knowledge; Status quo is the ultimate solution.

(Continued)

Table 4.1 (Cont.) Long-term implications of what we teach.

What we Teach	What it Leads to
Newton's first, second, and third laws	There is such a status called 'rest;' there can be transition between rest and motion only using external force (Might is Right); every action has equal and opposite reaction, even when it comes to intangibles, Justice = Revenge → Focus on external and tangibles.
Second law of thermodynamics. Entropy always positive (Lord Kelvin: "the universe would eventually reach a state of uniform temperature and maximum entropy from which it would not be possible to extract any work")	The transition from no-life to life is positive-entropy event? Human being is a liability. Men can only mess up things, while calling it 'natural science.' This would lead to the culture that transforms: Air → toxic fume (combustion engine, electricity) Water → pop (toxic chemical) Milk → ice cream Tomato → Ketchup
First approximation is steady state (implicit assumption is there is such a state for which time doesn't matter).	False expectation that steady state solutions have relevance to unsteady state → Short-term thinking
Equity shoulders debt (implicit assumption is there is no such thing as zero percent interest rate)	Perception that unless you have connections with World Bank, IMF, UNDP, you have no hope; raising funds (through taxes, and schemes) is the principal pre-occupation of business leaders.
Lowest-cost producers will make the most profit	Cut costs (more cuts for long-term investments, e.g., research), bulk produce; promote waste; dehumanize the environment → Focus on self-interest.

(Continued)

Table 4.1 (Cont.) Long-term implications of what we teach.

What we Teach	What it Leads to
Supply and demand dictate market prices	The most obvious answer is the only answer; Intangibles do not matter → Focus on tangibles and obvious.
History and the role of various players	Justification of the future policies based on the past; Resurfacing of past wrongs in the form of future wrongs (considering the transition from witch hunt to equal pay for women as a great progress while pension for raising children remains a laughable topic); Insensitivity to past wrongs (e.g., privateering for the King (or piracy for the pirate) → colonization → Veto powers in UN → globalization)
Mathematics of finite numbers	Intangibles (e.g., zero or infinity) don't matter
Regulations and standards	False perception that they are based on knowledge and are time-tested → Attempt to barely meet the regulation and view the enforcers as technology police → bureaucracy is the law and order, while creative thinking is trouble.
Known theories, postulates, and hypotheses	Knowledge is finite and remains within 'others'

system means a greater loss of creativity that is fundamental to scientific (non-doctrinal) cognition.

In previous chapters, we discussed how this process has impacted modern civilization. This chapter is dedicated to providing one with a clearer picture and point to the most prominent shortcoming of our education system and new science.

New science constantly introduces absurd concepts and answers logical questions with citation of paradox and doctrinal justification, misleading the general population in all aspects of human cognition.

4.2 Why Focus on Tangibles Makes it Impossible for us to Act on Conscience That is Needed to Increase our Knowledge of Truth

Science of Conscience: that one may be very difficult to render comprehensible by anyone who is focused on tangible. Anyone focused on tangible repeatedly follows the line of looking first for an acceptable, approved existing definition, having nothing to do with his own life experience, of what he "ought" to be seeing or perceiving BEFORE he will commit to "learning" anything. A simple example of 'focused on tangible' is this: Imagine there is a course on how to bake bread in a bakery. Because the person registered for this course aspires to work in a certain bakery that uses white bread ingredients with chemical additives (e.g., sugar, hydrogenated oil, and others, for taste and preservation), baked in an electrical oven, he has vested interest in finding out how this bakery operates and how his 'knowledge' will fit into a job prospect with the bakery. However, if he is focused on tangible, his eyes will roll during any discussion of what makes a good bread, why whole-wheat bread is better than white bread, the dangers of adding chemical additives, not to mention the long-term impact of electrical heating. In fact, such a person may not even last through the first few days, thinking this course is driving him crazy. He came here to 'learn' how to make bread in a bakery and he is being 'lectured' on nutritional values. He would be saying, "I need my job with the bakery, the heaven can wait. Just give me the ingredients of white bread, which setting the knob of the oven should be at and where is the timer button. I did not pay all of this money to 'increase my virtue' so I throw up next time I even eat white bread. This not a course on human health, man! If I needed such a course, I would go to a medical doctor, not a master chef!" This down-to-earth example serves as a basis for the first condition to increasing knowledge: one cannot be focused on tangibles and one cannot rush to find a number just so one can get back to the lazy lifestyle of robotic thinking. Other examples of this can be derived from: 1) Dessert making course; 2) Water engineering; 3) Food processing; 4) Pop-drink manufacturing; 5) Tobacco engineering; 6) Pharmaceutical sciences; 7) Genetic engineering; 8) Fluid flow; 9) Materials and manufacturing; and 10) Building design and architecture. With a focus on tangibles, every decision a person will make will be exactly opposite to what the decision should have been made based on true knowledge. Conscience is the driver of true knowledge.

Conscience is what an individual discovers by going with his own natural, unmediated reaction to events and surroundings, not assisted by or dependent upon any definition in some book somewhere. The last thing that conscience needs is activation by a dogma. As discussed in previous chapters, dogmatic thinking is akin to focusing on tangibles. Even the prophet Muhammad, the man widely believed to be the only person who acted on conscience all the time, did not get order from divine revelations on his daily decisions. He constantly made decisions based on conscience, and some of them were later discovered to be incorrect. One such example is cited in Chapter 80 of the Qur'an. This chapter begins with

عَبَسَ وَتَوَلَّىٰٓ (١) أَن جَآءَهُ ٱلۡأَعۡمَىٰ (٢) وَمَا يُدۡرِيكَ لَعَلَّهُۥ يَزَّكَّىٰٓ (٣) أَوۡ يَذَّكَّرُ فَتَنفَعَهُ ٱلذِّكۡرَىٰٓ (٤)

(1) He frowned and turned away (2) Because the blind man came unto him. (3) What could (possibly) inform thee but that he might grow (in grace) (4) Or take heed and so the reminder might avail him?

Obviously, the prophet himself is being chastised for ignoring a blind man whom he ignored in favor of elites, with whom he was busy discussing none other than 'conscience.' This passage shows that there is no escaping making decisions. One cannot rely on others' dictate and, more importantly, one cannot avoid responsibility of making decisions. One can never say, "This and that expert said, therefore I did it...My boss ordered me to do so...I wasn't quite thinking at that time..." There is no excuse for not acting on conscience. It is simply not logical to make up terms such as 'pathological psychopath.'

Even the Pharaohs would fall under this category. Indeed, not even the U.S. Supreme Court could be fooled on this one. The United States Supreme Court recognized decades ago that freedom of speech is inherently coupled with freedom of thought.

"Freedom of thought... is the matrix, the indispensable condition, of nearly every other form of freedom. With rare aberrations a pervasive recognition of this truth can be traced in our history, political and legal" (*Palko v. Connecticut*, 302 U.S. 319, 326–27 (1937)).

Even though it is only logical that freedom of speech is moot without freedom of thought because one can only speak what one cognizes, often referred to as 'internal dialogue,' the matter remains a hotly debated issue. Only recently, the United States Supreme court recognized that forcibly drugging a person would amount to altering his natural thought process, thereby interfering with his First Amendment rights (*Sell v. US*, Supreme Court, Case

Islam is credited to be the first economic system based on intention and not 'interest' (Zatzman and Islam, 2007), yet today Islam is portrayed as synonymous with Capitalism that is entirely based on interest and the modern banking system.

No. 02-5664). While administering anti-psychotic drugs to alter thought is recognized as an interference, use of other drugs, food, and energy sources with the specific intent of enforcing one party's interest remains an illusion to the general public and vast majority of the academics. For instance, most artificial fats and sweeteners are highly addictive (often compared with cocaine addiction, Klein, 2010), and light sources are so toxic that they are used to torture inmates. The new science does not see this, because history is disconnected from mass while energy source (mass) is disconnected from the expression of energy (light). A combined mass-energy analysis can account for all these events and explain also the science behind all pollutions ranging from light pollution to brain 'washing' (Taylor, 2004). Here, creating artificial light is equivalent to creating opacity for the brain by adding a 'strobe light'-like effect on the brain.

Humans are born into society, into collectives (family, followed by larger, different collectives) and this is where the sense of what is right and wrong becomes modulated. This cannot be taught. In fact, teaching anything, let alone 'conscience,' is an absurd idea. You cannot teach anything to anyone. Thinking about one's actions and their consequences further strengthens and defines conscience. It is a fact of living in this world that many things emerge to challenge the individual who would let his conscience be his guide.[1]

How the individual acts upon that realization: aye, there's the rub, as Hamlet says. Again: the issue becomes an individual choice. We do not mean by this the individuality of the choice, which is a thoroughly American idea, but rather the pathway by which the individual's actions become linked to their thought-process, and whether the long-term is in command or something else. Such choices cannot possibly be guided by some 'objective,' allegedly true-for-all-circumstances-cases type of checklist of 'Good' versus 'Bad.' That is why people focused on tangibles are constantly looking for a list of 'dos and don'ts.' They have no hope of acting on conscience. The individual, on the other hand, can have their

---

[1] For anyone familiar with Islam and the Qur'an, one could say: this is where jihad (literal meaning being sustained struggle, as in continuously climbing uphill or swimming against the current) must enter the picture. Conscience is the origin of jihad. In fact, if there is no jihad, there is no act of conscience. In general, it is the same for everyone: anyone who has been normally socialized knows exactly when they have acted in violation of conscience.

own checklist. If it is based on long-term, and not short-term, self-interest, such a checklist may even be valid at least in principle for other individuals who do not operate according to short-term self-interest. What there cannot be is any absolute checklist that works equally for those whose interest is based on short-term serving of the self and those whose interest is actually long-term. None of this forecloses using any or all of the mathematics and other findings of science to date. Rather, it imposes the requirement that the first assumption of one's chosen line or area of research is checked carefully before proceeding further with selecting the relevant or applicable mathematics and other modeling tools and deciding how far these are applied.

So, what is the single most important criterion for judging if an action has been based on conscience? Has one considered long-term implications of the action? "Long-term" here means infinity. In fact, one can argue, it is the only one that an individual has absolute control over. One cannot have any control over his short-term or anyone else's short-term and long-term, without violating natural laws. Violation of natural laws is aphenomenal – totally illogical. You cannot win a fight against nature.

4.3 New Science vs. Science of Intangibles

Consider Table 4.2 that lists various modern subjects and their meaning in the modern education system, as opposed to their root meaning. The departure of Western science down the path of tangibles-only can be more or less precisely dated. It was in western Europe, the late Middle Ages following the Second Crusade, as the scholars raised in the tradition of the One True Church (the Roman Catholic Holy See) began to appreciate that following the path of the "Saracen," i.e., the Moorish Empire, would probably demolish and further the authoritative role for the Church in arbitrating (actually, dictating the outcome of) disputes arising over truth and the scientific method. Table 4.2 illustrates this by comparing the philologies of key concepts emerging in medieval science and how meanings were "transformed."

Today, largely as a byproduct of the fact that so many phenomena can be recorded and measured mechanically or electronically by such a wide variety of means, intangibles may be differentiated from tangible phenomena more clearly than ever before as those

Table 4.2 Modern Subjects and their original meaning as compared to modern implications.

Subject	Popular Meaning in New Science	Most Probable Root Meaning
Algebra	Study rules of operations and relations and the construction of terms, polynomials (non-linear), and equations. Most commonly used algebra: Linear Algebra.	*Al-jabr* (Arabic) Reunion or restoration of broken parts (as in bone restoration)
Biology	Study of engineered bodies and artificial living.	Study of natural living bodies.
Calculus	Branch of mathematics focused on limits, derivative, integrals, and infinite series.	Small stone used for counting
Chemistry	Study of the properties and composition of artificial matter and their manufacturing (mass production)	Study of black (mineral rich) soil. Dates back to the knowledge of smelting in King David's time. *Al-Kimiya* (Arabic) means the art of transmuting metals.
Economics	Analysis of production, distribution, and consumption of goods and services – all regulated by the government or the establishment. 2011 Nobel Prize in Economics on "two-way relationship between the economy and policy – how policy affects the economy and vice versa."	Natural laws of family estate (household goods). Distinct from politics or policy making. In Arabic, the equivalent root word (qsd) stands for optimizing (economizing) and dynamic intention.
Geometry	Study of straight lines, angles, circles, planes, and other idealized (yet artificial) shapes.	Measurement of Earth, its content, and other celestial bodies

(Continued)

Table 4.2 (Cont.) Modern Subjects and their original meaning as compared to modern implications.

Subject	Popular Meaning in New Science	Most Probable Root Meaning
Medicine	Study of artificial chemicals that delay the symptoms.	The art of healing
Philosophy	Study of metaphysics, ethics, doctrinal logic, argument, etc.	Love of wisdom (that comes from logical thinking: homo sapiens means 'logical/thinking human;' in Arabic, it is *hekma* (wisdom) or *aql* (logic))
Physics	Study of artificial matter and energy in such fields as acoustics, optics, mechanics, thermo-dynamics, electro-magnetism, nuclear, and others.	Study of nature
Surgery	Study of radical procedures, specializing in controlling pain, bleeding, and infection, all through artificial means, such as laser surgery, antibiotic, anesthetic, etc. Preventive (breast, ovary, appendix) or cosmetic surgery is common.	Work of hand in order to heal

phenomena that resist ready quantification or verification by modern counting techniques.

Mathematically, the domains of the intangible include the entire fourth dimension, i.e., the time dimension. Cognized intellectually, the time dimension seems like a continuous space of infinite vastness, on which intangibles map a vast number of points and distances. Homing in for a closer look, on the other hand, would disclose an infinitude of discontinuities, many pairs of which would represent the durations of numerous events.

However, this should not be confused with time as a continuous function (e.g., digital). Intangibles refer to some continuous function of time, in the sense that every time function will also have a source, which is the origin of the function, its ultimate source and hence itself intangible. For any human activity, the source of any action is the intention. This is truly intangible because no one else would know about the intention of a person. Even though it has been long recognized in the modern justice system that the intention must be established prior to determining accountability, little is considered about this intangible in other disciplines such as science and engineering, or even social science.

Another aspect of the time function is the factor that depends on others. The dimensionality of intangibles is infinite, because:

a. the time function involved is not only continuous, but incorporates the interaction of all other elements, each possessing their own history and/or characteristic time; and
b. within the natural order, which itself has no boundary, whatever happens to one entity must have an effect on everything else.

In the post-Renaissance era, however, while progress was made to break free from doctrinal philosophy to new science, most intangible considerations have been discarded as pseudoscience or metaphysics that is beyond the scope of any engineering considerations.[2]

The notion of intangibles was in the core of various civilizations, such as Indian, Chinese, Egyptian, Babylonian, and others, for several millennia. Thousands of years ago, Indian philosophers commented about the role of time, as a space (or dimension), in unraveling the truth, the essential component of knowledge (Zatzman and Islam, 2007a). The phrase used by ancient Indian philosophers was that *the world reveals itself*. Scientifically, it would mean that time is the dimension in which all the other dimensions

---

[2] It can be argued that the lack of consideration of intangibles in the modern age is deliberate, due to focus on the short-term. Lord Keynes, who believed that historical time had nothing to do with establishing the truth or falsehood of economic doctrine, said, "In the long run, we are all dead" (cited by Zatzman and Islam, 2007).

New science has invested entirely in hardware, using the same old software and planning that has been proven to be dogmatic and successful only in promoting the new *shahada*: There is no god but Money, Maximum is the Prophet (as compared to There is no god but God, Muhammad is the Prophet)

completely unfold, so that truth becomes continuously known to humans, who use science (as in critical thinking). Another very well-known principle from Ancient India is the connection among *Chetna* (inspiration), *dharma* (inherent property), *karma* (deeds arising from *Chetna*), and *chakra* (wheel, symbolizing the closed loop of a sustainable lifestyle). Each of these concepts scientifically bears the intangible meanings, which cannot be expressed with conventional European mathematical approach (Joseph, 2000). Ketata *et al.* (2006a; 2006b; 2006c; 2006d) have recognized this fact, introducing a series of mathematical tools that can utilize the concept of meaningful zero and infinity in computational methods.

This itself points to one of the limitations in how far the earlier ancient civilizations were able to take their basically correct line about true sustainability at the level of thinking into actions, such as general construction and/or engineering practice. These ancient principles contain some of the most useful hints, extending far back into the oldest known human civilizations, of true sustainability as a state of affairs requiring the involvement of infinite time as a condition of maintaining a correct analysis, as well as ensuring positive pro-social conclusions (Khan and Islam, 2007b). However, the numerous limitations encountered in practice in the material conditions of those times, and the apparently much higher productivity per worker in modern times, made it easy to throw the baby out with the bathwater and dismiss all the valuable and hard-won insights achieved in the development of these societies' thought material, on the grounds that those societies themselves could not give rise to shopping malls and other "proofs" of advanced civilization. Moving from ancient India to ancient China, the Chinese philosophers provide one with useful insights into similar principles of sustainability and knowledge. As pointed out in the previous chapter, this notion can be related to Confucius's statement: *Strive for balance that remains quiet within*).

4.4 The Criterion of Truth and Falsehood

Table 4.3 summarizes the historical development in terms of scientific criterion, origin, pathway and consequences of the principal cultural approaches to reckoning, and reconciling, the tangible-intangible nexus.

With this table 4.3, we present the next table (Table 4.4) that shows the characteristic features of Nature. These are true features and are not based on perception. Each is true insofar as no example of their opposite has been sustained.

Nowhere in nature would one find real symmetry, homogeneity, rigidity, uniformity, and many other features that have been the fundamental traits of 'model,' 'ideal,' 'elemental particle,' 'beauty,' and other departure points in Eurocentric cognition.

Table 4.3 Criterion, origin, pathway and end of scientific methods in some of the leading civilizations of world history.

People	Criterion	Origin	Pathway	End
Zatzman and Islam (2007)	$\Delta t \to \infty$	Intention	$f(t)$	Consequences
Khan (2006)	$\Delta t \to \infty$	Intention	Natural	Sustainability
(Zatzman and Islam, 2007a)	$\Delta t \to \infty$	Intention	Natural	Natural (used $\Delta t \to \infty$ to validate intention)
Einstein	t as 4th-D	"God does not play dice…"	Natural	N/A
Newton	$\Delta t \to 0$	"external force" (1st Law)	No difference between natural & artificial	Universe will run down like a clock
Aquinas	Bible	Acceptance of Divine Order	All *knowledge* & *truth* reside in God; *choice* resides with Man	Heaven and Hell
Averröes	Al- Furqan الفرقان (meaning The Criterion, title of Chapter 25 of *The Qur'an*) stands for *Qur'an*	Intention (first *hadith*)	*Amal saliha* (good deed, de-pending on good intention)	Accomplished (as in *Muflehoon*, ٱلْمُفْلِحُونَ, 2:5), Good (+∞) Losers (as in *Khasheroon*, ٱلْخَـٰسِرُونَ, 58:19), Evil (-∞)

(Continued)

Table 4.3 (Cont.) Criterion, origin, pathway and end of scientific methods in some of the leading civilizations of world history.

People	Criterion	Origin	Pathway	End
Aristotle	*A* or *not-A* ($\Delta t = 0$)	Natural law	Natural or arti-ficial agency	(*Eudaimonia*, tr. "happiness," actually more like "Man in harmony with universe")
Ancient India	Serving others; "world reveals itself"	Inspiration (*Chetna*)	*Karma* (deed with inspiration, *chetna*)	Karma, salvation through merger with Creator
Ancient Greek (pre-Socratics)	*t* begins when Chaos of the void ended	the Gods can interrupt human intention at any time or place	N/A	N/A
Ancient China (Confucius)	N/A	Kindness	Quiet (intangible?)	Balance

Table 4.4 Typical features of natural processes as compared to the claims of artificial processes (Adapted from Khan and Islam, 2007a).

Features of Nature and Natural Materials	
Feature no.	**Feature**
1	Complex
2	Chaotic
3	Unpredictable
4	Unique (every component is different), i.e., forms may appear similar or even "self-similar," but their contents alter with passage of time
5	Productive
6	Non-symmetric, i.e., forms may appear similar or even "self-similar," but their contents alter with passage of time
7	Non-uniform, i.e., forms may appear similar or even "self-similar," but their contents alter with passage of time
8	Heterogeneous, diverse, i.e., forms may appear similar or even "self-similar," but their contents alter with passage of time
9	Internal
10	Anisotropic
11	Bottom-up
12	Multifunctional
13	Dynamic
14	Irreversible
15	Open system
16	True
17	Self healing
18	Nonlinear
19	Multi-dimensional

(Continued)

Table 4.4 (Cont.) Typical features of natural processes as compared to the claims of artificial processes (Adapted from Khan and Islam, 2007a).

Features of Nature and Natural Materials	
Feature no.	**Feature**
20	Infinite degree of freedom
21	Non-trainable
22	Infinite, never ending
23	Intangible
24	Open
25	Flexible
26	Continuous

Table 4.5 summarizes many currently used 'laws' and theories, all of which emerged from the new science after the Renaissance. Note how the first premises of nearly all of these theories violate fundamental features of nature. Only conservation of mass, that in fact has root in ancient times and theory of relativity, does not have an aphenomenal first premise. It is important to note that only recently Kwitko (2007) discredited Einstein's theory of relativity altogether. However, he did not elaborate on the first premise of the theory. Our contention is that Einstein's relativity theory appears to be spurious if processed through the science of tangibles. So far, there is no evidence that the first premise of the theory of relativity, as Einstein envisioned, is aphenomenal. Now, if new science has given us only theories and 'laws' that have a spurious first premise, Averröes' criterion would make new science aphenomenal. This is indeed found in modern technologies that have resulted in 'technological disasters,' reversing the originally declared 'intention' for every technology. This was shown in Table 2.1.

4.5 Effect of the Science of Tangibles

If all theories of new science are based on premises that violate fundamental traits of nature, such laws and theories should weaken considerably, or worse, implode, if applied as universal laws and theories. They can be applied only to certain fixed conditions that

Table 4.5 How the natural features are violated in the first premise of various 'laws' and theories of the science of tangibles.

Law or Theory	First Premise	Features Violated (see Table 4.2)
Conservation of mass	Nothing can be created or destroyed	None, but applications used artificial boundaries
Conservation of energy	No energy can be created or destroyed	None, but applications used artificial boundaries between mass and energy
Big Bang theory	14 billion years ago, there was an entity of infinite mass and zero volume that has been expanding after the Big Bang	1, 3, 6, 9, 14, 24, 26
Saul Perlmutter and Brian Schmidt (2011 Nobel Prize)	Universe is expanding with acceleration	1, 3, 6, 9, 14, 24, 26
Lavoisier's deduction	Perfect seal	15
Phlogiston theory	Phlogiston exists	16
Theory of relativity	Everything (including time) is a function of time	None (concept, but didn't identify time as a continuous function). Mathematical derivation 6, 7, 25
$E = mc^2$	Mass of an object is constant	13
	Speed of light is constant	13
	Nothing else contributes to E	14, 19, 20, 24

(Continued)

Table 4.5 (Cont.) How the natural features are violated in the first premise of various 'laws' and theories of the science of tangibles.

Law or Theory	First Premise	Features Violated (see Table 4.2)
Planck's theory	Nature continuously degrading to heat dead	5, 17, 22
Charles	Fixed mass (closed system), ideal gas, Constant pressure	24, 3, 7
Boyles	A fixed mass (closed system) of ideal gas at fixed temperature	24, 3, 7
Kelvin's	Kelvin temperature scale is derived from Carnot cycle and based on the properties of ideal gas	3, 8, 14, 15
Thermodynamics 1st law	Energy conservation (The first law of the thermodynamics is no more valid when a relationship of mass and energy exists)	None, but assumes existence of artificial boundary between and energy source and output
Thermodynamics 2nd law	Based on Carnot cycle which is operable under the assumptions of ideal gas (imaginary volume), reversible process, adiabatic process (closed system)	3, 8, 14, 15
Thermodynamics 0th law	Thermal equilibrium	10, 15
Poiseuille	Incompressible uniform viscous liquid (Newtonian fluid) in a rigid, non-capillary, straight pipe	25, 7

(Continued)

Table 4.5 (Cont.) How the natural features are violated in the first premise of various 'laws' and theories of the science of tangibles.

Law or Theory	First Premise	Features Violated (see Table 4.2)
Bernoulli	No energy loss to the sounding, no transition between mass and energy	15
Newton's 1st law	A body can be at rest and can have a constant velocity	Non-steady state, 13
Newton's 2nd law	Mass of an object is constant Force is proportional to acceleration External force exists	13 18
Newton's 3rd law	The action and reaction are equal	3
Newton's viscosity law	Uniform flow, constant viscosity	7, 13
Newton's calculus	Limit $\Delta t \rightarrow 0$	22
Fractal theory	Single pattern that repeats itself exists	1-4, 6, 8, 10

pertain to 'idealized' situations not extant in nature. For example, it can be said that the laws of motion developed by Newton cannot explain the chaotic motion of nature due to its assumptions which contradict the reality of nature. The experimental validity of Newton's laws of motion is limited to describing instantaneous macroscopic and tangible phenomena. However, microscopic and intangible phenomena are ignored. Classical dynamics, as represented by Newton's laws of motion, emphasize fixed and unique initial conditions, stability, and equilibrium of a body in motion (Ketata *et al.*, 2007). With the 'laws' and theories of Table 4.5, it is not possible to make a distinction between the products of the following 'engineering' processes. The same theories cannot be called upon to make the reversal.

Wood → plastic
Glass → PVC
Cotton → polyester
Natural fiber → synthetic fiber
Clay → cement
Molasses → Sugar
Sugar → Sugar-free sweeteners
Fermented flower extract → perfume
Water filter (Hubble bubble) → cigarette filter
Graphite, clay → chalk
Chalk → marker
Vegetable paint → plastic paint
Natural marble → artificial marble
Clay tile → ceramic tile
Ceramic tile → vinyl and plastic
Wool → polyester
Silk → synthetic
Bone → hard plastic
Organic fertilizer → chemical fertilizer
Adaptation → bioengineering

Millennia-old technologies were aesthetically superior, at the same time truly sustainable both in material and design.

Consider the above pictures, that show eye-pleasing designs that are the product of technologies several millennia old. None of them used artificial material or met the modern criteria of environmental sustainability, and yet each of them fulfills the true sustainability criteria (Khan and Islam, 2007a). Not that long ago, every technology used real (natural) materials and processes that did not violate natural traits of matter (Table 4.4). This raises the uncomfortable question: could it be that the modern 'science' is actually a disinformation machine, carefully doctored to obscure the difference between real and artificial, truly sustainable and inherently unsustainable? Zatzman (2007) examined this aspect of scientific progress. He argued that the 'chemicals are chemicals' mantra is not promoted out of ignorance, but out of necessity, the necessity to uphold the aphenomenal model that is incapable of existing or coexisting with knowledge, the truth. He showed how this mantra is the driver behind the aphenomenality of mass production, i.e., that 'more' must be 'better' simply because... it is more.

If this 'mass' is not the same mass that exists in nature, the implosive nature of the entire post-Renaissance model of 'new science' and the Industrial Revolution becomes very clear. Ironically, the same 'new science' that had no problem with Einstein's theories, all of which support 'mass is not mass' and 'heat is not heat' and

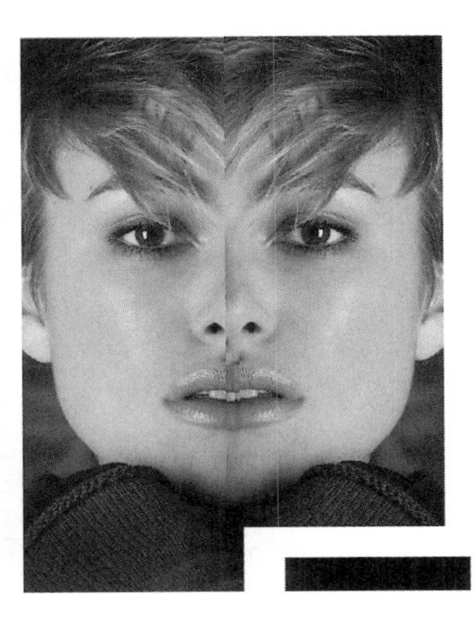

Just as we would create very ugly faces by imposing symmetry to human facial images, we have created an ugly environment by engineering nature with science of tangibles.

recognize the dependence on source and pathway, had tremendous problems with the notion that white light from the fluorescent light is not the same as the white light from the sun or that vitamin C from the organic orange is not the same as the vitamin C from the pharmaceutical plant. Most importantly, it did not see the difference between industrial CO_2 and organic CO_2, blaming modern-day global warming on carbon, the essence of organic matters. Rather than trying to discover the science behind these pathways, industries instead introduced more aphenomenal products that created an even darker opacity and obscured the difference between reality and the truth. The roller coaster was set in motion, spiraling down, bringing mankind to such a status that even a clear champion of new science, the Chemistry Nobel Laureate Robert Curl, called it a 'technological disaster.'

4.6 The Science of Matter and Energy

The existence of essential matter was known to all civilizations, from the prehistoric time. Artifacts exist from thousands of years ago that show the knowledge of engineering, dealing with materials needed for survival. This includes the knowledge of the carriage with wheels.

It is also reported that much sophisticated technology was present in the days of Pharaohs (some 6000 years ago) and with the

The usage of wheel and carriage may be a 9000-year-old technology (picture taken at Tripoli Museum).

Thamud. The Thamud created an entire city by carving on hard rock (Petra of Jordan). This is a technology that is unthinkable today. Pharaohs, whose masterpiece of civil engineering technology (in the form of pyramids) show craftsmanship unprecedented in today's time. Similarly, the chemical technology employed by the engineers of the Pharaoh is much more sophisticated and definitely sustainable. We know that the chemicals that were used to mummify are both non-toxic and extremely efficient. Today's technology that uses various known toxins is remotely as efficient and definitely harmful to the environment (the ones used in Lenin's mausoleum in Moscow or Pope's mausoleum in the Vatican). However, we have little information as to how those engineers designed their technological marvels. All one can say is that the so-called Stone Age and Bronze Age are not within the time frame of suggested 10,000 years of human existence.

The current era of the full order of social culture can date back to King David, who conquered Jerusalem some 3000 years ago, and established an empire that would last some 1300 years. This would be the first known society with remarkable law and order. He was also known to master material processing as the metal mines were being discovered. Only recently, an Israeli research team has discovered earliest-known 'Hebrew' text on a shard of pottery that dates to the time of King David. The following was reported by CNN.

The shard contains five lines of text divided by black lines (left picture) and 100 water pots similar to this one shown here were found in the same site (the right picture is from the Tripoli museum).

> "Professor Yosef Garfinkel of the Hebrew University of Jerusalem says that the inscribed pottery shard was found during excavations of a fortress from the 10th century BC. Archaeologists have yet to decipher the text, but initial

> interpretation indicates it formed part of a letter and contains the roots of the words "judge," "slave," and "king," according to the university. That may indicate it was a legal text, which archaeologists say would provide insights into the full order King David had created. The report also states that the text was clearly written by a trained scribe: "The shard was discovered at the Elah Fortress in Khirbet Qeiyafa, about 20 miles southwest of Jerusalem. Because the ostracon is similar to that found in other Israelite settlements, and because no pig bones were found at the site, archaeologists say the site was likely part of the Kingdom of Judea. Jewish dietary laws forbid the eating of pork."
>
> Among the artifacts found at the site are more than 100 jar handles bearing distinct impressions which may indicate a link to royal vessels, the university said. It is widely believed that the first set of Dead Sea Scrolls was discovered in 1947 by a Bedouin shepherd who ventured into a cave in the Judean Desert in search of a lost sheep or goat. The texts, written on crumbling parchment and papyrus, were found wrapped in linen inside earthenware jars. These discoveries show that a society of order and discipline was in place some 3000 years ago, and that they indeed had the ability to scribe and preserve with sustainable technologies.

His son, Solomon, would continue the expansion of the empire with very significant development in science and engineering. Only recently, Solomon's copper mine was reportedly discovered. On October 28, 2008, Discovery channel reported the following.

Solomon's mine was not a myth in the Middle East, it was a fact based on the Qur'an.

"The fictional King Solomon's Mines held a treasure of gold and diamond, but archaeologists say the real mines may have supplied the ancient king with copper. Researchers led by Thomas Levy of the University of California, San Diego, and Mohammad Najjar of Jordan's Friends of Archaeology, discovered a copper-production center in southern Jordan that dates to the 10th century B.C., the time of Solomon's reign. The discovery occurred at Khirbat en-Nahas, which means 'ruins of copper' in Arabic."

This mine was not a myth or fiction in the Middle East. Even the name of the location means 'ruins of copper,' and people knew for the last 1400 years at least that it was the location of the copper mine. In fact, Mohammad Najjar, the co-investigator of the above research project is personally known to the principal author and he was trying for some time to secure funding for the excavation. The funding came through collaboration with the University of California, and they discovered what was known for a long time. The rein of Solomon has been cited in the Qur'an a number of times, all in light of his knowledge regarding many aspects of science and technology that have been forgotten in the modern age. Consider the state of the mine depicted above, and compare that with today's mining technology. Below is a picture of the infamous Cape Breton tar pond of Nova Scotia. After mining the area for some 100 years, an entire cesspool of environmental disaster has been created. This would be equivalent to any of the Superfund sites of the United States. Similarly, over 100 million dollars have been spent just to assess the extent of the damage, while not a single drop of the tar

The infamous tar pond of Cape Breton, Nova Scotia.

pond has been cleaned. In the meantime, the mining industry is also in ruins, rendering the Cape Breton community the poorest in Nova Scotia, which is one of the poorest provinces of Canada. Even the recent finding of a gas reserve in Nova Scotia did not lift its economy, and the local community does not even get to use the gas that is extracted from their offshore. Contrast this with the Judea community of Solomon that thrived for a millennium after his death.

Solomon expanded the empire to control a wider area of the old world of the three continents. Its political and military power declined with the Israelite kingdom splitting into two regions, one in the north and the other in the south. The region of the north vanished in two centuries, whereas the region of the south, Judea, continued for another millennium (from there came the name Jew). During the 1000 years of Judea rule, this culture expanded to surrounding regions and exerted great influence among the Persian, Greek, Carthaginian, and even ancient Indian cultures. Subsequent political weakening did not stop other peoples from continuing to draw on the science and knowledge accumulated in ancient Judean culture, until the total destruction of the last semblance of Jewish rule in Jerusalem by the Roman occupation in 70 AD.

In our own time, dominated by a European and still Eurocentric culture, references to the ancient Judean source of knowledge are credited to the Greeks. There are several problems with such an assignment of credit. First, it ignores the fact that the vast annals of Judean civilization's knowledge directly benefited and enriched Greek cultural development. This "oversight" implicitly inserts a bias against the very idea of a source of knowledge that could have been shared with both the East and West and preserved in its original form for many centuries being revived and utilized during the Islamic era (7th–20th century A.D.) across a vast region extending from the Far East into central and southern Europe. Second, the pre-existing knowledge developed within the cultures of the East, extending from Persia to China and India (all of which enjoyed extensively elaborate social infrastructures, including access to ancient knowledge predating King Solomon) is entirely ignored or at least marginalized. Third, all of the work of the ancient Greeks that came to Europe was itself actually a result of a gigantic translation effort by Arab scholars. Some resided in Spain, the epicenter of knowledge for 400 years under Arab rule, and the home of Ibn Rushd or Averroes, most notably known as translator of most of Aristotle's work. These documents were only made available in Arabic, and were later translated into Latin before ending up ... in modern Greek. Reliance on this third-hand translation itself ensured that ancient Greek knowledge inevitably became severely distorted, as evidenced from the discussion below.

Fourth, there were powerful political forces that emerged in the last half of the 19th century, especially from Great Britain, possessed of a vested interest in fashioning a "scientific" justification for British hegemony or outright domination. The specific aim was to justify British domination of the African continent, primacy in all parts (especially those peopled by Arabs) of the Ottoman Empire lying outside or beyond the control of the Russian Tsar, and uncontestably exclusive control of the Indian subcontinent, from Ceylon to the Himalayas. As the late Conrad C. Reining, long associated with the Library of Congress, carefully documented, one of their favorite methods was to apply a distorted interpretation of the evolutionary principles of Charles Darwin, both to excuse racial abuse and terrorism meted out by British officialdom and/or their overseers, as well as to prepare the British and world public opinion to accept British interference and dictate in territories not yet under their full control as a boon to the future of humanity as a whole. The main target of this latter pose was to stomp out with hobnailed boots the very inkling of any idea that there was ever, or could ever be, any cultural link, continuum, or meeting point between the civilized values of the West (i.e., Europe) and the barbarism of the peoples of "the East": "East is East and West is West, and never the twain shall meet" was the least offensive manner in which Rudyard Kipling, the great poet of the glories of the British Empire, put it.

4.6.1 The European Knowledge Trail in Mass and Energy

4.6.1.1 Introduction

Why should we study history, particularly in the context of technology development? Is history useful for increasing our knowledge? The issue here is *not* whether new knowledge accumulates on the basis of using earlier established findings, with the entire body of knowledge then being passed on to later generations. The real issue is: *on what basis does an individual investigator cognize the existing state of knowledge*? If the individual investigator cognizes the existing state of knowledge on the basis of his/her own re-investigation of the bigger picture surrounding his/her field of interest, that is a conscious approach, one which shows that the investigator operating according to conscience.

If, on the other hand, one accepts as given the so-called conclusions reached up to now by others, such a consideration could introduce a problem: what were the *pathways* by which those earlier conclusions were reached? An investigator who declines to investigate those pathways is negating conscience.

Such negating of conscience is not a good thing for anyone to undertake. However, the fact is that there were for a long time external or surrounding conditions asserting an undue or improper influence on this front. What if, for example, there exists an authority (like the Church of Rome during the European Middle Ages) that steps into the picture as my-way-or-the-highway (actually: rack-and-thumbscrews) Knowledge Central, certifying certain conclusions while at the same time banishing all thinking or writing that leads to any other conclusions? Then the individual's scientific investigation itself and reporting will be colored and influenced by the looming threat of censorship and/or the actual exercise of that censorship. The latter could occur at the cost of one's career and 'pato' ["personal access to oxygen"].

Against this, mere interest or curiosity on the part of the investigator to find something out will not be enough. The investigator him/herself has to be driven by some particular consciousness of the importance for humanity of his/her own investigative effort. Of course, the Church agrees, but insists only that one have to have the Church's conscience ("everything we have certified is the Truth; anything that contradicts, or conflicts with, the conclusions we certified is Error; those who defend Error are agents of Satan who must be destroyed").

This would account for Galileo's resorting to defensive maneuvers (claiming he was not out to disprove scripture), a tactic of conceding a small Lie in order to be able to continue nailing down a larger, more important Truth. Why mix such hypocrisy into such matters? Because *it had worked for other investigators in the past*. What was new in Galileo's case was the decision of the Church of that time not to permit him that private space in which to maneuver, in order to make of him an example with which to threaten less-talented researchers coming after him. The worst we can say against Galileo after that point is that, once an investigator (in order to get along in life) goes along with this, s/he destroys some part of her/his usefulness as an investigator. This destruction is even more meaningful because it is likely to change the direction of the conscience pathway of the investigator, for example, leading him/her to pursue money instead of the truth.

The historical movement in this material illustrates the importance of retaining the earliest and most ancient knowledge. However, it leaves open the question of what was actually authoritative about earlier knowledge for later generations. The unstated but key point is that the authority was vested in the unchanging character of the key conclusions. That is to say, this authority was never vested in the integrity and depth of probing by earlier investigators and investigations into all the various pathways and possibilities.

There is another feature that is crucial regarding the consequences of vesting authority in a Central Knowledge-Certifier. For thousands of years, Indian mathematics was excelling in increasing knowledge, yet nobody knew about its findings for millennia outside of the villages or small surrounding territories, because there did not exist any notion of *publication of results and findings* for others. Contrast this with the enormous propaganda ascribing so many of the further advancements in the new science of tangibles to the system that emerged of scholarly publication and dissemination of fellow researchers' findings and results. This development is largely ascribed to "learning the lessons" of the burning of the libraries of Constantinople in 1453 (by those barbaric Ottomans, remember...), which deprived Western civilization of so much ancient learning.

The issue is publication, and yet, at the same time the issue is *not just* publication. Rather, it is: on what basis does publication of new findings and research take place? Our point here is that publication will serve to advance knowledge in rapid and great strides *if and only if* authority is vested in the integrity and depth of probing by earlier investigators and investigations into all the various pathways and possibilities. This issue has been discussed in Chapter 3. The issue is not just to oppose the Establishment in theory or in words. The issue is rather to oppose the Establishment in practice, beginning with vesting authority regarding matters of science and present state of knowledge in the integrity and depth of probing by earlier investigators and investigations to date into all the various pathways and possibilities of a given subject-matter.

4.6.1.2 Characterization of Matter and Energy

Around 450 B.C., a Greek philosopher, Empedocles, characterized all matter into earth, air, fire, and water. Note that the word 'earth' here implies clayey material or dirt; it is not the planet Earth. The origin of the word 'earth' (as a human habitat) originates from the Arabic word *Ardh*, the root meaning of which is the habitat of the human race children of Adam. Earth in Arabic is not a planet, as there are other words for planet. Similarly, the sun is not a star, it is precisely the one that sustains all energy needs of the earth. The word 'air' is *Hawa* in Arabic refers to air as in the atmosphere. Note that 'air' is not the same as oxygen (or even certain percentage of oxygen, nitrogen, and carbon dioxide, and so forth), it is the invisible component of the atmosphere that surrounds the earth. Air must contain all organic emission from earth for it to be 'full of life.' It cannot be reconstituted artificially. The term 'fire' is *naar* in Arabic, and refers to real fire, as when wood is burnt and both heat and light are produced. The word has the same root as light (*noor*), which however has a broader meaning. For instance, moonlight is called *noor*, whereas sunlight (direct light) is called *adha'a*. In Arabic, there is a different word for lightning (during a thunderstorm, for instance). In all, the characterization credited to Empedocles and known to modern Europe is in conformance with the criterion of phenomena as outlined in the previous section. It does not violate any of the fundamental properties of nature, as listed in Table 4.3. In fact, this characterization has the following strengths: 1) definitions are real, meaning they have a phenomenal first premise; 2) it recognizes the continuity in nature (including that between matter and energy); and 3) it captures the essence of natural lifestyle. With this characterization, nuclear energy would not emerge as an energy source. Fluorescent light would not qualify for natural light. In fact, with this characterization, none of the technologies (all of which are unsustainable and implosive) listed in Table 2.1 would come to existence.

In the context of characterization of matter, the concept of fundamental substance was introduced by another Greek philosopher named Leucippus, who lived around 478 B.C. Even though his original work was not accessible even to the Arabs who brought the annals of ancient Greek knowledge to the modern age, his student, named Democritus (420 B.C.) documented Leucippus' work, which was later translated in Arabic, then to Latin, followed by modern Greek and other European contemporary languages. That work contained the word 'atom' (ατομοζ in Greek) perpetrated as a fundamental unit of matter. This word created some discussion among Arab scientists some 900 years ago. They understood the meaning to be 'undivided' (this is different from the conventional meaning 'indivisible' used in Europe in the post-Renaissance era.

This would be consistent with Arab scholars because they would not assign any property (such as indivisible) that has the risk of being proven false (which is the case for the conventional meaning of atom)). Their acceptance of the word atom was again in conformance with the criteria listed in Table 2.1 and the fundamental traits of nature, as listed in Table 4.4. The atom was not considered to be indivisible, identical, uniform, or any of the other commonly asserted properties described in the contemporary Atomic theory. In fact, the fundamental notion of creating an aphenomenal basis or unit is a strictly European one. Arab annals of knowledge in the Islamic era starting from the 7th century do not have any such tradition (Zatzman, 2007). This is not to say that they did not know how to measure. On the contrary, they had yardsticks that were available to everyone. Consider in this, the unit of time as the blink of an eye (*tarfa*) for small scale and bushel of grain from medium scale (useful for someone who does the milling of grains using manual stone grinders), and the unit of matter as the dust particle (*dharra* means the dust particles that are visible when a window is opened to let the sunlight into a room: this word is erroneously translated as 'atom').

Heraclitus (540 B.C.) argued that all matter was in flux and vulnerable to change, regardless of its apparent solidity. This is obviously a more profound view, even though, like Democritus, he lacked any special lab-type facilities to investigate this insight further or otherwise to look into what the actual structure of atomic matter would be. It would turn out that the theory of Heraclitus would be rejected by subsequent Greek philosophers of his time. A further discussion follows.

A less elaborate 'atomic theory' as described by Democritus had the notion of atoms being in perpetual motion in a *void*. While being in constant motion (perpetual should not mean uniform or constant speed) is in conformance with natural traits, void is not something that is phenomenal. In Arabic, the word 'void' as in 'nothing' is used only with reference of original creation (Arabic word *'badia'* refers to creation from nothing – a trait attributed to God only). In Chapter 3, the Arabic word, *'cipher'* was introduced. For instance, a hand or a bowl can be empty because it has no visible content in it, but it would never be implied that it has nothing it (for instance, it must have air). The association of 'cipher' with zero was done much later when Arabs came to know about the role of zero from Indian mathematicians. One very useful application of zero was in its role as a filler: that alone made the counting system take a giant leap forward. However, this zero (or cipher or '*sunya*' in Sanskrit) never implies nothingness. In Sanskrit *Maha Sunya* (Great Zero) refers to the outer space, which is anything but void as in nothingness. Similarly, the equivalent word is *As-sama'a,* which stands for anything above the earth, including seven layers of stars in the entire universe (in conventional astronomical sense). In ancient Greek culture, however, void refers to the original status of the universe, which was thought to be filled with nothingness. This status is further confused with the state of chaos, Χαοσ, another Greek term that has void as its root. The word chaos does not exist in the Qur'an, as it is asserted there is no chaos in universal order that would not allow any state of chaos, signaling the loss of control of the Supreme Authority. It is not clear what notion Leucippus had regarding the nature of atomic particles, but from the outset, if it meant a particle (undivided) that is in perpetual motion, it would not be in conflict with the fundamental nature of natural objects. This notion would put everything in a state of flux. The mainstream Greek philosophy would view this negatively for its subversive implication that nature is essentially chaotic. Such an inference threatened the Greek mainstream view that Chaos was the Void that had preceded the coming into existence of the world, and that a natural order came into existence, putting an end to chaos. As stated earlier, this confusion arises from misunderstanding the origin of the universe. Even though this view was rejected by contemporary Greek scholars, this notion (nature is dynamic) was accepted by Arab scholars who did not see this as a conflict with natural order. In fact, their vision of the universe is that everything is in motion and there is no chaos. Often, they referred to a verse of the Qur'an (36:38) that actually talks about the sun as a constantly moving object, moving not just haphazardly, but in a precisely predetermined direction, assuring universal order. Another intriguing point that was made by Democritus is that the feel and taste of a substance is a function of ατομοσ (*atomos*) of the substance on the ατομοσ of our sense organs. This theory, advanced over one thousand years before Alchemists' revolutionary work on modern chemistry, was correct in the sense that it supports the fundamental traits of nature. This suggestion that everything that comes into contact contributes to the exchange of ατομοσ would have stopped us from making toxic chemicals, thinking that they are either inert (totally isolated from the system of interest) or that their concentration is so low that the leaching can be neglected. This would prevent us from seeing the headlines that we see everyday (the latest being on Nov. 8, 2008, see Website 1). This theory could have revolutionized chemical engineering 1000 years before the Alchemists (at least for Europe, as Egyptians already were much advanced in chemical engineering some 6000 years ago). This theory, however, was rejected by Aristotle (384–322B.C.), who became the most powerful and famous of the Greek scientific philosophers. Instead, Aristotle adopted and developed Empedocles's ideas of elemental substances, which were originally well-founded. While Aristotle took the fundamental concept of fire, water, earth, and air being the fundamental ingredients of all matter, he added qualitative parameters, such as hot, moist, cold, and dry. This is shown in Figure 4.1. This figure characterizes matter and energy in four elements, but makes them a function of only composition, meaning one can move from water (cold and moist) to fire (hot and dry) by merely changing the composition of various elements. Similarly, by changing the properties, one can introduce change in compositions. This description is the first known steady-state model that we have listed in Table 4.4. Nature, however, is not a steady state, and that is why this depiction is inherently flawed. In addition, the phase diagram itself has the symmetry imposed on it that is absent in nature. This theory of Aristotle was not picked up by the Arab scientists, even though many other aspects of Aristotle's philosophy were adapted after careful scrutiny, including the famous law of exclusion of the middle.

Democritus is indeed most often cited as the source of the atomic theory of matter, but there is a strong argument or likelihood that what he had in mind was a highly idealized notion, not anything based on actual material structure.

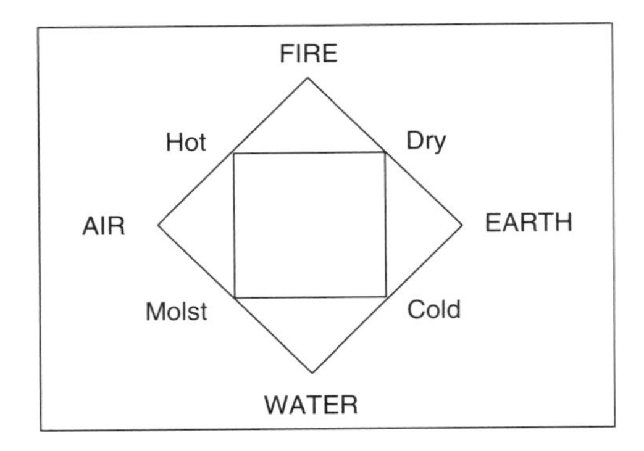

Figure 4.1 Aristotle's four-element phase diagram (steady-state).

For the Greeks, symmetry was believed to be good in itself, and was largely achieved by geometric rearrangement of [usually] two-dimensional space. There is an ambiguity as to whether Greek atomists thought of atoms as anything other than an infinite spatial subdivision of matter. Heraclitus' major achievement, which also marginalized him among the other thinkers of his time, unfortunately, was his incorporation of a notion of the effects of time as a duration of some kind, as some other kind of space in which everything played itself out.

Mainstream Greek philosophy following Plato was rigidly opposed to assigning any such role to time when it came to establishing what they called the essence of things. Plato and his school had held that all matter was physical representations of ideal forms. The task of philosophy was to comprehend these ideal forms in their essence. That essence was what the vast majority of Greek philosophers understood by "ideas."

Both Democritus and Heraclitus followed the main lines of Greek thought in accepting or assuming ideas as being something purer than immediate perception. These "ideas" had their purest form within human consciousness. In effect, although the Greeks never quite put it like this, the material world as we would understand it was deemed a function of our consciousness of ideas about the forms. The Greek philosophers were deeply divided over whether matter as an idea had to have any particular physical existence. Physical existence was something assigned largely to plants and animals. For both philosophers, the atom they had in mind was more of a fundamental idea than a starting point of actual material structure.

Of all the leading ancient Greek thinkers, in his lifelong wrestling with how to reconcile generally accepted notions and ideas about the world with that which could be observed beyond the surface of immediate reality in the short term, it was Aristotle who came closest to grasping the real world as both a material reality outside us and as a source of ideas. He himself never fully resolved the contradictions within his own position. However, he tended to side with the material evidence of the world outside over conjectures lacking an evidentiary or factual basis.

European literature is silent on the scientific progress by Arabs and other Muslim scientists who made spectacular progress in aspects of science, ranging from architectural mathematics and astronomy to evolution theory and medicine. Much of this was preserved, but by methods that precluded or did not include general or widespread publication. Thus, there could well have been almost as much total reliable knowledge 1400 years ago as today, but creative people's access and availability to that mass of reliable knowledge would have been far narrower. This aspect has been discussed in Chapter 3.

Science and technology as we see today would return to Europe in 16th century. However, much earlier, Thomas Aquinas (1225–1274 AD) adopted the logic of Averröes (derived from Aristotle's work), an Arab philosopher of Spain who was liked by Thomas Aquinas and was affectionately called 'The Interpreter.' However, Thomas Aquinas, whose fascination of Aristotle was well-known, introduced the logic of the Creator and His Book being the source of all knowledge to Europe, with a simple, yet highly consequential modification: he would color the (only) creator as God and define the collection of Catholic church documentation on what eventuated in the neighborhood of Jerusalem some millennium ago as the only communication of God to mankind (hence the title, *bible*, meaning the (only) Book). Further elaboration has been given in Chapter 3.

Even though astronomers and alchemists and scholars of many other disciplines were active experimental science in other part of the world for millennia, experimental science in continental Europe began only in the seventeenth century. Sir Francis Bacon (1561–1626) emphasized that experiments should be planned and the results carefully recorded so they could be repeated and verified. Again, there was no recognition of time as a dependent variable and a continuous function.

The work of Sir Isaac Newton (1643–1717) marks the most profound impact on modern European science and technology. Historically, what Thomas Aquinas' model did to European philosophy is the same as what Newton's model did to the new science. Various aspects of Newton's laws of motion, gravity and light propagation have recently been reviewed by Zatzman *et al.* (2008a, 2008b). In subsequent chapters of this book some of those discussions will be presented. Here, it suffices to indicate that Newton's laws suffered from the lack of a real first premise (see Table 4.4). With the exception of Einstein, every scientist took Newton's model as the ideal and developed new models based on the same, adding only factors thought to be relevant because experimental data were not matching with theoretical ones.

Boyle (1627–1691), an experimentalist, recognized the existence of constant motion in gas particles (*corpuscles*, in his word), the same idea that Heraclitus proposed over 2000 years before Boyle (the idea that was rejected by Aristotle and subsequent followers). While this recognition was in conformance with natural traits of matter (Table 4.4), his belief that the particles are: 1) in constant motion; and 2) uniform and rigid is in stark contradiction to real nature of matter. This fundamentally incorrect notion of matter continues to dominate kinetic molecular theory.

In the last half of the 18th century, John Dalton (1766–1844) reasserted the atomic view of matter, albeit now stripped of Heraclitus' metaphysical discussion and explanations. The essential observations of Dalton are:

1. Elements are composed of *atoms* (themselves being unbreakable).
2. All atoms of a given element have identical properties, and those properties differ from those of other elements.
3. Compounds are formed when atoms of different elements combine with one another in small whole numbers (this one emerges from previous assumption that atoms are unbreakable).
4. The relative numbers and kinds of atoms are constant in a given compound (this one asserts steady state, in contrast to notion of kinetic models).

Figure 4.2 shows Dalton's depiction of molecular structure. Note that in this figure, No. 28 denotes carbon dioxide. This depiction is fundamentally flawed because all four premises listed above are aphenomenal. This representation amounts to Aristotle's depiction of matter and energy as both being unique functions of composition and devoid of the time function (Figure 4.1). It should also be noted that today this is the same model that is used in all disciplines of new science. Consider the following figure (Figure 4.3) that shows depiction of non-organic molecules as well as a depiction of DNA.

This fundamentally flawed model of matter was used as the basis for subsequent developments in chemical engineering. In subsequent Europe-based studies, research in physic-chemical properties of matter was distinctly separate from research in energy and light. Even though Newton put forward theories for both, that

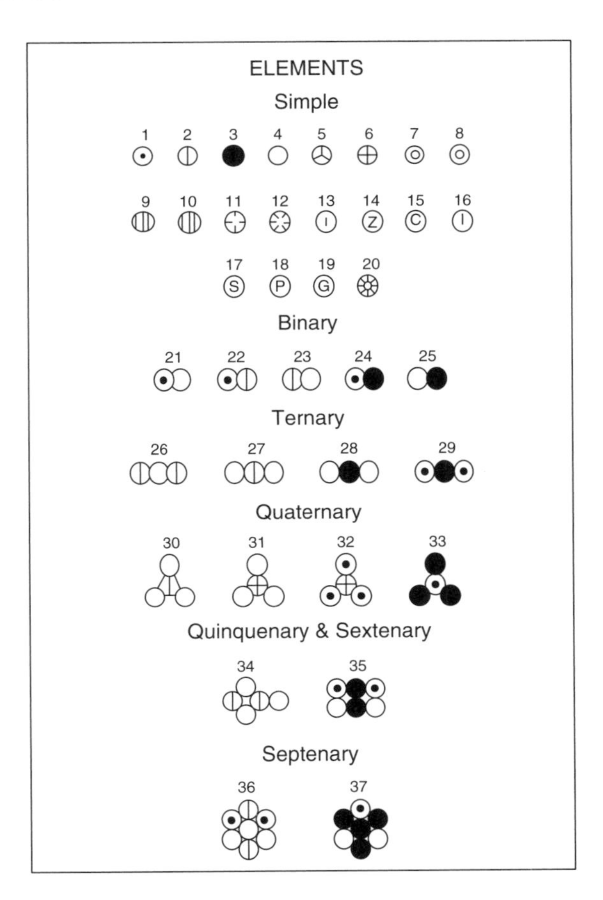

Figure 4.2 Depiction of Dalton's atomic symbols.

trend was clearly absent in subsequent research that saw different tracks of research, some focusing on chemistry, others on physics, astronomy, and numerous other branches of new science.

The law of conservation of mass was known to be true for thousands of years. In 450 B.C., Anaxagoras said, "Wrongly do the Greeks suppose that aught begins or ceases to be; for nothing comes into being or is destroyed; but all is an aggregation or secretion of pre-existing things; so that all becoming might more correctly be called becoming mixed, and all corruption, becoming separate." When Arabs translated this work from old Greek to Arabic, they had no problem with this statement of Anaxagoras. In fact, they were inspired by Qur'anic verses that clearly defined that the universe was created out of nothing, and that ever since its creation all

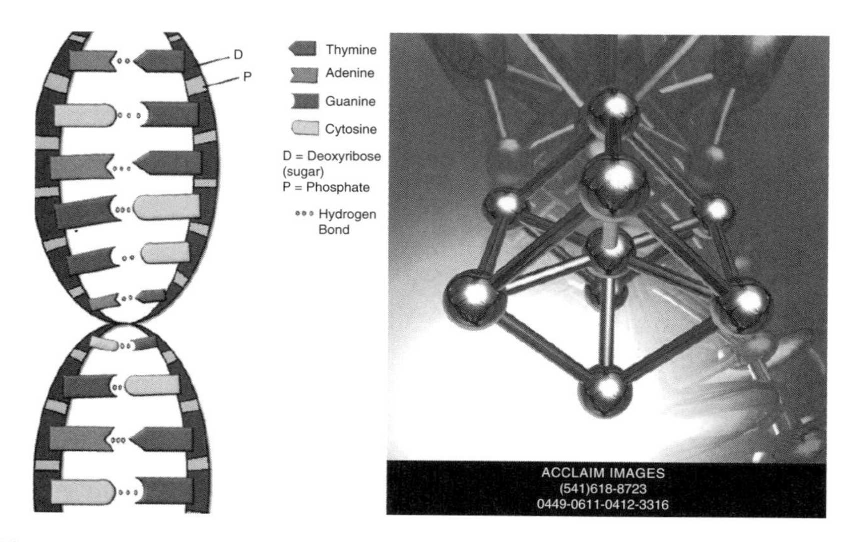

Figure 4.3 Symmetry and uniformity continue to be the main trait of today's scientific models (left, a DNA model; right, a molecular model).

has been a matter of phase transition, as no new matter or energy is created.

However in modern scientific literature, Antoine Laurent Lavoisier (1743–94) is credited to have discovered the law of conservation of mass. Lavoisier's first premise was that "mass cannot be created or destroyed." This assumption does not violate any of the features of nature. However, his famous experiment had some assumptions embedded in it. When he conducted his experiments, he assumed that the container was sealed perfectly. This would violate the fundamental tenet of nature that an isolated chamber cannot be created (Table 4.4). Rather than recognizing the aphenomenality of the assumption that a perfect seal can be created, he 'verified' his first premise (law of conservation of mass) 'within experimental error.' The error is not in the experiment, which remains real (hence, true) at all times, but it is in fact within the first premise that a perfect seal has been created. By avoiding the confrontation of this premise, and by introducing a different criterion (e.g., experimental error), which is aphenomenal and, hence, non-verifiable, Lavoisier invoked a European prejudice, linked to the pragmatic approach, that is 'whatever works is true.' This leads to the linking of measurement error to the outcome. What could Lavoisier have done with the knowledge of his time to link this to intangibles? For instance, if he left some room for leakage from the container, modern-day air conditioner design would have room for how much Freon is leaked to the atmosphere!

Lavoisier nevertheless faced extreme resistance from scientists who were still firm believers of the phlogiston theory (in Greek, *phlogios* means 'fiery'). This theory was first promoted by a German physician, alchemist, adventurer, and professor of Medicine Johann Joachim Becher (1635–1682). This theory recognizes a matter named phlogiston exists within combustible bodies. When burnt (energy added), this matter was thought to have been released to achieve its 'true' state. This theory enjoyed support of the mainstream European scientists for nearly 100 years. One of the proponents of this theory was Robert Boyles, the scientist who would gain fame for relating pressure with volume of gas. Mikhail Vasilyevich Lomonosov (Михаил Васильевич Ломоносов) (1711–1765) was a Russian scientist, writer, and polymath who made important contributions to literature, education, and science. He wrote in his diary: "Today I made an experiment in hermetic glass vessels in order to determine whether the mass of metals increases from the action of pure heat. The experiment demonstrated that the famous Robert Boyle was deluded, for without access of air from outside, the mass of the burnt metal remains the same."

Ever since the work of Lavoisier, the steady-state model of mass balance has been employed in all segments of chemistry and chemical engineering. These works focused on defining symbols and identifying new elements and classifying them. Current chemical symbols (formulas) are derived from the suggestions of Jöns Berzelius (1779–1848). He used oxygen as the standard reference for atomic mass (O = 16.00 AMU). In contrast to Dalton's assertion that water had a formula of HO, Berzelius showed it to be H_2O. For Dalton, all atoms had a valence of one. This made the atomic mass of Oxygen to be 8.

The consideration of mass being independent of time forced all chemical models to be steady or non-dynamic. More importantly, this model was embedded to the definition of time, coupling mass and energy in an intricate fashion that obscured the reality even from experts. This has been discussed in Chapter 3.

Albert Einstein came up with a number of theories, none of which are called 'laws.' The most notable theory was the theory of relativity. Unlike any other European scientists of modern times, this theory recognized the true nature of nature and does not have the first premise that violates the any fundamental feature of nature. Ironically, the very first scientific article that mentioned relativity after Einstein was by Walter Kaufmann, who 'conclusively' refuted the theory of relativity.

Even though this 'conclusive' refutation did not last very long, one point continues to obscure scientific studies, which is the expectation that something can be 'proven.' This is a fundamental misconception, as outlined by Zatzman and Islam (2007a). The correct statement in any scientific research should involve discussion of the premise the research is based on. The first premise represents the one fundamental intangible of a thought process. If the first premise is not true, because it violates fundamental feature(s) of nature, the entire deduction process is corrupted, and no new knowledge can emerge from this deduction.

Einstein's equally famous theory is more directly involved with mass conservation. He derived $E = mc^2$ using the first premise of Planck (1901). Einstein's formulation was the first attempt by European scientists to connect energy with mass. However, in addition to the aphenomenal premises of Planck, this famous equation has its own premises that are aphenomenal (see Table 4.4). However, this equation remains popular and is considered to be useful (in pragmatic sense) for a range of applications, including nuclear energy. For instance, it is quickly deduced from this equation that 100 kJ is equal to approximately 10^{-9} gram. Because no attention is given to the source of the matter, nor the pathway, the information regarding these two important intangibles is wiped out from the science of tangibles. The fact that a great amount of energy is released from a nuclear bomb is then taken as evidence that the theory is correct. By accepting this at face value (heat as the one-dimensional criterion), heat from nuclear energy, electrical energy, electromagnetic irradiation, fossil fuel burning, wood burning, or solar energy becomes identical. This has tremendous implication on economics, which is the driver of modern engineering.

4.6.2 Delinearized History of Mass and Energy Management in the Middle East

At the Petroleum Development Oman (PDO) Planetarium, Dr. Marwan Shwaiki recounted for us an arrestingly delinearized history of the Arab contribution to world scientific and technical culture. What follows is our distillation of some of the main outlines.

Human civilization is synonymous with working with nature. For thousands of years of known history, we know that man marveled in using mathematics to design technologies that created the basis of sustaining life on this planet. In this design, the natural system had been used as a model. For thousands of years, the sun was recognized as the source of energy that is needed to sustain life. For thousands of years, improvements were made over natural systems without violating natural principles of sustainability. The length of a shadow was used by ancient civilizations in the Middle East to regulate the flow of water for irrigation, a process still in present in some areas, known as the *fallaj* system. At night, stars and other celestial bodies were used to ascertain water flow. This is old, but by no means obsolete, technology. In fact, this technology is far superior to the irrigation implanted in the modern age that relies on deep-water exploitation.

For thousands of years of known history, stars were used to navigate. It was no illusion, even for those who believed in myths and legends: stars and celestial bodies are dynamic. This dynamic nature nourished poetry and other imaginings about these natural illuminated bodies for thousands of years. The Babylonians started these stories, as far as one can find out from known history. Babylonian civilization is credited with dividing the heavenly bodies into 12 groups, known as the Zodiac. The Babylonians are also credited with the sexagesimal principle of dividing the circle into 360 degrees and each degree into 60 minutes. They are not, however, responsible for the created confusion between the unit of time (second and minute) and space (Zatzman, 2007b). Their vision was more set on the time domain. The Babylonians had noticed that the sun returned to its original location among the stars once every 365 days. They named this length of time a "year." They also noticed that the moon made almost 12 revolutions during that period. Therefore, they divided the year into 12 parts and each of them was named a "month." Hence, the Babylonians were the first to conceive of the divisions of the astronomical clock.

Along came Egyptian civilization, which followed the path opened by the Babylonians. They understood, even in those days, that the sun is not just a star, and the earth is not just a planet. In a continuous advancement of knowledge, they added more constellations to those already identified by the Babylonians. They divided the sky into 36 groups, starting with the brightest star, Sirius. They believed (on the basis of their own calculations) that the sun took

10 days to cross over each of the 36 constellations. That was what they were proposing ***thousands of years before the Gregorian calendar fixed the number of days to some 365***. Remarkably, this latter fixation would actually violate natural laws; in any event, it was something of which the Egyptians had no part. The Gregorian "solution" was larded with a Eurocentric bias, one that solved the problem of the days that failed to add up by simply wiping out 12 days (Unix users can see this for themselves if they issue the command "cal 1752" in a terminal session).

It was the Greeks, some of whom, e.g., Ptolemy, traveled to Egypt to gather knowledge, who brought the total number of constellations to 48. This was a remarkable achievement. Even after thousands more years of civilization and the discovery of constellations in the southern sky, something previously inaccessible to the peoples to whose history we have access, the total number of constellations was declared to be 88 in 1930. Of course, the Greek version of the same knowledge contained many myths and legends, but it always portrayed as the eternal conflict between good and evil, between ugly and beautiful, and between right and wrong.

The emergence of Islam in the Arabian Peninsula catapulted Arabs to gather knowledge on a scale and at a pace unprecedented in its time. Even before this, they were less concerned with constellations as groups of stars, and far more focused on individual stars and using them effectively to navigate. Not by accident, star constellations' names are of Greek origin, while the names of individual stars are mostly of Arabic in origin. In the modern astronomical atlas, some 200 of the 400 brightest stars are given names of Arabic origin. Arabs, just like ancient Indians, also gave particular importance to the moon. Based on the movement of the moon among the stars, the Arabs divided the sky and its stars into 28 sections, naming them *manazil*, meaning the mansions of the moon. The moon is "hosted" in each mansion for a day and a night. Thus, the pre-Islamic Arabs based their calendar on the moon, although they noted the accumulating differences between the solar and lunar calendars. They also had many myths surrounding the sun, moon, and the stars. While Greek myths focused on kings and gods, however, Arab myths were more focused on individuals and families.

Prehistoric Indians and Chinese assumed that the Earth had the shape of a shell borne by four huge elephants standing on a gigantic turtle. Similarly, some of the inhabitants of Asia Minor envisaged that the Earth was in the form of a huge disk carried by three gigantic whales floating on water. The ancient inhabitants of Africa believed that the sun sets into a "lower world" every evening and that huge elephants pushed it back all night in order to rise the next morning. Even the ancient Egyptians imagined the sky in the shape of a huge woman surrounding the Earth and decorated from the inside with the stars. This was in sharp contrast to the ancient Greek belief that the stars were part of a huge sphere. Ptolemy refined the ancient Greek knowledge of astronomy by imagining a large sphere, with the stars located on the outer surface. He thought that all the planets known at the time (Mercury, Venus, Mars, Jupiter and Saturn) were revolving within this huge sphere, together with the sun and the moon.

The ancient Greeks, including Aristotle, assumed that the orbits of these celestial bodies were perfectly circular, and that the bodies would keep revolving forever. For Aristotle, such perfection manifested symmetric arrangements. His followers continue to use this model. Scientifically speaking, the spherical model is nothing different from the huge elephant on a gigantic turtle model, and so on. What precipitated over the centuries following Ptolemy is a Eurocentric bias that any models that the Greeks proposed are inherently superior than the models proposed by the ancient Indians, Africans, or Chinese. In the bigger picture, however, we know now that the pathways of celestial bodies are non-symmetric and dynamic. Only with this non-symmetric model can one explain retrograde motion of the planets, a phenomenon that even most ancient civilizations noticed. Eurocentric views, however, would continue to promote a single theory that saw the Earth as the center of the Universe. In Ptolemy's words: "During its rotation round the Earth, a planet also rotates in a small circle. On return to its orbit, it appears to us as if it is going back to the west." Of course, this assertion, albeit false, explained the observation of retrograde motion. Because it explains a phenomenon, it becomes true. The essence of pragmatic approach led to the belief that the Earth is indeed the center of the universe, a belief that would dominate the Eurocentric world for over one thousand years.

The knowledge gathered about astronomy by the ancient Chinese and Indians was both extensive and profound. The Chinese were particularly proficient in recording astronomical incidents. The Indians excelled in calculations and had established important astronomical observatories. It was the Arabs of the post-Islamic renaissance that would lead the world for many centuries, setting an example of how to benefit from knowledge of the previous civilizations. Underlying this synthesizing capacity was a strong motive to seek the truth about everything.

Among other reasons for this, a most important reason was that every practicing Muslim is required to offer formal prayer five times a day, all relating to the position of the sun on the horizon. They are also required to fast one month of the year and to offer pilgrimage to Mecca once in a lifetime, no matter how far they reside from Mecca (as long as they can afford the trip). Most importantly, they were motivated by the hadith of the prophet that clearly outlined, "It is obligatory for every Muslim man and woman to seek Knowledge through science (as in process)." This was a significant point of departure, diverging extremely sharply away from the Hellenized conception that would form the basis of what later became "Western civilization" at the end of the European Middle Ages. Greek thought, from its earliest forms, associated the passage of time not with the unfolding of new further knowledge about a phenomenon, but rather with decay and the onset of increasing disorder. Its conceptions of the Ideal and of the Forms are all entirely complete unto themselves, and, most significantly, they stand *outside* of time, the truth being identified with a point in which everything stands still. Even today, conventional models based on the "new science" of tangibles unfolded since the 17th century disclose their debt to these Greek models by virtue of the obsession with the steady state as the "reference point" from which to discuss many physical phenomena, as though there was such a state anywhere in nature. Implicitly, on the basis of such a standpoint, consciousness and knowledge exist in the here-and-now, after the past and before the future unfurls. Again, today, conventional scientific models treat time as the independent variable, in which one may go forward or backward, whereas time in nature cannot be made to go backward, even if a process is reversible. All this has a significant, but rarely articulated, consequence for how nature and its truths would be cognized. According to this arrangement, the individual's knowledge of the truth at any given moment, frozen outside of time, is *co-extensive* with whatever is being observed, noted, or studied.

The Islamic view diverged sharply by distinguishing belief, knowledge (i.e., some conscious awareness of the truth), and truth (or actuality). In this arrangement, the individual's knowledge of the truth or of nature is always fragmentary, and also time-dependent. Furthermore, how, whether, or even where knowledge is gathered cannot be subordinated to the individual's present state of belief(s), desires or prejudices. In the Islamic view, a person seeking knowledge of the truth cannot be biased against the source of knowledge, be it in the form of a geographical location or the tangible status of a people. Muslims felt compelled to become what we term as 'scientists' or independent thinkers, each person deriving their inspiration from the Qur'an and the hadith of prophet Muhammad. Hence, they had no difficulty gaining knowledge from the experience of their predecessors in different fields of science and mathematics. They were solely responsible for bringing back the writings of Greeks Aristotle and Ptolemy and the Indian Brahmagupta in the same breath. Neither were their role models: they were simply their ancestors whose knowledge Muslims did not want to squander. They started the greatest translation campaign in the history of mankind to convert the written works of previous civilizations into Arabic. In due course, they had gained all prior knowledge of astronomy, and that enabled them to become the world leaders in that field of science for five successive centuries. Even their political leaders were fond of science and knowledge. One remarkable pioneer of knowledge was Caliph Al-Mamoon, one of the Abbasite rulers. Some one thousand years before Europeans were debating how flat the Earth is, Al-Mamoon and his scholars already knew the earth is spherical (although, significantly, *not* in the European perfect-sphere sense), but he wanted them to find out the circumference of the Earth. Al-Mamoon sent out two highly competent scientific expeditions. Working independently, they were to measure the circumference of the Earth. The first expedition went to Sinjar, a very flat desert in Iraq. At a certain point, on latitude 35 degrees north, they fixed a post into the ground and tied a rope to it. Then they started to walk carefully northwards, in order to make the North Pole appear one degree higher in the sky. Each time the end of the rope was reached, the expedition fixed another post and stretched another rope from it, until their destination was reached: latitude 36 degrees north. They recorded the total length of the ropes and returned to the original starting point at 35 degrees north. From there, they repeated the experiment, heading south. They continued walking and stretching ropes between posts, until the North Pole dropped in the sky by one degree when they reached the latitude of 34 degrees.

The second of Al-Mamoon's expeditions did the same thing, but in the Kufa desert. When they had finished the task, both expeditions returned to Al-Mamoon and told him the total length of the rope used for measuring the length of one degree of the Earth's circumference. Taking the average of all expeditions, the length of one degree amounted to 56.6 Arabic miles. The Arabic mile is equal to 1973 meters. Therefore, according to the measurements made by the two expeditions, the Earth's circumference is equal to 40,252 kilometers. So, how does it compare with the circumference of the earth as we know today? Today, it is known to be 40,075 km if measured through the equator, a difference of less than 200 km. Contrast that with the debate that was taking place in Europe over the earth being flat many centuries later. Another important aspect is that this was the first time in known history that a state sponsored fundamental research. The motive of Caliph Al-Mamoon was not to capture more land, and history shows that these rulers were not the recipient of any tax. In fact, all rulers paid *zakat*, the obligatory charity, for the wealth they possessed, with the entire amount going to the poor. Also, the judicial system was separate from the administration. Judicial system being always in the hands of the 'most righteous', rather than most 'powerful.' In fact, during the entire Ottoman period, the state language was not Arabic. Arabic was the language of science. For the administration, it was Turkish for communication with the headquarters, and local languages for local communication. This attitude is starkly different from what we encountered in Europe. In the 16th century, Copernicus posited, "The Earth is not located in the center of the universe but the sun is. The earth and the planets rotate around the Sun." This simple observation of the truth could not be tolerated by the very Church that Galileo served during his entire life. Galileo saw the Earth moving and could not reconcile with any dogma that prohibited him from stating what he knew as the truth. In his words, "O people! Beware that your Earth, which you think stationary, is in fact rotating. We are living on a great pendulum." He discovered the four great moons of Jupiter. He was the inventor of the clock pendulum and the "Laws of Motion." The Church could not bear Galileo's boldness. He was put on trial. Confronted with such tyranny, Galileo, who was by then old and weak, yielded, and temporarily changed his mind. But while he was going out of the court, he stamped his feet in anger saying: "But you are still rotating Earth!" This was the beginning of new science that would dominate the world until today. Galileo marks the *eureka* moment in western 'science.' Science finally had broken out of the grip of the

Church, and, therefore, was free from the bias that had a chokehold on clear thinking. This is, unfortunately, is yet another misconception. The earth science that was unleashed after Galileo remains the science of tangibles. With this science, the earth is not flat or at a steady state, but it still is not the science of knowledge (Islam, 2007). The same mindset has led to a lot of scientists rejecting Darwin's theory, a topic that was actually handled by Muslim scholars centuries ago (see picture below).

Did Islamic scientists discover evolutionary theory before Darwin?
January 29, 2008

From Telegraph.co.uk :

Next year, we will be celebrating the 200th anniversary of Charles Darwin's birth, and the 150th of the publication of his On The Origin of Species, which revolutionised our understanding of biology.

But what if Darwin was beaten to the punch? Approximately 1,000 years before the British naturalist published his theory of evolution, a scientist working in Baghdad was thinking along similar lines.

For 700 years, the international language of science was Arabic

In the Book of Animals, abu Uthman al-Jahith (781-869), an intellectual of East African descent, was the first to speculate on the influence of the environment on species. He wrote: ŅAnimals engage in a struggle for existence; for resources, to avoid being eaten and to breed. Environmental factors influence organisms to develop new characteristics to ensure survival, thus transforming into new species. Animals that survive to breed can pass on their successful characteristics to offspring.Ó

During the Islamic era, science meant no restriction to knowledge gathering.

Also, consider the example of the case of the Earth itself. Ibn Kordathyah, an Arab scientist, mentioned the Earth is not flat in early in his books *Al-Masalik* and *Al-mamlik* in the 800s. So, what shape did he think the Earth was? It is the word بَيض أو بَيضاوي (*baidh* or *baidha*). In the modern Europe-dominated word, it is translated as elliptical. In reality, elliptical is an aphenomenal shape, meaning it doesn't exist anywhere in nature. The true meaning of this word is ostrich's egg or its nest, which, obviously, is not elliptical. The inspiration of Ibn Kordathyah came from the Qur'an (Chapter 79, verse 30). The ideal in Islamic culture is the Qur'an (Zatzman and Islam, 2007). Contrast this with western "science," for which the starting point would be the outlined circumference of a circle rendered as an ellipse which has "degenerated" into some kind of ovoid. Then the egg is elaborated as an extrusion into 3-D of a particular case or class of a non-spherical, somewhat ellipsoidal circumference. Why not just start with the egg itself, instead of with circles and ellipses? Eggs are real, actual objects. We can know all of their properties directly, including everything important to know about the strength and resilience of its shape as a container for its particular contents, without having to assume some so-called simple ideal and then extrapolate everything about and in an egg from these abstractions that exist solely in someone's imagination. Going the other direction, on the other hand, is the much richer scientific path. Once we have explored real eggs and generalized everything we find out, we can anticipate meaningfully what will happen in the relations between the form of other exterior surfaces found in nature and their interior contents.

The ostrich egg was the shape that was used to describe the earth in the 9th century.

What we see here is a difference in attitude between standpoints maintained pre- and post-Thomas Aquinas, the father of Europe-centric philosophy. Before his time, truth was bound up with knowledge, and could be augmented by subsequent inquiry. After that point, on the other hand, the correctness or quality of knowledge has been rendered as a function of its conformity with the experience or theories of the elite (called 'laws'). Before, personal experience was just 'personal.' After, the experience of the elite has become a commodity that can be purchased as a source of knowledge. Before, the source of knowledge was individual endeavor, research, and critical thinking. After, it became dogma, blind faith, and the power of external (aphenomenal) forces. After Thomas Aquinas, few Europeans have engaged in **increasing** knowledge *per se*. If they did, they were severely persecuted. Copernicus (1473–1543) is just one example. What was his offense? The Earth moves around a stationary sun. It was not complete knowledge (it is important to note that 'complete' knowledge is anti-knowledge), but it was knowledge in the right direction. His theory contradicted that of Ptolemy, and in general the Catholic Church. Yet, the following statement was made about him in Wikipedia: "While the heliocentric theory had been formulated by Greek, Indian and Muslim savants centuries before Copernicus, his reiteration that the sun — rather than the Earth — is at the center of the solar system is considered among the most important landmarks in the history of modern science."(Website 1) While there is some recognition that Copernicus's knowledge was not ***new*** knowledge, it did not prevent European scientists from making statements that would sanctify Copernicus. Goethe, for instance, wrote:

> "Of all discoveries and opinions, none may have exerted a greater effect on the human spirit than the doctrine of Copernicus. The world had scarcely become known as round and complete in itself when it was asked to waive the tremendous privilege of being the center of the universe. Never, perhaps, was a greater demand made on mankind — for by this admission so many things vanished in mist and smoke! What became of our Eden, our world of innocence, piety and poetry; the testimony of the senses; the conviction of a poetic — religious faith? No wonder his contemporaries did not wish to let all this go and offered every possible resistance to a doctrine which in its converts authorized and demanded a freedom of view and greatness of thought so far unknown, indeed not even dreamed of." (Website 1)

In the above statement, there are three items to note: 1) there is no reference to Copernicus's knowledge being prior knowledge; 2) there is no comment on what the problem was with Copernicus's theory; and 3) there is no explanation to why religious fanatics continued to stifle knowledge and how to handle them in the future.

What would be the knowledge-based approach here? To begin with, ask whether the theory contradicts the truth. European scholars did not ask this question. They compared with words in the Holy *Bible*, a standard whose authenticity, impossible to establish unambiguously, was itself subject to interpretation. When we ask the question, "Is such-and-such true?" we cannot simply define the truth as we wish. We have to state clearly the standard measure of this truth. For Muslim scientists prior to the European Renaissance, the Qur'an formed the standard. Here is the relevant passage of Chapter 36 (36-40) from the Qur'an addressing the matters of whether the sun is 'stationary,' the earth stands at the center of the solar system, or the moon is a planet:

وَٱلشَّمْسُ تَجْرِى لِمُسْتَقَرٍّ لَّهَا ذَٰلِكَ تَقْدِيرُ ٱلْعَزِيزِ ٱلْعَلِيمِ (٣٨)
وَٱلْقَمَرَ قَدَّرْنَـٰهُ مَنَازِلَ حَتَّىٰ عَادَ كَٱلْعُرْجُونِ ٱلْقَدِيمِ (٣٩)
لَا ٱلشَّمْسُ يَنۢبَغِى لَهَآ أَن تُدْرِكَ ٱلْقَمَرَ وَلَا ٱلَّيْلُ سَابِقُ ٱلنَّهَارِ ۚ وَكُلٌّ فِى فَلَكٍ يَسْبَحُونَ (٤٠)

One possible translation: "And the sun runs on its fixed course for a term (appointed). That is the Decree (the word comes from 'qadr' as in 'proportioned' or 'balanced') of the All-Mighty (Al-Aziz) and the All-Knowing (Al-Aleem, the root word being *ilm* or science). And the moon, we have measured (or proportioned, again coming from the root word, 'qadr') for it locations (literally meaning 'mansion') til it returns like the old dried curved date stalk. It is not for the sun to overtake the moon, nor does the night outstrip the day. They all float, each in an orbit."

When did you find out that sun is not stationary? What is the speed and how does the solar orbit look like? See the following table.

Bibliographic Entry	Result (w/Surrounding Text)	Standardized Result
Chaisson, Eric, & McMillan, Steve. *Astronomy Today*. New Jersey: Prentice-Hall, 1993: 533.	"Measurements of gas velocities in the solar neighborhood show that the sun, and everything in its vicinity, orbits the galactic center at a speed of about 220 km/s ...."	220 km/s

"Milky Way Galaxy." *The New Encyclopedia Britannica.* 15th ed. Chicago: Encyclopedia Britannica, 1998: 131.	"The Sun, which is located relatively far from the nucleus, moves at an estimated speed of about 225 km per second (140 miles per second) in a nearly circular orbit."	225 km/s
Goldsmith, Donald. *The Astronomers.* New York: St. Martin's Press, 1991: 39.	"If the solar system … were not moving in orbit around the center, we would fall straight in toward it, arriving a hundred million years from now. But because we do move (at about 150 miles per second) along a nearly circular path …."	240 km/s
Norton, Arthur P. *Norton's Star Atlas.* New York: Longman Scientific & Technical, 1978: 92.	"… the sun's neighborhood, including the Sun itself, are moving around the center of our Galaxy in approximately circular orbits with velocities of the order of 250 km/s."	250 km/s
Recer, Paul (Associated Press). "Radio Astronomers Measure Sun's Orbit Around Milky Way." *Houston Chronicle.* 1 June 1990.	"Using a radio telescope system that measures celestial distances 500 times more accurately than the Hubble Space Telescope, astronomers plotted the motion of the Milky Way and found that the sun and its family of planets were orbiting the galaxy at about 135 miles per second." "The sun circles the Milky Way at a speed of about 486,000 miles per hour."	217 km/s

With 20/20 hindsight, many write these days that the speed of the sun could be predicted using Newton's law. What is missing in this assertion is that Newton's law is absolute and that all hypotheses behind Newton's gravitational law are absolutely true. In addition, it also assumes that we know exactly how the gravitational attractions are imparted from various celestial bodies, a proposition that stands (not to put too fine a point on it) "over the moon"!

Only recently, the following depiction of the earth, other planets, and the sun has been envisioned.

Along came Galileo (1564–1642). Today, he is considered to be the "father of modern astronomy," the "father of modern physics," and the "father of science." As usual, the Church found reasons to ask Galileo to stop promoting his ideas. However, Galileo really was not a 'rebel.' He remained submissive to the Church and never challenged the original dogma of the Church that promotes the aphenomenal model. Consider the following quotations (Website 2):

> Psalm 93:1, Psalm 96:10, and Chronicles 16:30 state that "the world is firmly established, it cannot be moved." Psalm 104:5 says, "[the LORD] set the earth on its foundations; it can never be moved." Ecclesiastes 1:5 states that "the sun rises and the sun sets, and hurries back to where it rises."

Galileo defended heliocentrism and claimed it was not contrary to those Scripture passages. He took Augustine's position on Scripture, which was not to take every passage literally, particularly when the scripture in question is a book of poetry and songs, not a book of instructions or history. The writers of the Scripture wrote from the perspective of the terrestrial world, and from that vantage point the sun does rise and set. In fact, it is the earth's rotation which gives the impression of the sun in motion across the sky.

Galileo's trouble did not come from the Establishment because he contradicted Aristotle's principle. For instance, Galileo contradicted Aristotle's notion that the moon is a perfect sphere, or the heavy object would fall faster than lighter objects directly proportional to weight. Amazingly, both the Establishment and Galileo continued to be enamored with Aristotle, while bitterly fighting with each other. Could the original premise that Aristotle worked on be the same as that of the Church as well as Galileo's? Why didn't he rebel against this first premise?

Galileo's contributions to technology, as the inventor of the geometric and military compass used by gunners and surveyors, are notable. There, even Aristotle would agree, this was indeed τεχνε (techne) or "useful knowledge," meaning useful to the Establishment, of course.

The most remarkable technological development in the Middle East was during 8th to 1800th century. During that time, similar ancient Indian, Chinese, and Islamic philosophy outlined in the Qur'an did include intangibles, such as intention and the time function. The modern-day view holds that knowledge and solutions developed from and within nature might be either good, or neutral [zero net impact] in their effects, or bad, all depending on how developed and correct our initial information and assumptions are. The view of science in the period of Islam's rise was rather different.

Inventions from that era continue to amaze us today (Lu and Steinhardt, 2006). Recently, Al-Hassani (2006) documented 1001 inventions of that era. On reviewing those technologies, one discovers that none of them were unsustainable. We contend it was so because those technologies had a fundamentally phenomenal basis. Recall that phenomenal basis refers to intentions that are in conformance with natural laws and pathways that emulates nature. The following is a brief summary of some of the little known facts from that era.

Accounting

The following is a brief summary of historical recount of some of the major contributions of Middle Eastern scholars. For detailed information, readers are directed to Kline (1972), Struik (1967), and Logan (1986).

Originally accounting in the Islamic era was introduced to calculate obligatory charity and heritage (particularly of orphans) as prescribed by the Qur'an. Because any contract had to be written down with witnesses, it was obligatory for scholars of that time to come up with a numbering system. The currently used numbering system emerges from the Babylonians about 4000 years ago. Their system was base 60, or perhaps a combination of base 60 and base 10, and it was a positional or place-value system, that is, the relative position of a digit enters into determining its value. In our system we multiply by successive powers of 10 as we move to the left, whereas the Babylonian used powers of 60. Some argue that this system emerged from the sundial that had 360 degrees on it. All Arabic numbers we use today are ideograms created by Abu Ja'far Muhammad ibn Musa al-Khowarizmi (c.778–c.850). Using the abacus notations, he developed the manuscript decimal system. These numerals are a scientific display of the number of angles created. There is some ambiguities as to how exactly the numbers of his time looked, but it is certain that he had introduced both the numeral (including zero) and the modern positioning system in counting.

The same person wrote the book *Hidab al-jabr wal-muqubala*, written in Baghdad about 825 A.D. The title has been translated to mean "science of restoration (or reunion) and opposition" or "science of transposition and cancellation" and "The Book of Completion and Cancellation" or "The Book of Restoration and Balancing." The book essentially outlines restoration (*jabr*) and equivalent terms (the actual word, *Muqabalah* means 'comparison') to perform cancellation from two sides of an equation. The equality sign (=) in Arabic represents natural balance or optimization, including intangibles. The first operation, *Jabr*, is used in the step where $x - 2 = 12$ becomes $x = 14$. The left-side of the first equation, where x is lessened by 2, is "restored" or "completed" back to x in the second equation. *Muqabalah* takes us from $x + y = y + 7$ to $x = 7$ by "canceling" or "balancing" the two sides of the equation. Today's word *algebra* is a Latin variant of the Arabic word *al-jabr*. At the outset, this seems to be exactly the same as any 'equation' one would use in algebra or chemical equations. It is not so. Take for instance the simple equation, oxygen + hydrogen → water. It would be written in modern form as:

$$2H_2 + O_2 = 2H_2O \tag{4.1}$$

However, that above equality is not an equality: that symbol would be illegitimate. It is because the elemental forms of oxygen and hydrogen cannot be equal to the compound form on the right hand side. The most one can say about the two sides is they are equivalent, but that, too, is not quite correct. This sign (=) in original Arabic means it is an equivalence. For the above equation to be equivalent, one must have the following elements added:

$$2H_2 + O_2 + \Sigma = 2H_2O + \Sigma O + \Delta E\ (H,O,\Sigma) \tag{4.2}$$

Here, the symbol ΔE represents energy that in itself is a function of (H,O,Σ), where Σ is the summation of other matter present. This would be the minimum requirement for the legitimate equivalency represented by the equal sign. This simple accounting system keeps the pathway of all transformations and is equivalent to keeping track of the time function. This is necessary because matter and energy are both conservative and every matter is dynamic, a fact that was known even by Greek philosophers, dating back 500 B.C.

Arabs also merged the notion of zero in its use as a filler in accounting. The Arabic term *cipher* (this word was used by Arabs at least before 570 AD) means empty (not void). For instance, there could be a cipher hand, which would mean there is nothing visible, but it would not mean there is a void (meaning nothingness, as in the Greek word, Χαοζ that represents void). This is in conformance with the Sanskrit word, *Sunya*, which also means empty space. Indians were advanced in the sense that they were already using the positional and decimal system, using zero. When, Arabs adopted that notion, they called it *cipher*. Because the language of science remained Arabic for some 900 years after the 7th century, most of scientific and philosophical works of Muslim (not all Arab) scholars had to be translated into Latin before they were accessible to modern Europe. Medieval Latin's version of *cipher* became "ciphra." The Latin entered Middle English as "siphre" which eventually became "cypher" in English and "cipher" in American English. Until now, integers are referred to as "cyphers" in English, even though the usage is not common in American English. With "ciphra" taking on a new, more general meaning, a word derived from it, the Medieval Latin "zephirum" or "zepharino," came to be used to denote zero. This word eventually entered English as "zero." Interestingly, in Medieval Europe, some communities banned the positional number system. The bankers of Florence, for example, were forbidden in 1299 to use Indian-Arabic numerals. Instead, they had to use Roman numerals. Thus, the more convenient Hindu-Arabic numbers had to be used secretly. As a result "ciphra" came to mean a secret code, a usage that continues in English. Of course, resolving such a code is "deciphering," a very popular word in modern English (Peterson, 1998).

Fundamental Science and Engineering

Ibn al-Haitham (Alhacen) is known as the father of Modern Optics. Using an early experimental scientific method in his *Book of Optics*, he discovered that light has a finite speed. This is in sharp contrast to Aristotle's belief that the speed of light is infinity. While Al-Haitham and his contemporary, Persian Muslim philosopher and physicist Avicenna, demonstrated that light has finite speed, they did not seek a constant speed, as they were content with finite speed theory. This notion of constant speed is a European one and emerges from an aphenomenal first premise. In Avicenna's translated words, "if the perception of light is due to the emission of some sort of particles by a luminous source, the speed of light must be finite." The 'some sort of particles' were neither atoms nor photons. They were simply emissions from the source of illumination. Using this argument, along with the notion of equivalency (Eq. 4.2), let us consider the most known energy equation of today.

$$C + O_2 + \Sigma = CO_2 + \Sigma O + \Delta E\ (H,O,\Sigma) \tag{4.3}$$

Here, the rationale behind using ΣO is that every matter oxidizes because of the nature of matter. If one considers the Table 4.4 as the basis for describing natural material, this would become obvious. As usual, ΔE represents energy that in itself is a function of (C,O,Σ), where Σ is the summation of other matter present. The above equation conserves both *matter and energy balance*, terms familiar to human civilization for millennia. With this formulation, there would be no confusion between CO_2 coming from a toxic source or a natural source. For that matter, white light from a toxic source and white light from the sun would not be the same either. See Figure 4.5 that shows that the spectral analysis would show the optical toxicity can

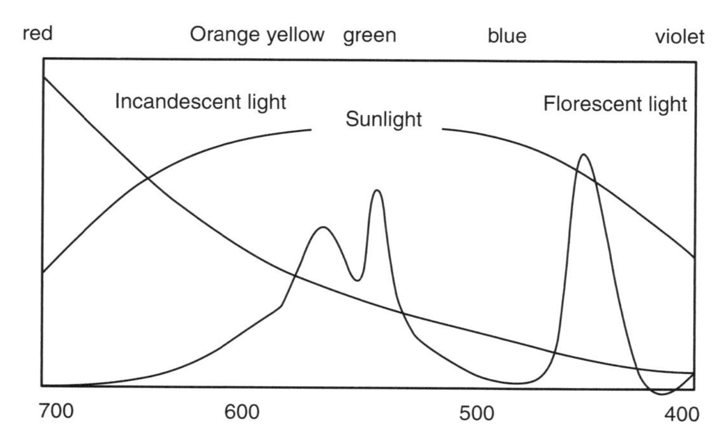

Figure 4.5 Spectral analysis shows the artificial white light is toxic (from Chhetri and Islam, 2008).

arise, but this is never depicted in modern-day chemical or physical accounting systems.

Wikepedia has some coverage of this 'first scientist' of our time. The following are some details:

Main interests: Anatomy, Astronomy, Engineering, Mathematics, Mechanics, Medicine, Optics, Ophthalmology, Philosophy, Physics, Psychology, Science.

Notable ideas: Pioneer in optics, scientific method, experimental science, experimental physics, experimental psychology, visual perception, phenomenology, analytic geometry, non-Ptolemaic astronomy, celestial mechanics.

Works: *Book of Optics, Doubts Concerning Ptolemy, On the Configuration of the World, The Model of the Motions, Treatise on Light, Treatise on Place.*

Influenced: Khayyam, al-Khazini, Averroes, Roger Bacon, Witelo, Pecham, Farisi, Theodoric, Gersonides, Alfonso, von Peuerbach, Taqi al-Din, Risner, Clavius, Kepler, John Wallis, Saccheri.

A moon crater is named after him, along with an asteroid (59239 Alhazen, Feb. 7, 1999).

For the moon crater, see Alhazen (crater). For the asteroid, see 59239 Alhazen.

In addition to being a nature scientist and philosopher, Ibn al-Haitham was also an experimentalist (in today's word) and even an engineer. He actually developed an extensive plan to build a dam on the river Nile, but later realized it was not sustainable. He also invented numerous gadgets, including the pinhole camera.

This line of activity was very common in those days in the Middle East. It was markedly different in two respects: their intention was doing long-term good to the society. Then, when they undertook a research task, they did not allow any dogma or false premises to interfere with their cognition process. They would freely access knowledge of other civilizations (e.g., Greek, Indian, and Chinese), and yet would not take them for granted. For instance, Aristotle and Ptolemy's work were vastly translated by Arab scholars, yet Ibn Haitham and others did not take them at face value, nor did they reject them outright. They carefully filtered the information and rejected notions that,

If we used Al-Hazen's theory, we would be able to discern between sunlight and moonlight, as well as between sunlight and artificial light. By using the photon as the unit of light, we have closed all avenues of discerning between real light and artificial light, let alone proper characterization of light pollution and noise pollution.

with contemporary knowledge, could not pass the reality check. This would explain why such explosive growth in science and technology took place in such a short span of time. Here is one example: Poiseuille's blood flow model is essentially a linearized form of Ibn Nafis' model.

Ibn Nafis first recorded observations on pulmonary blood circulation, a theory attributed to William Harvey 300 years later. Much later than that would come

Abbas Ibn Firnas, who made the first attempt of human flight in the 9th century, using adjustable wings covered with feathers. One thousand years later, the Wright brothers would attempt to fly, except that their flying machine was neither adjustable nor an emulation of birds. Leonardo da Vinci, who himself was inspired by Muslim scientists, is credited to have designed the first flying machine.

Zeng He, the Chinese Muslim admiral, used refined technologies to construct fleets of massive non-metal ship vessels five centuries ago. Until today, this technological marvel remained unsurpassed.

The Ottomans (the Muslim regime that ruled vast regions within Europe, Asia, and Africa) were known for their excellence in naval architecture and a powerful naval fleet, and continued to get recognized until the fall of the Ottoman Empire in the early 20th century.

Avicenna remains the most important name in medical science, with one clear distinction. He never believed in mass-producing 'nature-simulated' products that actually prove that artificial products do not work. Similarly, when Muslim scientists invented shampoo, they used olive oil and wood ash, without any artificial or toxic additions to the product. The word 'alkali' comes from Arabic and the word conserves the meaning of wood ash. Synthetically-made alkali would not qualify to be 'alkali' in Arabic.

The most visible contribution to engineering was in the areas of architecture and building design. Their marks are visible in many sites, ranging from the architectural icon of St. Paul's Cathedral in London, UK, to the horseshoe arches and gothic ribs of Al-Hamra in Granada, Spain, to the Taj Mahal, India. Many forget that buildings in those days did not need Newton's mechanics or Kelvin's thermodynamics, nor did they need ISO 900x standards to become the symbol of time-defying marvels of engineering design. They had natural designs that would eliminate the need of dependence on artificial fluid or electricity, and the structures themselves served multiple purposes, similar to what takes place in nature. These designs were in conformance with natural traits, as listed in Table 4.4. As an example consider the following building in Iran. Below is the schematic of air flow, showing how the design itself can create natural air conditioning (Figure 4.6).

Economics as a science is broadly associated with notions of producing, distributing or managing finite quantities of inputs and outputs. Often this finitude is interpreted as a relative scarcity. This can be misleading, even highly so. What really counts is whether a process is sustainable. Sustainability is to be measured by more than mere availability or unavailability of a supply of raw material components.

As already mentioned, the prime criterion of sustainability is whether a given process, and all the other processes to which it connects, are "natural," in the sense of *characteristic*, within whatever context the process normally unfolds. A secondary, derivative criterion is whether a process is truly "time-tested," i.e., capable of persisting indefinitely (assuming no other elements on which it depends are removed from the environment). The extent to which

A building in Iran that adopts natural air conditioning

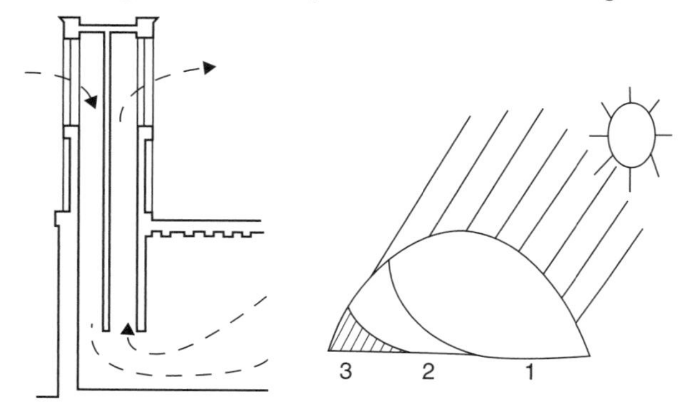

Arrows showing how natural air flow and heating/cooling occurs

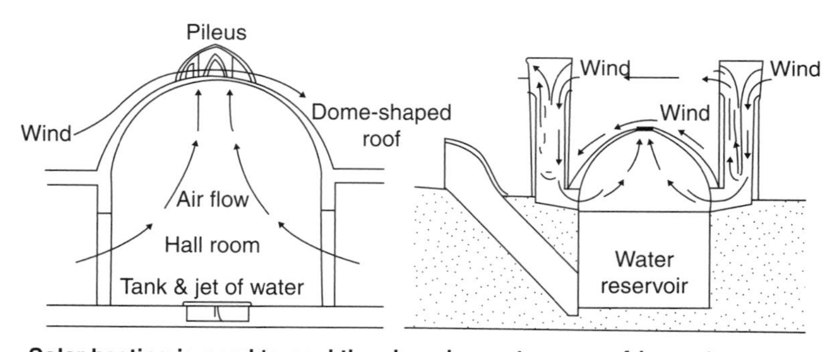

Solar heating is used to cool the air, using water as a refrigerant

Figure 4.6 Deconstructing Greek myths to disclose ancients' understanding of sustainable engineering.

the second criterion is fulfilled defines relative sustainability. The first criterion, which constitutes a definition of inherent sustainability, defines sustainable in an absolute sense. If the prime criterion is not met, no other criteria matter.

The question of sustainability addresses the matter of the pathway of a process. In our own day, this has emerged as the arena in which the supply-demand models of conventional economics fail most spectacularly. Under modern conditions, there are countless pathways by which potentially to finance, staff, and operate the production of almost any given commodity for any market. It is even entirely possible that all of them are unsustainable: nuclear energy production currently falls in that category, and it appears almost certain to remain in that category for some time to come. Back in the 18th and 19th centuries, when the foundations of scarcity-based economics were laid, economic space was far more open. Markets, far from being saturated, were frequently and chronically undersupplied at various times of the year. Competition, while fierce in markets between purveyors of the same goods or services, was almost non-existent between technologies, i.e., between pathways.

This matter of pathway is an especially rich vein. In practical terms, as discussed in depth at Chapter 4 *infra*, it can provide the entry point for introducing innovations in managerial and other practices, as well as policies that impinge on sustainability, including many underappreciated arrangements, such as green supply chains.[3]

The first level of observation and reflection employed by practitioners of what we have been calling the "nature-science" approach is something else that we call "root-pathway analysis."

Root-pathway analysis falls within a class of what may be deemed "organic" (as opposed to, or distinguishable from, "mechanical") analytical methods. Its greatest "weakness," viz., the lack of any elaboration of any pre-*existing* scientific laws or principles to account for phenomena observed in any portion of the time continuum proceeding from the root down to the present day, is in fact its greatest strength. Thus, phenomenon B may be observed following phenomenon A without assuming *any* inherent connection, be they 'a priori' assumptions about the existence of either phenomenon or assumptions about causation or even correlation between the phenomena.

---

[3] Advanced work around all these aspects of sustainability can be found in (Khan, 2006), (Khan & Islam, 2004), (Khan & Islam, 2005a), (Khan & Islam, 2005b), (Khan & Islam, 2005c), (Khan & Islam, 2006), (Khan & Islam, 2007a), (Khan & Islam, 2007b), (Khan *et al.*, 2005d), (Khan *et al.*, 2006) and (Khan *et al.*, 2008).

To demonstrate that awareness of the principles of sustainable engineering are far older than generally assumed, the author decided to apply root-pathway analysis to the utterly familiar and almost three-millennia-old mythology surrounding the "labors of Hercules," specifically the so-called Fifth Labor (to clean the Augean stables).

First, however, a comment about 'myths' & 'mythology' is necessary. Objectively considered, much of what is known in Anglo-American scholarship (and more generally throughout Eurocentric scholarship) as 'myth' actually comprises the recording and transmission for the present and future generations of important social or societal knowledge under conditions of general illiteracy and non-publication of research findings. Of course, this specific, objective, and highly material reality has often become blotted out by the excessive focus on elements from the narrative of the myths themselves that seem to stand in absolute contradiction to everything known or predicted by modern science.

Among almost all practicing scientists, engineers, and other educated people today, there is undeniably widespread ready acceptance of such a dismissive characterization of the content of the ancient Greek myths. But where others see nothing scientific in these narratives, the author senses rather an overpowering sense of Anglo-American and Eurocentric cultural hubris, the same hubris identified in the Greek myths themselves as the most fatal of sins among their gods. It is qualitatively the same hubris that was on display in all its ugliness following the destruction of the Twin Towers in New York City on 11 September 2001. Blaming this destruction of a 31-year old pair of skyscrapers on an allegedly Islamic hankering after mediaeval obscurantism and backwardness and hence a fanatical hatred of Western modernity, well-known public intellectuals across the political spectrum, ranging from Christopher Hitchens to Bernard Lewis and Samuel Huntington could not contain themselves to reflect, even if only for the briefest moment, on the fact that it was a non-Western, non-European, non-Anglo-American society that gave rise 31 centuries ago to the Egyptian pyramids which remain with us today.

Charles Darwin provided and published extremely disturbing proofs that neither stasis, nor gradual evolution, but transformative change by leaps was the norm throughout nature. During the last third of the 19th century, the impacts of this insight began to roil all areas of scholarly investigation in the social and natural sciences.

This was especially the case in Great Britain, the global imperial superpower of its day. In a conscious effort at least to blunt, if not entirely extinguish, these impacts, British imperial scholarship invented the entirely new field of anthropology, including social anthropology.

The study of ancient mythology was promptly transformed into a branch of the new "science." In the works of Edward Burnett Tylor at Oxford and more popular accounts such as James Frazier's *Golden Bough*, far from being recognized as resistance by the subject population in defense of their own thought material, "differences" between the British rulers and their subjects were ascribed to differences on the evolutionary scale in the development of tribal societies towards a reconciliation with, or acceptance of, modernity.

This notion of peaceful evolution towards modernity for less-civilized peoples provided a cosmetic screen, concealing a reality far less benign. On the North American continent in the 17th and 18th centuries, throughout the Asian subcontinent from the 17th to the 20th century, and on the African continent since the 19th century, the reality of British policy was always and everywhere the genocidal extermination of indigenous peoples and tribes as first principle. This was also the case even where the British were compelled by circumstances to rule through elite members from either these indigenous groups, or from among the colonial settler population living otherwise more or less peacefully alongside local indigenous peoples.

The British ruling classes' acceptance of a pseudo-Darwinian model of "civilization" as yet another organic evolutionary process from lower to higher stages was grounded in a crucial unstated assumption that civilization originated with the Greeks, sometime in the first millennia BCE. The two-fold truth, however, was in fact that:

1. neither Greece nor any other European part of the Eurasian land mass could provide the starting point or epicenter from which human social, economic, and political organization within Europe, beyond that of the cave, tribe, village, river-shore or lakefront, developed; and
2. the "Greek" foundations of European civilization were infused from the outset, from at least the midpoint of the 2nd millennia BCE, with the accomplishments and

> even mythological structures and thought-material of much older social formations in the Fertile Crescent and the Indus Valley, going back into the 3rd and 4th millennia BCE, as well as those of peoples along the southern littoral of the Mediterranean.

It was, therefore, no accident that the structures and narratives of the Greek myths shared so many points in common with the mythologies of these other social formations. Indeed, the Greek myth could be seen to embody the unity of human thought material across regions of the Eurasian continent lying to the west of China, including (most critically for the British ruling classes in general and the Raj in particular) the thought material of many peoples from northern parts of the Indian subcontinent. It was this very unity, however — something which would place their Indian subjects on an equal footing — that the British could never accept. As the imperial poet Rudyard Kipling would popularize: "East is East and West is West and never the 'twain shall meet."

Thus, by the end of the 19th century, within British scholarly discourse, two crucial foundational principles of human civilization-in-general would emerge:

1. that smashing this Eurasian-wide link by repeatedly asserting its nonexistence became the watchword and focus of everything British scholarship had to say concerning the Greek paternity of European culture; and
2. that human knowledge itself, as an overwhelmingly European cultural achievement, possessed an inherently hierarchical structure, comprising modern science, followed by revealed religion (headed by Christian religious forms), followed by mythic and/or literary traditions, followed by folklore, followed by animist belief.

According to the unstated principles of such a rigid hierarchy, the ancient Greek myths had a value already assigned by their place within this hierarchy. The important implication of such a position is that the Greek myths could never be usefully mined for insights into such recondite matters as the engineering practices of ancient Eastern civilizations. Nevertheless, looking out upon the world beyond the Anglosphere without any Eurocentric blinders, and applying the key understanding of the role of myth, garnered from root-pathway analysis above as a record and transmittal of engineering ideas and thinking in the ancient world, a careful re-examination of the actual content of the story of Hercules' Fifth Labor turns out to be most illuminating and instructive on this very point and its connection to notions of true sustainability:

> For the fifth labor, Eurystheus ordered Hercules to clean up King Augeas' stables. Hercules knew this job would mean getting dirty and smelly, but sometimes even a hero has to do these things. Then Eurystheus made Hercules' task even harder: he had to clean up after the cattle of Augeas in a single day.
>
> Now King Augeas owned more cattle than anyone in Greece. Some say that he was a son of one of the great gods, and others that he was a son of a mortal; whosever son he was, Augeas was very rich, and he had many herds of cows, bulls, goats, sheep and horses.
>
> Every night the cowherds, goatherds and shepherds drove the thousands of animals to the stables.
>
> Hercules went to King Augeas, and without telling anything about Eurystheus, said that he would clean out the stables in one day, if Augeas would give him a tenth of his fine cattle.
>
> Augeas couldn't believe his ears, but promised. Hercules brought Augeas's son along to watch. • First the hero tore a big opening in the wall of the cattle-yard where the stables were. Then he made another opening in the wall on the opposite side of the yard.
>
> • Next, he dug wide trenches to two rivers which flowed nearby. He turned the course of the rivers into the yard. The rivers rushed through the stables, flushing them out, and all the mess flowed out the hole in the wall on other side of the yard.
>
> When Augeas learned that Eurystheus was behind all this, he would not pay Hercules his reward. Not only that, he denied that he had even promised to pay a reward. Augeas said that if Hercules didn't like it, he could take the matter to a judge to decide.
>
> The judge took his seat. Hercules called the son of Augeas to testify. The boy swore that his father had agreed to give Hercules a reward. The judge ruled that Hercules would have to be paid. In a rage, Augeas ordered both his own son and Hercules to leave his kingdom at once. So the boy went to the north country to live with his aunts, and Hercules headed back to Mycenae. But Eurystheus said that this labour didn't count, because Hercules was paid for having done the work.

> – extracted from "The Augean Stables: Hercules Cleans Up," The Perseus Project, Tufts University. The bulleted (•) sentences describe the actual engineering acts of this labor. (This description, and the descriptions of the other labors of Hercules, may be viewed online at http://www.perseus.tufts.edu/Herakles/index.html)

The Greek myths have been deeply and extensively analyzed for centuries. These analyses never fail to mention or comment on the moral lessons which they have indeed been framed to extract and impress on the listener or reader. However, from a nature-science standpoint (meaning, in this particular context, comprising both the overwhelming presence of living human labor power unassisted by the dead labor of machinery and the singular absence of "high-tech"), the objective content of engineering interest here comes in the following five sentences:

> "First the hero tore a big opening in the wall of the cattle-yard where the stables were. Then he made another opening in the wall on the opposite side of the yard. Next, he dug wide trenches to two rivers which flowed nearby. He turned the course of the rivers into the yard. The rivers rushed through the stables, flushing them out, and all the mess flowed out the hole in the wall on other side of the yard."

Only as the byproduct of "something else" are the stables being cleaned, i.e., the obligatory task set by Eurystheus (in the name of the other gods) is accomplished. What Hercules' labor has also accomplished, however (indeed, it is actually the central accomplishment), has been to render Augeas' entire livestock-raising operation truly sustainable for some lengthy period extending indefinitely into the future.

All versions of the story make it more than clear that King Augeus was dealing with Hercules in bad faith. The King assumed from the outset that the stable-cleaning task could not possibly be accomplished in the assigned time period of one day, and therefore he would never have to worry about rendering Hercules any payment whatsoever, much less a one-tenth ownership share of his herd. Then, after the task was accomplished, King Augeus acted like the thief who yells 'Stop, thief!' Claiming he had dealt with Hercules without knowing that Hercules was in fact under orders from Eurystheus, he sought to wriggle out of his obligation.

Eurystheus saddles Hercules with the final indignity of refusing to "count" the accomplishment as one of the twelve labors because an independent arbitration process by a judge has ordered King Augeus to pay Hercules. This order appears to settle a conflict of opposing human-social interests, but only at the cost of transforming the character of this particular labor of Hercules from that of an obligation discharged at the behest of the gods into mere labor.

From the vantage point of our particular interest in this story, the inner meaning may be summarized thus: a major feat of sustainable engineering is reported faithfully and stands on its own merit. However, recognition of the significance of Hercules' using the opportunity presented by the immediate demand for stable cleaning to solve the far more significant problem of sustainable

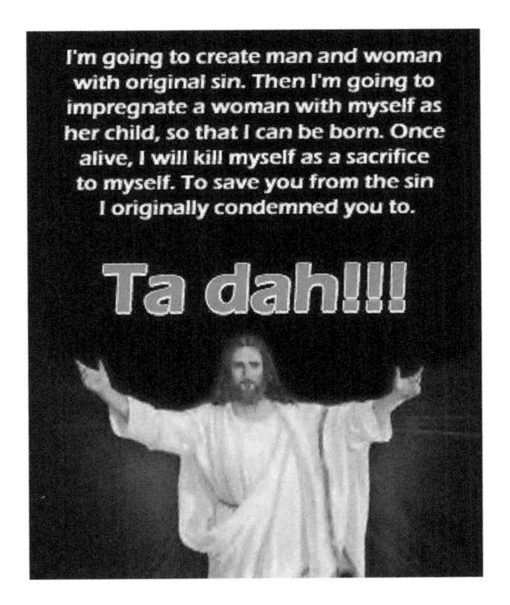

It has become fashionable in the 'enlightened' population to criticize doctrinal philosophy and its offshoots (e.g., creationism), but few realizes that the 'secular basis' of modern science has used the same dogmatic logic to introduce concepts of 'dark energy,' 'criminal gene,' 'gay gene,' 'selfish gene,' 'black hole,' and numerous others to justify whatever conclusion the 'enlightened population' would want the Establishment to believe (in order to maximize their funding). It is noteworthy that scientists who promoted symmetry in atomic level and those who promoted non-symmetry both received the Nobel Prize. Scientists who invented DDT and the CT Scan, and those who determined the vile of them, both won the Nobel Prize.

livestock raising is hindered and obscured by the hopeless tangle of human-social relations attending the working-out of conflicting lines of self-interest.

Reconsidered thus, the issue of scarcity raised at the opening of this section is in fact irrelevant. From a nature-science standpoint, at the dawn of the 21st century (as compared to the time of Hercules and King Augeus several millennia ago), unless there is a broad social movement and/or intellectual consensus standing behind them and occupying the space for change opened up by such accomplishments, how differently should present-day authorities be expected to respond to successful sustainable engineering solutions that appear as "one-offs"?

4.7 Paradigm Shift in Scientific and Engineering Calculations

The first problem with the laws and theories of new science has been the aphenomenal basis. This means the first premise is not real. As mentioned earlier, this problem is compounded by the coupling of the fundamental units of time and space. In modern science, several mitigating factors were at work. From Bernoulli at the end of the 1600s into the mid-1700s, the method emerged of considering physical mass, space, and time as homogeneous at some abstract level, and hence, divisible into identical units. This trend had a positive starting point with a number of French mathematicians who were seeking ways to apply Newton's calculus to analyzing change within physical phenomena generally (and not just problems of motion). This became something of a back-filling operation, intended to maneuver around the troublesome problem of nailing down Leibniz's extremely embarrassing differentials, especially all those *dx*'s in integration formulas.

In England, by the last third of the 1700s, the philosophy of Utilitarianism (often described as 'the greatest good for the greatest number') was spread widely among engineers and others involved in the new industrial professions of the budding Industrial Revolution. Underlying this thinking is the notion that the individual is, at some abstract level, the unitary representative of the whole. Of course, this means all individuals at that level are identical, i.e., they have lost any individuality relative to one another, and are merely common units of the whole.

Generalizations of this kind made it trivially easy to put a single identifying number on anything, including many things where a single identifying number would be truly inappropriate, anti-nature, misleadingly incomplete, and false. For example, 'momentum' is one of the physical phenomena that is conserved throughout nature — not possible to create or destroy — and being able to draw a proper momentum balance is part of analyzing and discussing some natural phenomena. But if we want to compare momenta, the 'single identifying number' approach limits and reduces us to comparing the numerical magnitude of the speed component of their respective velocities.

Some units, like that of force, are entirely concocted to render certain equations dimensionally consistent, like the so-called constant of proportionality that allows us to retain Newton's second law, which computes force as the product of mass and acceleration. Essentially: more back-filling...

So, the atom indeed turns out to be quite breakable, but breakable into what? Quarks, photons, all of them unit-ized! That is what quantized means, but this is then "qualified" (i.e., rendered as opaque as possible) with elaborate hedges about exact time or place being a function of statistical probability.

With the time function being the dependent continuous function, having an aphenomenal basis (the essence of unit) would falsify subsequent engineering calculations. Zatzman (2008) analyzed the merits (or demerits) of the currently used time standard. The time unit known as 'the second' (that standard unit of time measurement for both Anglo-American and SI systems) can be multiplied or divided infinitely, but that is the only real aspect of this unit. Recent research has inserted some strikingly modern resonances into much ancient wisdom that has long been patronized, yet largely disregarded as insufficiently scientific for modern man. Everyone knows, for example, that the 'second' is a real unit of time, whereas the phrase 'blink of an eye,' an actual time unit that has been used in many cultures for expressing short-term events, is considered to be poetic enough, but relatively fanciful. The latest research has established what happens, or more precisely what does not happen, in the blink of an eye. The visual cortex of the brain resets, and when the lens of the retina reopens to the outside world, the information it now takes in is new and not a continuation of the information it took in before the eye-blink. Hence, visual perception which was long assumed to be essentially continuous for the brain, regardless of the blink, is in fact discontinuous. Hence, the natural unit of time is the blink of an eye: it is the shortest unit of time in which the perceptual apparatus of the individual cognizes the continuity of the external world. Consider yet another natural unit of time, this time for a factory worker. Imagine he is told that after he packs 100 boxes of a certain product, he can go for a coffee break. This is a real unit, and it is immediately connected to direct productivity, and hence, economic growth. This manner of characterizing time with real units is both convenient and productive for an individual. Of course, this was known in many cultures for the longest time, even though the science behind this wisdom was not legitimized through standardization. What better way to measure time than the rise and setting of the sun, when a farmer knows he must finish sowing seeds during the daylight? This method is fundamentally sustainable.

If it is accepted that standards ought to have a natural basis, then the notion that a standard must be strictly quantifiable and measurable, or capable of calibration, in order to be "objective," and hence, acceptable is more a matter for Kabbalists, Triskaidekaphobes or some other numerological cult than for engineers or serious students of science. The giveaway is the obsession with "objectivity," which is intimately bound up yet again with "disinterestedness." A natural basis must be applied where the phenomenon actually is, as it actually is. The reason is simply that there must occur a loss of information in applying standards developed for some macro level to a micro/nano level. The determination of a ceiling or a floor for a phenomenon is the very essence of any process of defining a nature-based standard. A ceiling or floor is a region in space-time, but it is not a number. We can readily distinguish effects above a floor or below a ceiling from effects below the floor or above the ceiling. However, from a reference frame based on time t = 'right now,' there is no way we can know or appreciate what 60 Hz, or 9 ppb ± 0.3%, or whatever is going to mean over time and space within nature.

Underlying the thinking that a standard can be reduced to a number is some notion of equilibrium-points, represented by a 'real (sic) number', enjoying some sort of existence anywhere in nature. This assumption is unwarranted. The 'ceiling' / 'floor' approach that characterizes naturally-based standards makes no such assumption. The 'ceiling' / 'floor' approach incorporates everything attaching to and surrounding the phenomenon in its natural state. Standards based on nature must be as four-dimensional as nature itself, whereas a point on a real-number line is a one-dimensional standard. Scientifically speaking, this alone should guarantee that, when it comes to the real, i.e., natural, world, such a standard must be as useless as it will be meaningless.

The notion of using some exact number as a "standard" is put forward as being objective and not dependent on any subjective factor, but it is inherently aphenomenal. Assurances that the standard would undergo only "changes in the light of new knowledge" (or higher-precision measurement technology) do nothing to reduce this aphenomenality. For example, we have all heard this concession from the Establishment: the moment we can detect effects of so-called low-level radiation with better detection equipment, then and only then will it be acceptable to rewrite the standards. Again the idea is that we just move the standard to the new equilibrium-point on the spectrum.

Let us say we knew at this time what the true so-called background level is of naturally occurring radiation from uranium ore in the earth before it is extracted and processed. Let us say we were able to account definitively for the separate and distinct effects of background radiation in the atmosphere in the aftermath of all the various kinds of nuclear testing and bombings since 1945, plus all radiative effects of nuclear power plants, uranium mining and processing, and any other anthropogenically-induced source. Is it likely that the consequences of radiation not augmented by human intervention are anywhere near as dangerous as these anthropogenically-induced sources?

There is another deep-going error that attaches to any and every aphenomenal pathway. As a result of being subordinated by our practices, nature becomes marginalized in our thinking, to the point that we develop and propagate standards that are utterly alien to how anything in nature actually works. The very notion of elaborating a standard that will "keep" us safe itself finesses asking the first most obvious and necessary question. That question is: are we in fact safe ***in the present environment***, in general, or are we menaced by it and if so, from what direction(s)? Because it has become so 'natural' for us to look at this matter from the standpoint of $\Delta t \geq 0$, the future beyond $t + \Delta t$ does not arise even as a consideration. Hence, this first, most obvious and necessary question will not even be posed.

If we have set out a standard that was not based on nature to begin with, changes to that standard over time will improve neither its usefulness nor the process(es) it is supposed to regulate. Just ask the question: why are we beset in the first place by so many standards, and so many amendments to them? Here is the rub: there is no natural standard that could justify any aphenomenal pathway. The presence and further proliferation of so many one-dimensional standards, then, must be acknowledged for what they actually are. They are a function of the very presence and further proliferation of so many aphenomenal pathways in the first place. Any given standard may seem to protect the relative positions into the foreseeable future of competing short-term interests of the present. Its defenders and regulators may be truly disinterested in theory or practice about maintaining and enforcing it. Regardless, the setting of any standard that is "objective" only in the sense that its key criteria accord with something that enjoys no actual existence in nature (and therefore possesses no characteristic non-zero time/existence of its own) can never protect the interests of nature or humanity in the long-term.

Standards based on natural phenomena as they are, and not as eternal or constant under all circumstances, would provide an elegant solution and response. For too long, there is this discourse about "standards' and the necessity to measure mass according to some bar kept in an evacuated bell jar somewhere in Paris, or to set or measure the passage of time with the highest precision according to some clock on a computer in Washington. Now they admit the movement of the cesium atoms for the clock was a bit off and that the bar in Sevres has lost weight over the last few centuries. Surprise!

If, for instance, the unit of time, the second, were replaced by the blink of an eye, or bushel of grains for a Grameen bank farmer, or packages for a factory worker, all calculations would have to change. Each person would have an independent conclusion based on his/her own characteristic time unit. This characteristic time unit will lead to characteristic solutions, the only one that is natural. Any solution that is imposed from others is external and, hence, aphenomenal. This would result in one solution per person, and would honor the fundamental trait of nature, which is uniqueness. There are no two items that are identical. When the manufacturers claim 'nature-identical flavors,' they are simply making a false statement. False is aphenomenal or anti-nature.

There must be a general line of solution that incorporates all individually characteristic forms of solution. This solution-set is infinite. Hence, even though we are talking about a general line or direction of solution, it is not going to produce a unique one-size-fits-all answer. The individual has to verify the general line of solution for him/herself. This is where the individually characteristic part must emerge.

The notion of unit comes from the need of basing on an unbreakable departure point. The word 'atom' indeed means 'unbreakable' in Greek. Atoms are not 'unbreakable' to begin with. 'Second,' on the other hand, is a unit that is not only non-unbreakable, people take pride in the fact that it can broken indefinitely. The unit of time violates the fundamental notion of unit. Unless the unit problem is solved, standards cannot be selected because aphenomenal units will lead to aphenomenal standards, much like Aristotle's notion of 'beauty' being that of symmetry and, after several millennia, we now give the Nobel Prize to someone who discovers 'symmetry' breaks down (Nobel Prize in Physics, 2008).

4.8 Summary and Conclusions

We might have the false impression that our cognition skills have improved ever since the notion of 'flat earth' was discredited. However, as this chapter shows, the illogical mindset that once propelled the religious establishment to promote 'original sin' still reigns supreme in the scientific community. It is not in the topic of theology or even social science, it is in the topic of hard science that allowed every assumption of Newton to remain unchallenged. Other scientists took Newton's assumptions as 'laws' and never bothered to question the non-existence of these assumptions in nature. This happened even though the modern age equipped us with greater observation skills.

In this process, every promise made at the onset of a 'revolution' has been broken and opposite happened. "Green revolution" killed the greeneries, Communism that promised no-government created the biggest government, Capitalism that promised biggest capital produced the greatest debt, sex revolution that promised liberation of women created the largest slavery scheme, War on terror has created the greatest terror mankind ever witnessed. The mistake we continue to make is to ask the system that created the disaster to fix it.

The approach taken in Modern Europe can only be characterized as the myopic approach. This approach is inherently implosive and anti-conscience. The model is applied in both social and technological matters. This approach converts everything natural to everything artificial in order to make it appealing. On the social side, this model translates into the following conversion of good into evil.

History, culture → entertainment (e.g., fashion show, 'beauty' context)
Smiles → Laughter (e.g., stand-up comedy)
Love → Lust
Love of children → Pedophilia
Passion → Obsession
Contentment → Gloating
Quenching thirst → Bloating
Feeding hunger → Gluttony
Consultation → Politicking
Freedom → Panic
Security → Fear
Liberation → Enslavement
Dream → Fantasy
Justice → Revenge
Science → "Technological development"
Social progress → "Economic development"
Positive internal change (true civilization) → Negative external change (true savagery)

On the technology side, this approach converts everything natural and free into artificial and toxic, with only the profit margin going up. In order to deceive a promise is made that creates public appeal and people are drawn to Consumerism. See the following transition.

Air → cigarette smoke, air pollution [Combustion engine; the promise was to give sex appeal, power and comfort]
Protective atmospheric layer → Radiation deflecting layer [Chemical technology, the promise was to clean and provide safety]
Water → Carbonated drink (dehydrating) [the promise was to quench thirst]
Rain → Acid rain [the promise was to give clean air/water system]
Milk → Ice Cream [the promise was to give nutrition]
Cocoa → Chocolate [the promise was to make nutrition available]
Tomato → Ketchup [the promise was to give healthy food]
Corn, potato → chips, popcorns (trans fats that increase bad cholesterol while reducing good ones!) [the promise was to provide good food]
Human asset → Human liability [The promise was to civilize]

Figure 4.7 Some natural pathways negated by human-engineered intervention.

Historical time in social development and characteristic time in natural processes each exist, and operate, objectively and independently of our will, or even our perception. They are certainly not perceived as such by us humans living in the present. We cognize these phenomena, and their objectivity, only in the process of summing up matters on the basis of looking back from the vantage point of the present.

We may idealize the arc of change, development and/or motion of a process. This idealization can be as tractable or complex as we desire, with a view to being reproducible in experiments of various kinds. What weight is to be assigned, however, to any conclusions drawn from analysis of this idealization and how it works? Can those conclusions apply to what is actually happening in the objective social or natural process? The nub of this problem is that the input-state and output-state of an actual process can be readily simulated in any such idealization or its experimental reproduction. The actual pathway (meaning: how matters actually proceeded from input to output) is very likely another matter entirely, however.

When it comes to things that are human-engineered (the fashioning of some process or product, for example), the pathway of the natural version may not seem or even be particularly important. But the pragmatic result of simulating an idealization cannot be confused with actual understanding of the science of how the natural process works.

Essentially, that idealization takes the form of a First Assumption. The most dangerous such First Assumptions are the most innocent-seeming. Consider, for example, the notion of the speed of light taken as a constant — in a vacuum. Where in nature is there a vacuum? Since no such location is known to exist anywhere in nature, if the speed of light is observed to vary, i.e., not be constant, does this mean any observed non-constant character can be ascribed to the absence of a vacuum, so therefore the original definition remains valid? Or does it mean, rather, that we need better measuring instruments?

This notion of the speed of light being constant in a vacuum has been retrofitted to make it possible to bridge various gaps in our knowledge of actually-observed phenomena. It is an example of an idealization. By fitting a "logically necessary" pathway of steps between input and output, however, on the basis of applying conclusions generated by an idealization of some social or natural process to the social or natural process itself, it becomes trivially easy to create the appearance of a smooth and gradual development or evolution from one intermediate state to another intermediate state. In such linearizing and smoothing, some information-loss, perhaps even a great deal, necessarily occurs. Above all, however, what is being passed off as a scientific explanation of phenomena is in fact an aphenomenal construction on the actual social or natural process. This aphenomenal modeling of reality closes all loops and bridges all gaps with fictions of various kinds.

One necessary corrective to this hopeless course should rely instead on the closest possible observation of input-state (i.e., historical origin), pathway, and output-state (i.e., the present condition, as distinct from a projection) of the actual social or natural process, starting with the present, meaning: that current output-state. Whatever has been clearly established, and whatever still remains incompletely understood are then summed up. A process of elimination is launched. This is based on abstracting absence to advance an hypothesis that might both account for whatever gaps remain in the observer's knowledge, and is possible to test. The observer plans out some intervention(s) that can establish in practice whether the hypothesized bridging of the gap in knowledge indeed accounts for what has been "missing."

All processes explained up to now rather simplistically only insofar as their change, development, and motion conformed to known laws of social or natural development can be reviewed by these same methods and their conventional explanation replaced with these essentially "delinearized" histories.

5

Fundamentals of Mass and Energy Balance

5.1 Introduction

Nature is balanced, yet continuously dynamic. Unless nature were balanced, any disturbance would result in chain reactions that would lead to total implosion. Unfortunately, as an outgrowth of the linearizations that are developed and applied throughout currently used models of nature, scientists and engineers become myopic. A widespread presumption has emerged that assumes balance is a point of steady-state equilibrium and that the retention of balance takes place mechanically, thus a given steady state is perturbed, chaos ensues, a new steady-state equilibrium emerges. If one attempted to publish an article with detailed 500-ms by 500-ms recordings of data about someone he or she observed holding out his empty hat and a handkerchief, tossing the handkerchief into the hat, shaking them and finally pulling out a rabbit, should he be applauded as a careful scientific observer of the phenomena unfolding before his eyes or dismissed and mocked as a thorough dupe of the magician?

Those who defend the observation of cycles of "steady state," perturbation, chaos, and "new steady state" as the best evidence we can collect regarding any natural balance are operating under an unwarranted assumption, that some displacement(s) in space and time which correlate with some or any stage(s) of organic growth and development must register somehow, somewhere. This is a defect, not only in the quantity of the data collected, but especially in the quality of the data collected. However, certain qualities in the reasoning that is applied to summarize the available data tend to obscure this latter defect from view. Only if one already knows the actual internal signatures of the stages of organic development and growth and finds some displacement(s) in space and time external to the organism that corresponds, then is it reasonable to assert the correlation. But to reason in the other direction – "I saw this external sign; the internal development must therefore be thus" – is only acceptable if one is certain from the outset that no other correlation is possible, because all other possibilities in the field of available observations have been accounted for.

When it comes to sorting out natural processes, the quality of the reasoning is no substitute for a clear and warranted understanding of what is actually going on within the process, as well as in its relations to phenomena that surround it and may interact with it. Inductive reasoning about an effect from the evidence of a cause to a conclusion is always safe, although limited in its generality, whereas deductive reasoning from observed effects backwards to its causes is seductively breathtaking in its generality but justifiable only within a fully defined field of observed causes and effects. The ongoing value of deductive reasoning lies in the questions that require further investigation, while the ongoing value of inductive reasoning ought to be its demonstration of how modest and limited our understanding presently is.

5.2 The Difference Between a Natural Process and an Engineered Process

For purposes of study and investigation, any process may be reduced to an indefinitely extended sequence of intervals of any length. All such processes may then be considered "equal." However, as a result of this first-order abstraction, a process in which the observer can definitely be separated from the subject matter under observation may be treated no differently than a process in which the observer cannot necessarily be separated, or considered truly separate, from the subject matter under observation. In other words, to the extent that both involve physical or chemical laws in some manner, a natural process (a process unfolding in the natural or social environment) and an engineered process (something in a laboratory or under similarly controlled conditions) may be treated the same insofar as they involve the same physical-chemical laws, governing equations, and so forth.

There are reasons for wishing to be able to treat a natural process and an engineered process the same. This can lend universal authority to scientific generalizations of results obtained from experiments, e.g., in a laboratory. However, this is where many complications emerge regarding how science is cognized among researchers on the one hand, and how those involved with implementing the findings of research cognize science on the other. Furthermore, in the literature of the history of science, and especially of experimental and research methods, there are sharp differences of approach between what happens with such cognition in the social sciences and what happens with such cognition in the engineering and so-called "hard" sciences. In either case, these differences and complications are shaped by the ambient intellectual culture. How the science is cognized is partly a function of discourse, terminology, and rhetoric, and partly a function of the attitudes of researchers in each of these fields toward the purposes and significance of their work for the larger society. Zatzman *et al.* (2007a) examined these issues at considerable length.

5.3 The Measurement Conundrum of the Phenomenon and its Observer

5.3.1 Background

The single most consequential activity of any scientific work is probably that which falls under the rubric of "measurement." There is a broad awareness among educated people in the general public of the endless and unresolved debates among social scientists over what it is that they are actually counting and measuring. This awareness is stimulated by a sense that, because the phenomena they examine cannot be exactly reproduced and experiments can only examine narrowly selected pieces of the social reality of interest, the social sciences are scientific in a very different way than the engineering and "hard" sciences. This outlook conditions and frames much of the discussion of measurement issues in the social science literature up to our present day.

During the 1920s and 1930s, when the social sciences in North America were being converted into professions based on programs of post-graduate-level academic formation, many dimensions of these problems were being discussed fully and frankly in the literature. In a memorable 1931 paper, Prof. Charles A. Ellwood, who led in professionalizing sociology, chaired the American Sociological Association in 1924, and produced (before his death in 1946) more than 150 articles and standard textbooks in the field that sold millions of copies, weighed in strongly on these matters:

> A simple illustration may help to show the essence of scientific reasoning or thinking. Suppose a boy goes out to hunt rabbits on a winter morning after a fresh fall of snow. He sees rabbit tracks in the fresh snow leading toward a brush pile. He examines the snow carefully on all sides of the brush pile and finds no rabbit tracks leading away from it. Therefore he concludes that the rabbit is still in the brush pile.
>
> Now, such a conclusion is a valid scientific conclusion if there is nothing in the boy's experience to contradict it, and it illustrates the nature of scientific reasoning. As a matter of fact, this is the way in which the great conclusions of all sciences have been reached-all the facts of experience are seen to point in one direction and to one conclusion. Thus the theory of organic evolution has been accepted by biological scientists because all the facts point in that direction-no facts are known which are clearly against this conclusion. Organic evolution is regarded as an established scientific fact, not because it has been demonstrated by observation or by methods of measurement, but rather because all known facts point to that conclusion.
>
> This simple illustration shows that what we call scientific method is nothing but an extension and refinement of common sense, and that it always involves reasoning and the interpretation of the facts of experience. It rests upon sound logic and a common-sense attitude toward human experience. But the hyper-scientists of our day deny this and say that science rests not upon reasoning (which cannot be trusted), but upon observation, methods of measurement, and the use of instruments of precision. Before the boy concluded that there was a rabbit in the brush pile, they say, he should have gotten an x-ray machine to see if the rabbit was really there, if his conclusion is

> to be scientific; or at least he should have scared bunny from his hiding place and photographed him; or perhaps he should have gotten some instrument of measurement, and measured carefully the tracks in the snow and then compared the measurements with standard models of rabbit's feet and hare's feet to determine whether it was a rabbit, a hare, or some other animal hiding in the brush pile. Thus in effect does the hyper-scientist contrast the methods of science with those of common sense.
>
> Now, it cannot be denied that methods of measurement, the use of instruments of precision, and the exact observation of results of experiment are useful in rendering our knowledge more exact. It is therefore, desirable that they be employed whenever and wherever they can be employed. But the question remains, in what fields of knowledge can these methods be successfully employed? No doubt the fields in which they are employed will be gradually extended, and all seekers after exact knowledge will welcome such an extension of methods of precision. However, our world is sadly in need of reliable knowledge in many fields, whether it is quantitatively exact or not, and it is obvious that in many fields quantitative exactness is not possible, probably never will be possible, and even if we had it, would probably not be of much more help to us than more inexact forms of knowledge.
>
> It is worthy of note that even in many of the so-called natural sciences quantitatively exact methods play a very subordinate role. Thus in biology such methods played an insignificant part in the discovery and formation of the theory of organic evolution. (16)

A discussion of this kind is largely absent in current literature of the engineering and "hard" sciences, and is symptomatic of the near-universal conditioning of how narrowly science is cognized in these fields. Many articles in this journal, for example, have repeatedly isolated the "chemicals are chemicals" fetish, with the insistent denial of what is perfectly obvious to actual common sense, namely, that what happens to chemicals and their combinations in test tubes in a laboratory cannot be matched 1:1 with what happens to the same combinations and proportions of different elements in the human body or anywhere else in the natural environment.

During the 20th century and continuing today in all scientific and engineering fields pursued in universities and industry throughout North America and Europe, the dominant paradigm has been that of pragmatism: the truth is whatever works. This outlook has tended to discount, or place at a lower level, purely "scientific" work, meaning experimental or analytical-mathematical work that produces hypotheses and/or various theoretical explanations and generalizations to account for phenomena or test what the researcher thinks he or she cognizes about phenomena. The pressure has been for some time to first, make something work, and second, to explain the science of why it works later, if ever.

This brings up the question, "Is whatever has been successfully engineered actually the truth, or are we all being taken for a ride?" The first problem in deconstructing this matter is the easy assumption that technology, i.e., engineering, is simply "applied science," a notion that defends engineering and pragmatism against any theory and any authority for scientific knowledge as more fundamental than practical.

Ronald Kline is a science historian whose works address various aspects of the relationship of formal science to technologies. In a 1995 article, he points out:

> A fruitful approach, pioneered by Edwin Layton in the 1970s, has been to investigate such "engineering sciences" as hydraulics, strength of materials, thermodynamics, and aeronautics. Although these fields depended in varying degrees on prior work in physics and chemistry, Layton argued that the groups that created such knowledge established relatively autonomous engineering disciplines modeled on the practices of the scientific community. (195)

Kline does go on to note that "several historians have shown that it is often difficult to distinguish between science and technology in industrial research laboratories; others have described an influence flowing in the opposite direction – from technology to science – in such areas as instrumentation, thermodynamics, electromagnetism, and semiconductor theory." Furthermore, he points out that although a "large body of literature has discredited the simple applied-science interpretation of technology – at least among historians and sociologists of science and technology – little attention has been paid to the history of this view and why it (and similar beliefs) has [*sic*] been so pervasive in American culture. Few, in other words, have [examined]... how and why historical actors described the relationship between science and technology the way they did and to consider what this may tell us about the past." It is important to note, however, that all this still leaves the question of how the relationship of engineering rules of thumb (by which the findings of science are implemented in some technological form or other) might most usefully apply to the source findings from "prior work in physics and chemistry."

One of the legacies of the pragmatic approach is that any divergence between predictions about, and the reality of, actual outcomes of a process is treated usually as evidence of some shortcoming in the practice or procedure of intervention in the process. This usually leapfrogs any consideration of the possibility that the divergence might actually be a sign of inadequacies in theoretical understanding and/or the data adduced in support of that understanding. While it may be simple in the case of an engineered process to isolate the presence of an observer and confine the process of improvement or correction to altering how an external intervention in the process is carried out, matters are somewhat different when it comes to improving a demonstrably flawed understanding of some natural process. The observer's reference frame could be part of the problem, not to mention the presence of apparently erratic, singular, or episodic epiphenomena that are possible signs of some unknown sub-process(es). This is one of the greatest sources of confusion often seen in handling so-called data scatter, presumed data 'error,' or anomalies generated from the natural version of a phenomenon that has been studied and rendered theoretically according to outcomes observed in controlled laboratory conditions.

Professor Herbert Dingle (1950), more than half a century ago, nicely encapsulated some of what we have uncovered here:

> Surprising as it may seem, physicists thoroughly conversant with the ideas of relativity, and well able to perform the necessary operations and calculations which the theory of relativity demands, no sooner begin to write of the theory of measurement than they automatically relapse into the philosophical outlook of the nineteenth century and produce a system of thought wholly at variance with their practice. (6)

He goes on to build the argument thus:

> It is generally supposed that a measurement is the determination of the magnitude of some inherent property of a body. In order to discover this magnitude we first choose a sample of the property and call it a 'unit.' This choice is more or less arbitrary and is usually determined chiefly by considerations

> of convenience. The process of measurement then consists of finding out how many times the unit is contained in the object of measurement. I have, of course, omitted many details and provisos, for I am not criticising the thoroughness with which the matter has been treated but the fundamental ideas in terms of which the whole process is conceived and expressed. That being understood, the brief statement I have given will be accepted, I think, as a faithful account of the way in which the subject of measurement is almost invariably approached by those who seek to understand its basic principles. Now it is obvious that this is in no sense an 'operational' approach. 'Bodies' are assumed, having 'properties' which have 'magnitudes.' All that 'exists', so to speak, before we begin to measure. Our measurement in each case is simply a determination of the magnitude in terms of our unit, and there is in principle no limit to the number of different ways in which we might make the determination. Each of them–each 'method of measurement,' as we call it-may be completely different from any other; as operations they may have no resemblance to one another; nevertheless they all determine the magnitude of the same property and, if correctly performed, must give the same result by necessity because they are all measurements of the same independent thing. (6–7)

Then Prof. Dingle gives some simple but arresting examples:

> Suppose we make a measurement – say, that which is usually described as the measurement of the length of a rod, AB. We obtain a certain result-say, 3. This means, according to the traditional view, that the length of the rod is three times the length of the standard unit rod with which it is compared. According to the operational view, it means that the result of performing a particular operation on the rod is 3. Now suppose we repeat the measurement the next day, and obtain the result, 4. On the operational view, what we have learnt is unambiguous. The length of the rod has changed, because 'the length of the rod' is the name we give to the result of performing that particular operation, and this result has changed from 3 to 4. On the traditional view, however, we are in a dilemma, because we do not know which has changed, the rod measured or the standard unit; a change in the length of either would give the observed result. Of course, in practice there would be no dispute; the measurements of several other rods with the same standard, before and after the supposed change, would be compared,

> and if they all showed a proportionate change it would be decided that the standard had changed, whereas if the other rods gave the same values on both occasions, the change would be ascribed to the rod AB; if neither of these results was obtained, then both AB and the standard would be held to have changed. If an objector pointed out that this only made the adopted explanation highly probable but not certain, he would be thought a quibbler, and the practical scientist would (until recently, quite properly) take no notice of him. (8–9)

The "operational view" is the standpoint according to which the reference frame of the observer is a matter of indifference. The conventional view, by way of contrast, assumes that the observer's standard(s) of measurement corresponds to actual properties of the object of observation. However, when the physical frame of reference in which observations are made is transformed, the assumption about the reference frame of the observer that is built into the "operational view" becomes dysfunctional. Such a change also transforms the reference frame of the observer, making it impossible to dismiss or ignore:

> But with the wider scope of modern science he can no longer be ignored. Suppose, instead of the length of the rod AB, we take the distance of an extra-galactic nebula, N. Then we do, in effect, find that of two successive measurements, the second is the larger. This means, on the traditional view, that the ratio of the distance of the nebula to the length of a terrestrial standard rod is increasing. But is the nebula getting farther away or is the terrestrial rod shrinking? Our earlier test now fails us. In the first place, we cannot decide which terrestrial rod we are talking about, because precise measurements show that our various standards are changing with respect to one another faster than any one of them is changing with respect to the distance of the nebula, so that the nebula may be receding with respect to one and approaching with respect to another. But ignore that: let us suppose that on some grounds or other we have made a particular choice of a terrestrial standard with respect to which the nebula is getting more distant. Then how do we know whether it is 'really' getting more distant or the standard is 'really' shrinking? If we make the test by measuring the distances of other nebulae we must ascribe the change to the rod, whereas if we make it by measuring other 'rigid' terrestrial objects we shall get no consistent result at all. We ought,

> therefore, to say that the probabilities favour the shrinking of the rod. Actually we do not; we say the universe is expanding. But essentially the position is completely ambiguous.

As long as the transformation of the frame of reference of the phenomenon is taken properly into account, the operational view will overcome ambiguity. However, the comparison of what was altered by means of the preceding transformation serves to establish that there is no such thing as an absolute measure of anything in physical reality:

> Let us look at another aspect of the matter. On the earth we use various methods of finding the distance from a point C to a point D: consider, for simplicity, only two of them-the so-called 'direct' method of laying measuring rods end to end to cover the distance, and the indirect method of 'triangulation' by which we measure only a conveniently short distance directly and find the larger one by then measuring angles and making calculations. On the earth these two methods give identical results, after unavoidable 'experimental errors' have been allowed for, and of course we explain this, as I have said, by regarding these two processes as alternative methods of measuring the same thing. On the operational view there are two different operations yielding distinct quantities, the ' distance' and the 'remoteness,' let us say, of D from C, and our result tells us that, to a high degree of approximation, the distance is equal to the remoteness. Now let us extend this to the distant parts of the universe. Then we find in effect that the 'direct' method and the triangulation method no longer give equal results. (Of course they cannot be applied in their simple forms, but processes which, according to the traditional view, are equivalent to them, show that this is what we must suppose.) On the view, then, that there is an actual distance which our operations are meant to discover-which, if either, gives the 'right' distance, direct measurement or triangulation? There is no possible way of knowing. Those who hold to the direct method must say that triangulation goes wrong because it employs Euclidean geometry whereas space must be non-Euclidean; the correct geometry would bring the triangulation method into line with the other, and the geometry which is correct is then deduced so as to achieve just this result. Those who hold to triangulation, on the other hand, must say that space is pervaded by a field of force which distorts the measuring rods, and again the strength

> of this field of force is deduced so as to bring the two measurements into agreement. But both these statements are arbitrary. There is no independent test of the character of space, so that if there is a true distance of the nebula we cannot know what it is. On the operational view there is no ambiguity at all; we have simply discovered that distance and remoteness are not exactly equal, but only approximately so, and then we proceed to express the relation between them.

The nature-science approach takes what Prof. Dingle has outlined to the final stage of considering nature four-dimensional. If time is taken as a characteristic measure of any natural phenomenon, and not just of the vastnesses of distance in cosmic space, then the observer's frame of reference can be ignored if and only if it is identical to that of the phenomenon of interest. Otherwise, and most if the time, it must be taken into account.

That means, however, that the assumption that time varies independently of the phenomena unfolding within it may have to be relaxed. Instead, any and all possible (as well as actual) non-linear dependencies need to be identified and taken explicitly into account. At the time Prof. Dingle's paper appeared in 1950, such a conclusion was neither appealing nor practicable. The capabilities of modern computing methods since then, however, have reduced previously insuperable computational tasks to almost routine procedures.

In mentioning the "operational view," Prof. Dingle has produced a hint of something entirely unexpected regarding the problem-space and solution-space of reality according to Einstein's relativistic principles. Those who have been comfortably inhabiting a three-dimensional sense of reality have nothing to worry about. For them, time is just an independent variable. In four-dimensional reality, however, it is entirely possible that the formulation of a problem may never completely circumscribe how we could proceed to operationalize its solution. In other words, we have to anticipate multiple possible, valid solutions to one and the same problem formulation. It is not a 1:1 relationship, so linear methods that generate a unique solution will be inappropriate. The other possibilities are one problem formulation with multiple solutions (1:many) or multiple formulations of the problem having one or more solutions in common (many:1, or many:many). In modelers' language, we will very likely need to solve non-linear equation descriptions of the relevant phenomena by non-linear methods, and we may still never know if or when we have all possible solutions. (This matter of solving non-linear equations with non-linear methods is developed further at §§4–5 *supra.*) At this moment, it is important to establish what operationalizing the solution of a problem, especially multiple solutions, could mean or look like. Then, contrast this to how the scenario of one problem with multiple solutions is atomized and converted into a multiplicity of problems based on different sub-portions of data and data-slices from which any overall coherence has become lost.

5.3.2 Galileo's Experimental Program: An Early Example of the Nature-Science Approach

Although science and knowledge are not possible without data measurements, neither science nor knowledge is reducible to data, techniques of measurement, or methods of data analysis. Before operationalizing any meaningful solution(s) to a research problem, there is a crucial prior step, the stage of "experimental design," of formulating the operational steps or the discrete sub-portions of the problematic, in which our knowledge is incomplete. For this, there is no magic formula or guaranteed road to success. This is where the true art of the research scientist comes into play. In adopting the standpoint and approach of nature-science, in which actual phenomena in nature or society provide the starting point, the different roles of the investigator as participant on the one hand and as observer on the other can be clearly delineated as part of the research program.

Galileo's experiments with freely falling bodies are well known and widely discussed in the literature of theoretical physics, the history of science and technology, the history of the conflicts between science and religion, and a number of other areas. There is a widespread consensus about his principled stance in defense of the findings in his own research, coupled with some ongoing disputes as to how consistently he could defend principles against the pressures of the Inquisition. In a departure from that path in the literature, rather than adding or passing judgment on Galileo's conduct vis-à-vis the Church authorities, this paper addresses the backbone of Galileo's stand, namely, his conviction that his method was a more reliable guide to finding the truth of the nature of freely falling bodies than any guesswork by Aristotle, a Greek who had been dead for two millennia. The standpoint adopted here is that Galileo's research program represents an early model of the nature-science approach — the first by a European, in any event. Its "correction" by those who came after him, on the other hand, represents a corruption of his method by the mandates of "new science," mandates whereby subsequent investigators would become so preoccupied with tangible evidences to the point of excluding other considerations.

The following information about Galileo's approach to experimental methods and the controversy that continued around it, taken from the summary from the Wikipedia entry "Two New Sciences," summarizes the most significant developments on which the claim of the first European proponent of the nature-science approach is based:

> *The Discourses and Mathematical Demonstrations Relating to Two New Sciences* (*Discorsi e dimostrazioni matematiche, intorno a due nuove scienze*, 1638) was Galileo's final book and a sort of scientific testament covering much of his work in physics over the preceding thirty years.
>
> Unlike the *Dialogue Concerning the Two Chief World Systems* (1632), which led to Galileo's condemnation by the Inquisition following a heresy trial, it could not be published with a license from the Inquisition. After the failure of attempts to publish the work in France, Germany, or Poland, it was picked up by Lowys Elsevier in Leiden, The Netherlands, where the writ of the Inquisition was of little account.
>
> The same three men as in the Dialogue carry on the discussion, but they have changed. Simplicio, in particular, is no longer the stubborn and rather dense Aristotelian; to some extent he represents the thinking of Galileo's early years, as Sagredo represents his middle period. Salviati remains the spokesman for Galileo.
>
> Galileo was the first to formulate the equation for the displacement s of a falling object, which starts from rest, under the influence of gravity for a time t:
>
> $$s = \tfrac{1}{2}gt^2$$
>
> He (Salviati speaks here) used a wood molding, "12 cubits long, half a cubit wide and three finger-breadths thick" as a ramp with a straight, smooth, polished groove to study rolling balls ("a hard, smooth and very round bronze ball"). He lined

> the groove with "parchment, also smooth and polished as possible." He inclined the ramp at various angles, effectively slowing down the acceleration enough so that he could measure the elapsed time.
>
> He would let the ball roll a known distance down the ramp, and used a water clock to measure the time taken to move the known distance. This clock was "a large vessel of water placed in an elevated position; to the bottom of this vessel was soldered a pipe of small diameter giving a thin jet of water, which we collected in a small glass during the time of each descent, whether for the whole length of the channel or for a part of its length; the water thus collected was weighed, after each descent, on a very accurate balance; the differences and ratios of these weights gave us the differences and ratios of the times, and this with such accuracy that although the operation was repeated many, many times, there was no appreciable discrepancy in the results." (Website 1)

It is critical to add that, instead of clocking standardized "seconds" or minutes, this method of time measurement calibrates one natural motion by means of another natural duration.

> The water clock mechanism described above was engineered to provide laminar flow of the water during the experiments, thus providing a constant flow of water for the durations of the experiments. In particular, Galileo ensured that the vat of water was large enough to provide a uniform jet of water.
>
> Galileo's experimental setup to measure the literal flow of time, in order to describe the motion of a ball, was palpable enough and persuasive enough to found the sciences of mechanics and kinematics. (ibid.)

Although Galileo's procedure founded "time" in physics, in particular, on the basis of uniformity of flow in a given interval, this would later be generalized as the notion of a linear flow of time. Einstein would later overthrow this notion with regard to the vastnesses of space in the universe and what the nature-science approach proposes to correct in all investigations of processes unfolding in the natural environment of the earth.

> The law of falling bodies was discovered in 1599. But in the 20th century some authorities challenged the reality of Galileo's experiments, in particular the distinguished French historian

> of science Alexandre Koyré. The experiments reported in *Two New Sciences* to determine the law of acceleration of falling bodies, for instance, required accurate measurements of time, which appeared to be impossible with the technology of 1600. According to Koyré, the law was arrived at deductively, and the experiments were merely illustrative thought experiments. Later research, however, has validated the experiments. The experiments on falling bodies (actually rolling balls) were replicated using the methods described by Galileo, and the precision of the results was consistent with Galileo's report. Later research into Galileo's unpublished working papers from as early as 1604 clearly showed the reality of the experiments, and even indicated the particular results that led to the time-squared law. (ibid.)

Of interest here is the substance of Koyré's challenge, that the time it would take objects to fall to the ground from the top of the Tower of Pisa could never have been measured precisely enough in Galileo's day to justify his conclusion. Of course, subsequent experimental verification of Galileo's conclusions settles the specific question, but Koyré's objection is important here for another reason.

What if, instead of following Galileo's carefully framed test, there was a series of increasingly precise measurements of exactly how long it took various masses in free fall to reach the ground from the same height? The greater the precision, the more these incredibly small differences would be magnified. One could hypothesize that air resistance accounted for the very small differences, but how could that assertion then be positively demonstrated? If modern statistical methods had been strictly applied to analyzing the data generated by such research, magnificent correlations might have been demonstrated. None of these correlations, however, would point conclusively to the uniformity of acceleration of the speed at which these freely falling objects descend over any other explanation. As long as the focus remained on increasing the precision of measurement, the necessity to drop Aristotle's explanation entirely (that object fall freely at speeds proportional to their mass) would never be established unambiguously.

The following possible representation of how the data resulting from such a falsified experimental approach reinforces this conclusion (Figure 5.1).

Galileo's act of publishing the *Discorsi* in 1638 as the main summary of the most important part of his life's work was not an

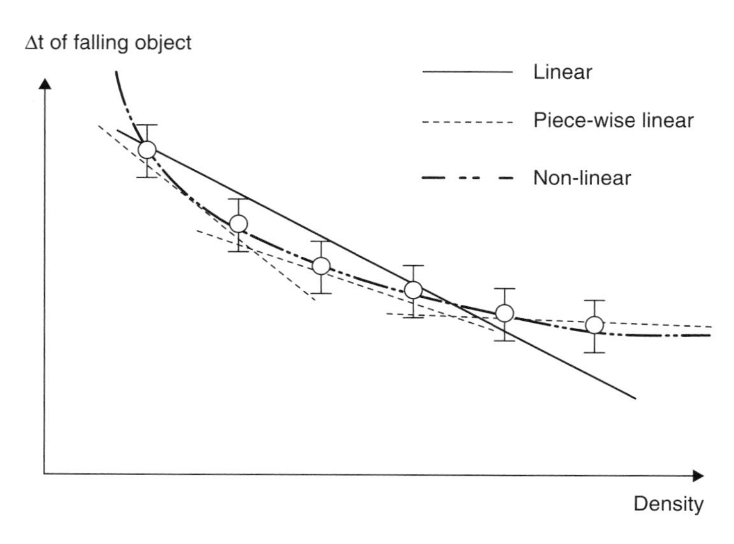

Figure 5.1 If the model were true, then the theory would be verified. So, what if the model were false from the outset, betrayed by retention of a first assumption that was not more carefully scrutinized? With more observations and data, this modeling exercise could carry on indefinitely, e.g., a 5-parameter univariate non-linear "function" like $y = ax + bx^2 + cx^3 + px^4 + qx^5$... that continues towards ever higher degrees of "precision."

ordinary act of defiance. It was his affirmation before the entire world that he had stuck to his original research program and never took the easy way out. The nature-science approach settles the question of whether one is on the right path to begin with, and that is Galileo's primary accomplishment. Today, the time has arrived to take matters to the next stage. In order to capture and distinguish real causes and real effects as a norm in all fields of scientific and engineering research, it is necessary to apply what Albert Einstein elaborated, from examining the extremities of space, namely that time is a fourth dimension, to the observation of all phenomena, including those that are natural/social and those that are more deliberately engineered.

Proponents of the "new science" of tangibles have long insisted that Galileo is truly their founding father. In fact, their premise has been that the collection and sifting of the immediate evidence of nature and its processes is the sum total of reliable scientific knowledge. Without conscious investigation of nature and critical questioning of conclusions that no longer fit the available evidence, how does science advance? This contribution is

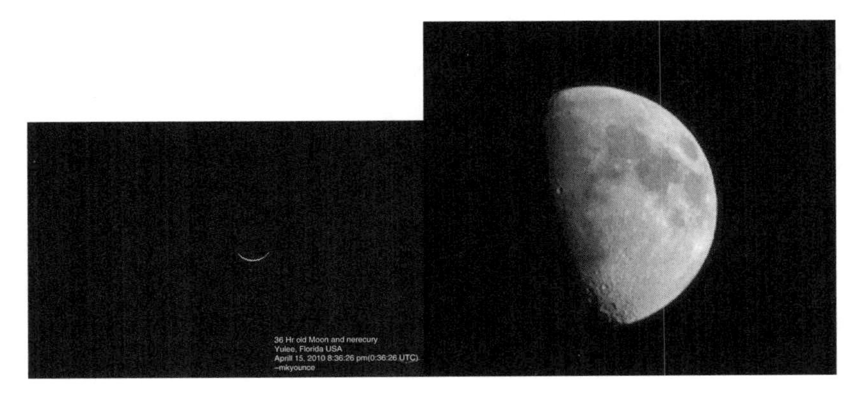

In modelization, the first premise must be true at all time, it cannot be subjective or a matter of perception at a particular time. Any model based on perception will make the entire process aphenomenal.

acknowledged by recognizing Galileo as the first European practitioner of nature-science.

Although science and knowledge are bound to give rise to data and measurement, the reverse is not necessarily true. In the scramble to produce models that generate information about phenomena with increasing specificity and precision, it is easy to lose sight of this elementary insight and identify advances in data gathering and management with advances in scientific understanding. The coherence of research efforts undertaken in all fields has come under threat from this quarter, conventional research methods and approaches have come into question, and novel alternatives excite growing interest. As the review of the "engineering approach" serves to reconfirm, what decides the usefulness of any mathematical modeling tool is not the precision of the tool itself, but rather the observer's standpoint of a real process unfolding in nature. In fact, this ultimately limits or determines its actual usefulness. As Galileo's example clearly shows, regardless of the known value of improved precision of measurement, greater precision cannot overcome problems that arise from having embarked on the wrong path to begin with. The same goes for how precisely, i.e., narrowly, one specifies the reference frame of an observer or what is being observed. Conventional methods of modeling have long since passed the point of being able to specify to the point of excluding the general context of the reality surrounding the observer or the phenomena of interest. The apparent short-term gain in specificity may end up being vastly overcompensated by such things as the impossibility of attaching any meaningful physical interpretation to elaborately computed mathematical results and, with that, the subsequent realization that a research effort may have begun by looking in the wrong direction in the first place.

5.4 Implications of Einstein's Theory of Relativity on Newtonian Mechanics

The underlying problems that the philosophy of science addressed before World War II were linked to deciding what weight to give Einstein's disturbance of the Newtonian worldview of "mechanism." There was considerable concern over the principle of relativity and how far some of its new concepts might be taken, e.g., where the idea of time as a fourth dimension might be taken next.

Alongside this came pressing concerns after World War II to eliminate the powerful influence of the advances in science coming out of the Soviet Union. These concerns were based on a paradigm shift that was universally rejected in Western countries but nevertheless produced powerful results in many engineering fields throughout the Soviet system that could not be ignored. The notion on which all scientific research in the Soviet Union came to be based was that the core of scientific theory-building depended on sorting out the necessary and sufficient conditions that account for observed phenomena. In the West, this was allowed only for mathematics, and that, too, with limited scope. Introducing this notion into investigations of, and practical engineering interventions in, natural processes was dismissed as "excessive determinism." Instead, an entire discussion of "simplicity" as a diversionary discourse was developed.

This discourse had an aim. It was intended, first and foremost, to avoid dealing with the entire matter of necessary and sufficient conditions. This approach reduced the process of selecting the more correct, or most correct, theoretical explanation accounting for one's data to the investigator's personal psychology of preference for "the simple" over "the complex." This was put forward in opposition to the notion that a scientific investigator might want to account for what is in line with his, her, or others' observations or counter others' observations of the same phenomenon (Ackermann 1961; Bunge 1962; Chalmers 1973b; Feuer 1957, 1959; Goodman 1961; Quine 1937; Rudner 1961; Schlesinger 1959, 1961).

As presented in Chapter 1 in the context of historical development, the notion that nature is dynamic has been known for at least two and a half millennia. However, Einstein's revelation of the existence of the fourth dimension was a shock, because all scientific and engineering models used in the post-Renaissance world were steady-state models. The analysis presented in Chapter 1 shows that the acceptance of these models, without consideration of their existence in the real world, led to subsequent models that remain aphenomenal. In this, it is important to note that an aphenomenal hypothesis will entirely obscure the actual pathway of some natural phenomena that anyone with a correct starting point would set out to study. Unless this is recognized, resorting to curve-fitting and/or other retrofitting of data to highly linearized preconceptions and assumptions concerning the elementary physics of the phenomenon in its natural environment will not validate the aphenomenal model. Zatzman *et al.* (2008a, 2008b) re-asserted this stance in a recent series of articles on Newtonian mechanics. They highlighted the following:

1. All natural phenomena at all stages and phases are four-dimensional (time being the fourth dimension).
2. All models based on conceiving the time factor as "the independent variable" rather than as a dimension in its own right are aphenomenal (unreal) and bound to fail.
3. Any engineering design based on these aphenomenal bases can only simulate nature for a very short period of time, and never emulate it.
4. Our present-day "technological disaster" is essentially a litany of the inevitable failure that aphenomenally-theorized engineering, beyond the shortest of short-terms, will always produce.

With the revolutionary paper of Albert Einstein, the understanding of light was placed on a fully scientific basis for the first time as a form of matter that radiates as an energy wave (Website 1). To establish this, it was necessary to breach the wall in human thought that was created as a result of following the misconceived starting point that "light" was the opposite of "dark" (in Arabic language, the word for light is *noor*, which is inherently singular, where as the word for darkness is *dhulumat*, which is inherently plural, thus making is easy to understand opposite to darkness cannot be light).

How had an everyday notion become such an obstacle to scientific understanding? Investigation unexpectedly disclosed the culprit, lurking in what was universally acknowledged for the three hundred and twenty-seven years preceding Einstein's paper, to represent one of the most revolutionary scientific advances of all time, namely, the affirmation by Isaac Newton that "motion" was the opposite of "rest" (Website 3). Nowhere does Newton's mechanism explicitly deny that motion is the very mode of existence of all matter. However, his system crucially fails to affirm this as its first principle. Instead, his First Law of Motion affirms that a body in motion remains in motion (disregarding drag effects due to friction), and an object at rest remains at rest unless acted upon by an external force. The corollary flowing immediately from this law is not that motion is the mode of existence of all matter, but only that motion is the opposite of rest.

In Newton's day, no one doubted that the earth and everything in the heavens had been created according to a single all-encompassing clock. Among Christians (mainly Catholics and Protestants during the Reformation), all that was disputed at this time was whether the clock was set in motion by a divine creator and then carried on heedless of human whims, or whether humans, by their individual moral choices, could influence certain operations of this clock. The most important implication is that any notions regarding the universe as a mechanism operating according to a single all-encompassing clock lose all coherence. Newton's great antagonist, Bishop George Berkeley, excoriated Newton's failure to close the door to such heretically anti-Christian views (Stock, 1776).

In the same moment that the unquestioning belief in a single all-encompassing clock is suspended or displaced, it becomes critical to affirm the existence of, and a role for, the observer's frame of reference. Newton, his contemporaries, and other men and women of science before Einstein's theory of relativity were certain that Newton's laws of motion defined the mechanism governing force of any kind, motion of any type, as well as any of their possible interactions. What we now understand is that, in fact, Newton's laws of motion actually defined only the possibilities for a given system of forces, one that would moreover appear "conservative" only because any observer of such a system was assumed to stand outside it.

To seventeenth-century European contemporaries of Isaac Newton, this matter of frame of reference was of no moment whatsoever. For them, the flora, humans, and non-human fauna of the known world had never been anywhere other than where they currently were. Nor was it generally known or understood that people lived in human social collectives other than those already known since "ancient times," an era beginning in western Asia some finite but unknown number of years before the birth of Jesus Christ. There was no reason to wrestle with the prospect of any other frame of reference, either in historical time or in spaces elsewhere in the universe. Only by means of subsequent research, of a type and on a scale that could not have been undertaken in Newton's day, did it become possible to establish that in any space-time coordinates anywhere in the universe, from its outermost cosmic reaches to the innermost sub-atomic locale, mass, energy, and momentum would be, and must be, conserved, regardless of the observer's frame of reference.

In a universe defined by a single clock and common reference frame, three principal physical states of matter (vapor, solid, and liquid) could be readily distinguished (by driving externally applied energy, bounded by powerful electro-magnetic force fields, and still without having to readjust any other prevailing assumptions about a single clock and common reference-frame, a fourth highly transient plasma state could be further distinguished). Overall, motion could and would still be distinguished as the opposite of rest. What, however, can be said to connect matter in a vapor, solid, or liquid state to what happens to a state of matter at either sub-atomic or cosmic spatial scales? These are the regions for which the conservation of matter, energy, and momentum must still be accounted. However, in these regions, matter cannot possibly be defined as being at rest without introducing more intractable paradoxes and contradictions. It is only when motion is recognized as the mode of existence of all matter in any state that these paradoxes become removable.

5.5 Newton's First Assumption

Broadly speaking, it is widely accepted that Newton's system, based on his three laws of motion, accounting for the proximate physical reality in which humans live on this Earth coupled with the elaboration of the principle of universal gravitation to account for motion in the heavens of space beyond this Earth, makes no special axiomatic assumptions about physical reality outside what any human being can observe and verify.

For example, Newton considers velocity, v, as a change in the rate at which a mass displaces its position in space, s, relative to the time duration, t, of the motion of the said mass. That is:

$$v = \frac{\partial s}{\partial t} \tag{5.1}$$

This is no longer a formula for the average velocity, measured by dividing the net displacement in the same direction as the motion impelling the mass by the total amount of time that the mass was in motion on that path. This formula posits something quite new, actually enabling us to determine the instantaneous velocity at any point along the mass's path while it is still in motion.

The velocity that can be determined by the formula given in Equation 5.1 above is highly peculiar. It presupposes two things. First, it presupposes that the displacement of an object can be derived relative to the duration of its motion in space. Newton appears to cover that base already by defining this situation as one of what he calls "uniform motion." Secondly, what exactly is the time duration of the sort of motion Newton is setting out to explain and account for? It is the period in which the object's state of rest is disturbed, or some portion thereof. This means the uniformity of the motion is not the central or key feature. Rather, the key is the assumption in the first place that motion is the opposite of rest.

In his first law, Newton posits motion as the disturbance of a state of rest. The definition of velocity as a rate of change in spatial displacement relative to some time duration means that the end of any given motion is either the resumption of a new state of rest, or the starting point of another motion that continues the disturbance of the initial state of rest. Furthermore, only to an observer external to the mass under observation can motion appear to be the disturbance of a state of rest and a state of rest appear to be the absence or termination of motion. Meanwhile, within nature, is anything ever at rest? The struggle to answer this question exposes the conundrum implicit in the Newtonian system: everything "works" and all systems of forces are "conservative" if and only if the observer stands outside the reference frame in which a phenomenon is observed.

In Newton's mechanics, motion is associated not with matter-as-such, but only with force externally applied. Inertia, on the other hand, is definitely ascribed to mass. Friction is considered only as a force equal and opposite to that which has impelled some mass into motion. Friction in fact exists at the molecular level as well as at all other scales, and it is not a force externally applied. It is a property of matter itself. It follows that motion must be associated fundamentally not with force(s) applied to matter, but rather with matter itself. Although Newton nowhere denies this possibility, his first law clearly suggests that going into motion and ceasing to be in motion are equally functions of some application of force external to the matter in motion; motion is important relative to some rest or equilibrium condition.

Examination of developments in European science and what prominent historians of this era of new science have had to say about Newton's mathematical treatment of physical problems compels the conclusion that the failure to ascribe motion to matter in general is implicit in, and built into, Newton's very approach to these problems (Cohen 1995; Grabiner 2004).

For example, Grabiner (2004) explains Newton's approach thus:

> ...Newton first separated problems into their mathematical and physical aspects. A simplified or idealized set of physical assumptions was then treated entirely as a mathematical system. Then the consequences of these idealized assumptions were deduced by applying sophisticated mathematical techniques. But since the mathematical system was chosen to duplicate the idealized physical system, all the propositions deduced in the mathematical system could now be compared with the data of experiment and observation. Perhaps the mathematical system was too simple, or perhaps it was too general and choice had to be made. Anyway, the system was tested against experience. And then — this is crucial — the test against experience often required modifying the original system. Further mathematical deductions and comparisons with nature would then ensue... What makes this approach non-trivial is the sophistication of the mathematics and the repeated improvement of the process. It is sophisticated mathematics, not only a series of experiments or observations, that links a mathematically describable law to a set of causal conditions. (842)

What is this initial "simplified or idealized set of physical assumptions" but the isolation from its surrounding environment of the phenomenon of interest? Immediately, before the assumptions are even tested against any mathematical approximation, this must narrow the investigative focus in a way that is bound to impose some loss of connected information of unknown significance. No amount of the "sophistication of the mathematics" can overcome such an insufficiency. On the contrary, the very "sophistication of the mathematics" can be so exciting to the point of blinding the investigator's awareness of any other possible connections. No doubt this produces some answer, and no doubt "the sophistication of the mathematics" renders the approach "non-trivial" as well. However, nowhere in this is there any guarantee that the answer will be either physically meaningful or correct. In a Newtonian physical system, however, the logic and standard of proof is the following: if a phenomenon can be cognized by everyone, e.g., the motion of an object or mass, and if some mathematical demonstration is developed that confirms the hypothesis of a law purporting to account for the said phenomenon, then the law is considered to have been verified.

Is this about science in the sense of establishing knowledge of the truth by exposing and eliminating error, or is it about something else? The key to solving this problem is to answer the question, is the scientific authority of Newton's approach "knowledge-based," or is it based on something else? In discussing the career of the 18th century Scottish mathematician Colin Maclaurin, who discovered the Maclaurin series, or the more general form of the Taylor series used extensively throughout all fields of engineering and applied science, Grabiner hints at an answer when she writes, "Maclaurin's career illustrates and embodies the way mathematics and mathematicians, building on the historical prestige of geometry and Newtonianism, were understood to exemplify certainty and objectivity during the eighteenth century. Using the Newtonian style invokes for your endeavor, whatever your endeavor is, all the authority of Newton... The key word here is 'authority.' Maclaurin helped establish that..." (2004, 841). Unlike the conditions attached to the publication of Galileo's and Copernicus' works, Newton's struggle was at no time a fight over the authority of the Church, not even with the Church of England, a Protestant Church very much at odds with the Roman Catholic Vatican on various Christian theological doctrines. Grabiner notes, for example, that Newton argued that the Earth was not perfectly spherical because the forces of its own rotation led to flattening at its poles over time. Note how artfully this line of argument dodges questioning the basis of the religious authorities' long-standing assertion that the Earth had to be spherical in order to fit with "God's plan." Grabiner's descriptions of Maclaurin's career-long coat tailing in Newton's reputation make more than clear that the issue became whether the scientists and mathematicians could govern themselves under their own secular priesthood, one that would no longer be accountable before any theological censor but that would prove to be no less ruthless in attacking any outsider challenging their authority. This is the meaning of Maclaurin's efforts to refute and discredit Bishop George Berkeley's serious questioning of Newton's calculus (Grabiner 2004). Berkeley and those who agreed with his criticisms were ridiculed for daring to challenge Newton the Great. During the 18th century, Maclaurin, and his example to others, ensured that the authority of Newtonian mathematics directly replaced the previous authority (either that of Christian scripture or the arbitrary exercise of monarchical power) across a wide range of commercial and other fields.

A knowledge-based approach to science explains and accounts for actual phenomena not mainly or only in themselves, but also in relation to other phenomena, and especially those characteristically associated with the phenomenon of interest. Then, how knowledge-based was Newton's mathematization of physics as a form of science? Simply by virtue of how the investigator has isolated the phenomenon, e.g., looked at the motion, but only the motion, of one or more tangible temporally finite object-masses, the phenomenon may appear cognizable by all. When it comes, however, to establishing anything scientifically valid about actual phenomena characteristic of some part of the natural environment, this act of isolation is the starting-point of great mischief. For any phenomenon considered in its natural or characteristic environment, the basis (or bases) of any change of the phenomenon is/are internal to the phenomenon. The conditions in which that change may manifest are all external to the phenomenon. However, when only some particular part of the phenomenon is isolated, what has actually happened? First, almost "by definition" so to speak, some or any information about the root-source of the phenomenon and about its pathway up to the isolated point or phase is already discounted. As a result, consideration of any conditions of change external to the phenomenon has become marginalized. What weight should then be assigned to any mathematical demonstration of the supposed law(s) of operation of the phenomenon thus isolated? Among other things, such demonstrations are substituted for any physical evidence surrounding the phenomenon in its characteristic state-of-nature, which would serve to corroborate the likeliest answer(s) to the question of what constitutes the internal basis (or bases) of change(s) within the phenomenon. Such mathematical demonstrations effectively marginalize any consideration of an internal basis (or bases). In other words, isolating only the tangible and accessible portion of a phenomenon for observation and mathematical generalization transforms what was a phenomenon, characteristic of some portion of the natural environment, into an aphenomenon. Whereas the basis of any change in a real, natural, characteristic phenomenon is internal and its conditions of change are external, neither any external conditions of change nor any idea of what might be the internal basis of change for the phenomenon attaches to the aphenomenon.

One alternative approach to the problem of motion encapsulated in Newton's first law would be to consider "rest" as a relative or transient condition, rather than as something fundamental that gives way to motion only as the result of disturbance by some external force (Newton's second law, often summarized by his $F = ma$ equation-relationship, represents in effect a dynamic case of his first law). Newton's schema has made it simple to take for granted the idea of rest and static equilibriums in all manners of physical situations. Clearly, however, by the very same token according to which motion should not be viewed as "absence" or disturbance of "rest," no equilibrium state should be considered static, permanent, or anything other than transitional. Is it not absurd, not to mention embarrassingly elementary, yet no less necessary, to ask how any "steady-state" reproducible under laboratory conditions, which are always controlled and selected, could ever be taken as definitive of what would occur "in the field," or in the phenomenon's native environment within nature? Yet, it would seem that an excessive focus on reproducing some measurement, a measurement, moreover, that would approximate the prediction of a governing equation developed from an idealized model, seems to have obscured this fundamental absurdity of infinitely reproducing an equilibrium state.

This was precisely the point at which the "authority" of Newton's mathematics could become a source of scientific disinformation. As long as the existence of uniform motion (first law) or constant accelerated (second law) motion is taken for granted, one loses the ability to see the role of this disinformation. On the other hand, mathematically speaking, Newton's third law says that $\sum \mathbf{F} = 0$, i.e., that the algebraic sum of all the forces acting on some object-mass "at rest" is zero. Physically speaking, however, "= 0" does not mean there are no forces acting. Rather, "=" means that there is something of a dynamic equilibrium in effect between what is represented on the left-hand and right-hand sides of this expression. One dynamic state of affairs may have given way to another, differently dynamic state of affairs, but does this "rest" truly constitute absence of motion? That is the first level at which disinformation may enter the picture. No single mathematical statement can answer this question or encompass the answer to this question in any particular case.

Mathematically speaking, in general:

LHS: expression *(simple or complex)*
= **RHS**: value *(number or function)*

In the sciences of physics and chemistry and the engineering associated with their processes, that same "=" sign, which often translates into some sort of balance between a process (or processes) described in one state on the left-hand side and an altered condition, state of matter, or energy on the right-hand side, is also used to describe data measurements (expressed as a number on the right-hand side) of various states and conditions of matter or energy (described symbolically on the left-hand side). The equivalence operator as some kind of balance, however, is meaningful in a different way than the meaning of the same equivalence operator in a statement of a numerical threshold reached by some measurement process. Confusing these "=" signs is another potential source of scientific disinformation. The problem is not simply one of notational conventions; "the sophistication of the mathematics" becomes a starting point for sophistries of various kinds.

5.6 First Level of Rectification of Newton's First Assumption

On the one hand, in a Newtonian mechanical system, time-duration remains tied to the motion of a mass inside a referential frame of active forces. On the other hand, momentum is preserved at all scales, from the most cosmic to the nano- and inter-atomic level, and at none of these scales can time or motion stop. Although Newton posited gravitation as a universally acting force, we now know that electromagnetic forces predominate in matter at the nano- or inter-atomic level. Electromagnetic forces, like frictional forces, can exist and persist without ever having been externally applied. Reasoning thus "by exhaustion," Newton's three laws of motion plus the principle of universal gravitation are actually special cases of "something else." That "something else" is far more general, like the universal preservation of mass-energy balance and conservation of momentum. The connecting glue of this balance that we call nature is that motion is the mode of existence of all matter. This is what renders time a characteristic of matter within the overall context of mass-energy-momentum conservation.

By considering motion as the mode of existence of all matter, it also becomes possible at last to treat time, consistently, as a true fourth dimension, and no longer as merely the independent variable. In other words, time ceases to be mainly or only a derivative of some spatial displacement of matter. Also, if time is characteristic of matter (rather than characteristic particularly or only of its spatial displacement), then transformation of matter's characteristic scale must entail similarly transforming the scale on which time and its roles are accounted.

From this, it follows as well that whatever we use to measure time cannot be defined as fixed, rigid, or constant, e.g., a standard like the "second," or selected or imposed without regard to the reference-frame of the phenomenon under observation. It is possible to assign physical meaning to $\partial s/\partial t$ (velocity), so long as, and only so long as, time is associated with matter only indirectly, e.g., in reference to a spatial displacement rather than directly to matter.

However, it seemed impossible and unthinkable to assign any physical meaning to $\partial t/\partial s$. Only with Einstein's conception of relativity does this become possible. However, Einstein's work confines this possibility to applications at vast distances in space measurable in large numbers of light-years. Comparing the scale of everyday reality accessible to human cognition (a terrestrial scale, so to speak) to nano-scale or any similarly atomic-molecular scale is not unlike comparing terrestrial scales of space and time to thousands of light-years removed in space and time. It would, therefore, seem no less necessary to look at time as a fourth dimension in all natural processes. Of course, without also transforming the present arbitrary definitions of space and time elaborated in terms of a terrestrial scale, the result would be, at best, no more informative than retaining the existing Newtonian schema and, at worst, utterly incoherent. Subdividing conventional notions of space or time to the milli-, micro-, or nano- scale has been unable to tell us anything meaningful about the relative ranking of the importance of certain phenomena at these scales (including, in some cases, their disappearance). These phenomena are common and well-known either on the terrestrial scale or on the non-terrestrial scales, but rarely seen and less understood on the terrestrial scale. Relative to the terrestrial scale, for example, the electron appears to be practically without mass. It does possess what is called "charge," however, and this feature has consequences at the atomic and subatomic scale that disappear from view at the terrestrial scale.

Einstein's tremendous insight was that, at certain scales, time becomes a spatial measurement, while quantum theory's richest idea was that, at certain other scales, space becomes a temporal measurement. However, the effort to explain the mechanics of the quantum scale in terrestrially meaningful terms led to a statistical interpretation and a mode of explanation that seemed to displace any role for natural laws of operation that would account for what happens in nature at that scale. Reacting against this, Einstein famously expostulated that "God does not play dice with the world." This comment has been widely interpreted as a bias against statistical modes of analysis and interpretation in general. Our standpoint, however, is different. The significance here of these areas of contention among scientists is not about whether any one of these positions is more or less correct. Rather, the significance is that the assertion that all of nature, at any scale, is quintessentially four-dimensional accords with, and does not contradict, profoundly different and even opposing observations of, and assertions about, similar phenomena at very different scales.

There may be any number of ways to account for this, including mathematical theories of chaos and fractal dimension. For example, between qualitatively distinct scales of natural phenomena, there may emerge one or more interfaces characterized by some degree of mathematical chaos and multiple fractal dimensions. Statements of that order are a matter of speculation today and research tomorrow. Asserting the four-dimensionality of all nature, on the other hand, escapes any possibility of 0 mass, 0 energy, or 0 momentum. Simultaneously, this bars the way to absurdities like a mass-energy balance that could take the form of 0 mass coupled with infinite energy or 0 energy coupled with infinite mass. Nature up to now has mainly been characterized as flora, fauna, and the various processes that sustain their existence, plus a storehouse of other elements that play different roles in various circumstances and have emerged over time periods measured on geological and intergalactic scales. According to the standpoint advanced in this paper, nature, physically speaking, is space-time completely filled with matter, energy, and momentum. These possess a temporal metric, which is characteristic of the scale of the matter under observation. Speaking from the vantage point of the current state of scientific knowledge, it seems highly unlikely that any such temporal metric could be constant for all physically possible frames of reference.

5.7 Second Level of Rectification of Newton's First Assumption

Clarification of gaps in Newton's system makes it possible to stipulate what motion is and is not. However, this still leaves open the matter of time. If time is considered mainly as the duration of motion arising from force(s) externally applied to matter, then it must cease when an object is "at rest." Newton's claim in his first law of motion, that an object in motion remains in (uniform) motion until acted upon by some external force, appears at first to suggest that, theoretically, time is physically continual. It is mathematically continuous, but only as the independent variable, and according to Eq. 5.1, velocity v becomes undefined if time-duration t becomes 0. On the other hand, if motion ceases, in the sense of ∂s, the rate of spatial displacement, going to 0, then velocity must be 0. What has then happened, however, to time? Where in nature can time be said to either stop or come to an end? If Newton's mechanism is accepted as the central story, then many natural phenomena have been operating as special exceptions to Newtonian principles. While this seems highly unlikely, its very unlikelihood does not point to any way out of the conundrum.

This is where momentum, p, and, more importantly, its "conservation," come into play. In classically Newtonian terms:

$$p = mv = m\frac{\partial s}{\partial t} \tag{5.2}$$

Hence,

$$\frac{\partial p}{\partial t} = \frac{\partial}{\partial t}m\frac{\partial s}{\partial t} + m\frac{\partial^2 s}{\partial t^2} \tag{5.3}$$

If the time it takes for a mass to move through a certain distance is shortening significantly as it moves, then the mass must be accelerating. An extreme shortening of this time corresponds, therefore, to a proportionately large increase in acceleration. However, if the principle of conservation of momentum is not to be violated, either (a) the rate of increase for this rapidly accelerating mass is comparable to the increase in acceleration, in which case the mass itself will appear relatively constant and unaffected; (b) mass will be increasing, which suggests the increase in momentum will be greater than even that of the mass's acceleration; or (c) mass must

By introducing a real elementary particle that is real (exists in nature), unique, non-uniform, flexible, non-rigid, non-zero mass, and so forth, we can retain both quantity and quality of the iceberg. The same model would be able to distinguish between organic matter and non-organic matter, natural light and artificial light, natural sound and artificial sound, and all other natural and artificial matter and energy.

diminish with the passage of time, which implies that any tendency for the momentum to increase also decays with the passage of time.

The rate of change of momentum ($\partial p/\partial t$) is proportional to acceleration (the rate of change in velocity, expressed as $\partial^2 s/\partial t^2$) experienced by the matter in motion. It is proportional as well to the rate of change in mass with respect to time ($\partial m/\partial t$). If the rate of change in momentum approaches the acceleration undergone by the mass in question, i.e., if $\partial p/\partial t \rightarrow \partial^2 s/\partial t^2$, then the change in mass is small enough to be neglected. On the other hand, a substantial rate of increase in the momentum of a moving mass on any scale much larger than its acceleration involves a correspondingly substantial increase in mass.

The analytical standpoint expressed in Equations 5.2 and 5.3 above work satisfactorily for matter in general, as well as for Newton's highly specific and peculiar notion of matter in the form of discrete object-masses. Of course, here it is easy to miss the "catch." The "catch" is the very assumption in the first place that matter is an aggregation of individual object-masses. While this may be true at some empirical level on a terrestrial scale, say, 10 balls of lead shot or a cubic liter of wood sub-divided into exactly 1,000 one-cm by one-cm by one-cm cubes of wood, it turns out, in fact, to be a definition that addresses only some finite number of properties of specific forms of matter that also happen to be tangible and, hence, accessible to us on a terrestrial scale.

Once again, generalizing what may only be a special case, before it has been established whether the phenomenon is a unique case, a special but broad case, or a characteristic case, begets all manner of mischief.

To appreciate the implications of this point, consider what happens when an attempt is made to apply these principles to object-masses of different orders and/or vastly different scales but within the same reference-frame. Consider the snowflake, a highly typical piece of natural mass. Compared to the mass of an avalanche of which it may come to form a part, the mass of any individual component snowflake is negligible. Negligible as it may seem, however, it is not zero. Furthermore, the accumulation of snowflakes in an avalanche's mass of snow means that the cumulative mass of snowflakes is heading towards something very substantial, infinitely larger than that of any single snowflake. To grasp what happens for momentum to be conserved between two discrete states, consider the starting-point $p = mv$. Clearly in this case, that would mean that in order for momentum to be conserved,

$$p_{avalanche} = p_{snowflakes\text{-}as\text{-}a\text{-}mass} \tag{5.4}$$

which means

$$m_{\text{avalanche}} v_{\text{avalanche}} = \sum\nolimits_{snowflake=1}^{\infty} m_{snowflake} v_{snowflake} \tag{5.5}$$

On a terrestrial scale, an avalanche is a readily observed physical phenomenon. At its moment of maximum (destructive) impact, an avalanche looks like a train wreck unfolding in very slow motion. However, what about the energy released in the avalanche? Of this we can only directly see the effect, or footprint, and another aphenomenal absurdity pops out: an infinitude of snowflakes, each of negligible mass, have somehow imparted a massive release of energy. This is a serious accounting problem; not only momentum, but mass and energy as well, are conserved throughout the universe. This equation is equivalent to formulations attributed to knowledge of Avicenna as well as Ibn-Haithan (Equations 5.3 and 5.4) who both recognized that any form of energy must be associated with a source. Philosophically, this was also seen by Aristotle, and later confirmed and extended by Averroës, whose model permeated to modern Europe through the work of Thomas Aquinas (Chhetri and Islam 2008).

The same principle of conservation of momentum enables us to "see" what must happen when an electron (or electrons) bombards a nucleus at a very high speed. Now we are no longer observing or operating at terrestrial scale. Once again, however, the explanation conventionally given is that since electrons have no mass, the energy released by the nuclear bombardment must have been latent and entirely potential stored within the nucleus.

Clearly, then, in accounting for what happens in nature (as distinct from a highly useful toolset for designing and engineering certain phenomena involving the special subclass of matter represented by Newton's object-masses), Newton's central model of the object-mass is insufficient. Is it even necessary? Tellingly on this score, the instant it is recognized that there is no transmission of energy without matter, all the paradoxes we have just elaborated are removable. Hence, we may conclude that, for properly understanding and being able to emulate nature on all scales, mass-energy balance and the conservation of momentum are necessary and sufficient. On the other hand, neither constancy of mass, nor constancy of speed of light, nor even uniformity in the passage and measure of time is necessary or sufficient.

In summary, the above analysis overcomes several intangibles that are not accounted for in conventional analysis. It includes a) a source of particles and energy, b) particles that are not visible or measurable with conventional means, and c) tracking of particles based on their sources (the continuous time function).

5.8 Fundamental Assumptions of Electromagnetic Theory

Once the confining requirement that phenomena be terrestrially tangible and accessible to our perception is removed, it quickly becomes evident that the appearance of energy radiating in "free space" (electromagnetic phenomena such as light energy, for instance) is an appearance only. (As for the transmission of any form of energy at a constant speed through a vacuum, this may signal some powerful drug-taking on the observer's part, but otherwise it would seem to be a physical impossibility since nature has no such thing as a vacuum.) Nowhere in nature can there be such a thing as energy without mass or mass without energy. Otherwise, the conservation of mass and energy both come into question. Matter at the electronic, inter-atomic, inter-molecular level cannot be dismissed as inconsequential by virtue of its extremely tiny amounts of mass. Electromagnetic phenomena would appear to demonstrate that whatever may be lacking in particulate mass at this level is made up for by electron velocity, or the rate at which these charged particles displace space. The inter-atomic forces among the molecules of a classically Newtonian object-mass sitting "at rest," so to speak, must be at least a match for, if not considerably stronger than, gravitational forces. Otherwise the object-mass would simply dissolve upon reaching a state of rest as molecular matter is pulled by gravity towards the center of the earth. The fact, that in the absence of any magnetic field, either applied or ambient, "Newtonian" object-masses seem electrically neutral, means only that they manifest no net charge. Lacking the Bohr model or other subsequent quantum models of matter at the electronic level, nineteenth-century experimenters and theorists of electricity had no concept of matter at the electronic level comprising extremely large numbers of very tiny charged particles. They struggled instead to reconcile what was available to their limited means for observing electrical phenomena with Newtonian mechanics. Information as to what was at the input of an electrical flow and what was measured or observed at an output point of this flow was available for their observation. The actual composition of this flow, however, remained utterly mysterious. No amount of observation or careful measurement of this flow could bring anybody closer to discovering or hypothesizing the electronic character of matter, let alone bring anybody closer to discovering or hypothesizing that the manner in which electricity flowed was a function of this fundamental electronic character. As discussed in the first part of this chapter, absent Galileo's careful deconstruction of the Aristotelian notion that objects falling freely reached the earth at different times dependent on their mass, the fundamental fact of the existence and operation of gravitational attraction would have been missed. In the absence of an alternative hypothesis, Newton's mechanics were assumed to apply to electrical flow.

Among the leading developers in European science, there were many disputes about the theory of electrical phenomena, their experimental verification, or both, but these were all within the camp of what might be broadly called "Newtonianism." Before Maxwell there had been those, such as Ampère, Oersted, and Berzelius, who proposed to model electrical phenomena as a Newtonian kind of action-at-a-distance (Mundy, 1989). It was their line of thought that inspired Faraday's experimental program at the start of his career towards the end of the 1810s, as Sir Humphrey Davy's assistant in the laboratories of the Royal Institution. That line of thought also ended up raising questions whose answers, sought and obtained by Faraday, ultimately refuted such explanations of current flow. Faraday's experiments showed that there were other electrical effects that did not operate at a distance, and that there could not be two distinct kinds of electricity or "electrical fluids." Maxwell adopted a compromise position that electricity manifested the characteristics of an "incompressible fluid" but was not itself a fluid:

> "The approach which was rejected outright was that of 'purely mathematical' description, devoid of 'physical conceptions;' such an approach, Maxwell felt, would turn out to be unfruitful. More favorably viewed, and chosen for immediate use in 1856, was the method of physical analogy. Physical analogies

> were not only more physical and suggestive than purely mathematical formulae; they were also less constraining than physical hypotheses. Their use, then, constituted a desirable middle way, and Maxwell proceeded to treat electric fields, magnetic fields, and electric currents each by analogy with the flow of an incompressible fluid through resisting media. There was no suggestion here that in an actual electric field, for example, there was some fluid-flow process going on; rather, an analogy was drawn between the two different physical situations, the electric field and the fluid flow, so that with appropriate changes of the names of the variables the same equations could be applied to both" (Siegel 1975, 365).

Meanwhile, although Maxwell became the head of the Cavendish Laboratory, the world-famous research center at Cambridge University, neither he nor his students would ever undertake any directed program of experiments to establish what electricity itself might be (Simpson 1966). Instead, they remained supremely confident that systematic reapplication of Newtonian principles to all new data forthcoming regarding electrical effects would systematically yield whatever electricity was not.

The common understanding among engineers is that Maxwell's equations of electromagnetism established the notion that light is an electromagnetic phenomenon. Broadly speaking, this is true, but Maxwell had a highly peculiar notion of what constituted an electromagnetic phenomenon. First and foremost, it was a theoretical exercise not based on any actual experimental or observational program of his own regarding any electrical or electromagnetic phenomena at all. Secondly, and most tellingly in this regard, when Maxwell's equations are examined more closely, his original version includes an accounting for something he calls "displacement current" whose existence he never experimentally verified (Simpson 1966, 413; Bork 1963, 857; Chalmers 1973a, 479). Furthermore, the version of Maxwell's equations in general use was actually modified by Hertz; this was the version on which Einstein relied.

Some historical background helps illuminate what was going on within this development before, during, and following Maxwell's elaboration of his equations. Notably, Maxwell seemed to have felt no compelling need to further establish, for his own work, what electricity might be (Bork 1967). As a strong proponent of the experimental findings of Michael Faraday, he felt no need to "reinvent the wheel." Faraday's brilliance lay in his design and execution of experimental programs that systematically eliminated false or unwarranted inferences from the growing body of knowledge of electrical phenomena one by one (Williams 1965). Maxwell saw a need to furnish Faraday's work with a mathematical basis so that the theoretical coherence of mankind's knowledge in this field could be presented with the same elegance that the rest of physics of that time was presented, relying on a foundation of Newtonian mechanics: "Maxwell's objective was to establish Faraday's theory on a surer physical basis by transforming it into a mechanical theory of a mechanical aether, that is, an aether whose behavior is governed by the principles of Newtonian mechanics" (Chalmers 1973b, 469). One of Faraday's biographers has questioned whether he had a general theory about electrical phenomena as opposed to experimentally demonstrable explanations of specific electrical phenomena, many of them linked (Williams 1965). Notwithstanding that issue, however, Maxwell firmly accepted the existence of "a mechanical aether" as something required for fitting a Newtonian theoretical framework in order to render existing knowledge of electromagnetic phenomena coherent.

What is known, but not well understood, is the degree to which the general dependence among scientists on a mathematical, i.e., aphenomenal, framing of a natural phenomenon like electricity (a phenomenon not normally readily accessible in complete form to the five senses) exercised so much influence over those scientists whose sense of physics was not initially trained in the Newtonian mold. During the 1870s, one of the most important developers of insights opened by Maxwell's work, for example, was Hermann von Helmholtz. Helmholtz came to physics via physiology, in which he had become interested in the electrical phenomena of the human body. The task of "squaring the circle," so to speak, fell to Helmholtz. He reconciled Maxwell's equations with the "action-at-a-distance" theories of Ampère and of Weber especially, who formulated an equation in 1847 predicting dielectric effects of charged particles as a form of electrical potential. In order to keep the analytical result consistent with the appropriate physical interpretations of observed, known phenomena of open and closed circuits, charged particles, dielectrics, and conductors, Helmholtz was compelled to retain the existence of "the aether." However, his analysis set the stage for his student, Hertz, to predict and extrapolate the source of electromagnetic waves propagating in "empty space," a shorthand for a space in which "the aether" did not seem to play any role (Woodruff 1968).

What is much less well known is that Maxwell's mentor, Faraday, rejected the assumption that such an "aether" existed. He maintained this position, albeit unpublished, for decades. One of his biographers reproduced for the first time in print a manuscript from the eight-volume folio of Faraday's diary entitled "The Hypothetical Ether," which establishes, in the drily understated words of his biographer, that "the ether quite obviously did not enjoy much favor in Faraday's eyes" (Williams 1965, 455). This was long before the famous Michelson-Morley experiment failed to measure its "drift" and placed the asserted existence of the aether in question among other men of science (Williams 1965; Holton 1969). The real shocker in all this, moreover, is the fundamental incoherence that Maxwell ended up introducing into his theoretical rendering of electromagnetic phenomena. Maxwell was struggling to remain theoretically in conformity with a Newtonian mechanical schema: "The major predictions of Maxwell's electromagnetic theory, namely, the propagation of electromagnetic effects in time and an electromagnetic theory of light, were made possible by Maxwell's introduction of a displacement current" (Chalmers 1973a, 171). The "major predictions" were correct; the justification, namely displacement current, was false.

In the first part of this chapter, an absurdity was deliberately extrapolated of scientists and engineers. The example showed that in the absence of Galileo's point of departure, 21st century research would only refine the precision of demonstrations proving Aristotle's assumption about the relationship of mass to rate of free fall for heavier-than-air object-masses. The history of the issues in dispute among experimenters and theorists of electrical phenomena, before the emergence of modern atomic theory, serves to illustrate, with factual events and not imaginary projections, the same difficulty that seized the development of scientific investigation in the shadow of Newtonian "authority."

This is something seen repeatedly in what we have identified elsewhere as part of aphenomenal modeling of scientific explanations for natural phenomena (Zatzman and Islam 2007). What Maxwell in effect erected was the following false syllogism:

- Any deformation of matter in space, including wave-like action, must fulfill requirements of Newtonian mechanism.

- Postulating electrical flow as a displacement due to electromagnetic waves propagating in space at the speed of light and causing mechanical deformation of an "aether" across vast distances anywhere in space fulfills this requirement.
- Therefore, electromagnetic waves must propagate anywhere in space at the speed of light.

In the 19th century, at a time when models of matter on an electronic scale were sketchy to non-existent, Newton's mechanics, developed for object-masses on a terrestrial scale, were assumed to apply. Once the "mechanical aether" was found not to exist, however, light and other electromagnetic phenomena as forms of energy became separated from the presence of matter.

Einstein disposed of the "mechanical aether" also without the benefit of more fully developed modern atomic theory. He instead retained the aphenomenal idea that light energy could travel through a vacuum, i.e., in the complete absence of matter. Meanwhile, practical (that is, physically meaningful) interpretations of modern atomic theory itself today, however, persist in retaining a number of aphenomenal assumptions that make it difficult to design experiments that could fully verify, or falsify, Einstein's general relativity theory. Hermann Weyl, one of Einstein's close collaborators in elaborating relativistic mathematical models with meaningful physical interpretations, summarized the problem with stunning clarity 64 years ago in an article suggestively entitled "How Far Can One Get With a Linear Field Theory of Gravitation in Flat Space-Time?" He wrote, "Our present theory, Maxwell + Einstein, with its inorganic juxtaposition of electromagnetism and gravitation, cannot be the last word. Such juxtaposition may be tolerable for the linear approximation (L) but not in the final generally relativistic theory" (Weyl 1944, 602).

As an example of the persistence of aphenomenal models in many areas of practical importance, starting with the assertion of a scale of compressibility-incompressibility for example, matter is classified as either purely and entirely incompressible (also known as solid), slightly compressible (also known as liquid), or compressible (also known as gas). In other words, one and the same matter is considered to possess three broad degrees of compressibility. This counterposes the idea that matter could exist characteristically mostly in a solid, liquid, or vapor state, but between each of these states there is some non-linear point of bifurcation. Before such a bifurcation point, matter is in one state, and after that point it is distinctly in another state. The underlying basis of the compressibility-scale reasoning is not hard to spot. A little reflection uncovers the notion of discrete bits of matter, conceived of as spherical nuclei consisting of neutrons, and positively charged discrete masses orbited by negatively-charged, much smaller, faster-moving balls called electrons.

Newton's laws of motion are insufficient for determining where in space at any point in time any piece of electronic-scale matter actually is. A statistical version of mechanics, known as quantum theory, has been developed instead to assert probable locations, taking into account additional effects that do not normally arise with object-masses on terrestrial scale, such as spin and charge. It is a system whose content is radically modified from that of Newtonian mechanism, but whose form resembles a Newtonian system of planets and satellite orbiting bodies. These arrangements are often pictured as electron or atom "clouds" purely for illustrative purposes. Mathematically, these are treated as spherical balls, interatomic forces are computed in terms of spherical balls, porosity at the molecular level is similarly computed according to the spherical-balls model, and so forth. Just as the curl and divergence in Maxwell's famous equations of electromagnetism purport to describe Newtonian mechanical deformations of a surface arising from externally acting forces, the aphenomenal assumption of a spherical-balls model underpins the compressibility-incompressibility scale already mentioned. The mathematics to deal with spheres and other idealized shapes and surfaces always produces some final answer for any given set of initial or boundary conditions. It prevails in part because the mathematics needed to deal with "messy," vague things, such as clouds, are far less simple. It is the same story when it comes to dealing with linearized progression on a compressibility scale, as opposed to dealing with nonlinear points of bifurcation between different phases of matter.

All the retrofitting introduced into the modern electronic theory cannot hide the obvious. The act of deciding "yes or no" or "true or false" is the fundamental purpose of model-building. However, as a result of adding so many exceptions to the rule, the logical discourse has become either corrupted or rendered meaningless. For example, there is no longer any fundamental unit of mass. Instead we have the atom, the electron, the quark, the photon, etc. The idea is to present mass "relativistically," so to speak, but the upshot is that it becomes possible to continue to present the propagation of light energy in the absence of mass. What was discussed above, regarding conservation of momentum, hints at the alternative. This alternative is neither to finesse mass into a riot of sub-atomic fragments nor to get rid of it, but rather to develop a mathematics that can work with the reality in which mass undergoes dynamic change.

Outside those scales in which Newton either actually experimented (e.g., object-masses) or about which he was able to summarize actual observational data (e.g., planetary motion), the efforts to retain and reapply a Newtonian discourse and metaphor at all scales continue to produce models, analysis, and equations that repeatedly and continually diverge from what is actually observed. For example, mass and energy are analyzed and modeled separately and in isolation. In nature, mass and energy are inseparable. The conservation of each is a function of this physical (phenomenal) inseparability. Hence, within nature, there is some time function that is characteristic of mass. Once mass and energy are conceptually separated and isolated, however, this time function disappears from view. This time function accounts for the difference between artificial, synthesized versions of natural products or phenomena on the one hand, and the original or actual version of the material or phenomenon in its characteristic natural environment. It is why results from applying organic and chemical fertilizers (their pathways) are not and can never be the same, why fluorescent light will not sustain photosynthesis that sunlight very naturally produces and sustains, why the antibacterial action of olive oil and antibiotics cannot be the same. Wherever these artificial substitutes and so-called "equivalents" are applied, the results are truly magical, in the full sense of the word, since magic is utterly fictional.

It is easy to dismiss such discourse as "blue sky" and impractical. Such conclusions have been the reaction to a number of obstacles that have appeared over the years. One obstacle in adapting engineering calculations that account for natural time functions and their consequences has been the computational burden of working out largely non-linear problems by non-linear methods. These are systems in which multiple solutions will necessarily proliferate. However, modern computing systems have removed most of the practical difficulties of solving such systems. Another obstacle is that time measurement itself appears to have been solved long ago, insofar as the entire issue of natural time functions has been finessed by the introduction of artificial clocks. These have been introduced in all fields of scientific and engineering research work, and they have varying degrees of sophistication. All of them measure time in precisely equal units (seconds, or portions thereof). Since Galileo's brilliant demonstration (using a natural clock) of how heavier-than-air objects in free fall reach the earth at the same time regardless of their mass, scientists have largely lost sight of the meaning, purpose, and necessity of using such natural time measurement to clock natural phenomena (Zatzman *et al.* 2008). Another obstacle is that many natural time functions have acquired undeserved reputations as being subjective and fanciful. That is what has happened, for example, to the concept of "the blink of an eye." Yet, contemporary research is establishing that "the blink of an eye" is the most natural time unit (Zatzman 2008). Another obstacle is the trust that has been (mis)placed in conventional Newtonian time functions. Classically the Newtonian calculus allows the use of a Δt of arbitrary length, something that causes no problems in an idealized mathematical space. However, any Δt that is longer than the span in which significant intermediate occurrences naturally occur out in the real world, i.e., in the field, is bound to miss crucially important moments, such as the passage of a process through a bifurcation point to a new state. In petroleum engineering, an entire multi-million-dollar sub-industry of retrofitting and history matching has come into existence to assess and take decisions regarding the divergence of output conditions in the field from the predictions of engineering calculations. If, on the other hand, the scientifically desirable natural time function were the basis of such engineering calculations in the first place, the need for most, if not all, history matching exercises would disappear. This is a source of highly practical, large savings on production costs.

5.9 Aims of Modeling Natural Phenomena

The inventor of the Hamming code (one of the signal developments in the early days of information theory) liked to point out in his lectures on numerical analysis that "the purpose of computing is insight, not numbers" (Hamming 1984). Similarly, we can say the aim in modeling natural phenomena, such as nature, is direction (or, in more strictly mathematical-engineering terms, the gradient). That is, this aim is not, and cannot be, some or any precise quantity.

There are three comments to add that help elaborate this point. First, with nature being the ultimate dynamical system, no quantity, however precisely measured, at time t_0 will be the same at time $t_0 + \Delta t$, no matter how infinitesimally small we set the value of that Δt. Secondly, in nature, matter in different forms at very different scales interacts continually, and the relative weight or balance of very different forces (intermolecular forces, interatomic forces of attraction and repulsion, and gravitational forces of attraction) cannot be predicted in advance. Since nature operates to enable and sustain life forms, however, it is inherently reasonable to confine and restrict our consideration to three classes of substances that are relevant to the maintenance or disruption of biological processes. Thirdly, at the same time, none of the forces potentially or actually acting on matter in nature can be dismissed as negligible, no matter how "small" their magnitude. It follows that it is far more consequential for a practically useful nature model to be able to indicate the gradient/trend of the production, and conversion or toxic accumulation of natural biomass, natural non-biomass, and synthetic sources of biomass respectively.

As already discussed earlier, generalizing the results for physical phenomena observed at one scale to fit all other scales has created something of an illusion, one reinforced moreover by the calculus developed by Newton. That analytical toolset included an assumption that any mathematical extension, x, might be infinitely subdivided into an infinite quantity of Δx–es which would later be (re-) integrated back into some new whole quantity. However, if the scales of actual phenomena of interest are arbitrarily mixed, leapfrogged, or otherwise ignored, then what works in physical reality may cease to agree with what works in mathematics. Consider in this connection the extremely simple equation:

$$y = 5 \tag{5.6}$$

Taking the derivative of this expression, with respect to an independent variable x, yields:

$$\frac{dy}{dx} = 0 \tag{5.7}$$

To recover the originating function, we perform:

$$\int dy = c \tag{5.8}$$

Physically speaking, Eq. 5.8 amounts to asserting that "something" of indefinite magnitude, designated as c (it could be "5" as a special case (with proper boundaries or conditions), but it could well be anything else) has been obtained as the result of integrating Eq. 5.7, which itself had an output magnitude of 0, i.e., nothing. This is scientifically absurd. Philosophically, even Shakespeare's aging and crazed King Lear recognized that "nothing will come of nothing: speak again" (Shakespeare 1608). The next problem associated with this analysis is that the pathway is obscured, opening the possibility of reversing the original whole. For instance, a black (or any other color) pixel within a white wall will falsely create a black (or any other color corresponding to the pixel) wall if integrated without restoring the nearby pixels that were part of the original white wall. This would happen even though, mathematically, no error has been committed. This example serves to show the need for including all known information in space as well as in time. Mathematically, this can be expressed as:

$$\int_{t=0}^{t=\infty} \int_{s=1}^{s=\infty} mv = constant \tag{5.9}$$

The aim of a useful nature model can be neither to account for some "steady state," an impossibility anywhere in nature, nor to validate a mechanical sub-universe operating according to the criteria of an observer external to the process under observation. Dynamic balances of mass, energy, and momentum imply conditions that will give rise to multiple solutions, at least with the currently available mathematical tools. When it comes to nature, a portion of the space-time continuum in which real physical boundary conditions are largely absent, a mathematics that requires $\Delta t \to 0$ is clearly inappropriate. What is needed are non-linear algebraic equations that incorporate all relevant components (unknowns and other variables) involved in any of these critical balances that must be preserved by any natural system.

5.10 Challenges of Modeling Sustainable Petroleum Operations

Recently, Khan and Islam (2007a, 2007b) outlined the requirements for rending fossil fuel production sustainable. This scientific study shows step by step how various operations ranging from exploration to fuel processing can be performed in such a manner that resulting products will not be toxic to the environment. However, modeling such a process is a challenge because the conventional characterization of matter does not make any provision for separating sustainable operations from unsustainable ones. This description is consistent with Einstein's revolutionary relativity theory, but does not rely on Maxwell's equations as the starting point. The resulting equation is shown to be continuous in time, thereby allowing transition from mass to energy. As a result, a single governing equation emerges. This equation is solved for a number of cases and is shown to be successful in discerning between various natural and artificial sources of mass and energy. With this equation, the difference between chemical and organic fertilizers, microwave and wood stove heating, and sunlight and fluorescent light can be made with unprecedented clarity. This analysis would not be possible with conventional techniques. Finally, analysis results are shown for a number of energy and material related prospects. The key to the sustainability of a system lies within its energy balance. Khan and others recast the combined energy-mass balance equation in the following form as in Equation 5.9.

Dynamic balances of mass, energy, and momentum imply conditions that will give rise to multiple solutions, at least with the currently available mathematical tools. In this context, Equation 5.9 is of utmost importance. This equation can be used to define any process to which the following equation applies:

$$Q_{in} = Q_{acc.} + Q_{\text{out}} \tag{5.10}$$

In the above classical mass balance equation, Q_{in} expresses mass inflow matter, Q_{acc} represents the same for accumulating matter, and Q_{out} represents the same for outflow matter. Q_{acc} will have all terms related to dispersion/diffusion, adsorption/desorption, and chemical reactions. This equation must include all available information regarding inflow matters, e.g., their sources and pathways, the vessel materials, catalysts, and others. In this equation, there must be a distinction made among various matter based on their source and pathway. Three categories are proposed: 1) biomass (BM); 2) convertible non-biomass (CNB); and 3) non-convertible non-biomass (NCNB). Biomass is any living object. Even though conventionally dead matters are also called biomass, we avoid that denomination, as it is difficult to scientifically discern when a matter becomes non-biomass after death. The convertible non-biomass (CNB) is the one that, due to natural processes, will be converted to biomass. For example, a dead tree is converted into methane after microbial actions, the methane is naturally broken down into carbon dioxide, and plants utilize this carbon dioxide in the presence of sunlight to produce biomass. Finally, non-convertible non-biomass (NCNB) is a matter that emerges from human intervention. These matters do not exist in nature and their existence can only be considered artificial. For instance, synthetic plastic matters (e.g., polyurethane) may have similar composition as natural polymers (e.g., human hair, leather), but they are brought into existence through a very different process than that of natural matters. Similar examples can be cited for all synthetic chemicals, ranging from pharmaceutical products to household cookware. This denomination makes it possible to keep track of the source and pathway of a matter. The principal hypothesis of this denomination is that all matters naturally present on Earth are either BM or CNB with the following balance:

$$\text{Matter from natural source} + CNB_1 = BM + CNB_2 \qquad (5.11)$$

The quality of CNB_2 is different from or superior to that of CNB_1 in the sense that CNB_2 has undergone one extra step of natural processing. If nature is continuously moving toward a better environment (as represented by the transition from a barren Earth to a green Earth), CNB_2 quality has to be superior to CNB_1 quality. Similarly, when matter from natural energy sources comes in contact with BMs, the following equation can be written:

$$\text{Matter from natural source} + B_1M = B_2M + CNB \qquad (5.12)$$

Applications of this equation can be cited from biological sciences. When sunlight comes in contact with retinal cells, vital chemical reactions take place that result in the nourishment of the nervous system, among others (Chhetri and Islam 2008a). In these mass transfers, chemical reactions take place entirely differently depending on the light source, of which the evidence has been reported in numerous publications (Lim and Land 2007). Similarly, sunlight is also essential for the formation of vitamin D, which is essential for numerous physiological activities. In the above equation, vitamin D would fall under B_2M. This vitamin D is not to be confused with the synthetic vitamin D, the latter one being the product of an artificial process. It is important to note that all products on the right-hand side have greater value than the ones on the left-hand side. This is the inherent nature of natural processing – a scheme that continuously improves the quality of the environment and is the essence of sustainable technology development.

The following equation shows how energy from NCNB will react with various types of matter:

$$\text{Matter from unnatural source} + B_1M = NCNB_2 \tag{5.13}$$

An example of the above equation can be cited from biochemical applications. For instance, if artificially generated UV is in contact with bacteria, the resulting bacteria mass would fall under the category of NCNB, stopping further value addition by nature. Similarly, if bacteria are destroyed with synthetic antibiotic (pharmaceutical product, pesticide, etc.), the resulting product will not be conducive to value addition through natural processes, and instead becomes a trigger for further deterioration and insult to the environment.

$$\text{Matter from unnatural source} + CNB_1 = NCNB_3 \tag{5.14}$$

An example of the above equation can also be cited from biochemical applications. The $NCNB_1$, which is created artificially, reacts with CNB_1 (such as N_2, O_2) and forms NCNB3. The transformation will be in a negative direction, meaning the product is more harmful than it was earlier. Similarly, the following equation can be written:

$$\text{Matter from unnatural source} + NCNB2_1 = NCNB_2 \tag{5.15}$$

An example of this equation is that the sunlight leads to photosynthesis in plants, converting NCBM to MB, whereas fluorescent lighting would freeze that process and never convert natural non-biomass into biomass.

5.11 Implications of a Knowledge-based Sustainability Analysis

The principles of the knowledge-based model proposed here are restricted to those of mass (or material), balance, energy balance, and momentum balance. For instance, in a non-isothermal model, the first step is to resolve the energy balance based on temperature as the driver for some given time-period, the duration of which has to do with characteristic time of a process or phenomenon. Following the example of the engineering approach employed by Abou-Kassem (2007) and Abou-Kassem *et al.* (2006), the available temperature data are distributed block-wise over the designated time-period of interest. Temperature being the driver of the bulk process of interest, a momentum balance may be derived. Velocity would be supplied by local speeds, for all known particles. This is a system that manifests phenomena of thermal diffusion, thermal convection, and thermal conduction without spatial boundaries, but giving rise nonetheless to the "mass" component.

The key to the system's sustainability lies with its energy balance. Here is where natural sources of biomass and non-biomass must be distinguished from non-natural, non-characteristic, industrially synthesized sources of non-biomass.

5.11.1 A General Case

Figure 5.2 envisions the environment of a natural process as a bioreactor that does not and will not enable conversion of synthetic non-biomass into biomass. The key problem of mass balance in this process, as in the entire natural environment of the earth as a whole, is set out in Figure 5.3, in which the accumulation rate of synthetic non-biomass continually threatens to overwhelm the natural capacities of the environment to use or absorb such material.

In evaluating Equation 5.10, it is desirable to know all the contents of the inflow matter. However, it is highly unlikely to know all the contents, even on a macroscopic level. In the absence of a technology that would find the detailed content, it is important to know the pathway of the process in order to have an idea of the source of impurities. For instance, if de-ionized water is used in a system, one would know that its composition would be affected by the process of de-ionization. Similar rules apply to products of organic sources. If we consider combustion reaction (coal, for instance) in a burner,

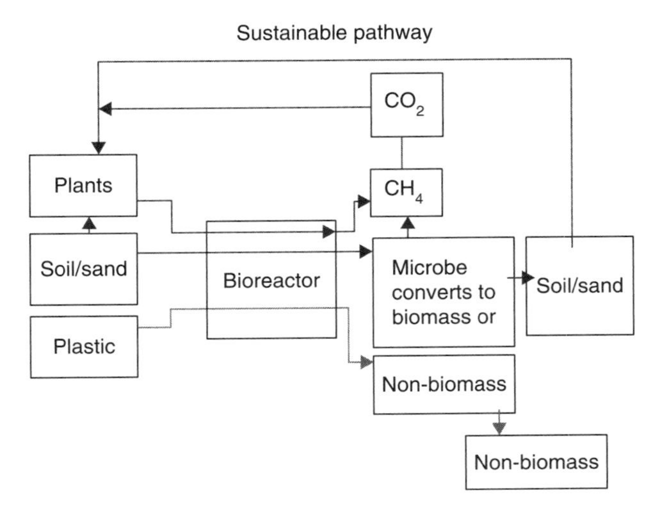

Figure 5.2 Sustainable pathway for material substance in the environment.

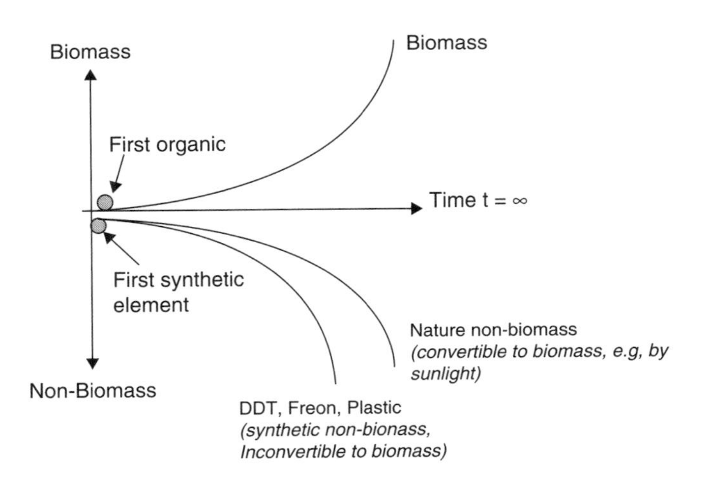

Figure 5.3 Transitions of natural and synthetic materials.

the bulk output will likely be CO_2. However, this CO_2 will be associated with a number of trace chemicals (impurities) depending on the process it passes through. Because Equation 5.10 includes all known chemicals (e.g., from source, absorption/desorption products, catalytic reaction products), it is possible to track matters in terms of CNB and NCNB products. Automatically, this analysis will lead to differentiation of CO_2 in terms of pathway and the composition of the environment, the basic requirement of Equation 5.11. According to Equation 5.11, charcoal combustion in a burner made of clay will release CO_2 and natural impurities of the charcoal and the materials from the burner itself. Similar phenomena can be expected from a burner made of nickel plated with an exhaust pipe made of copper.

Anytime CO_2 is accompanied with CNB matter, it will be characterized as beneficial to the environment. This is shown in the positive slope of Figure 5.3. On the other hand, when CO_2 is accompanied with NCNB matter, it will be considered harmful to the environment, as this is not readily acceptable by the ecosystem. For instance, the exhaust of the Cu or Ni-plated burner (with catalysts) will include chemicals, e.g., nickel, copper from pipe, trace chemicals from catalysts, besides bulk CO_2 because of adsorption/desorption, catalyst chemistry, and so forth. These trace chemicals fall under the category of NCNB and cannot be utilized by plants (negative slope from Figure 5.3). This figure clearly shows that the upward slope case is sustainable as it makes an integral component of the ecosystem. With the conventional mass balance approach, the bifurcation graph of Figure 5.3 would be incorrectly represented by a single graph that is incapable of discerning between different qualities of CO_2 because the information regarding the quality (trace chemicals) is lost in the balance equation. Only recently, the work of Sorokhtin *et al.* (2007) has demonstrated that without such distinction, there cannot be any scientific link between global warming and fossil fuel production and utilization. Figure 5.4 shows how the mere distinction between harmful CO2 and beneficial CO2 can alter the environmental impact analysis. In solving Equation 5.10, one will encounter a set of non-linear equations. These equations cannot be linearized. Recently, Moussavizadegan *et al.* (2007) proposed a method for solving non-linear equations. The principle is to cast the governing equation in engineering formulation, as outlined by Abou-Kassem *et al.* (2006), whose principles were further elaborated in Abou-Kassem (2007).

The non-linear algebraic equations then can be solved in multiple solution mode. Mousavizadegan (2007) recently solved such an equation to contemporary, professionally acceptable standards of computational efficiency. The result looked like what is pictured in Figure 5.5.

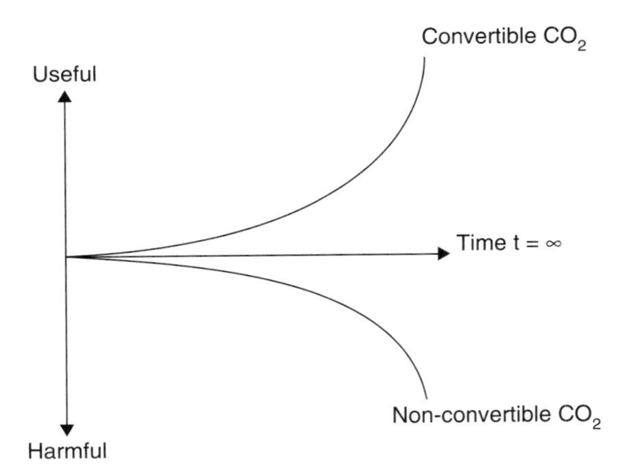

Figure 5.4 Divergent results from natural and artificial.

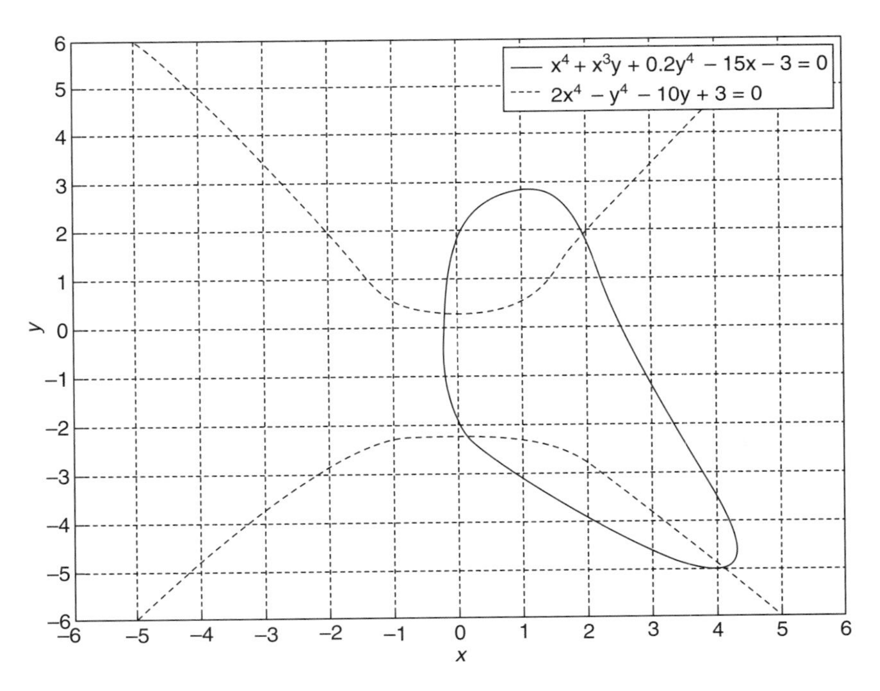

Figure 5.5 The solution behavior manifested by just two non-linear bivariate equations, $x^4 + x^3y + 0.5y^4 - 15x - 3 = 0$ and $2x^4 - y^4 - 10y + 3 = 0$, suggests that a "cloud" of solutions would emerge.

5.11.2 Impact of Global Warming Analysis

In light of the analysis shown in the above section, consider the problem we encounter in evaluating global warming and its cause, as considered by Chhetri and Islam (2008b).

The total energy consumption in 2004 was equivalent to approximately 200 million barrels of oil per day, which is about 14.5 terawatts, over 85% of which comes from fossil fuels (Service 2005). Globally, about 30 billion tons of CO_2 is produced annually from fossil fuels, which includes oil, coal, and natural gas (EIA 2004). The industrial CO_2 produced from fossil fuel burning is considered solely responsible for the current global warming and climate change problems (Chhetri and Islam 2007). Hence, burning fossil fuels is not considered to be a sustainable option. However, this "sole responsibility" is not backed with science (in absence of our analysis above). The confusion emerges from the fact that conventional analysis does not distinguish between CO_2 from natural processes (e.g., oxidation in national systems, including breathing) and CO_2 emissions that come from industrial or man-made devices. This confusion leads to making the argument that man-made activities cannot be responsible for global warming. For instance, Chilingar and Khilyuk (2007) argued that the emission of greenhouse gases by burning of fossil fuels is not responsible for global warming and, hence, is not unsustainable. In their analysis, the amount of greenhouse gases generated through human activities is scientifically insignificant compared to the vast amount of greenhouse gases generated through natural activities. The factor that they do not consider, however, is that greenhouse gases that are tainted through human activities (e.g., synthetic chemicals) are not readily recyclable in the ecosystem. This means when "refined" oil comes in contact with natural oxygen, it produces chemicals (called non-convertible, non-biomass, NCNB). (See Equations 5.13–5.15.)

At present, for every barrel of crude oil, approximately 15% additives are added (California Energy Commission 2004). These additives, with current practices, are all synthetic and/or engineered materials that are highly toxic to the environment. With this "volume gain," the following distribution is achieved (Table 5.1).

Each of these products is subject to oxidation, either through combustion or low-temperature oxidation, which is a continuous process. Toward the bottom of the table, the oxidation rate is decreased, but the heavy metal content is increased, making each

Table 5.1 Petroleum products yielded from one barrel of crude oil in california (From California Energy Commission, 2004).

Product	**Percent of Total**
Finished Motor Gasoline	51.4%
Distillate Fuel Oil	15.3%
Jet Fuel	12.3%
Still Gas	5.4%
Marketable Coke	5.0%
Residual Fuel Oil	3.3%
Liquefied Refinery Gas	2.8%
Asphalt and Road Oil	1.7%
Other Refined Products	1.5%
Lubricants	0.9%

product equally vulnerable to oxidation. The immediate consequence of this conversion through refining is that one barrel of naturally occurring crude oil (convertible non-biomass, CBM) is converted into 1.15 barrel of potential non-convertible non-biomass (NCNB) that would continue to produce more volumes of toxic components as it oxidizes either though combustion or through slow oxidation. Refining is by and large the process that produces NCNB, similar to the process described in Equations 5.12–5.15. The pathways of oil refining illustrate that the oil refining process utilizes toxic catalysts and chemicals, and the emission from oil burning also becomes extremely toxic. Figure 5.6 shows the pathway of oil refining. During the cracking of the hydrocarbon molecules, different types of acid catalysts are used along with high heat and pressure. The process of employing the breaking of hydrocarbon molecules is thermal cracking. During alkylation, sulfuric acids, hydrogen fluorides, aluminum chlorides, and platinum are used as catalysts. Platinum, nickel, tungsten, palladium, and other catalysts are used during hydro processing. In distillation, high heat and pressure are used as catalysts. As an example, just from oxidation of the carbon component, 1 kg of carbon, which was convertible non-biomass, would turn into 3.667 kg of carbon dioxide (if completely

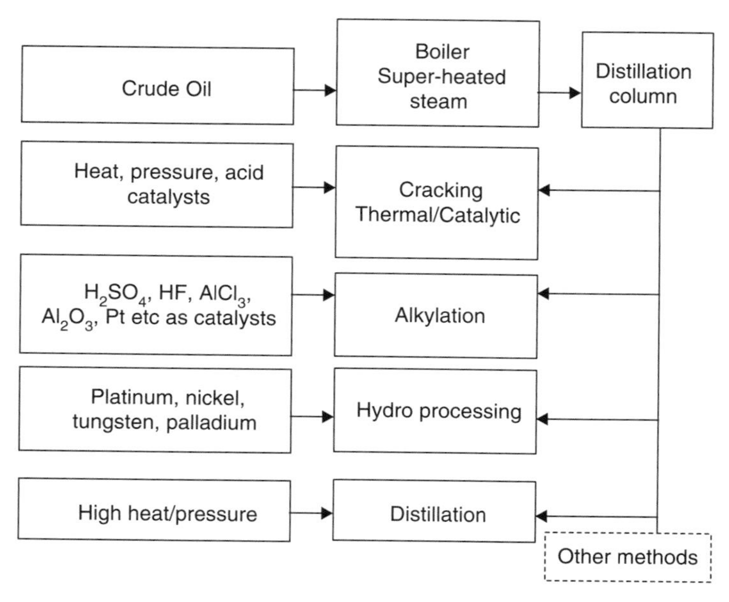

Figure 5.6 Pathway of oil refining process.

burnt) that is now no longer acceptable by the ecosystem, due to the presence of the non-natural additives. Of course, when crude oil is converted, each of its numerous components would turn into such non-convertible non-biomass. Many of these components are not accounted for or even known, let alone a scientific estimation of their consequences. Hence, the sustainable option is either to use natural catalysts and chemicals during refining, or to design a vehicle that directly runs on crude oil based on its natural properties. The same principle applies to natural gas processing (Chhetri and Islam 2008).

5.12 Concluding Remarks

It has long been established that Einstein's work on relativity displaced and overcame known limitations in the applicability of Newton's laws of motion at certain physical scales. Less considered, however, has been another concurrent fact. The implications of certain pieces of Einstein's corrections of Newton, especially the role of time functions, opened up a much larger question. Perhaps the Newtonian mechanism and accounting for motion, by way of the laws of motion and universal gravitation, would have to be adjusted with regard to natural phenomena in general at other scales.

Careful analysis suggests that conservation of mass, energy, and momentum are necessary and sufficient in accounting for natural phenomena at every scale, whereas the laws of motion are actually special cases and their unifying, underlying assumption (that there can be matter at rest anywhere in the universe) is aphenomenal. The sense or assumption that it is necessary to fit all physical cases to the Newtonian schema seems to have led scientists and engineers on a merry chase, particularly in dealing with phenomena observed at the molecular, atomic, electronic, and (today) the nano scale. The history of what actually happened with Faraday's experimental program, Maxwell's equations, and Einstein's generalization of Maxwell, by means of Lorentz's transformations, illustrates the straitjacket in which much applied science was placed, as a result of insisting all observed phenomena fit Newton's schema and rendering these phenomena aphenomenally, such that the logic of the chain of causation became twisted into syllogisms expressing a correct conclusion, the phenomenon, as the result of two or more falsified premises.

The nature-science standpoint provides a way out of this impenetrable darkness created by the endless addition of seemingly infinite layers of opacity. We have no case anywhere in nature where the principle of conservation of mass, energy, or momentum has been violated. The truly scientific way forward, then, for modern engineering and scientific research would seem to lie on the path of finding the actual pathway of a phenomenon from its root or source to some output point by investigating the mass-balance, the energy balance, the mass-energy balance, and the momentum balance of the phenomenon.

This seems to be a fitting 21st century response to the famous and, for its time, justified lament from giants of physical science, such as Pierre Duhem, who was a leading opponent of Maxwell's method. Drawing an important distinction between modeling and theory, he pointed out, even after Einstein's first relativity papers had appeared, that, in many cases, scientists would employ models (using Newtonian laws) about physical reality that were "neither an explanation nor a rational classification of physical laws, but a model constructed not for the satisfaction of reason, but for the pleasure of the imagination" (Duhem, P. 1914, 81; Ariew and Barker 1986). The engineering approach has profound implications both for modeling phenomena and for the mathematics that are used to analyze and synthesize the models.

6

A True Sustainability Criterion and its Implications

6.1 Introduction

"Sustainability" is a concept that has become a buzzword in today's technology development. Commonly, the use of this term infers that the process is acceptable for a period of time. True sustainability cannot be a matter of definition. In this chapter, a scientific criterion for determining sustainability is presented. In this chapter, a detailed analysis of different features of sustainability is presented in order to understand the importance of using the concept of sustainability in every technology development model.

As seen in previous chapters, the only true model of sustainability is nature. A truly sustainable process conforms to the natural phenomena, both in source and process. Scientifically, this means that true long-term considerations of humans should include the entire ecosystem. Some have called this inclusion "humanization of the environment'" and put this phenomenon as a pre-condition to true sustainability (Zatzman and Islam 2007). The inclusion of the entire ecosystem is only meaningful when the natural pathway for every component of the technology is followed. Only such design can assure both short-term (tangible) and long-term (intangible) benefits.

However, tangibles relate to short-term and are very limited in space, whereas the intangibles relate either to long-term or to other elements of the current time frame. Therefore, a focus on tangibles will continue to obscure long-term consequences. The long-term consequences will not be uncovered until intangible properties are properly analyzed and included. Recently, Chhetri and Islam (2008) have established that by taking a long-term approach, the outcome that emerges from a short-term approach is reversed. This distinction is made in relation to energy efficiency of various energy sources. By focusing on just heating value, one comes up with a ranking that diverges into what is observed as the global warming phenomenon. On the other hand, if a long-term approach were taken, none of the previously perpetrated technologies would be considered "efficient" and would have been replaced long ago with truly efficient (global efficiency-wise) technologies, thereby avoiding the current energy crisis. This chapter emphasizes intangibles due to their inherent importance, and shows how tangibles should link with intangibles. This has opened up the understanding of the relationship between intangible and tangible scales, from microscopic to macroscopic properties.

It has long been accepted that nature is self-sufficient and complete, rendering it as the true teacher of how to develop sustainable technologies. From the standpoint of human intention, this self-sufficiency and completeness is actually a standard for declaring nature perfect. "Perfect" here, however, does not mean that nature is in one fixed, unchanging state. On the contrary, nature has the capacity to evolve and sustain, which is what makes it such an excellent teacher. This perfection makes it possible and necessary for humanity to learn from nature, not to fix nature, but to improve its own condition and prospects within nature in all periods, and for any timescale. The significance of emulating nature is subtle but crucial; technological or other development undertaken within the natural environment only for a limited, short term must necessarily, sooner or later, end up violating something fundamental or characteristic within nature. Understanding the effect of intangibles, and the relations of intangibles to tangibles, is important for reaching appropriate decisions affecting the welfare of society and nature as well. A number of aspects of natural phenomena have been discussed here to find out the relationship between intangibles and tangibles. The target of this study is to provide a strong basis to the sustainability model. The combined mass and energy balance equation has provided sufficient and necessary support for the role of intangibles in developing sustainable technologies.

6.2 Importance of the Sustainability Criterion

Few would disagree that we have made progress as a human race in the post-Renaissance world. Empowered with new science, led by Galileo and later championed by Newton, modern engineering is credited to have revolutionized our lifestyle. Yet, centuries after those revolutionary moves of new science pioneers, modern day champions (e.g., Nobel Laureate Chemist Robert Curd) find such technology development modes more akin to "technological disaster" than technological marvels. Today's engineering, which is driven by economic models that have been criticized as inherently unsustainable by Nobel Laureates with notable economic theories (Joseph Stiglitz, Paul Krugman, and Mohamed Yunus), has positioned itself as environmentally unsustainable and socially unacceptable. Figure 6.1 shows the nature of the problem that human habitats are facing today. All value indices that would imply improvement of human social status have declined, whereas per capita energy consumption has increased. This figure clearly marks a problem of directionality in technology development. If

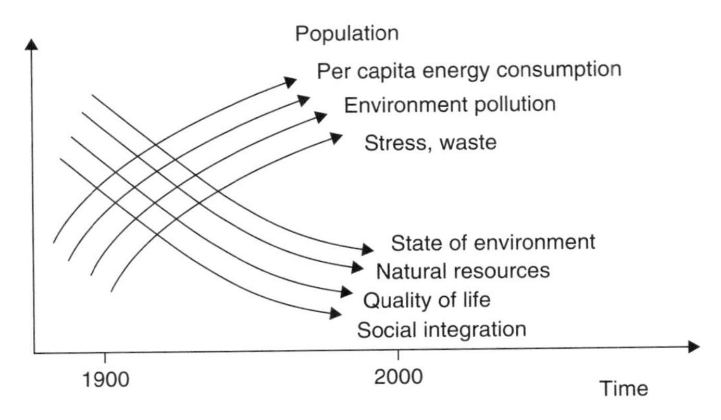

Figure 6.1 Rising costs and declining values during the post-Renaissance world.

increasing per capita energy consumption is synonymous with economic growth (this is in line with the modern-day definition of gross domestic product, GDP), further economic growth would only mean an even worse decline in environmental status. This is not merely a technological problem, it is also a fundamental problem with the outlook of our modern society and how we have evolved in the modern age leading up to the Information Age. Table 6.1 summarizes various man-made activities that have been

Table 6.1 Energy consumed through various human activities.

Activity	Btu	Calorie
Amatch	1	252
An apple	400	100,800
Making a cup of coffee	500	126,000
Stick of dynamite	2,000	504,000
Loaf of bread	5,100	1,285,200
Pound of wood	6,000	1,512,000
Running a TV for 100 hours	28,000	7,056,000
Gallon of gasoline	125,000	31,500,000
20 days cooking on gas stove	1,000,000	252,000,000
Food for one person for a year	3,500,000	882,000,000
Apollo 17's trip to the moon	5,600,000,000	1,411,200,000,000
Hiroshima atomic bomb	80,000,000,000	20,160,000,000,000
1,000 transatlantic jet flights	250,000,000,000	63,000,000,000,000
United States in 1999	97, 000,000,000,000,000	24,444,000,000,000,000,000

synonymous with social progress (in the modern age) but have been the main reason why we are facing the current global crisis. The current model is based on conforming to regulations and reacting to events. It is reactionary because it is only reactive, and not fundamentally proactive. Conforming to regulations and rules that may not be based on any sustainable foundation can only increase long-term instability. Martin Luther King, Jr. famously pointed out, "We should never forget that everything Adolf Hitler did in Germany was 'legal.'" Environmental regulations and technology standards are such that fundamental misconceptions are embedded in them; they follow no natural laws. A regulation that violates natural law has no chance to establish a sustainable environment. What was "good" and "bad" law for Martin Luther King, Jr., is actually sustainable (hence, true) law and false (hence, implosive) law, respectively. With today's regulations, crude oil is considered to be toxic and undesirable in a water stream, whereas the most toxic additives are not. For instance, a popular slogan in the environmental industry has been, "Dilution is the solution to pollution." This is based on all three misconceptions that were discussed in Chapter 2, yet all environmental regulations are based on this principle. The tangible aspect, such as the concentration, is considered, but not the intangible aspect, such as the nature of the chemical, or its source. Hence, "safe" practices initiated on this basis are bound to be quite unsafe in the long run. Environmental impacts are not a matter of minimizing waste or increasing remedial activities, but of humanizing the environment. This requires the elimination of toxic waste altogether. Even non-toxic waste should be recycled 100%. This involves not adding any anti-nature chemical to begin with, then making sure each produced material is recycled, often with value addition. A zero-waste process has 100% global efficiency attached to it. If a process emulates nature, such high efficiency is inevitable. This process is the equivalent of greening petroleum technologies. With this mode, no one will attempt to clean water with toxic glycols, remove CO_2 with toxic amides, or use toxic plastic paints to be more "green." No one will inject synthetic and expensive chemicals to increase EOR production. Instead, one would settle for waste materials or naturally available materials that are abundantly available and pose no threat to the eco-system.

The role of a scientific sustainability criterion is similar to the bifurcation point shown in Figure 6.2. This figure shows the importance

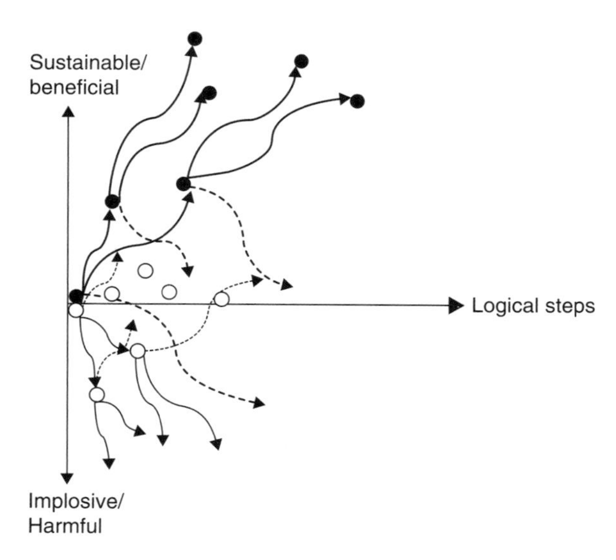

Figure 6.2 The role of the first criterion premise in determining the pathways to sustainable and implosive developments.

of the first criterion. The solid circles represent a natural (true) first premise, whereas the hollow circles represent an aphenomenal (false) first premise. The thicker solid lines represent scientific steps that would increase overall benefits to the whole system. At every phenomenal node, spurious suggestions will emerge from an aphenomenal root (e.g., bad faith, bottom-line driven, or myopic models), as represented by the dashed thick lines. However, if the first premise is sustainable, no node will appear ahead of the spurious suggestions. Every logical step will lead to sustainable options. The thinner solid lines represent choices that emerge from aphenomenal or false sustainability criteria. At every aphenomenal node, there will be aphenomenal solutions; however, each will lead to further compounding of the problem, radically increasing environmental costs in the long run. This process can be characterized as the antithesis of science (as a process). At every aphenomenal node, anytime a phenomenal solution is proposed, it is deemed spurious, because it opposes the first criterion of the implosive model. Consequently, these solutions are rejected. This is the inherent conflict between sustainable and unsustainable starting points. In the implosive mode, phenomenal plans have no future prospect, as shown by the absence of a node.

Albert Einstein famously said, "The thinking that got you into the problem is not going to get you out." Figure 6.2 shows the two different thought processes that dictate diverging pathways. Once launched in the unsustainable pathway, there cannot be any solution other than to return to the first bifurcation point and re-launch in the direction of the sustainable pathway. This logic applies equally to technology development as well as economics.

6.3 The Criterion: The Switch that Determines the Direction at a Bifurcation Point

It is long understood that the decision-making process involves asking "yes" or "no" questions. Usually, this question is thought to be posed at the end of a thought process or logical train. It is less understood that the "yes" or "no" question cannot lead to a correct answer if the original logical train did not start with a correct first premise, and if the full time domain (defining logical train) is not fully considered. Consider the following question. Is whole-wheat bread better than white bread? One cannot answer this question without knowledge of the past history. For instance, for organic flour (without chemical fertilizer, genetic alteration, pesticide, metallic grinder, or artificial heat sources), whole-wheat bread is better. However, if the flour is not organic, then more questions need to be asked in order to first determine the degree of insult caused to the natural process, and to determine what would be the composition of the whole wheat if all ingredients (including trace elements from grinder, chemical fertilizers, pesticide, heat source, and others) were considered. In this analysis, one must include all elements in space at a given time. For instance, if trace elements from pesticide or a metallic grinder are neglected, the answer will be falsified. In this particular case, whole-wheat non-organic bread is worse than white non-organic bread, but it will not be shown as such if one does not include all elements in time and space (mass and energy). In summing up these two points, one must consider the full extent of time from the start of a process (including the logical train), and one must include all elements in space (for both mass and energy sources), in line with the theory advanced by Khan *et al.* (2008). Each of these considerations will have a question regarding the diversion of the process from a natural process. At the end, anything that is natural is sustainable, and therefore, is good.

Let us rephrase the question:

Q: Is whole-wheat bread better than white bread? The conventional answer sought would be either yes or no, or true or false. However, without a proper criterion for determining true or false, this question cannot be answered. In order to search for the knowledge-based answer to this question, the following question must be asked:

Q_k: Are both the white bread and the whole-wheat bread organic? If the answer is yes, then the answer to Q is yes. If the answer is no, then the following knowledge-based question has to be asked:

Q_{k1}: Are both non-organic? If the answer is yes, then the answer to Q becomes no, meaning whole-wheat bread is not better than white bread. If the answer to Q_{k1} is no, then another knowledge-based question has to be asked:

Q_{k2}: Is the white bread organic? If the answer is yes, then the answer to Q becomes no, meaning whole-wheat non-organic bread is not better than white organic bread. If the answer to Q_{k2} is no, then the answer to Q is yes, meaning whole-wheat organic bread is better than white non-organic bread.

In the above analysis, the definition of "organic" has been left to the imagination of the reader. However, it must be stated that a 100% scientific definition of organic cannot be achieved. Scientifically, organic means something that has no anti-conscious intervention of human beings. Obviously, by nature being continuous in space and time, there is no possibility of having a 100% organic product. However, this should not stop one from searching for the true answer to a question. At the very least, this line of analysis will raise new questions that should be answered with more research, if deemed necessary. For this particular question, Q, we have only presented the mass balance aspect. For instance, organic bread also means that it is baked in a clay oven with natural fuel. Now, what happens if this energy balance is not respected? This poses another series of questions. Let us call them energy-related questions, QE. This question must be asked in the beginning, meaning before asking the question Q_k.

QE: Are both the whole-wheat and non-whole wheat breads organically baked? If the answer is yes, then the previous analysis stands. If the answer is no, then the following knowledge-seeking question must be asked:

QE_K: Are both breads baked non-organically (e.g., electricity, microwave, processed fuel, recombined charcoal, or steel stove)? If the answer is yes, then the previous analysis stands. If the answer is no, then it is a matter of more research. To date, we do not have enough research to show how whole-wheat flour would react with non-organic energy sources as compared to white flour.

It is clear from the above analysis that we come across many knowledge-seeking questions and each question demarks a bifurcation point. At each bifurcation point, the question to ask is, "Is the process natural?" The time frame to investigate is many times the characteristic time of a process. For environmental sustainability, the characteristic time of the process is the duration of human species in existence. This can easily transform into infinity, as originally proposed by Khan and Islam (2007). The process, then, involves taking the limit of a process as time goes to infinity. If the process is still sustainable, it can be considered a natural process, and is good for the environment. Otherwise, the process is unnatural, and therefore, unsustainable. This analysis shows that the most important role of the time dimension is in setting the direction. In Figure 6.2, we see that a real starting point would lead to knowledge, whereas an unreal starting point will lead to prejudice. If the time dimension is not considered on a continuous ("continual" is not enough) basis, even the logical steps cannot be traced back in order for one to verify the first premise. It is not enough to back up a few steps; one must back up to the first premise that led to the bifurcation between sustainable and implosive pathways. Zatzman *et al.* (2008) have recently highlighted the need for such considerations of the time domain in order to utilize the time dimension as a switch. It turns out that, with such considerations, scientists cannot determine the cause of global warming with the science that assumes all molecules are identical, thereby making it impossible to distinguish between organic CO_2 and industrial CO_2. Similarly, scientists cannot determine the cause of diabetes unless there is a paradigm shift that distinguishes between sucrose in honey and sucrose in Aspartame® (Chhetri and Islam 2007).

6.3.1 Some Applications of the Criterion

The same logic would indicate that, unless the science includes intangibles, the cause(s) of global warming could not be determined either. What remain uncharted are the role of pathways and the passage of time, something that cannot be followed meaningfully in lab-controlled conditions, in transforming the internal basis of changes in certain natural phenomena of interest. One example has been given by Khan and Islam (2007b) in the context of the use of catalysts. Tangible science says catalysts play no role in the chemical reaction equation because they do not appear in the result or outcome. No mass balance accounts for the mass of the catalyst lost during a reaction, and no chemical equation accounts for what happens to the "lost" catalyst molecules when they combine with the products during extremely unnatural conditions. By using the science of tangibles, one can argue that the following patent is a technological breakthrough (El-Shoubary *et al.* 2003). This patented technology separates Hg from a contaminated gas stream using $CuCl_2$ as the main catalyst. At a high temperature, $CuCl_2$ would react with Hg to form Cu-Hg amalgam. This process is effective when combined with fire-resistant Teflon membranes.

1. Patent #6,841,513 – "Adsorption powder containing cupric chloride." January 11, 2005.
2. Patent #6,589,318 – "Adsorption powder for removing mercury from high temperature, high moisture stream." July 8, 2003.
3. Patent #6,5824,97 – "Adsorption powder for removing mercury from high temperature high moisture gas stream." June 24, 2003.
4. Patent #6,558,642 – "Method of adsorbing metals and organic compounds from vaporous streams." May 6, 2003.
5. Patent #6,533,842 – "Adsorption powder for removing mercury from high temperature, high moisture gas stream." March 18, 2003.
6. Patent #6,524,371 – "Process for adsorption of mercury from gaseous streams." February 25, 2003.

This high level of recognition for the technology is expected. After all, what happens to Teflon at high temperature and what happens to Cu-Hg amalgam is a matter of long term, or at least of time being beyond the "time of interest." Khan (2006) describes this as "time = right now."

However, if longer-term time is used for the analysis and a bigger area is considered for the mass balance, it would become clear that the same process has actually added more waste to the environment, in the form of dioxins released from Teflon and Cu-Hg. The dioxins from both would be in a more harmful state than their original states in Teflon, $CuCl_2$, and gas stream, respectively. In the efficiency calculation, nearly 90% efficiency is reported within the reactor. This figure makes the process very attractive. However, if the efficiency calculation is conducted including the entire system, in which the heater resides, the efficiency drops drastically. In addition, by merely including more elements, the conversion of Hg in a natural gas stream and Cu in $CuCl_2$ solution into Cu-Hg sludge, as well as the addition of chlorine in the effluent gas, poses the difficult question as to what has been accomplished overall.

Another example can be given from the chemical reaction involving honey and Aspartame®. With the science of tangibles, the following reactions take place:

$$\text{Honey} + O_2 \rightarrow \text{Energy} + CO_2 + \text{Water}$$

$$\text{Aspartame}^{®} + O_2 \rightarrow \text{Energy} + CO_2 + \text{Water}$$

In fact, a calorie-conscious person would consider Aspartame® a better alternative to honey, as the energy produced in Aspartame is much less than that of honey for the same weight burnt. An entirely different picture emerges if all components of honey and Aspartame® are included. In this case, the actual compositions of water as a product are very different for the two cases. However, this difference cannot be observed if the pathway is cut off from the analysis and if the analysis is performed within an arbitrarily set confine. Similar to confining the time domain to the "time of interest," or time = right now, this confinement in space perverts the process of scientific investigation. Every product emerging after the oxidation of an artificial substance will come with long-term consequences for the environment. These consequences cannot be included with the science of tangibles. Zatzman and Islam (2007) detailed the following transitions in commercial product development, and argued that this transition amounts to an increased focus on tangibles in order to increase the profit margin in the short-term. The quality degradation is obvious, but the reason behind such technology development is quite murky. At present, the science of tangibles is totally incapable of lifting the fog out of this mode of technology development.

A third example involves natural and artificial vitamin C. Let us use the example of lemon and vitamin C. It has been known for the longest time that the lemon has both culinary and medicinal functions in ancient cultures, ranging from the Far East to Africa. However, in European literature, there is a confusion that certain fruits (e.g., the orange in the following example) are only for pleasure (culinary) and others are for medicinal applications (e.g., the lemon). Apart from this type of misconception, the point to note is that lemon was known to cure scurvy, a condition that arises from lack of vitamin C. So, lemons can cure scurvy: this is premise number one. Reference to this premise is made in old literature. For instance, the following site states (Anonymous, 2008):

> "An Italian Jesuit and full professor in Rome, Ferrari was an incredible linguist and broad scholar ... A lover of flowers, he authored four volumes on the culture of flowers (1632) illustrated by some of the same engravers as was *Hesperides*. He was the 'first writer to collect all the evidence available on the location of the garden of the Hesperiden and on the stealing of the Apples.' Ferrari described numerous medicinal preparations based upon citrus blossoms or fruits. He notes that the orange is usually eaten for pleasure alone; the lemon, citron and pomegranate as medicine. He mentions only in passing using citrus to cure scurvy since his frame of reference is the Mediterranean world in which this disease was not a problem."

Wikipedia discusses this same premise and states, "In 1747, James Lind's experiments on seamen suffering from scurvy involved adding Vitamin C to their diets through lemon juice" Wikipedia, 2008). Now with that first premise, if one researches what the composition of a lemon is, one would encounter the following types of comments readily. As an example, note this statement on the website, "I am a chemist and I know that lemon juice is 94% water, 5% citric acid, and 1% unidentifiable chemicals." Of course, other chemists would use more scientific terms to describe the 1% "unidentified chemicals." For instance, the website of Centre national de la recherche scientifique of France, http://cat.inist.fr/, states:

> "This interstock grafting technique does not increase the flavonoid content of the lemon juice. Regarding the individual flavonoids, the 6,8-di-C-glucosyl diosmetin was the most affected flavonoid by the type of rootstock used. The interstock used is able to alter the individual quantitative flavonoid order of eriocitrin, diosmin, and hesperidin. In addition, the HPLC-ESI/MSn analyses provided the identification of two new flavonoids in the lemon juice: Quercetin 3-O-rutinoside-7-O-glucoside and chrysoeriol 6,8-di-C-glucoside (stellarin-2). The occurrence of apigenin 6,8-di-C-glucoside (vicenin-2), eriodictyol 7-O-rutinoside, 6,8-di-C-glucosyl diosmetin, hesperetin 7-O-rutinoside, homoeriodictyol 7-O-rutinoside and diosmetin 7-O-rutinoside was also confirmed in lemon juice by this technique."

The entire exercise involves determining the composition using the steady-state model, meaning the composition does not change with time. In addition, this line of science also assumes that only composition (and not the dynamics of matter) matters, much like the model Aristotle used some 2,500 years ago. One immediate "useful" conclusion of this is that 5% of lemon juice is ascorbic acid. With this isolated aphenomenal first premise, one can easily proceed to commercialization by producing ascorbic acid with techniques that are immediately proven to be more efficient and, therefore, more economical. This is because when ascorbic acid is manufactured, the pills have much higher concentration of ascorbic acid than lemon juice ordinarily would have. In addition, the solid materials used to manufacture vitamin C pills are also cheaper than lemons and are definitely more efficient in terms of preservation and commercialization. After all, no lemon would last for a year, whereas vitamin C pills will. Overall, if vitamin C is ascorbic acid, then manufactured vitamin C can be marketed at a much lower price than vitamin C from lemons. Now, if a clinical test can show that the manufactured vitamin C indeed cures scurvy, no one can argue that real lemons are needed, and they would obviously be a waste of money. This is the same argument put forward by Nobel Laureate chemist Linus Pauling, who considered that synthetic vitamin C is identical to natural vitamin C and warned that higher-priced "natural" products are a "waste of money." Some thirty years later, we now know synthetic vitamin C causes cancer, while natural vitamin C prevents it (Chhetri and Islam 2008). How could this outcome be predicted with science? Remove the fundamental misconception that lemon juice is merely 5% ascorbic acid, which is independent of its source or pathway.

The above false conclusions, derived through conventional new science, could be avoided by using a criterion that distinguishes between real and artificial. This can be done using the science of intangibles that includes all phenomena that occur naturally, irrespective of what might be detectable. For the use of catalysis, for instance, it can be said that if the reaction cannot take place without the catalyst, clearly it plays a role. Just because at a given time (e.g., time = right now) the amount of catalyst loss cannot be measured, it does not mean that it (catalyst loss and/or a role of catalysts) does not exist. The loss of catalyst is real, even though one cannot measure it with current measurement techniques. The science of intangibles does not wait for the time when one can "prove" that catalysts are active. Because nature is continuous (without a boundary in time and space), considerations are not focused on a confined "control" volume. For the science of tangibles, on the other hand, the absence of the catalyst's molecules in the reaction products means that one would not find that role there.

The science of tangibles says that you cannot find it in the reaction product, so it does not count. The science of intangibles says that obviously it counts, but, just as obviously, not in the same way as what is measurable in the tangible mass-balance. This shows that the existing conventional science of tangibles is incomplete. To the extent that it remains incomplete, on the basis of disregarding or discounting qualitative contributions that cannot yet be quantified in ways that are currently meaningful, this kind of science is bound to become an accumulating source of errors.

6.4 Current Practices in Petroleum Engineering

In very short historical time (relative to the history of the environment), the oil and gas industry has become one of the world's largest economic sectors, a powerful globalizing force with far-reaching impacts on the entire planet that humans share with the rest of the natural world. Decades of continuous growth of the oil and gas operations have changed, and in some places transformed, the natural environment and the way humans have traditionally organized themselves. The petroleum sectors draw huge public attention due to their environmental consequences. All stages of oil and gas operations generate a variety of solids, liquids, and gaseous wastes (Currie and Isaacs 2005; Wenger *et al.* 2004; Khan and Islam 2003a; Veil 2002; Groot 1996; Wiese *et al.* 2001; Rezende *et al.* 2002; Holdway 2002). Different phases of petroleum operations and their associated problems are discussed in the following sessions.

6.4.1 Petroleum Operations Phases

In the petroleum operations, different types of wastes are generated. They can be broadly categorized as drilling wastes, human-generated wastes, and other industrial wastes. There are also accidental discharges, for example via air emission, oil spills, chemical spills, and blowouts.

During the drilling of an exploratory well, several hundred tons of drilling mud and cuttings are commonly discharged into the marine environment. Though an exploratory activity, such as seismic exploration, does not release wastes, it nevertheless also has a potential negative impact (Cranford *et al.* 2003; Putin 1999). According to a report (SECL 2002; Putin 1999), seismic shooting kills plankton, including eggs, larvae of many fish and shellfish species, and juveniles that are very close to the airguns. The most important sub-lethal effect on adult organisms exposed to chronic waste discharges, from both ecological and fisheries perspectives, is the impairment of growth and reproduction (GESAMP 1993; Putin 1999). Growth and reproduction are generally considered to be the most important sub-lethal effects of chronic contaminant exposure (Cranford *et al.* 2003). Seabirds aggregate around oil drilling platforms and rigs in above-average numbers, due to night lighting, flaring, food, and other visual cues. Bird mortality has been documented due to impact on the structure, oiling, and incineration by the flare (Wiese *et al.* 2001).

Khan and Islam (2005) reported that a large quantity of water discharge takes place during petroleum production, primarily resulting from the discharge of produced water, which includes injected fluid during drilling operations as well as connate water of high salinity. They also reported that produced water contains various contaminants, including trace elements and metals from formations through which the water passed during drilling, as well as additives and lubricants necessary for proper operation. Water is typically treated prior to discharge, although historically this was not the case (Ahnell and O'Leary 1997).

Based on the geological formation of a well, different types of drilling fluids are used. The composition and toxicity of these drilling fluids are highly variable, depending on their formulation. Water is used as the base fluid for roughly 85% of drilling operations internationally, and the remaining 15% predominantly use oil (Reis 1996). Spills make up a proportionately small component of aquatic discharges (Liu 1993).

CO_2 emissions are one of the most pressing issues in the hydrocarbon sector. There are direct emissions, through flaring and burning fossil fuels, from production sites (Figure 6.3). For example, during exploration and production, emissions take place due to control venting and/or flaring and the use of fuel. Based on 1994 British Petroleum figures, it is reported that emissions by mass were 25% volatile organic compounds (VOCs), 22% CH4, 33% NOx, 2% SOx, 17% CO, and 1% particulate matter. Data on CO_2 are not provided (Ahnell and O'Leary 1997). Until now, flaring is considered a production and refining technique, which wastes a huge amount of valuable resources through burning. The air emissions during petroleum processing are primarily due to uncontrolled volatilization

Figure 6.3 Flaring from an oil refinery.

and combustion of petroleum products in the modification of end products to meet consumer demand (Ahnell and O'Leary 1997). Oils, greases, sulphides, ammonia, phenols, suspended solids, and chemical oxygen demand (COD) are the common discharges into water during refining (Ahnell and O'Leary 1997). Natural gas processing generally involves the removal of natural gas liquid (NGLs), water vapor, inert gases, CO_2, and hydrogen sulphide (H_2S). The by-products from processing include CO_2 and H_2S (Natural Resources Canada 2002a).

The oil sector contributes a major portion of CO_2 emissions. Figure 6.4 presents the world historical emissions and projected emissions of CO_2 from different sectors. About 29 billion tons of CO_2 are released into the air every year by human activities, and 23 billion tons come from industry and burning fossil fuels (IPCC 2001; Jean-Baptiste and Ducroux 2003), which is why this sector is blamed for global warming. The question is, how might these problems best be solved? Is there any possible solution?

Throughout the life cycle of petroleum operations, there are accidental discharges, e.g., via air emission, oil spills, chemical spills, and blowouts. Figure 6.5 shows the annual total amount of oil release in the marine environment. Crude oil is one of the major

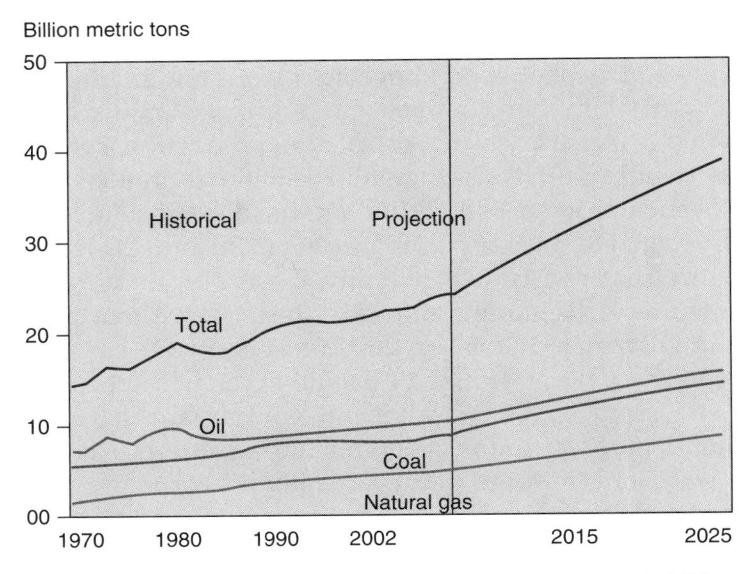

Figure 6.4 World CO_2 Emissions by oil, coal, and natural gas, 1970–2025 (EIA 2004).

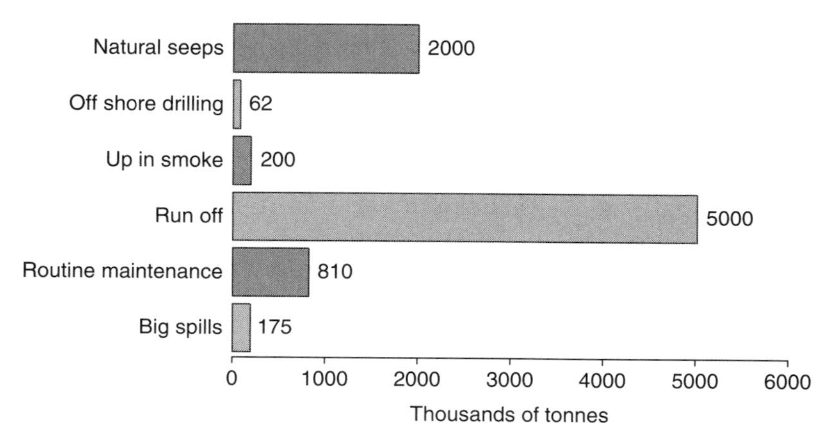

Figure 6.5 Annual total amount of oil release in the marine environment (Oil in the Sea 2003).

toxic elements released in the marine environment by the oil industry. On average, 15 million gallons of crude oil are released yearly from offshore oil and gas operations into the marine environment. There are in total 700 million gallons of oil discharged from other sources into the sea (United States Coast Guard 1990; Khan and Islam 2004). Other sources of oil release are routine maintenance of shipping, domestic/urban runoff, up in smoke, and natural seepages.

The examples mentioned above are only a few examples of the current technology development mode in petroleum sector. They are likely to increase as the petroleum production continues to increase (Figure 6.6). It is hard to find a single technology that does not have such problems. In addition to the use of technologies that are unsustainable, the corporate management process is based on a structure that resists sustainability. Generally, corporate policy is oriented towards gaining monetary benefits without producing anything (Zatzman and Islam 2006b). This model has imploded spectacularly in the aftermath of the fall of the world energy giant, Enron, in December 2001 (Deakin and Konzelmann 2004; Zatzman and Islam 2006). Post-Enron events, including the crisis that afflicted World Dot Com, indicate that practically all corporate structures are based on the Enron model (Zatzman and Islam 2005).

It is clear from above discussion that there are enormous environmental impacts from current petroleum operations, but with high market demand and technological advancement in exploration and development, the petroleum operations continue to

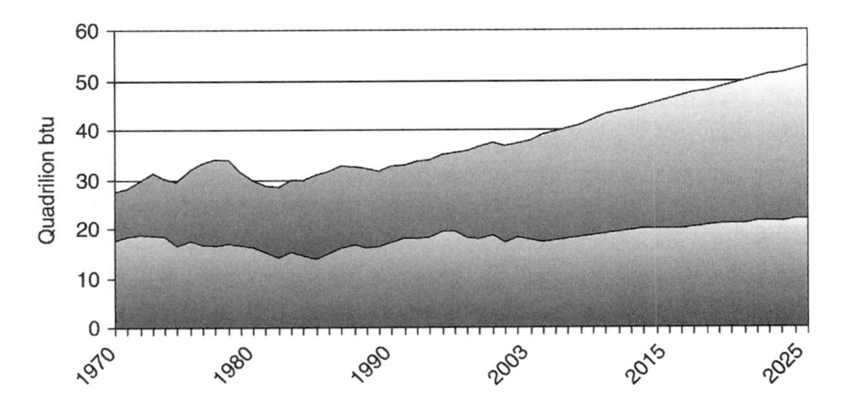

Figure 6.6 Estimated Total World Production and Demand of Oil (EIA 2003).

spread all around the world and even into remote and deeper oceans (Wenger *et al.* 2004; Pinder 2001). Due to the limited supply of onshore oil and gas reserves, and the fact that these reserves have already been exploited for a long term, there is an increasing pressure to explore and exploit offshore reserves. As a result of declining onshore reserves, offshore oil and gas operations have increased dramatically within the last two decades (Pinder 2001). This phenomenon has already been evident in many parts of the world. For example, the gas reserves on the Scotian Shell, Canada that were unsuitable or unfeasible in the 1970s are found to be economically attractive at present (Khan and Islam 2006).

6.4.2 Problems in Technological Development

The technologies promoted in the post-industrial revolution are based on the aphenomenal model (Islam 2005). This model is a gross linearization of nature ("nature" in this context includes humanity in its social nature). This model assumes that whatever appears at $\Delta t = 0$ (or time = right now) represents the actual phenomenon. This is clearly an absurdity. How can there be such a thing as a natural phenomenon without a characteristic duration and/or frequency? When it comes to what defines a phenomenon as truly natural, time, in one form or another, is of the essence.

The essence of the modern technology development scheme is the use of linearization, or reduction of dimensions, in all applications. Linearization has provided a set of techniques for solving equations that are generated from mathematical representations of observed physical laws – physical laws that were adduced correctly, and whose mathematical representations as symbolic algebra have proven frequently illustrative, meaningful, and often highly suggestive. However, linearization has made the solutions inherently incorrect. This is because any solution for t = "right now" represents the image of the real solution, which is inherently opposite to the original solution. Because this model does not have a real basis, any approach that focuses on the short-term may take the wrong path. Unlike common perception, this path does not intersect the true path at any point in time other than t = "right now." The divergence begins right from the outset. Any natural phenomenon or product always travels an irreversible pathway that is never emulated by the currently used aphenomenal model of technology development. Because, by definition, nature is non-linear and "chaotic" (Glieck 1987), any linearized model merely represents the image of nature at a time, t = "right now," in which their pathways diverge. It is safe to state that all modern engineering solutions (all are linearized) are anti-nature. Accordingly, the black box was created for every technology promoted (Figure 6.7).

This formulation of a black box helped keep "outsiders" ignorant of the linearization process that produced spurious solutions for every problem solved. The model itself has nothing to do with knowledge. In a typical repetitive mode, the output (B) is modified by adjusting input (A). The input itself is modified by redirecting (B). This is the essence of the so-called "feedback" mode that has become very popular in our day. Even in this mode, nonlinearity may arise as efforts are made to include a real object in the black box. This nonlinearity is expected. Even a man-made machine would generate chaotic behavior that becomes evident only if we have the means of detecting changes over the dominant frequency range of the operation.

We need to improve our knowledge of the process. Before claiming to emulate nature, we must implement a process that allows us to observe nature (Figure 6.8). Research based on observing nature

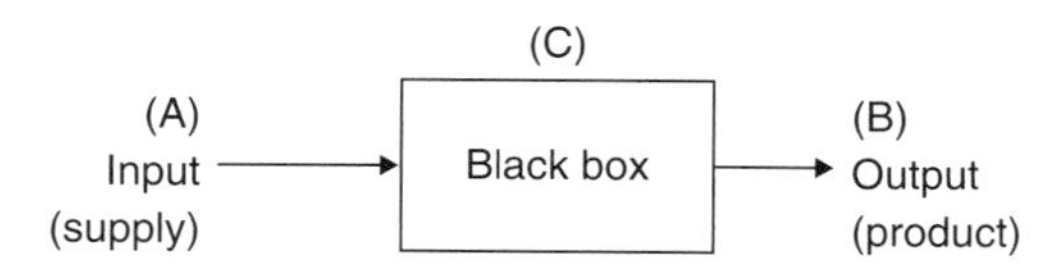

Figure 6.7 Classical "engineering" notion (redrawn from Islam 2005a).

is the only way to avoid spurious solutions due to linearization or elimination of a dimension.

Sustainable development is characterized by certain criteria. The time criterion is the main factor in achieving sustainability in technological development. However, in the present definition of sustainability, a clear time direction is missing. To better understand sustainability, we can say that there is only one alternative to sustainability, namely, unsustainability. Unsustainability involves a time dimension, but it rarely implies an immediate existential threat. Existence is threatened only in the distant future, perhaps too far away to be properly recognized. Even if a threat is understood, it may not cause much concern now, but it will cumulatively work in its effect in the wider time scale. This problem is depicted in Figure 6.9.

In Figure 6.9, the impact of the wider time scale is shown where A and B are two different development activities that are undertaken

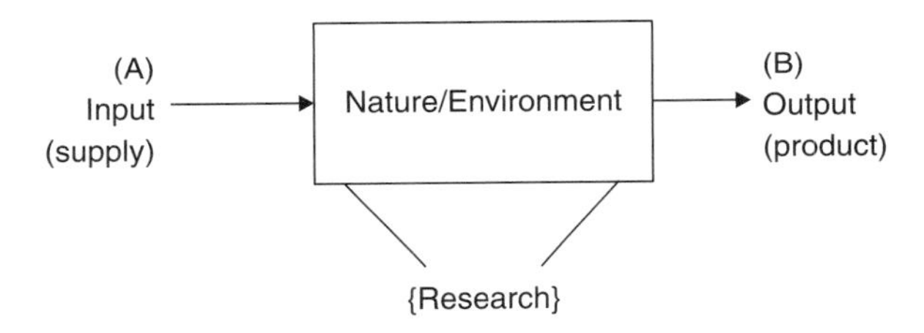

Figure 6.8 Research based on observing nature intersects classical "engineering" notion (redrawn from Islam 2005a).

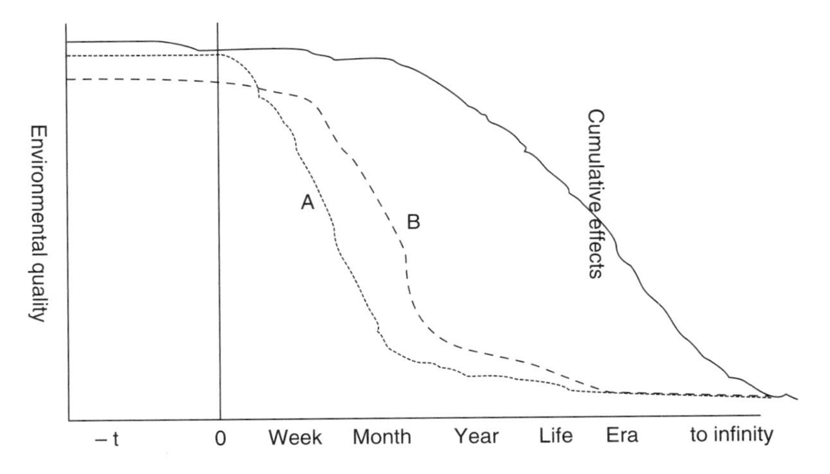

Figure 6.9 Cumulative Effects of Activities A and B within different Temporal Periods.

in a certain time period. According to the conventional environmental impact assessment (EIA), or sustainability assessment process, each project has insignificant impacts in the environment in the short time scale. However, their cumulative impacts will be much higher and will continue under a longer time scale. The cumulative impacts of these two activities (A and B) are shown as a dark line.

6.5 Development of a Sustainable Model

The sustainability model developed by Khan and Islam (2007a) provides the basis for the direction of sustainable technology. According to this model, a process is sustainable if and only if it travels a path that is beneficial for an infinite span of time. Otherwise, the process must diverge in a direction that is not beneficial in the long run. Pro-nature technology is the long-term solution. Anti-nature solutions come from schemas that comprehend, analyze, or plan to handle change on the basis of any approach, in which time-changes, or Δt, are examined only as they approach 0 (zero), that has been designed or selected as being good for time t = "right now" (equivalent to the idea of $\Delta t \rightarrow 0$). Of course, in nature, time "stops" nowhere, and there is no such thing as steady-state. Hence, regardless of the self-evident tangibility of the technologies themselves, the "reality" in which they are supposed to function usefully is non-existent, or "aphenomenal," and cannot be placed on the graph (Figure 6.10). "Good" technology can be developed if and only if it travels a path that is beneficial for an infinite span of time. In Figure 6.10, this concept is incorporated in the notion of "time tending to Infinity," which (among other things) implies also that time-changes, instead of approaching 0 (zero), could instead approach Infinity, i.e., $\Delta t \rightarrow \infty$. In this study, the term "perception" has been introduced, and it is important at the beginning of any process. Perception varies from person to person. It is very subjective, and there is no way to prove if a perception is true or wrong, or if its effect is immediate. Perception is completely one's personal opinion developed from one's experience without appropriate knowledge. That is why perception cannot be used as the base of the model. However, if perception is used in the model, the model would look as follows (Figure 6.11).

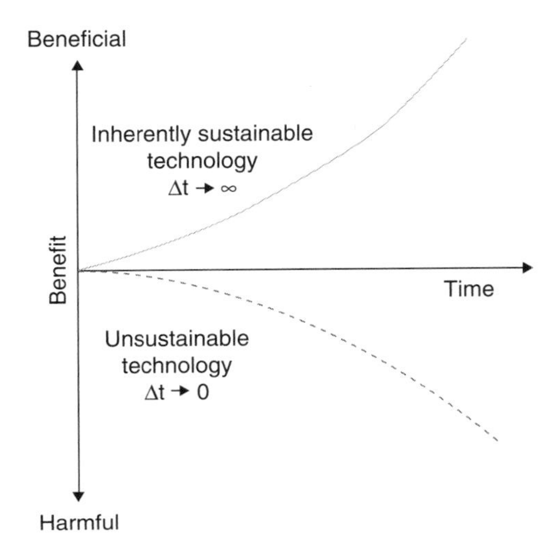

Figure 6.10 Direction of sustainability (Redrawn from Khan and Islam 2007a).

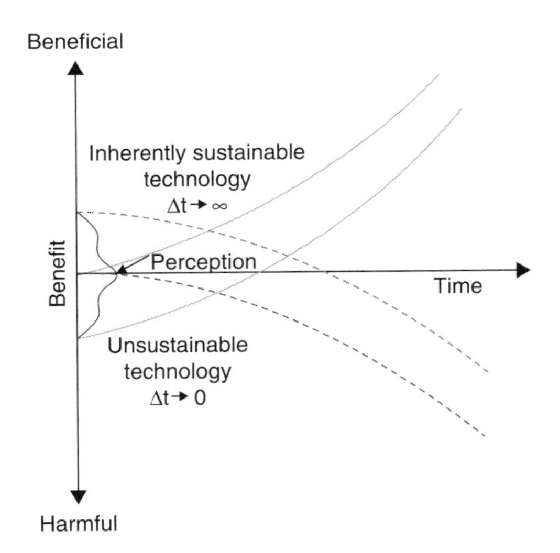

Figure 6.11 Direction of sustainability (Modified from Khan and Islam 2007a).

6.6 Violation of Characteristic Time Period or Frequency

Another problem with current technology is that it violates the natural, characteristic time. To understand the concept, let us use the following example. Visualize an electric fan rotating clockwise.

This is actuality: the truth, not a matter of perception or blind faith. If someone perceives the fan to be rotating counter-clockwise and asserts his observation as the truth, it would be equivalent to doctrinal cognition. Even if one perceives it as rotating counter-clockwise, he should make the background information available so that the truth (fan turning clockwise) can be revealed.

An electric fan can be made to appear as rotating the opposite way, to stand still, or to rotate faster by using a strobe light or by closing and opening one's eyes, matching desired frequency.

For instance, if observation with the naked eye is assumed to be complete, and no other information is necessary to describe the motion of the electric fan, a false conclusion will be validated. Using a strobe light, the motion of the electric fan can be shown to have reversed, depending on the applied frequency of the strobe. If the information that the observation of the electric fan was carried out under a strobe light is omitted, knowledge about the actual, true motion of the fan would be obscured (to say the least): simply by changing the frequency, perception has been rendered the opposite of reality. Similar obscurity to the truth can be imposed by opening and closing one's eyes to match a certain frequency. Such movement of the eyes is not sustainable in the same way covering up truth is not sustainable, *as truth unravels itself*.

How can it be ensured that any prediction of counter-clockwise fan rotation is discarded? If mention of the frequency of the light under which the observation was being made is included, it becomes obvious that the frequency of the strobe is responsible for perceiving the fan as rotating counter-clockwise. Even the frequency of the light would be insufficient to recreate the correct image, since only sunlight can guarantee an image closest to the truth and any other light distorts what the human brain will process as the image. Frequency is the inverse of time, but as this example serves to make more than clear, time is the single most important parameter in revealing the truth.

So, what is natural frequency? For the eyes, it is the blink of an eye; for light, it is sunlight with all of its components. The very use of artificial light or artificial blinking eyes violates the natural frequency, thereby creating obstacles to sustainability and environmental integrity.

Characteristic time is similar to the natural life cycle of any living being. However, characteristic time does not include any modification of life cycle time due to non-natural human intervention. For instance, the life span of an unconfined natural chicken can be up to 10 years, yet table fowls or broilers reach adult size and are slaughtered at six weeks of age (PAD 2006). The characteristic time for broiler chickens has been violated due to human intervention. This study has emphasized on characteristic time because of its pro-nature definition. Anything found in nature, grown and obtained naturally, has been optimized as a function of time. However, anything produced either by internal, genetic intervention or external, chemical fertilizer along with pesticide utilization is guaranteed to have gone through imbalances. These imbalances are often justified in order to obtain short-term tangible benefits, by trading off with other intangible benefits that are more important. In the long run, such systems can never produce long-term good.

6.7 Observation of Nature: Importance of Intangibles

Nature is observed and recorded only in tangible aspects detectable with current technologies. Accordingly, much of what could only be taking place as a result of intangible, but very active, orderliness within nature is considered "disorder" according to the tangible standard. The greatest confusion is created when this misapprehension is then labeled "chaotic" and its energy balance on this basis is portrayed as headed towards "heat death," "entropy," or the complete dissipation of any further possibility of extracting "useful work."

Reality is quite different. In nature, there is not a single entity that is linear, symmetric, or homogeneous. On Earth, there is not a single process that is steady, or even periodic. Natural processes are chaotic, but not in the sense of being arbitrary or inherently tending towards entropy. Rather, they are chaotic in the sense that what is essentially orderly and characteristic only unfolds with the passage of time within the cycle or frequency that is characteristic of the given process at a particular point. What the process looks like at that point is neither precisely predictable, previous to that point, nor precisely reconstructible or reproducible after that point. The path of such a process is defined as chaotic on the basis of it being aperiodic, non-linear, and non-arbitrary.

Nature is chaotic. However, the laws of motion developed by Newton cannot explain the chaotic motion of nature, due to their assumptions that contradict the reality of nature. The experimental validity of Newton's laws of motion is limited to describing instantaneous macroscopic and tangible phenomena. Microscopic and intangible phenomena are ignored, however. The classical dynamics, as represented by Newton's laws of motion, emphasize fixed and unique initial conditions, stability, and equilibrium of a body in motion (Ketata *et al.* 2007a). However, the fundamental assumption of constant mass is adequate to conflict Newton's laws of motion. Ketata *et al.* (2007a) formulated the following relation to describe the body in continuous motion in one space:

$$m = \frac{F}{\left((6t+2)+\left(3t^2+2t+1\right)^2\right)ce^u} \tag{6.1}$$

where F is the force on the body;

$$u = t^3 + t^2 + t + 1;$$

and c is a constant.

The above relation demonstrates that the mass of a body in motion depends on time, whether F varies over time or not. This is absolutely the contradiction of the first law of motion. Similarly, the acceleration of a body in motion is not proportional to the force acting on the body, because mass is not constant. Again, this is a contradiction of the second law of motion.

Here it is found that time is the biggest issue which, in fact, dictates the correctness of Newton's laws of motion. Considering only instantaneous time ($\Delta t \rightarrow 0$), Newton's laws of motion will be experimentally valid with some error. However, considering the infinite time span ($\Delta t \rightarrow \infty$), the laws cannot be applicable. That is why sustainable technologies that include short-term to long-term benefits cannot be explained by Newton's laws. To overcome this difficulty, it is necessary to break out of "$\Delta t \rightarrow 0$" in order to include intangibles, which is the essence of pro-nature technology development.

In terms of the well-known laws of conservation of mass (m), energy (E), and momentum (p), the overall balance, B, within nature may be defined as a function of all of them:

$$B = f(m, E, p) \tag{6.2}$$

The perfection without stasis that is nature means that everything that remains in balance within it is constantly improving with time. That is:

$$\frac{dB}{dt} > 0 \tag{6.3}$$

If the proposed process has all concerned elements, such that each element is following this pathway, none of the remaining elements of the mass balance will present any difficulty. Because the final product is considered as time extends to infinity, the positive ("> 0") direction is assured.

Pro-nature technology, which is non-linear, increases its orderliness on a path that converges at infinity, after providing maximum benefits over the intervening time. This is achievable only to the extent that such technologies employ processes as they operate within nature. They use materials whose internal chemistry has been refined entirely within the natural environment and whose subsequent processing has added nothing else from nature in any manner other than its characteristic form. Any and every other technology is anti-nature. The worst among them are self-consciously linear, "increasing" order artificially by means of successive superpositions that supposedly take side-effects and negative consequences into account as they are detected. This enables the delivery of maximum power, or efficiency, for an extremely short term. It does so without regard to coherence or overall sustainability, and at the cost of detrimental consequences carrying on long after the "great advances" of the original anti-nature technology have dissipated. Further disinformation lies in declaring the resulting product "affordable," "inexpensive," "necessary," and other self-serving and utterly false attributes, while increasing only very short-term costs. Any product that is anti-nature would turn out to be prohibitively costly if long-term costs are included. A case in point is tobacco technology. In Nova Scotia alone, 1,300 patients die each year of cancer emerging directly from smoking (Islam 2003). These deaths cost us 60 billion dollars in body parts alone. How expensive should cigarettes be? If intangibles are included in any economic analysis, a picture very different from what is conventionally portrayed will emerge (Zatzman and Islam 2007).

Any linearized model can be limited or unlimited, depending on the characteristics of the process (Figure 6.12). The "limited linearized model" has two important characteristics, more tangible features than intangible, and a finite, limited amount of disorder or imbalance. Because only linearized models are man-made, nature has time to react to the disorder created by this limited model, and

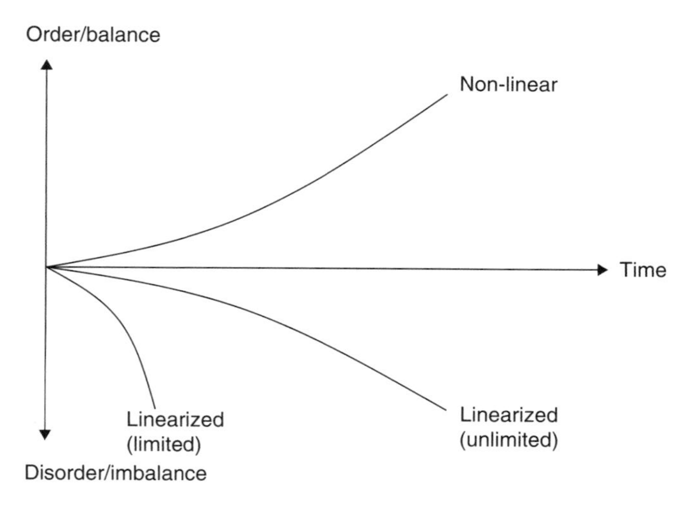

Figure 6.12 Pathway of nature and anti-nature (Modified from Khan and Islam 2006).

it may, therefore, be surmised that such models are unlikely to cause irreparable damage.

With more intangible features than tangible and an unlimited degree of disorder, or imbalance, the unlimited linearized model is characterized by long-term effects that are little understood but far more damaging. Contemporary policy-making processes help conceal a great deal of actual or potential imbalance from immediate view or detection, a classic problem with introducing new pharmaceuticals, for example. Since a drug has to pass the test of not showing allergic reactions, many drugs make it into the market after being "tweaked" to delay the onset of what are euphemistically called "contra-indications." An elaborate and tremendously expensive process of clinical trials is unfolded to mask such "tweaking," mobilizing the most heavily invested shareholders of these giant companies to resist anything that would delay the opportunity to recoup their investment in the marketplace. The growing incidences of suicide among consumers of Prozac® and other SSRI-type anti-depressant drugs and of heart-disease "complications" among consumers of "Cox-2" type drugs for relief from chronic pain are evidence of the consequences of the unlimited linearized model and of how much more difficult any prevention of such consequences is (Miralai 2006). In forms of concentrations, unlimited pertains to intangible.

Here is another example of how the unlimited linearized model delays the appearance of symptoms. If food is left outside, in two to three days it will cause food poisoning, which provokes diarrhea. However, if the food is placed in artificial refrigeration, the food will retain some appearance of "freshness" even after several weeks, although its quality will be much worse than the "rotten" food that was left outside. Another more exotic, but non-industrial example can be seen in the reaction to snake venom. The initial reaction is immediate. If the victim survives, there is no long-term negative consequence. Used as a natural source or input to a naturally-based process, snake venom possesses numerous long-term benefits and is known for its anti-depressive nature.

Repositioning cost-benefit analysis away from short-term considerations, such as the cheapness of synthesized substitutes, to the more fundamental tangible/intangible criterion of long-term costs and benefits, the following summary emerges: tangible losses are very limited, but intangible losses are not.

6.8 Analogy of Physical Phenomena

Mathematicians continue to struggle with the two entities "0" and "∞," whose full meanings and consequences continue to mystify (Ketata *et al.* 2006a, 2006b). However, these two entities are most important when intangible issues are counted, as the following simple analogy from well-known physical phenomena (Figure 6.13) can demonstrate. As "size," i.e., space occupied (surface area or volume) per unit mass, goes down, the quantity of such forms of matter goes up. This quantity approaches infinity as space occupied per unit mass heads towards zero. However, according to the law of conservation of mass and energy, mass can neither be created nor destroyed, and it only can transform from one form to another form. This contradiction was resolved in the early 20th century, when it was proven that as mass decreased, its quantity could increase as particles of mass were converted into quanta of energy.

Infinity means that a quantity is too large to count exactly, but that it enjoys practical existence. Conventionally, zero on the other hand denotes non-existence, posing another paradox that is nonetheless removable when the intangible aspect is considered. Something that is infinite in number is present everywhere but has no size. As Figure 6.13 shows, mass turns into energy at the end and loses "size," a transition of the tangible into the intangible. This also signifies that the number of intangibles is much more than that of tangibles. We can measure the tangible properties, but it is difficult to measure the intangible. Yet, inability to measure the intangible hardly demonstrates non-existence. Happiness, sorrow, and so forth are all clearly intangible, and they possess no tangible properties whatsoever, no matter how tangible their causes. As Figure 6.13 suggests, the scale of the intangible is potentially far more consequential than that of the tangible.

6.9 Intangible Cause to Tangible Consequence

Short-term intangible effects are difficult to understand, but consideration of the treatment procedures employed by homeopaths may serve to elucidate. The most characteristic principle of homeopathy is that the potency of a remedy can be enhanced by dilution, an inconsistency with the known laws of chemistry (Homeopathy 2006). In some cases, the dilution is so high that it is extremely

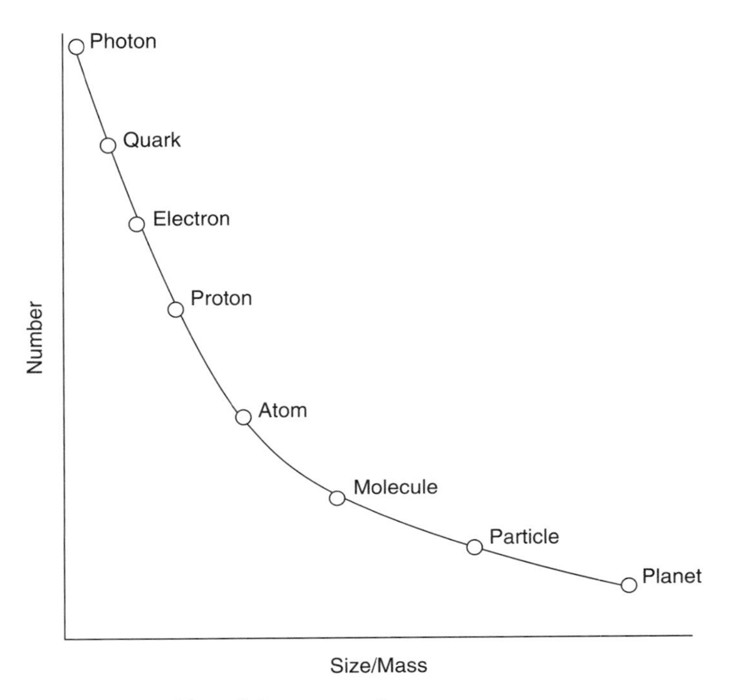

Figure 6.13 Relation of Size/Mass to number.

unlikely that one molecule of the original solution would be present in that dilution. As there is no detectable mechanism to this, the effect of the molecule cannot always be understood, and that is why the homeopathy still remains controversial to the modern science of tangibles. However, the trace ingredient of dilution is not always ignorable. Recently, Rey (2003) studied the thermoluminescence of ultra-high dilution of lithium chloride and sodium chloride, and found the emitted light specific of the original salts dissolved initially. The dilution was beyond Avogadro's number ($\sim 6.0 \times 10^{23}$ atoms per mole), but its effect was visible. In other words, when concentration of a substance descends to below detection level, it cannot be ignored, as its effects remain present. This is where greater care needs to be taken in addressing the harmful potential of chemicals in low concentrations. Lowering the concentration cannot escape the difficulty, a significant consideration when it comes to managing toxicity. Relying on low concentration as any guarantee of safety defeats the purpose when the detection threshold used, to regulate what is "safe," is higher than the lowest concentrations at which these toxins may be occurring or accumulating in the environment. Although the science that will identify the accumulation of effects from toxic concentrations before they reach the threshold of regulatory detection remains to be established, the point is already clear. Tangible effects may proceed from causes that can remain intangible for some unknown period of time.

Mobile phones are considered one of the biggest inventions of modern life for communication. So far, the investigation of the harmful effects of using mobile phones has been limited only to the human brain damage from non-natural electro magnetic frequency. An official Finnish study found that people who used the phones for more than 10 years were 40 percent more likely to get a brain tumor on the same side that they held the handset (Lean and Shawcross 2007). However, recently it has been observed that mobile frequency also causes serious problems for other living beings of nature, which are very important for the balance of the ecological system. Recently, an abrupt disappearance of the bees that pollinate crops has been noticed, especially in the United States, as well as some other countries of Europe (Lean and Shawcross 2007). The plausible explanation of this disappearance is that radiation from mobile phones interferes with bees' navigation systems, preventing the famously home-loving species from finding their way back to their hives. Most of the world's crops depend on pollination by bees. That is why a massive food shortage has been anticipated due to the extinction of these bees, which is due to radiation given off by mobile phones. Albert Einstein once said that if bees disappeared, "man would have only four years of life left" (Lean and Shawcross 2007). This is how a non-natural high-tech instrument poses tangible effects in the long run due to its intangible causes.

6.10 Removable Discontinuities: Phases and Renewability of Materials

By introducing time-spans of examination unrelated to anything characteristic of the phenomenon being observed in nature, discontinuities appear. These are entirely removable, but they appear to the observer as finite limits of the phenomenon, and as a result, the possibility that these discontinuities are removable is not even considered. This is particularly problematic when it comes to phase transitions of matter and the renewability or non-renewability of energy.

The transition between solid, liquid, and gas in reality is continuous, but the analytical tools formulated in classical physics are anything but continuous. Each P-V-T model applies to only one phase and one composition, and there is no single P-V-T model that is applicable to all phases (Cismondi and Mollerup 2005). Is this an accident? Microscopic and intangible features of phase transitions have not been taken into account, and as a result of limiting the field of analysis to macroscopic, tangible features, modeling becomes limited to one phase and one composition at a time.

When it comes to energy, everyone has learned that it comes in two forms, renewable and nonrenewable. If a natural process is being employed, however, everything must be "renewable" by definition, in the sense that, according to the law of conservation of energy, energy can be neither created nor destroyed. Only the selection of the time-frame misleads the observer into confounding what is accessible in that finite span with the idea that energy is therefore running out. The dead plant material that becomes petroleum and gas trapped underground in a reservoir is being added to continually. However, the rate at which it is extracted has been set according to an intention that has nothing to do with the optimal timeframe in which the organic source material could be renewed. Thus, "non-renewability" is not any kind of absolute fact of nature. On the contrary, it amounts to a declaration that the pathway on which the natural source has been harnessed is anti-nature.

6.11 Rebalancing Mass and Energy

Mass and energy balance inspected in depth discloses intention as the most important parameter and sole feature that renders the individual accountable to, and within, nature. This draws serious consequences for the black box approach of conventional engineering, because a key assumption of the black box approach stands in contradiction to one of the key corollaries of the most fundamental principle of all, the law of conservation of matter.

Conventionally, the mass balance equation is represented as "mass-in equals mass-out" (Figure 6.14). In fact, however, this is only possible if there is no leak anywhere and no mass can flow into the system from any other point, thereby rendering the entire analysis a function of tangible, measurable quantities, or a "science" of tangibles only.

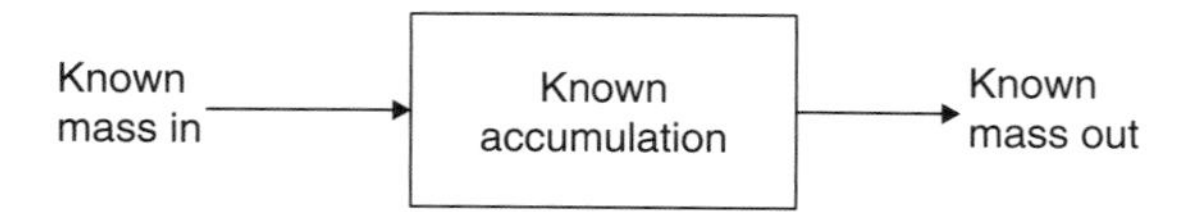

Figure 6.14 Conventional Mass Balance equation incorporating only tangibles.

The mass conservation theory indicates that the total mass is constant. It can be expressed as follows:

$$\Sigma_0^\infty m_i = \text{Constant} \tag{6.4}$$

where m = mass and i is the number from 0 to ∞.

In a true sense, this mass balance encompasses mass from the very macroscopic to microscopic, detectable to undetectable, or in other words, from tangible to intangible. Therefore, the true statement should be as illustrated in Figure 6.15:

"***Known*** mass-in" + "***Unknown*** mass-in"
= "***Known*** mass-out + "***Unknown*** mass-out"
+ "***Known*** accumulation" + "***Unknown*** accumulation"
(6.5)

The unknowns can be considered intangible, yet they are essential to include in the analysis, as they incorporate long-term and other elements of the current timeframe.

In nature, the deepening and broadening of order is continually observed: many pathways, circuits, and parts of networks are partly or even completely repeated, and the overall balance is further enhanced. Does this actually happen as arbitrarily as conventionally assumed? A little thought suggests this must take place principally in response to human activities and the response of the environment to these activities and their consequences. Nature itself has long established its immediate and resilient dominion over every activity and process of everything in its environment, and there is no other species that can drive nature into such modes of response. In the absence of the human presence, nature would not be provoked into having to increase its order and balance, and everything would function in "zero net waste" mode.

An important corollary of the law of conservation of mass is that no mass can be considered in isolation from the rest of the universe.

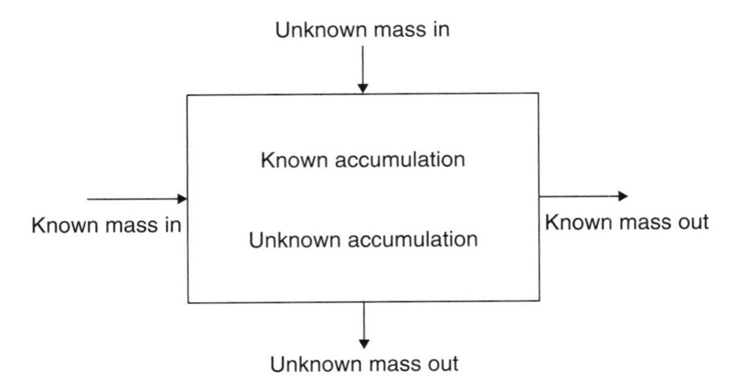

Figure 6.15 Mass-balance equation incorporating tangibles and intangibles.

Yet, the black box model clearly requires such an impossibility. However, since human ingenuity can select the time frame in which such a falsified "reality" would be, the model of the black box can be substituted for reality, and the messy business of having to take intangibles into account is foreclosed once and for all.

6.12 Energy: The Current Model

A number of theories have been developed in the past centuries to define energy and its characteristics. However, none of the theories are enough to describe energy properly. All of the theories are based on many idealized assumptions that have never existed practically. Consequently, the existing model of energy and its relation to others cannot be accepted confidently. For instance, the second law of thermodynamics depends on Carnot's cycle in classical thermodynamics, and none of the assumptions of Carnot's cycle exist in reality. The definitions of ideal gas, reversible process, and adiabatic process used in describing the Carnot's cycle are imaginary. In 1905, Einstein came up with his famous equation, $E=mc^2$, which shows an equivalence between energy (E) and relativistic mass (m) in direct proportion to the square of the speed of light in a vacuum (c^2). However, the assumption of constant mass and the concept of a vacuum do not exist in reality. Moreover, this theory was developed on the basis of Planck's constant, which was derived from black body radiation. Perfectly black body does not even exist in reality. Therefore, the development of every theory has depended on a series of assumptions that do not exist in reality.

6.12.1 Supplements of Mass Balance Equation

For whatever else remains unaccounted, the energy balance equation supplements the mass balance equation, which in its conventional form necessarily falls short in explaining the functionality of nature coherently as a closed system.

For any time, the energy balance equation can be written as:

$$\int_0^\infty \Sigma a_i = \text{Constant}\,,\ i \text{ going from 1 to infinity} \qquad (6.6)$$

where a is the activity equivalent to potential energy.

In the above equation, only potential energy is taken into account. Total potential energy, however, must include all forms of activity, and once again a large number of intangible forms of activity, e.g., the activity of molecular and smaller forms of matter, cannot be "seen" and accounted for in this energy balance. The presence of human activity introduces the possibility of other potentials that continually upset the energy balance in nature. There is overall balance but some energy forms, e.g., electricity (either from combustion or nuclear sources), which would not exist as a source of useful work except for human intervention, continually threaten to push this into a state of imbalance.

In the definition of activity, both time and space are included. The long term is defined by time reaching to infinity. The "zero waste" condition is represented by space reaching infinity. There is an intention behind each action, and each action plays an important role in creating overall mass and energy balance.

The role of intention is not to create a basis for prosecution or enforcement of certain regulations. Rather, it is to provide the individual with a guideline. If the product, or the process, is not making things better with time, it is fighting nature, a fight that cannot be won and is not sustainable. Intention is a quick test that will eliminate the rigorous process of testing feasibility, long-term impact, and so forth. Only with "good" intentions can things improve with time. After that, other calculations can be made to see how fast the improvement will take place.

In clarifying the intangibility of an action or a process, with reference to the curve of Figure 6.12, the equation has a constant, which is actually an infinite series:

$$a = \Sigma_0^\infty a_i = a_0 + a_1 + a_2 + a_3 + ... \tag{6.7}$$

If each term of Equation 6.6 converges, it will have a positive sign, which indicates intangibility; hence, the effect of each term thus becomes important for measuring the intangibility overall. On this path, it should also become possible to analyze the effect of any one action and its implications for sustainability overall, as well.

In can be inferred that man-made activities are not enough to change the overall course of nature. Failure until now, however, to account for the intangible sources of mass and energy has brought about a state of affairs in which, depending on the intention attached to an intervention, the mass-energy balance can either be restored and maintained over the long term, or increasingly threatened and compromised in the short term. In the authors' view, it would be better to develop the habit of investigating nature and the possibilities it offers to humanity's present and future by considering time t at all scales, reaching to infinity. This requires eliminating the habit of resorting to time scales that appear to serve an immediate ulterior interest in the short term, but that in fact have nothing to do with the natural phenomena of interest, and therefore lead to something that will be anti-nature in the long term and short term.

The main obstacle in discussing and positioning human intentions within the overall approach to the laws of conservation of mass, energy, and momentum stems from notions of the so-called "heat death" of the universe predicted in the 19th century by Lord Kelvin and enshrined in his second law of thermodynamics. In fact, this idea, that the natural order must "run down" due to entropy, eliminating all sources of "useful work," naively assigns a permanent and decisive role for negative intentions in particular, without formally fixing or defining any role whatsoever for human intentions in general. Whether failures arise out of the black box approach to the mass-balance equation, or out of the unaccounted, missing potential energy sources in the energy-balance equation, failures in the short term become especially consequential when made by those who defend the *status quo* to justify anti-nature "responses," the kind well-described as typical examples of "the roller coaster of the Information Age" (Islam 2003).

6.13 Tools Needed for Sustainable Petroleum Operations

Sustainability can be assessed only if technology emulates nature. In nature, all functions or techniques are inherently sustainable, efficient, and functional for an unlimited time period. In other words, as far as natural processes are concerned, "time tends to infinity." This can be expressed as *t* or, for that matter, $\Delta t \rightarrow \infty$.

By following the same path as the functions inherent in nature, an inherently sustainable technology can be developed (Khan and Islam 2005b). The "time criterion" is a defining factor in the sustainability and virtually infinite durability of natural functions. Figure 6.18 shows the direction of nature-based, inherently sustainable technology contrasted with an unsustainable technology. The path of sustainable technology is its long-term durability and environmentally wholesome impact, while unsustainable technology is marked by Δt approaching 0. Presently, the most commonly used theme in technology development is to select technologies that are good for t = "right now," or $\Delta t = 0$. In reality, such models are devoid of any real basis (termed "aphenomenal" by Khan *et al.* 2005) and should not be applied in technology development if we seek sustainability for economic, social, and environmental purposes.

In terms of sustainable technology development, considering pure time (or time tending to infinity) raises thorny ethical questions. This "time-tested" technology will be good for nature and good for human beings. The main principle of this technology will be to work toward, rather than against, natural processes. It would not work against nature or ecological functions. All natural ecological functions are truly sustainable in this long-term sense. We can take a simple example of an ecosystem technology (natural ecological function) to understand how it is time-tested (Figure 6.16).

In nature, all plants produce glucose (organic energy) through utilizing sunlight, CO_2, and soil nutrients. This organic energy is then transferred to the next higher level of organisms which are small animals (zooplankton). The next higher (tropical) level organism (high predators) utilizes that energy. After the death of all organisms, their body masses decompose into soil nutrients, which again take plants to keep the organic energy looping (Figure 6.16). This natural production process never malfunctions and remains constant for an infinite time. It can be defined as a time-tested technique.

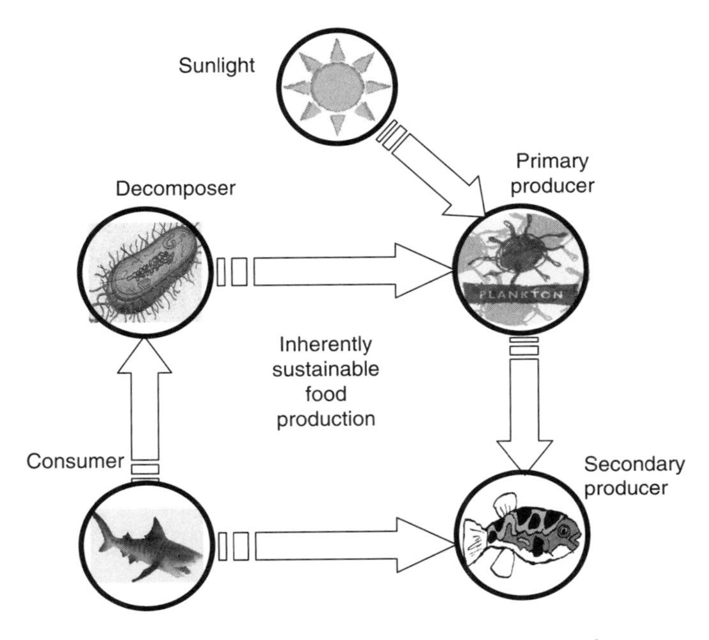

Figure 6.16 Inherently sustainable natural food production cycle.

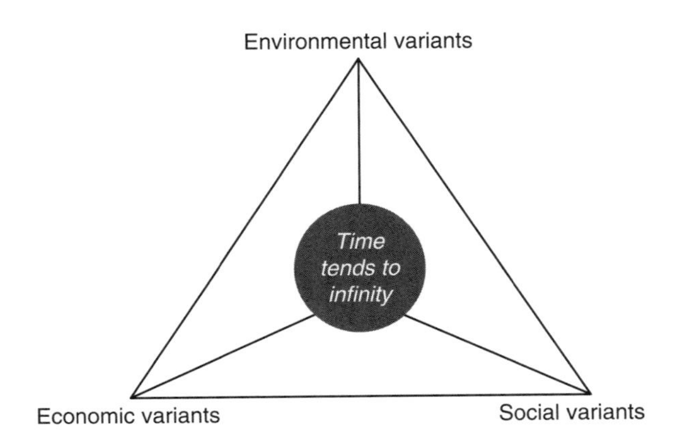

Figure 6.17 Pictorial view of the major elements of sustainability in technology development.

This time-tested concept can equally apply to technology development. New technology should be functional for an infinite time. This is the only way it can achieve true sustainability (Figure 6.17). This idea forms the new assessment framework that is developed and shown in Figures 6.17 and 6.18. The triangular sign of

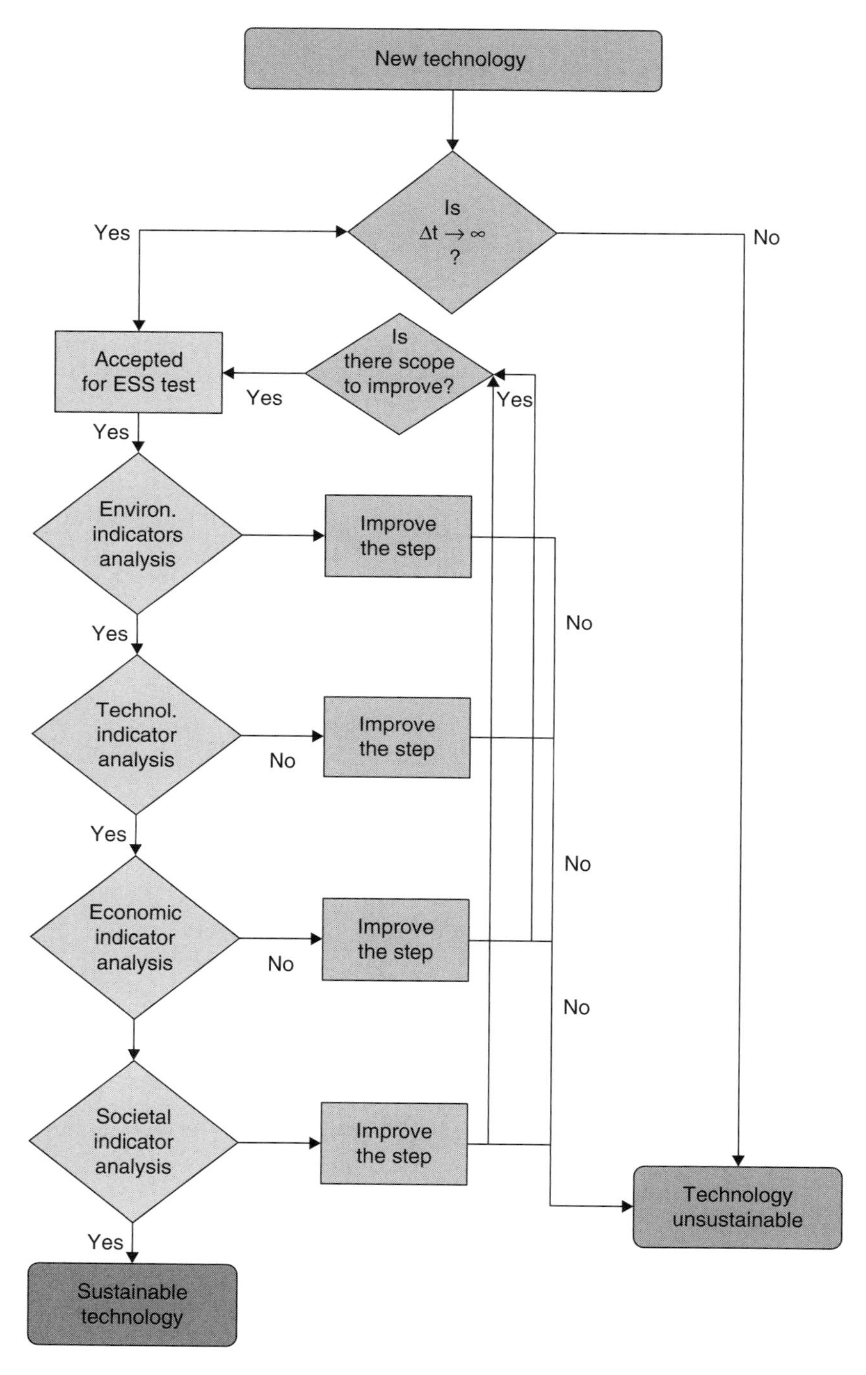

Figure 6.18 Proposed sustainable technology flowchart (modified Khan and Islam 2005a).

sustainability in Figure 6.17 is considered the most stable sign. This triangle is formed by different criteria that represent a stable sustainability in technology development. Any new technology could be evaluated and assessed by using this model. There are two selection levels, the primary level and the secondary level. A technology must fulfill the primary selection criterion, "time," before being taken to the secondary level of selection.

For a simulation test, we imagine that a new technology is developed to produce a product named "Ever-Rigid." This product is non-corrosive, non-destructive, and highly durable. The "Ever-Rigid" technology can be tested using the proposed model to determine whether it is truly sustainable or not. The first step of the model is to find out if the "Ever-Rigid" technology is "time-tested." If the technology is not durable over infinite time, it is rejected as an unsustainable technology and would not be considered for further testing. For, according to the model, time is the primary criterion for the selection of any technology.

If the "Ever-Rigid" technology is acceptable with respect to this time criterion, then it may be taken through the next process to be assessed according to a set of secondary criteria. The initial set of secondary criteria analyzes environmental variants. If it passes this stage, it goes to the next step. If the technology is not acceptable in regard to environmental factors, then it might be rejected, or further improvements might be suggested to its design. After environmental evaluation, the next two steps involve technological, economic, and societal variants analyses, each of which follows a pathway similar to that used to assess environmental suitability. Also, at these stages, either improvement on the technology will be required, or the technology might be rejected as unsustainable.

6.14 Assessing the Overall Performance of a Process

In order to break out of the conventional analysis introduced through the science of tangibles, we proceed to discuss some salient features of the time domain and present how using time as the fourth dimension can assess the overall performance of a process. Here, time *t* is not orthogonal to the other three spatial dimensions. However, it is no less a dimension for not being mutually orthogonal. Socially available knowledge is also not orthogonal either with respect to time t, or with respect to the other three spatial dimensions. Hence, despite the training of engineers and scientists in higher mathematics that hints, suggests, or implies that dimensionality must be tied up somehow in the presence of orthogonality, orthogonality is not a relationship built into dimensionality. It applies only to the arrangements we have invented to render three spatial dimensions simultaneously visible, i.e., tangible.

We start off with the fundamental premise that real is sustainable, whereas artificial is not sustainable. The word 'real' here applies to natural matters, both tangible and intangible. There can be comparison between two real processes in order to determine marketability, feasibility, etc. but there cannot be any comparison between real and artificial matters. Let us recap the notion of real and artificial.

Real: Root natural and processing natural (for matter, natural material must be the root; for cognition, good intention must be the root)

Artificial: Root artificial and processing artificial

Scientifically, the above analysis is equivalent to long-term criterion (time approaching infinity), as developed by Khan (2006). Applying this analysis to recently developed materials listed in Table 6.2 shows that the intention behind developing the artificial products was not to meet long-term sustainability.

The rankings show a clear pattern, that is, all technologies were put in place to increase profitability as the first criterion. Instead of using true sustainability as the first single most important criterion, the profit margin has been the single most important criterion used for developing a technology ever since the Renaissance saw the emergence of a short-term approach in an unparalleled pace. As Table 6.2 indicates, natural rankings generally are reversed if the criterion of profit maximization is used. This affirms, once again, how modern economics has turned pro-nature techniques upside down (Zatzman and Islam 2007). This is the onset of the economics of tangibles, as shown in Figure 6.19. As processing is done, the quality of the product is decreased. Yet, this process is called value addition in the economic sense. The price, which should be proportional to the value, in fact goes up inversely proportional to the real value (opposite to perceived value, as promoted through advertisement). Here, the value is fabricated. The fabricated value is made synonymous with real value or quality without any further discussion of what constitutes quality. This perverts the entire

Table 6.2 Synthesized and natural pathways of organic compounds as energy sources, ranked and compared according to selected criteria.

Natural (Real) Ranking ("Top" Rank Means Most Acceptable)	**Aphenomenal Ranking by the following Criteria**			
	Bio-degradability	**Efficiency[1],** e.g., $\eta = \frac{\text{Outp} - \text{Inp}}{\text{Inp}} \times 100$	**Profit Margin**	**Heating Value (cal/g)**
1. Honey	*2*	*4* "sweetness /g"	*4*	*1*
2. Sugar	*3*	*3*	*3*	*2*
3. Sacchharine	*4*	*2*	*2*	*3*
4. Aspartame	*1*	*1*	*1*	*4*
1. Organic wood	*1* REVERSES	*4* REVERSES if	*4* REVERSES if	*4*
2. Chemically-treated wood	*2* depending	*3* toxicity is	*3* organic wood	*3*
3. Chemically grown, Chemically treated wood	*3* on applic'n,	*2* considered	*2* treated with	*2*
4. Genetically-altered wood	*4 e.g.*, durability	*1*	*1* organic chemicals	*1*
1. Solar	Not applicable	*5* # Efficiency	*5*	*5* # - Heating
2. Gas		*4* cannot be	*4*	*4* value cannot
3. Electrical		*3* calculated for	*3*	*3* be
4. Electromagn etic		*2* direct solar	*2*	*2* calculated
5. Nuclear		#	*1*	# for direct solar

(Continued)

Table 6.2 (Cont.) Synthesized and natural pathways of organic compounds as energy sources, ranked and compared according to selected criteria.

Natural (Real) Ranking ("Top" Rank Means Most Acceptable)	Aphenomenal Ranking by the following Criteria			
	Bio-degradability	Efficiency[1], e.g., $\eta = \frac{Outp - Inp}{Inp} \times 100$	Profit Margin	Heating Value (cal/g)
1. Clay or wood ash	1 Anti-	6 REVERSES if	6	4 # 1 cannot
2. Olive oil + wood ash	3 bacterial	5 global is	5	6 be ranked
3. Veg oil + NaOH	4 soap won't	4 considered	4	5
4. Mineral oil + NaOH	5 use olive oil;	3	3	3
5. Synthetic oil + NaOH	6 volume	2	2	2
6. 100% synthetic soap-free soap)	2 needed for cleaning unit area	1	1	#
1. Ammonia	1	Unknown	3	Not applicable
2. Freon	2		2	
3 Non-Freon synthetic	3		1	

1. Methanol 2. Glycol 3. Synthetic polymers (low dose)	*1* 2 3	*1* For hydrate 2 control 3	3 2 *1*	Not applicable
1. Sunlight 2. Vegetable oil light 3. Candle light 4. Gas light 5. Incandescent light 6. Fluorescent light	Not applicable	6 5 4 3 2 *1*	6 5 4 3 2 *1*	Not applicable

[1]This efficiency is local efficiency that deals with an arbitrarily set size of sample.

*Calorie/gm is a negative indicator for "weight watchers" (interested in minimizing calorie) and a positive indicator for energy drink makers (interested in maximizing calorie).

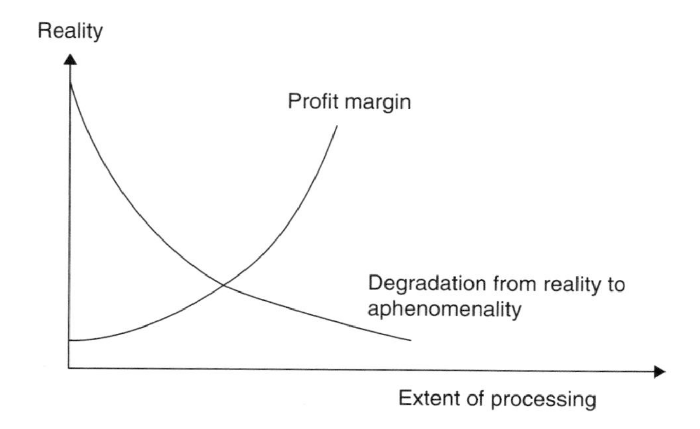

Figure 6.19 The profit margin increased radically with external processing.

value addition concept and falsifies the true economics of commodity (Zatzman and Islam 2007). Only recently, the science behind this disinformation has begun to surface (Shapiro *et al.* 2006).

6.15 Continuous Time Function as the Inherent Feature of a Comprehensive Criterion

The most serious, most important, most significant, and truest acid test of a proposed scientific criterion is that it accounts for everything necessary and sufficient to explain the phenomenon (its origin, its path, and its endpoint), thereby rendering it positively useful to human society. The same criterion was used in previous civilizations to distinguish between real and artificial. Khan (2007) introduced a criterion that identifies the endpoint, by extending time to infinity. This criterion avoids scrutiny of the intangible source of individual action (namely, intention). However, Zatzman and Islam (2007a) pointed out that the endpoint at time t = infinity can be a criterion, but it will not disclose the pathway unless a continuous time function is introduced. Mustafiz (2007) used this concept and introduced the notion of knowledge dimension, a dimension that arises from introducing time as a continuous function. In all of these deductions, it is the science of intangibles that offers some hope. It is important to note that the insufficiency just mentioned is not overcome by doing "more" science of tangibles "better." It is already evident that what is not being addressed are intangible components that cannot be wrinkled out or otherwise measured by existing means available within the realm of the science of tangibles.

Intangibles, which essentially include the root and pathway of any phenomenon, make the science suitable for increasing knowledge, as opposed to increasing confidence in a conclusion that is inherently false (Zatzman and Islam 2007a). Zatzman and Islam (2007) introduced the following syllogism to make this point about the science of intangibles:

> All Americans speak French. (major premise)
> Jacques Chirac is an American. (minor premise)
> Therefore, Jacques Chirac speaks French.
> (conclusion-deduction)

If, in either the major or minor premise, the information relayed above is derived from a scenario of what is merely probable (as distinct from what is actually known), the conclusion (which happens to be correct in this particular case) would not only be acceptable as something independently knowable, but also reinforced as something statistically likely. This, then, finesses determining the truth or falsehood of any of the premises, and, eventually, someone is bound to "reason backwards" to deduce the statistical likelihood of the premises from the conclusion. This latter version, in which eventually all the premises are falsified as a result of starting out with a false assumption asserted as a conclusion, is exactly what has been identified and labeled elsewhere as the aphenomenal model (Khan *et al.* 2005). How can this aphenomenal model be replaced with a knowledge model? Zatzman and Islam (2007a) emphasized the need of recognizing the first premise of every scientific discourse. They used the term "aphenomenality" (contrasted to truth) to describe, in general, the non-existence of any purported phenomenon or any collection of properties, characteristics, or features ascribed to such a purported but otherwise unverified or unverifiable phenomenon. If the first premise contradicts what is true in nature, then the entire scientific investigation will be false. Such investigation cannot lead to reliable or useful conclusions.

Consider the following syllogism (the concept of "virtue" intended here is that which holds positive value for an entire collectivity of people, not just for some individual or arbitrary subset of individual members of humanity):

> All virtues are desirable.
> Speaking the truth is a virtue.
> Therefore, speaking the truth is desirable.

Even before it is uttered, a number of difficulties have already been built into this seemingly noncontroversial syllogism. When it is said that "all virtues are desirable," there is no mention of a time factor (pathway) or intention (source of a virtue). For instance, speaking out against an act of aggression is a virtue, but is it desirable? A simple analysis would indicate that unless the time is increased to infinity (meaning something that is desirable in the long run), practically all virtues are undesirable. (Even giving to charity requires austerity in the short-term, and defending a nation requires self-sacrifice, an extremely undesirable phenomenon in the short-term). The scientifically correct reworking of this syllogism should be:

> All virtues (both intention and pathway being real) are desirable for time approaching infinity.
> Speaking the truth is a virtue at all times.
> Therefore, speaking the truth is desirable at all times.

The outcome of this analysis is the complete disclosure of source, pathway (time function), and final outcome (t approaching ∞) of an action. This analysis can and does restore to its proper place the rational principle underlying the comparison of organic products to synthetic ones, free-range animals to confined animals, hand-drawn milk to machine-drawn, thermal pasteurization with wood fire compared to microwave and/or chemical Pasteurization®, solar heating compared to nuclear heating, the use of olive oil compared to chemical preservatives, the use of natural antibiotics compared to chemical antibiotics, and so forth. When it comes to food or other matter ingested by the human body, natural components ought to be preferred because we can expect that the source and pathway of such components, already existing in nature, will be beneficial (assuming nontoxic dosages of medicines and normal amounts of food are being ingested). Can we have such confidence when it comes to artificially simulated substitutes? The pathway of the artificial substitute's creation lies outside any process already given in nature, the most important feature of food.

6.16 Conclusions

If the criterion used to characterize a technology is not scientifically true, the resulting ranking would make unsustainable technologies more appealing than the sustainable ones. This chapter demonstrates that currently prevalent criteria are set to maximize profit margin while creating environmental disaster. In order to reverse the current trend of environmental insult, a paradigm shift is necessary, which can only occur by reversing the entire process, from root to the external. Instead of creating an economic feasibility filter first, followed with creating perception of superiority of the scheme that would maximize profit of the desired scheme, the sustainability question has to be asked first, followed by implementation of natural economics.

7

What is Truly Green Energy?

For the last 100 years, fossil fuels have become the major driving factors of modern economic development. However, we are always told that fossil fuel is bad for the environment and is inherently unsustainable. Added to that is the fact that the oil price has gone up sharply over the recent years. Because no one knows or cares find out what is equitable price of oil today, everyone seems convinced that oil price hikes are the reason we have an economic crisis. This perception of global energy crisis is sharpened by the widening gap between the discovery of the petroleum reserves and the rate of production. Environmentally, fossil fuel burning is the major cause of producing greenhouse gases, which are believed to be major precursors for the current global warming problem. This chapter analyzes the shortcomings of conventional energy development, deconstructs the conventional energy models, and proposes the energy model with technology options that are innovative, economically attractive, environmentally appealing, and socially responsible. It is shown that crude oil and natural gas are compatible with organic processes that are known to produce no harmful oxidation products.

7.1 Introduction

The modern economic development is largely dependent on the consumption of large amounts of fossil fuels. Due to this reason, fossil fuel resources are sharply depleting. With the current consumption rate, the oil use will reach to its highest level within this decade. For example, humans today collectively consume the equivalent of a steady 14.5 trillion watts of power and 80% of that comes from fossil fuel (Smalley, 2005). Moreover, oil prices have skyrocketed and have shown severe impacts on all economic sectors. Yet, oil is expected to remain the dominant energy resource in decades to come with its total share of world energy consumption (Figure 7.1). This analysis indicates that except for hydropower resources, the consumption of other resources will still continue to rise.

Worldwide oil consumption is expected to rise from 80 million barrels per day in 2003 to 98 million barrels per day in 2015 and then to 118 million barrels per day in 2030 (EIA, 2006a). Transportation and industries are the major sectors for oil demand in the future (Figure 7.2). The transportation sector accounts for about 60% of the total projected increase in oil demand in the next two decades, followed by the industrial sector. Similarly, natural gas demand is

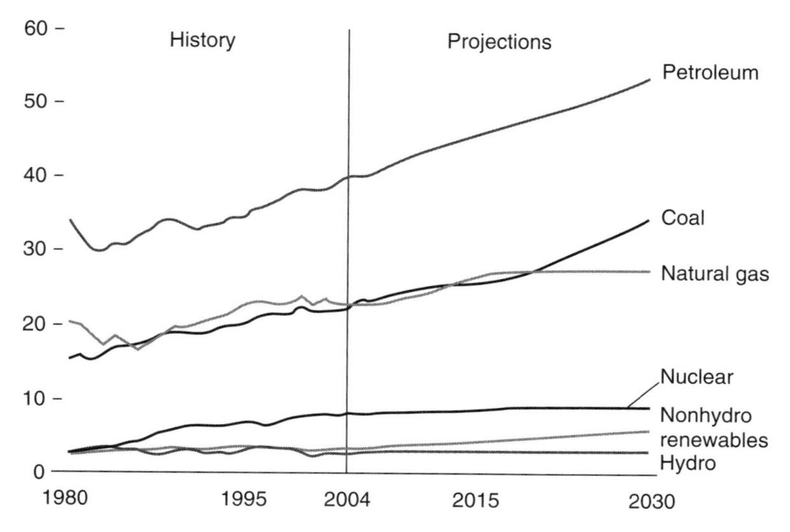

Figure 7.1 Global energy consumption by fuel type (Quadrillion Btu) (EIA, 2006a).

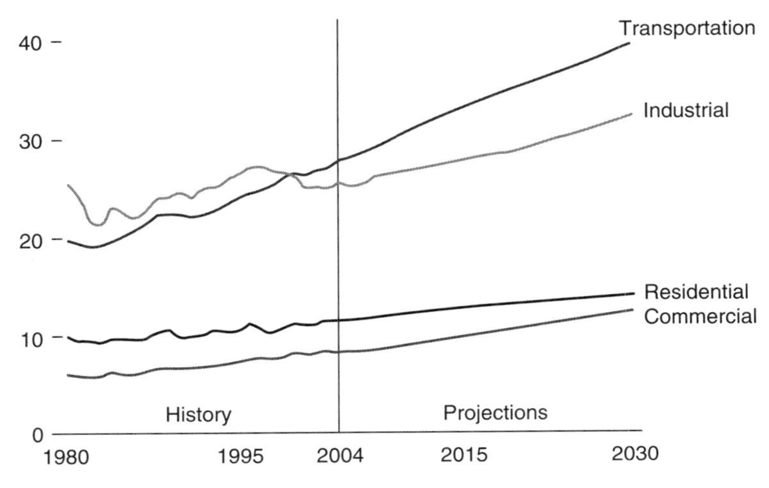

Figure 7.2 Delivered Energy Consumption by Sector (quadrillion Btu) (EIA, 2006a).

expected to rise by an average of 2.4% per year over the 2003–2030 period, and coal use by an average of 2.5% per year. Total world natural gas consumption is projected to rise from 95 trillion cubic feet in 2003 to 134 trillion cubic feet in 2015 and 182 trillion cubic feet in 2030 (EIA, 2006a). The oil demand in residential and commercial sectors will also increase constantly. The residential oil consumption demand increases much lower than other sectoral oil demand, which means almost half of the world's population with access to modern form of energy will continue to depend on the traditional fuel resources.

Burning of fossil fuel has several environmental problems. Due to the increased use of fossil fuels, the world carbon dioxide emission is increasing severely and is expected to grow continuously in the future. Currently, the total CO_2 emission from all fossil fuel sources is about 30 billion tons per year (Figure 7.3). The total CO_2 emission from all fossil fuels is projected to be almost 44 billion tons by 2030, which exceeds 1990 levels by more than double (EIA, 2006b). At present, the CO_2 emission is at the highest level in 125,000 years (Service, 2005).

The current technology development mode is completely unsustainable (Khan and Islam, 2006a). Due to the use of unsustainable technologies, the energy production and consumption usually have an environmental downside, which may in turn threaten human

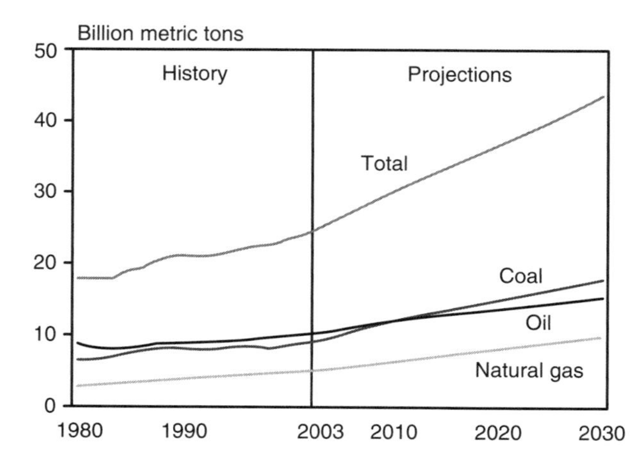

Figure 7.3 World CO_2 Emissions by oil, coal and natural gas, 1970–2025 (adopted from EIA, 2005).

health and quality of life. Impacts on atmospheric composition, deforestation leading to soil erosion and siltation of water bodies, the disposal of nuclear fuel waste, and occasional catastrophic accidents such as Chernobyl and Bhopal are some of the widely recognized problems.

The price of petroleum products is constantly increasing due to two reasons. First, the global oil consumption is increasing, due to the increased industrial demand, higher number of vehicles, increase in urbanization, and higher population. The industrial and transportation sector demand is rapidly increasing, and is expected to increase in the future. The current oil consumption rate is much higher than the discovery of new oil reserves. Secondly, the fossil fuel use is subjected to meet strict environmental regulations such as low sulfur fuel, unleaded gasoline, and so forth. This increases the price of the fuel. Figure 7.4 shows the rising trend of regular gasoline prices in the United States from early 2006 to date. The U.S. energy demand is significantly increasing. To meet the increasing demand, the U.S. imports will also increase. Figure 7.5 indicates the net import of energy on a Btu basis projected to meet a growing share of total U.S. energy demand (EIA, 2006a). It is projected that the net imports are expected to constitute 32 percent and 33 percent of total U.S. energy consumption in 2025 and 2030, respectively, which is up from 29 percent in 2004.

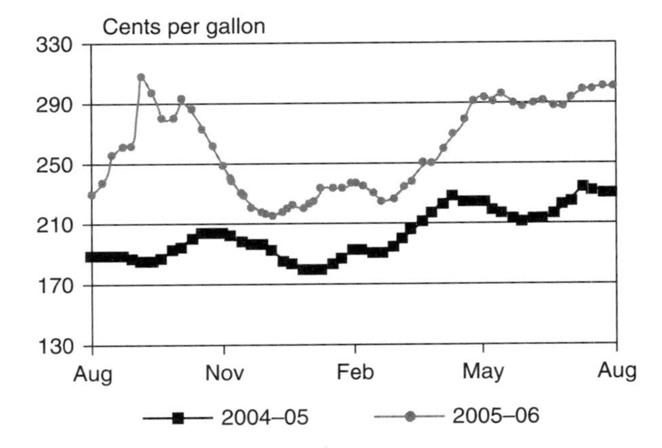

Figure 7.4 Regular gasoline price (EIA, 2006c).

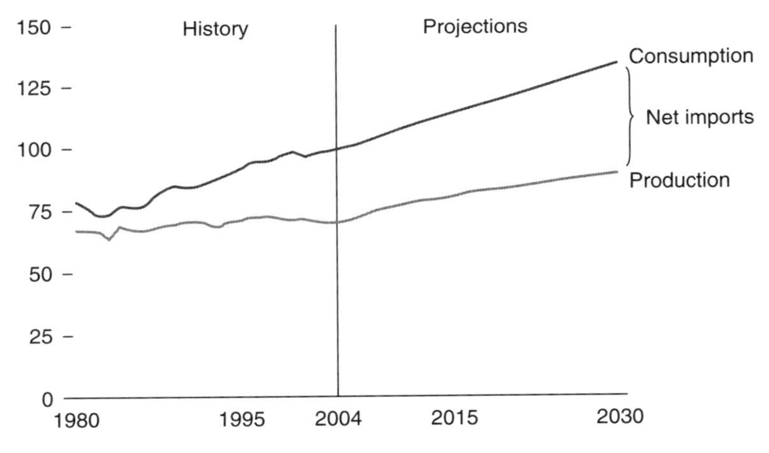

Figure 7.5 Total energy production, consumption and imports for US from 1980–2030 (quadrillion Btu) (EIA, 2006a).

The modern economic development is highly dependent on the energy resources and their effective utilization. There is a great disparity between the rate at which fossil fuel is being used up and the rate at which the new reserves are found. Moreover, the new and renewable energy resources are not being developed at the pace necessary to replace fossil fuel. Most of the energy resources, which are claimed to replace fossil fuel, are again based on fossil fuel for their primary energy resources. The alternatives being developed are based on chemical technologies which are hazardous to the environment. The technology development follows the degradation of chemical technologies as honey → sugar → saccharine → aspartame syndrome (Islam *et al*, 2006). Artificial light which is alternative to the natural light has several impacts to human health. Schernhammer (2006) reported a modestly elevated risk of breast cancer after longer period of rotating night work. Melatonin-depleted blood from premenopausal women exposed to light at night stimulates growth of human breast cancer xenografts in rats (Blask, *et al.*, 2005). The gasoline engine, which replaced the steam engine, became worse than its earlier counterpart. Modern gasoline and diesel engine use fuel which is refined by using highly toxic chemicals and catalysts. Biodiesel, which is touted to replace petroleum diesel, uses similar toxic chemicals and catalysts, such as methanol and sodium hydroxide, producing similar exhausts gas as that of petroleum diesel. This principle applies in every technological development. The major problem facing the energy development is that the conventional policies are meant to maintain the status quo, and all of the technological developments taking place are anti-nature. This chapter provides a comprehensive analysis of global energy problems and possible solutions to meet the global energy problems in sustainable way.

7.2 Global Energy Scenario

The global energy consumption share from different sources is shown in Table 7.1. The analysis carried out by EIA (2006a) showed that oil remains the dominant energy source, followed by coal and natural gas. It is projected that the nuclear energy production will also increase by more than two times by the year 2030. Renewable energy sources, such as biomass, solar, and hydro, will not be increased significantly compared to the total energy consumption. Renewable energy sources supply 17% of the world's primary energy. They include traditional biomass, large and small hydropower, wind, solar geothermal, and biofuels (Martinot, 2005).

The total global energy consumption today is approximately 14.5 terawatts, equivalent to 220 million barrels of oil per day (Smalley, 2005). The global population rise will settle somewhere near 10 billion by 2050, based on the current average population increase

Table 7.1 World total energy consumption by source from 1990–2030 (Quadrillion Btu).

Source	**History**			**Projections**					**Ave. Annual % Change**
	1990	**2002**	**2003**	**2010**	**2015**	**2020**	**2025**	**2030**	
Oil	136.1	158.7	162.1	191.0	210.5	229.8	254.9	277.5	2.0
Natural Gas	75.20	95.90	99.10	126.6	149.1	170.1	189.8	218.5	3.0
Coal	89.40	96.80	100.4	132.2	152.9	176.5	202.8	231.5	3.1
Nuclear	20.40	26.70	26.50	28.90	31.10	32.90	34.00	34.70	1.0
Others	28.30	32.20	32.70	45.80	50.60	58.80	64.10	73.20	3.0
Total	47.30	410.3	420.7	524.2	594.2	666.0	745.6	835.4	2.6

Source: EIA, 2006b

(WEC, 2005). The per-capita energy consumption is still rising in the developing countries as well as in developed countries. However, almost half of the world's population in the developing countries still relies on traditional biomass sources to meet their energy needs. Smalley (2005) argued that to meet the energy requirements for almost 10 billion people on the Earth by 2050, approximately 60 terawatts of energy is required, which is equivalent to some 900 million barrels of oil per day. According to him, major reservoirs of oil will have been used up by that time. Thus, there should be some alternatives to fulfill such huge energy requirements to maintain the growing economic development.

The point that Smalley does not make is that the need for an alternate source does not arise from the apparent depletion of petroleum basins. In fact, some 60% of a petroleum reservoir is left unrecovered even when a reservoir is called 'depleted' (Islam, 2000). It is expected that with more appropriate technologies, the current recoverable will double. Similarly, it is well-known that the world has more heavy oil and tar sands than 'recoverable' light oil (Figure 7.6). It is expected that technologies will emerge so that heavy oil, tar sands, and even coal are extracted and processed economically without compromising the environment. This will need drastic changes in practically all aspects of oil and gas operations (Khan and Islam, 2006b). However, the immediate outcome would be that all negative impacts of petroleum operations and usage will be eradicated, erasing the boundary between renewable and non-renewable energy sources.

Figure 7.6 indicates the similar trend for natural gas reserves. The conventional technology is able to recover lighter gases, which are at relatively lower depths. Recently, more focus is being placed to develop technologies in order to recover the coalbed methane. There are still large reserves of tight gas, Devonian shale gas, and

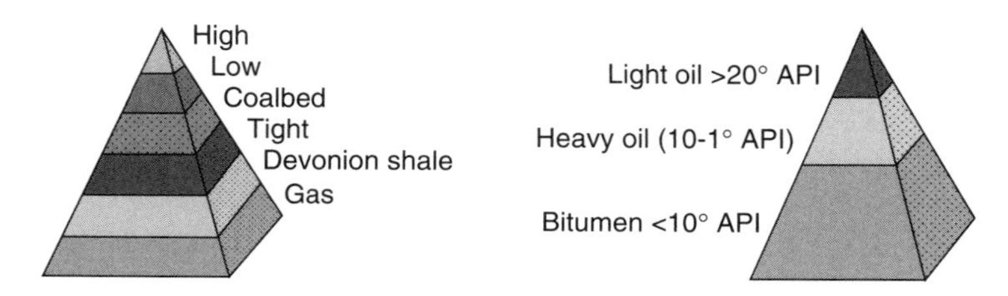

Figure 7.6 Worldwide Oil and Natural Gas Resource Base (after Stosur, 2000).

gas hydrates. It is expected that new technologies will emerge to economically recover the deeper gas and hydrates, so that natural gas can contribute significantly to the global fuel scenario. The problem with the natural gas industry currently is that it uses highly toxic glycol for dehydration, and amines for the removal of carbon dioxide and hydrogen sulfide (Chhetri and Islam, 2006a). Oxidation of glycol produces carbon monoxide, which is poisonous, and amines form carcinogens with other oxidation products. Hence, the use of processed natural gas is hazardous to human health and the environment. Chhetri and Islam (2006a) proposed the use of natural clay material for the dehydration of natural gas and use of natural oil to remove the carbon dioxide and hydrogen sulfides from the natural gas streams.

Smalley (2005) also missed the most important reason why the correctly used *modus operandi* in energy management is unsustainable. With current practices, thousands of toxic chemicals are produced at every stage of the operation. Many of these toxic products are manufactured deliberately in name of value addition (Globe and Mail, 2006). These products would have no room to be circulated in a civic society if it was not for the lack of long-term economic considerations (Zatzman and Islam, 2006). These products are routinely touted as cheap alternatives to natural products. This has two immediate problems associated to it. First, natural products used to be the most abundant and, hence, the cheapest. The fact that they became more expensive and often rare has nothing to do with free market economy and natural justice. In fact, this is the testimony of the type of manipulation and market distortion that have become synonymous with Enron, WorldCom and others, which failed due to corrupt management policies. The second problem with touting toxic materials as cheap (hence, affordable) is that the long-term costs are hidden. If long-term costs and liabilities were incorporated, none of the products would emerge as remotely cheap.

Most of the economic hydropower sources have already been exploited. There is huge potential to generate power from ocean thermal and tidal energy sources. These resources have great promise, but have yet to be commercialized. These sources are more feasible in isolated islands, as power transportation in such islands is difficult. Tidal energy has also a great potential, but depends on the sea tides, which are intermittent. Biomass is a truly renewable source of energy, however, sustainable harvesting and replenishment is a major challenge in utilizing this resource. This is the major energy source for almost half of the world's population residing in the developing countries. It is also argued that increasing the use of biomass energy would limit the arable land to grow food for increased population. The global share of biomass energy is less than 1% (Martinot, 2005). Wind is a clean energy source. Wind energy development is increasing rapidly. However, this is highly location-specific. This will be an effective supplement for other renewable energy resources to meet the global energy requirement for the long term.

Nuclear energy has been the most debated source of energy. It has been argued that the development of nuclear energy reduces greenhouse gas emission. The detailed analysis showed that nuclear energy creates unrecoverable environmental impacts because of nuclear radiations, which have very long half-lives, millions and billions of years. The nuclear disasters which the world has already witnessed can not be afforded anymore. The safe disposal of nuclear waste has yet to be worked out, and is proving to be an absurd concept. Bradley (2006) reported that the disposal of spent fuel has been in debate for a long time and has not been solved.

Hydrogen energy is considered as the major energy carrier for the 21st century, but the current mode of hydrogen production is not sustainable. Use of electricity to electrolyze to produce electricity becomes a vicious cycle which has very little or no benefit. The search for hydrogen production from biological methods is an innovative idea, which is yet to be established commercially. The most debated problem in hydrogen energy is the storage and transportation of energy. However, hydrogen production using solar heating has good promise. Geothermal energy involves high drilling costs, reducing its economic feasibility. The geothermal electricity generation is characterized by low efficiency. However, the application of direct heat for industrial processes or other uses would contribute significantly. The problems and prospect of each of the energy sources are discussed below.

7.3 Ranking with Global Efficiency

Energy sources are generally classified based on the various sources from which they are developed, such as hydropower, solar, biomass, or fossil fuels. Moreover, the conversion of energy into usable forms, such as electricity, light or heat, is an important activity of the energy industry. During this energy conversion from one form to another, efficiency comes into play that determines how efficient an energy system is, based on output received from certain energy input. Energy efficiency is the ratio of output (energy released by any process) and input (energy used as input to run the process). Hence, efficiency, η = output/input. The efficiency of a particular unit measured based on the ratio of output to input is also considered as the local efficiency. However, the conventional local efficiency does not include the efficiency of the use of the by-products of the system and the environmental impacts caused during processing, or after the disposal of the systems. It is likely that conventional efficiency calculation does not represent the real efficiency of the systems. The economic evaluation of any energy projects carried out with the local efficiency alone may be either undervalued or overvalued, due to the fact that several impacts to systems are ignored. Energy efficiency is considered to be a cost-effective strategy to improve the global economies without increasing the consumption of energy.

This chapter introduced the term 'global efficiency,' which is calculated not only with the ratio of output and input, but also by taking into account the system's impact to the environment, with the potential reuse of the by-products of the whole life cycles of the system in consideration. The concept of calculating the global efficiency of various types of energy systems has been discussed. Global efficiency not only considers the energy sources, but also the pathway of the complete conversion process from sources to the end use. Provided the global efficiency is considered, conventional economic evaluation of energy systems would appear differently than what is observed today.

7.4 Global Efficiency of Solar Energy to Electricity Conversion

7.4.1 PV Cells

Solar energy is the most abundant energy source available on Earth. Service (2005) wrote that Earth receives 170,000 TW of energy every moment, one-third of which is reflected back to the atmosphere. The fact is that the Earth received more energy in an hour than is consumed by humans in a year. However, utilizing such a huge energy source is not easy. Even though the cost of power generation from photovoltaic is decreasing, it is still the most expensive power generation option, compared to wind, natural gas coal, nuclear and others. In terms of pollutant avoidance as compared to fuels burning, it is argued that PV can avoid significant amount of pollutant emission, such as CO_2, NO_x, SO_2 and particulates. However, in terms of other environmental aspects, wide-scale deployment of solar photovoltaic technologies has several potential long-term environmental implications (Tsoutsos *et al.*, 2005). Bezdek (1991) argued that, given the current technologies on a standardized energy unit basis, solar energy systems may initially cause more greenhouse gas emissions and environmental degradation than do the conventional nuclear and fossil energy systems. They further argued that it is important to recognize the substantial costs, hazardous wastes, and land-use issues associated with solar technologies. Solar cells are manufactured using silica. Even though silica is considered nontoxic in its elemental form, formation of silicon dioxide in the environment can not be avoided. Silicon dioxide is considered to be a potent respiratory hazard. Figure 7.7 shows a schematic of the components of a complete solar photovoltaic system.

EDD guidelines (2002) reported about the environmental and social impacts of PV cells, from manufacturing to decommissioning. It was reported that the manufacturing of solar cells uses toxic and hazardous materials, such as phosphine (phosphorous hydride:PH3), during the manufacturing of solar cells. The PH3 used during the manufacturing of the amorphous silicon cell poses a severe fire hazard through spontaneous chemical

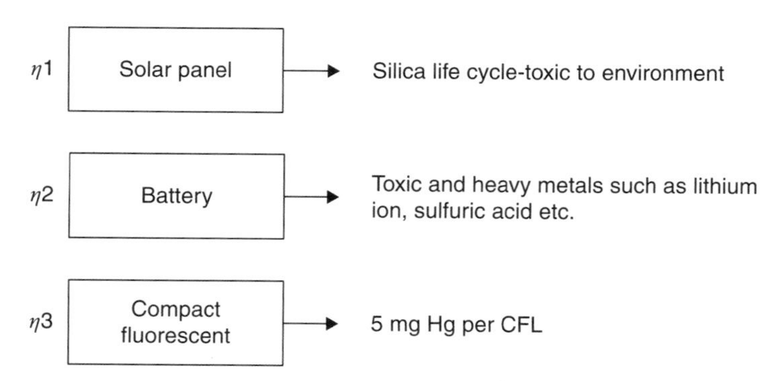

Figure 7.7 Evaluation of global efficiency of solar PV system.

reaction. This poses an occupational and public health hazard during manufacturing and operation. The decommissioning of PV cells releases atmospheric emission of toxic substance, leading to land and ground water contamination. (EDD guidelines, 2002). Table 7.2 is the summary of some of the chemicals used in the manufacturing of solar cells. The use of hydrofluoric acid (HF), nitric acid (HNO3) and alkalis (e.g., NaOH) for wafer cleaning, removing dopant oxides, and reactor cleaning poses occupational health issues related to chemical burns and inhalation of fumes. These chemicals are also released into the atmosphere during the process. The process also generates toxic P2O5 and Cl2 gaseous effluents, which are hazardous to health (Fthenakis, 2003). Hence, the life cycle of the solar cell shows that it has several environmental issues which need to be addressed to avoid the long-term environmental impacts.

Battery Life Cycle in PV System

The battery consists of various heavy metals, such as lead, cadmium, mercury, nickel, cobalt, chromium, vanadium, lithium, manganese and zinc, as well as acidic or alkaline electrolytes (Morrow, 2001). The exposure of such metals and electrolytes may have adverse impacts to humans and the natural environment. Even though the recycling and reuse of batteries has been practiced, they can not be recharged forever, and these metals and electrolytes leak to the environment during the life cycle of the operation of recycling. Moreover, the other factors, such as voltage, ampere-hour rating, cycle life, charging efficiency and self-discharge characteristics are also important in evaluating the total amounts of hazardous waste generated per unit of battery use. The use of corrosive electrolytes and toxic heavy metals needs to be addressed before disseminating the large number of batteries through PV systems.

Rydh and Sande (2005) reported that the overall battery efficiency, including direct energy losses during operation and the energy requirements for production and transport of the charger, the battery, and the inverter is 0.41–0.80. For some batteries, the overall battery efficiency is even lower than the direct efficiency of the charger, the battery, and the inverter (0.50–0.85). Nickel Metal hydrate (NiMH) battery has usually lower efficiency (0.41) compared to Li-ion battery (0.8 maximum). However, if we consider the global efficiency of the battery, the impact of heavy metals

Table 7.2 Some hazardous materials used in current PV manufacturing (Fthenakis, 2003).

Material	Source	TLV-TWA (ppm)	STEL (ppm)	IDLH (ppm)	ERPG2 (ppm)	Critical Effects
Arsine	GaAs CVD	0.05	–	3	0.5	Blood, kidney
Arsenic compounds	GaAs	0.01 mg/m^3	–	–	–	Cancer, lung
Cadmium compounds	CdTe and CdS deposition CdCl2 treatment	0.01 mg/m^3 (dust) 0.002 mg/m^3 (fumes)			NA	Cancer, kidney
Carbon tetrachloride	Etchant	5	10		100	Liver, cancer, greenhouse gas
Chloro-silanes	a-Si and x-Si deposition	5	–	800	–	Irritant
Diborane	a-Si dopant	0.1	–	40	1	pulmonary
Hydrogen sulfide	CIS sputtering	10	15	100	30	Irritant, flammable
Lead	Soldering	0.05 mg/m^3	–	–	–	blood, kidney, reproductive
Nitric acid	Wafer cleaning	2	4	25	–	Irritant, Corrosive
Phosphine	a-Si dopant	0.3	1	50	0.5	Irritant, flammable
Phosphorous oxychloride	x-Si dopant	0.1	–	–	–	Irritant, kidney
Tellurium compounds	CIS deposition	0.1 mg/m^3				cyanosis, liver

Note: TLV-TWA: The Threshold Limit Value, Time Weighted Average, STEL: Short Term Exposure Level, IDLH: The Immediately Dangerous to Life or Health, ERPG-2: The Emergency Response Planning Guideline-2.

and corrosive electrolytes on the environment needs to be considered, which significantly lowers the global efficiency of the battery system itself.

Compact Fluorescent Lamp

Compact fluorescent lamps (CFLs) are considered to be the most popular lamps recently. Each CFL is reported to prevent the emission of 500–1000 kg of carbon dioxide and 4–8 kg of sulfur dioxide every year in the United States (Polsby, 1994). It is considered that a CFL consumes 4–5 times less energy and could last up to 13 times longer than the standard incandescent lamp producing the same lumen (Kumar *et al.*, 2003). However, CFLs have some other environmental concerns. Energy Star (2008) reported that CFLs contain an average of 5 milligrams of mercury, as it is one of the essential components of CFL. The CFLs also contain phosphor, which is copper-activated zinc sulfide or the silver-activated zinc sulfide, the exposure of which has environmental impacts in the long-term. Hence, despite being energy-efficient, CFL still needs improvement to avoid the environmental impacts.

The global efficiency of the solar cell system including its pathways is shown in Figure 7.8. The average efficiency of such solar panels is 15%. The solar panels are made of silicon cells, which are very inefficient in hot climate areas, and are inherently toxic, as they consist of heavy metals, such as silicon, chromium, lead, and others. The energy is stored in batteries for nighttime operations. These batteries are exhaustible and have a short life even if they are rechargeable.

The batteries have a maximum efficiency of some 41–80% (Rydh and Sande, 2005). The compact fluorescent lamp generating

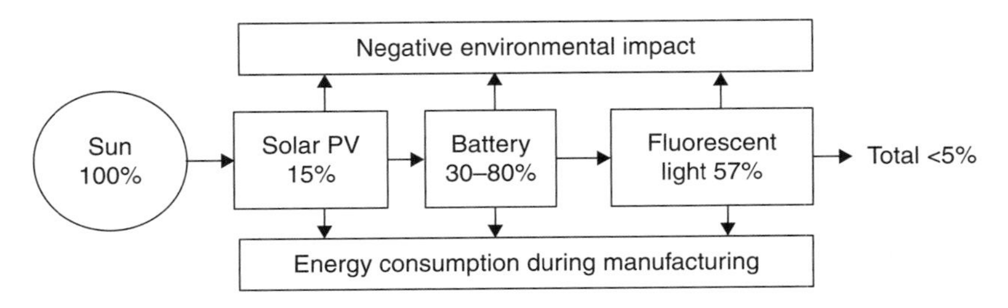

Figure 7.8 Flow chart showing pathway of generating artificial light from natural sun light.

800 lumens of light uses 14 watts of electricity and has lumens-per-watt efficiency of over 57% (Adler and Judy, 2006). Thus considering the local efficiency of all the components in the system in Figure 7.15, the global efficiency of the overall PV system is less than 5%. Natural and 100% efficient light is converted to less than 5% efficient artificial light.

$$\text{Global efficiency of PV system } (\eta_G) = \eta_1 \times \eta_2 \times \eta_3 \times \eta_4$$

Solar panels are widely used for lighting in isolated areas where other sources of energy are not available. Moreover, the embodied energy of the solar cells is very high and emits huge amount of CO_2, due to fossil fuel use during manufacturing of the solar cells. Toxic materials in the batteries and CFLs are one of the most environmentally polluting components (Islam *et al.*, 2006). The severity is particularly intense when they are allowed to oxidize. Note that oxidation takes place at any temperature.

7.4.2 Global Efficiency of Direct Solar Application

In the case of direct solar heating using fluid heat transfer, there are no such toxic chemicals used in the systems. However, there is a significant efficiency loss from source to the end use (Figure 7.9). There is heat loss in reflecting surfaces, such as the parabolic reflector during the heat transfer to the fluid heating from the heat exchanger, from fluid to steam conversion in the Einstein cycle, and in steam turbine to generators and transmission. These losses all decrease the global efficiency of the systems. Hence the global efficiency (η_G) of the direct solar application is:

$$\eta_G = \eta_1 \times \eta_2 \times \eta_3 \times \eta_4 \times \eta_5 \times \eta_6 \times \eta_7$$

The total solar radiance known as global solar irradiance on the earth's surface is made up of direct and diffuse components. However, for the solar collector, the global solar irradiation ($I\beta_G$) on a slope with an inclination angle β can be as follows (Li and Lam, 2004):

$$\beta_G = I\beta_B + I\beta_D + I\beta_R$$

Where $I\beta_B$ is the direct solar radiation propagating along the line joining the receiving surface and the sun, $I\beta_D$ is the scattered solar

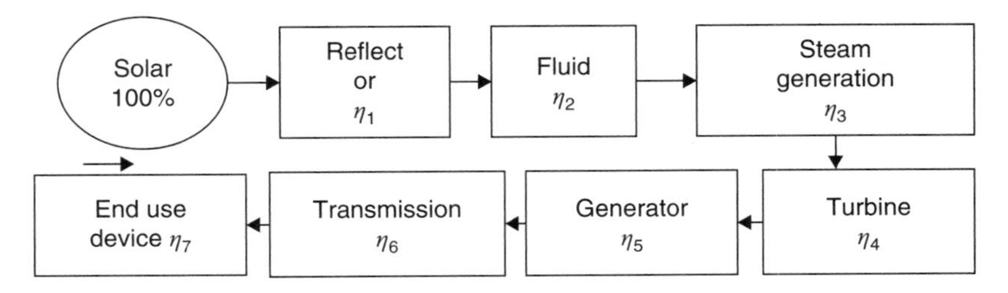

Figure 7.9 Global efficiency of solar to electricity conversion system.

radiation, and $I\beta_R$ is the ground reflected irradiance plane with surface inclination β.

After incidence of the radiation energy to the surface, some heat is lost, which can be summarized as follow:

Surface Assembly (Tiwari, 2002)

1. Variation in Shape
2. Ambient temperature
3. Heat diffusivity and conductivity of absorber
4. Optical consideration of the reflective surface

Receiver Assembly

5. Placement of receiver
6. Heat loss in the receiver
7. Behavior of receiver
8. Heat transfer fluid in the receiver

For a parabolic surface, ambient temperature is considered as one of the important parameters for heat loss from the surface. The lower the ambient temperature, the higher the temperature difference (between the temperature of the absorbed material and the ambient temperature), and the higher the convection heat loss. When the solar beam incidents on the parabolic surface, a portion of the heat from the beam is diffused through the surface materials. The parabolic trough collector (PTC) is compiled by a number of materials, and each material has a different heat conductivity, which is related to conduction heat loss. Some heat loss takes place after reflection, due to receiver assembly. If the receiver is not placed properly, it cannot receive all the reflection in the line of focus. The material of the receiver itself absorbs some heat, depending on the material of the receiver. From the receiver, some heat is transmitted to the ambient air by convection. Selection of heat transfer fluid is also important because thermal capacity of heat transfer fluid dictates the performance of a thermal fluid to energy transfer.

Figure 7.10 shows an energy balance over a parabolic trough collector.

If x is the total reflection on the surface and y is the heat transfer to the fluid in the receiver, then

$$\text{Total loss of radiation, } L = (x - y)$$

For a parabolic surface with the same receiver assembly and parabolic surface assembly (except the incident surface) and the same ambient temperature, the loss is a function of the incident surface:

$$L = f\,(\text{incident surface})$$

Incident surface is the most important for a parabolic surface which actually dictates the reflection of the solar beam from the surface. The reflectivity of a surface depends on the color and the gloss of the surface. It is already known that the white surface is the most reflective, where the black surface has the least reflectivity. So, the closer the surface is to white, the more reflective it is. On the other hand, surface smoothness depends on how polished it is. The polished surface is glossy, and reduces the fracture or coarseness of the surface, and thus reduces the diffused reflection on those coarse

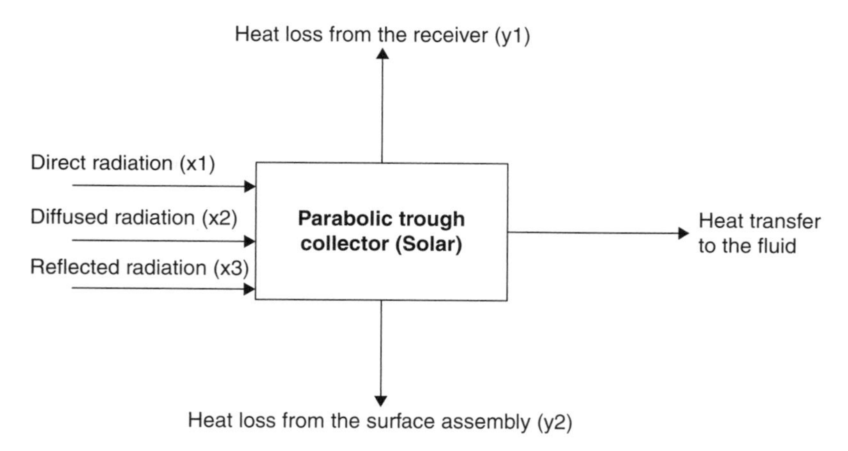

Figure 7.10 Global efficiency of a solar trough.

surfaces. Polished surface exhibits specular reflection, which has the same incidence and reflection angles. The reflective surface can be a thin layer of any material on top of the collector surface assembly. Due to the thin nature of the reflective surface, the diffusivity and conductivity of heat of the surface is not that important. That is why it can be speculated that the white polished surface can be used as the reflective surface for a parabolic trough collector.

The use of natural materials as reflecting material can enhance the global efficiency by minimizing the energy input. For example, a natural mineral surface, such as limestone surfacing, can be used. Erdogan (2000) reported that rocks made up of a single mineral, such as marble or limestone, show increasing value of surface reflectance with decreasing grain size. Moreover, small crystal size and dense crystallization have brightness-enhancing effects on the polished rock surfaces. The reflection coefficient K depends on the mineral composition of rocks, grain size, color, and so forth. It is reported that for rich massive ore with an admixture of pyrite, K = 16.0–18.5%, and for rich massive ore with an admixture of chalcopyrke, K = 11.5–13.5%, coarse-grained pyrite has value of K = 10.5–19.0%, but for fine grained pyrite, the value of K = 29.0–56.5% (Shekhovtsov and Shekhovtsov, 1970). The same study reported that the dark-gray limestone has K = 15.5–16.5%, but white limestone has K = 72.0–78.0%.

From the viewpoint of energy efficiency, it is found that solar energy is very suitable and efficient to be used directly to the water heating system. However, it can also be applied to a number of direct heating applications. Figure 7.11 shows the global efficiency of direct solar heating application.

$$\eta_G = \eta_1 \times \eta_2 \times \eta_3$$

Global Efficiency of a Steam Power Plant to Cooling System

The global efficiency from primary heating to the cooling input includes the efficiencies of several units (Figure 7.12). This global efficiency calculated by Khan *et al* (2007) is as following:

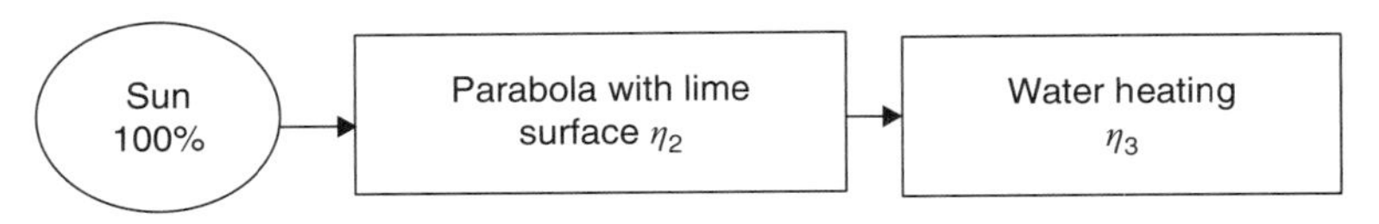

Figure 7.11 Global efficiency of the direct solar to electricity conversion system.

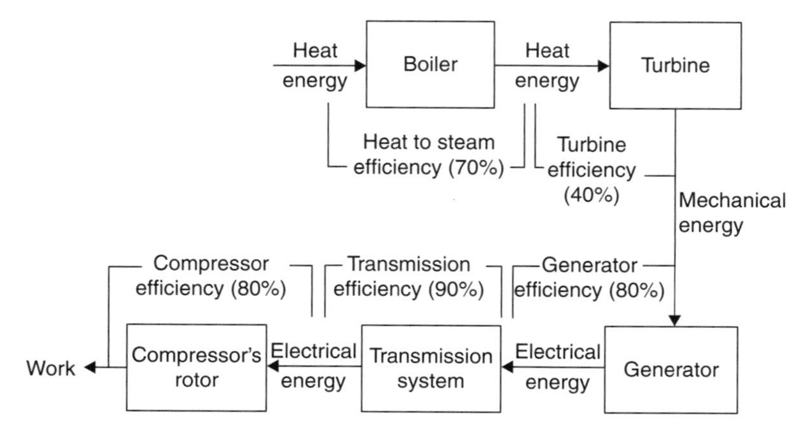

Figure 7.12 Global efficiency of a steam power plant to cooling system (Khan *et al.*, 2007a).

Global heat transfer efficiency (η_{Global}) = Heat-to-Steam efficiency (70%) × Turbine efficiency or thermal efficiency (ηt) × Generator efficiency (80%) × Transmission efficiency (90%) × Compressor's rotor efficiency (80%). Hence,

$$\textit{Global heat transfer efficiency } (\eta_{Global}) = 40\% \times (\eta \mathrm{t})$$

However, this global efficiency helps to find out the co-efficient of performance (COP) of a vapor compression cooling/refrigeration system. So far, COP for a vapor cooling/refrigeration system is calculated by the ratio of heat removed to the net work (output of a compressor), disregarding the efficiency of the units to bring the energy from primary heat to the compressor's input. That is why Khan *et al.* (2007) proposed to include the global efficiency so that the real COP is obtained as follows:

$$COP_v = \eta_{Global} \times \frac{\text{heat removed}}{\text{net work}}$$

Due to inclusion of global efficiency, the true COP of the vapor compression system is found to decrease a lot. However, the scenario is different for the absorption cooling/refrigeration system, because that system includes the primary heat from the source.

According to Khan *et al.* (2007a), for the same level of surrounding and cooling temperatures, the COP of absorption system is almost 2.5 times greater than that of the vapor system.

It is noted that the source of heat was not included in the calculation of global efficiency or COP. The heat source could be from renewable sources of non-renewable sources. The extraction efficiency from fossil fuel (non-renewable source) is actually the combustion efficiency of the fossil fuel, which can vary from 50% to 90%, depending upon fuel specification (Khan *et al.*, 2007a). On the other hand, the extraction of solar energy from a parabolic solar collector (PTC) is the combined efficiency of receiving energy to the heat transfer fluid in the receiver and the transmission line of that fluid, which can vary from 50% to 70% (Khan *et al.*, 2007a).

7.4.3 Combined-Cycle Technology

In combined heat and power (CHP) technology, also called co-generation, heat and power are sequentially generated from a single primary energy source. The two different forms of energy could be electrical energy and thermal energy, or mechanical energy and thermal energy. The sequence of generation could also be in any combination of different forms of energy. If an industry that needs both electrical energy as well as low pressure process steam, CHP could be ideally beneficial. It has an advantage of reducing the primary energy use, reducing the overall cost of the system. Even though CHP technology is considered as one of the most efficient energy technologies, there is significant loss in global efficiency of the system. Figure 7.13 is the schematic of combined heat and power generation technology. The efficiency for cycle 1 and cycle 2 are calculated separately and added to get the overall global efficiency of the CHP system.

Efficiency of Cycle 1$(\eta_{G1}) = \mu_1 \times \mu_2 \times \mu_6 \times \mu_7 \times \mu_8$ where μ_1 is the local efficiency of combustion chamber in which heat loss through flue gas, incomplete combustion of the fuels and the loss through boiler or chamber wall should be deducted.

$$\text{Efficiency of Cycle } 2(\eta_{G2}) = \mu_3 \times \mu_4 \times \mu_5 \times \mu_8$$

$$\text{Overall Global Efficiency } (\eta_G) = \eta_{G1} + \eta_{G2}$$

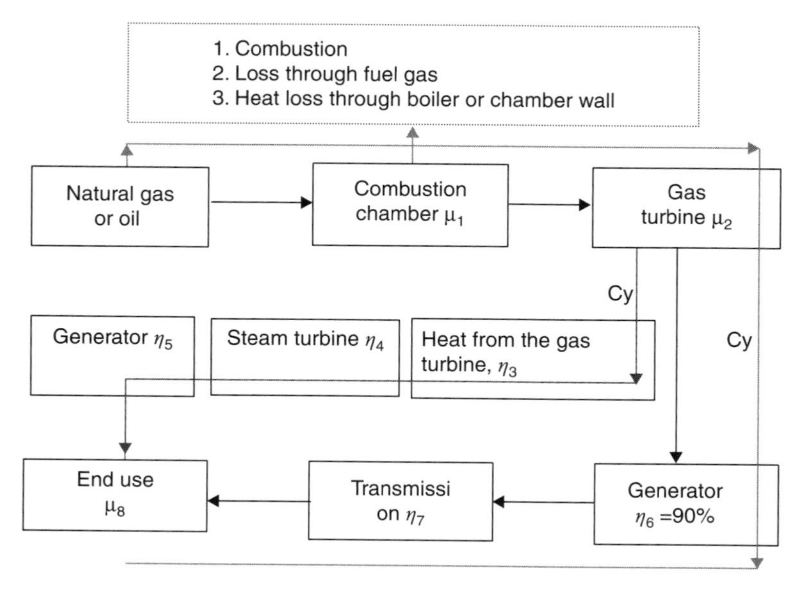

Figure 7.13 Global efficiency from natural gas burning to electricity.

7.4.4 Hydroelectricity to Electric Stove

Hydroelectricity is generated by utilizing the energy of falling water. Electricity from hydropower is a renewable form of energy, and is considered an environment-friendly energy source. Hydro electric power plants can either be 'run-of-the-river'-type or storage reservoir-type. In 'run-of-the-river'-type power plants, it is not necessary to build large dams to store water, but the water is simply diverted from the river into the channel carrying the water, and then to penstock pipe. However, in storage-type reservoirs, a high dam is constructed to store water that increases the water head, as well as supplies water at the peak load requirements.

Power development from water resources utilizes valuable natural resources. Hence, increasing the efficiency of power production will have significant contributions to the environment as well as the economy. Calculating the global efficiency, which is the summation of individual efficiency of each unit, is very important. The storage reservoir will lose its water through evaporation in the reservoir, as well as from the conveyance channel. There is a huge loss in penstock pipe due to friction, bend, and joint loss. The overall loss in this system could reach up to 15%, leaving the total efficiency

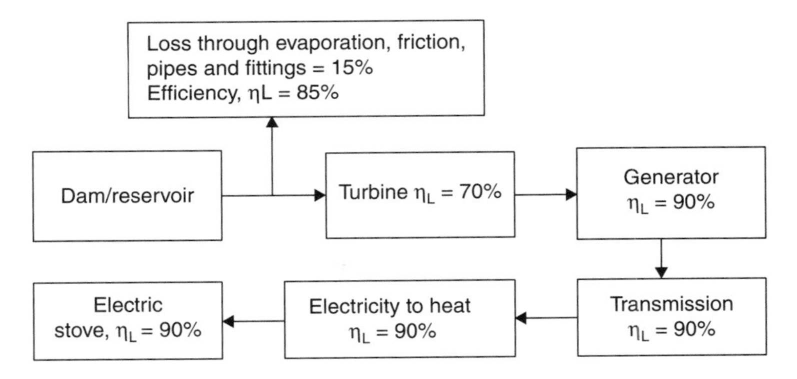

Figure 7.14 Global efficiency of hydroelectricity to electric stove.

before turbine as 85%. Different types of turbines have different efficiencies; however, pelton turbine efficiency could range from 70–90% (DOE, 2001). Natural Resource Canada (2004) published the average efficiencies of different impulse turbines (Pelton 80–90%, Turgo 80–95% and cross flow 65–85%) and reaction turbines (Francis 80–90%, pump as turbine 60–90%, Propeller 80–95% and Kaplan 80–90%). The same publication reported that the efficiency of synchronous generators varies from 75 to 90% depending on the size, and that of induction generators is approximately 75% at full load, where it reduces up to 65% part load. Hence, for calculation, average turbine efficiency is taken as 70% and generator efficiency is taken as 90%. Green (2004) reported that the average loss in electrical transmission lines is approximately 10%. The electric heating stove efficiency is approximately 90%. Hence, the global efficiency of hydro power to cooking stove is calculated as shown in Figure 7.14.

$$\text{Global Efficiency of Hydroelectricity to Electric Stove}$$
$$= 85\% \times 70\% \times 90\% \times 90\% \times 90\% \times 90\% = 39.06\%$$

7.4.5 Global Efficiency of Biomass Energy

Biomass energy is one of the most sustainable sources of energy. It originates from sunlight, and continues to be in one of the forms of biomass in a system. The key to the system's sustainability lies with its energy balance. Here is where natural sources of biomass and non-biomass must be distinguished from non-natural, non-characteristic industrially synthesized sources of non-biomass. In the same way that sunlight photosynthesizes plant material into living material, whereas fluorescent lighting would freeze that process, synthetic/naturally non-characteristic non-biomass, no matter how much solar energy is available anywhere in the system, can never convert natural non-biomass into biomass.

What is envisioned here is the atmosphere as a bioreactor, along the outlines of Figure 7.15, that does not, and will not, enable conversion of synthetic non-biomass into biomass. The key problem of mass balance in the atmosphere, as in the entire natural environment of the earth as a whole, is set out in Figure 7.16: the accumulation rate of synthetic non-biomass continually threatens to overwhelm the natural capacities of the environment to use or absorb such material. Hence, such analysis could form the basis of calculating the global efficiency of biomass combustion technologies.

Combustion of wood in traditional stoves has relatively low efficiency, in the ranges of 10–15% (Shastri *et al.*, 2002). Chhetri (1997) reported from the experimental investigation that some of the traditional stoves designed precisely reached the efficiency of up to 20%. Some improved cook stoves have efficiency up to 25% (Kaoma *et al.*,1994). However, the conventional efficiency calculation is

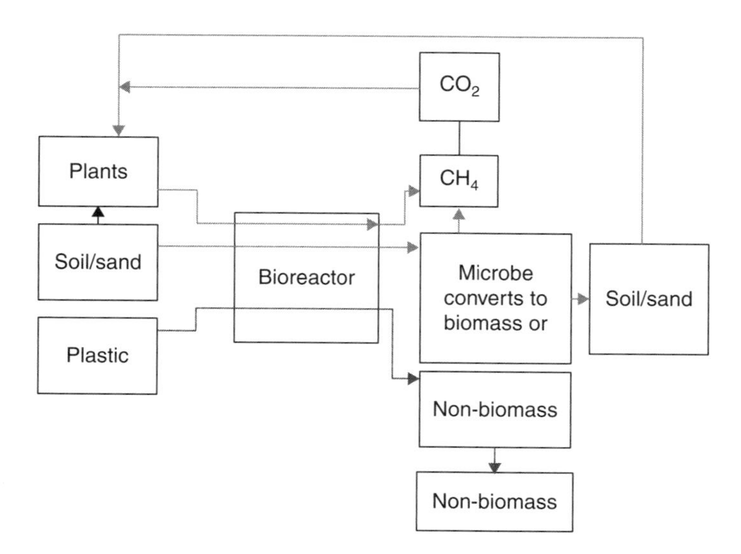

Figure 7.15 Sustainable pathway for material substance in the environment.

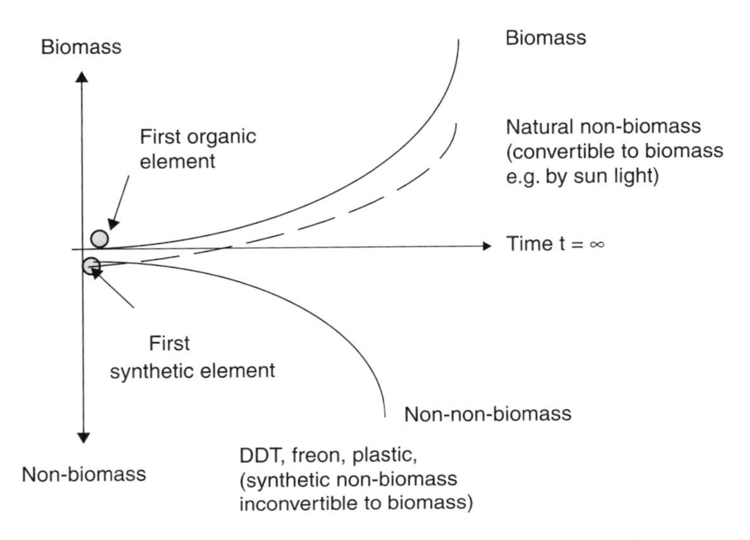

Figure 7.16 Synthetic non-biomass that cannot be converted into biomass will accumulate far faster than naturally sourced non-biomass, which can potentially converted to biomass.

based on the local efficiency, considering only the fuel input and heat output in the system itself. This method does not consider the utilization of byproducts, such as the fresh CO_2, which is essential for the plant photosynthesis, use of exhaust heat for household water heating using a heat exchanger, use of ash as surfactant for enhanced oil recovery or fertilizer, and good sources of natural minerals such silica, potassium, sodium, calcium and others.

Wood ash is a very rich source of silica, which is an important source for industrial applications. The ash also contains various minerals, such as potassium, sodium, magnesium, calcium, and others. Conventionally, ash has been in use as a source of fertilizer because of its high mineral content. Moreover, it is also used as a natural detergent. Wood ash has also been in use for centuries as saponification agent for soap production from vegetable oils and animal or fish fats. A fine wood ash is a very good raw material for making nontoxic toothpaste. Chhetri and Islam (2008) argued the possibility of extracting potassium or sodium to use as a natural catalyst for the transesterification of vegetable oil to produce biodiesel.

Rahman *et al.* (2004) reported that maple wood ash has the potential to adsorb both arsenic (III) and arsenic (V) from contaminated aqueous streams at low concentration levels, without any chemical treatment. Static tests showed up to 80% arsenic removal, and in various dynamic column tests, the arsenic concentration was reduced from 500 ppb to lower than 5ppb. Moreover, ash was traditionally used as a water disinfecting agent, possibly because of some mineral content in it.

Khan *et al.* (2007b) developed an energy-efficient stove, fueled by compacted sawdust, that utilizes exhaust heat coming from flue gas for household water heating (Figure 7.24). A simple heat exchanger can be used to transfer heat from flue gas to cold water. They also designed an oil-water trap to trap all particulate matters emitted from wood combustion. The particulates or the carbon soot collected from oil-water mixture are a very good nano-material which has very high industrial demand. The soot can also used as a nontoxic paint. When the particulates are trapped into the oil-water trap and the heat is extracted for water heating, the CO_2 emitted is a 'fresh' or 'new' CO_2 which is most favored by plants during photosynthesis (Islam, 2004; Khan *et al*, 2007b). Combustion of wood fuel thus does not contribute to the greenhouse effect.

Figure 7.17 is the pictorial representation of a zero-waste stove model. In the figure, four rectangular boxes represent the outputs from the stove, which are energy, CO_2, particulates, and heat from exhaust gas. To make this stove sustainable, these outputs should be completely utilized. Indeed, the output of this system will be the input for others. In the earlier discussion, it was shown that most wastes will be utilized. For example, heat trapped from exhaust will be used for home water heating, and particulates as nano-materials. It is also reported by Saastamoinen *et al.* (2005) that wood burning is considered to have zero net release of CO_2, since an equal amount of CO_2 is captured during the growing of firewood. As a result, considering the above factors, the proposed stove has achieved zero-waste objectives. Considering the ecological aspects and pollution hazards, the wood stoves are recommended to use primarily for residential heating (Olsson and Kjallstrand, 2004).

Figure 7.18 shows that burning wood in a stove produces by-products, such as CO_2, that are readily recycled by another natural process called photosynthesis. Other by-products, such as alkali from wood ash, can be utilized to make another product, such as soap. Some studies (Rahman, 2007; Chhetri *et al.*, 2008) reported

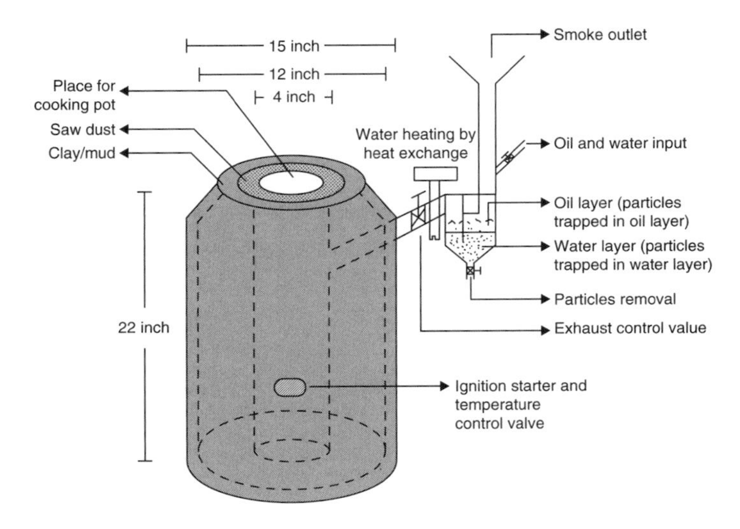

Figure 7.17 Sawdust packed stove (after Khan *et al.*, 2007b).

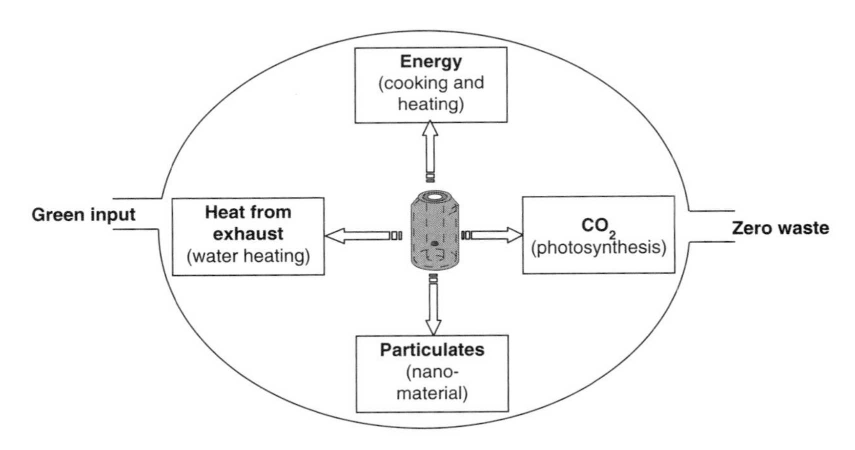

Figure 7.18 Pictorial view of zero-waste models for wood stove.

that alkali from wood ash can be used for the enhanced oil recovery application instead of sodium hydroxide. Similarly, the carbon soots collected can be used to make oil-based paints. Moreover, wood ash contains varieties of minerals which are essential for plants to grow, and hence, can be used as fertilizer. Wood ash has been in use for a long time to treat acidic soil.

Hence, considering the byproducts (Fig. 7.19), including wood ash for surfactant or catalysts for soap making, hot flue gas,

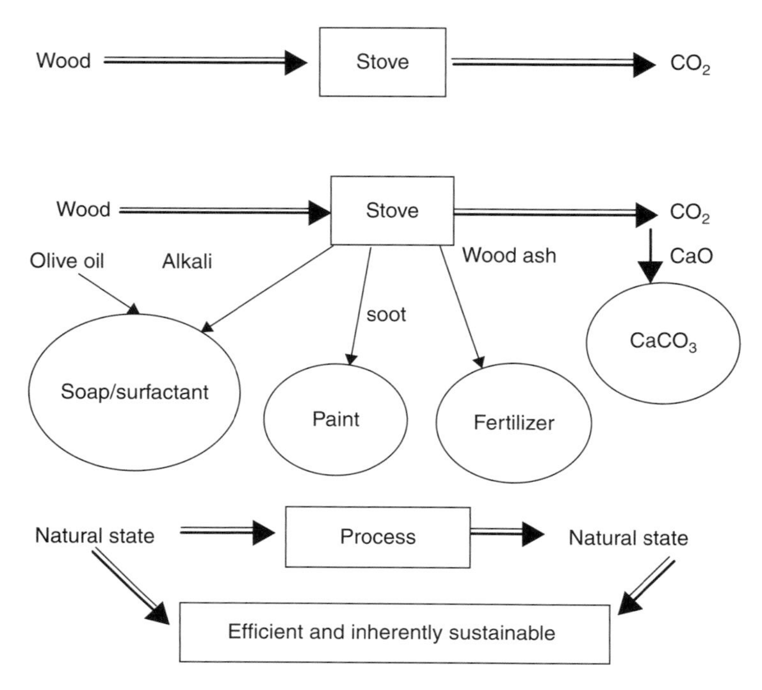

Figure 7.19 Wood burning and value addition of by-products.

nano-particles from the carbon soot, and the fresh CO_2 for plants for photosynthesis, the global efficiency of wood burning reaches to a highest level. The heat loss in the stove surface itself will contribute a small amount of heat that lowers some efficiency. Assuming the 5–10% radiation and conduction loss, the global efficiency of wood combustion in the stove is considered to be more than 90%. The most important point to make here is that there are no negative environmental impacts during this process. In terms of efficiency, every by-product can be used for another process, and follows the natural cycle. This process is entirely a natural process, and is inherently sustainable. Thus, wood combustion in effectively-designed stoves has one of the highest efficiency among the combustion technologies.

7.4.6 Global Efficiency of Nuclear Power

Nuclear power is considered as one of the most efficient technology for power generation. This is true if the criteria for evaluation are the local efficiency. The efficiency of thermal to net electric conversion from a nuclear power plants is considered to be over 50%.

However, if we consider the global efficiency, the scenario can be entirely different. Conversion of uranium ore from natural state to UF_6 and UO_2, enrichment, processing, and power generation involves emission of radioactive radiation that has very a long half-life. Hence, it takes infinite time for uranium to return to the natural state because of the violation its characteristic time. See Figure 7.20 for the breakdown of various components.

The spent fuel is the major concern in nuclear power generation. Even though it is argued that it is feasible to store the spent fuel in geological storage for thousands of years, it is highly unlikely that this can be a solution for the long term. Since the radiation continues for millions of years, the current design of storage systems for some thousand years will not solve the problem.

Mortimer (1989) reported that a nuclear power system releases 4–5 times more CO_2 from its life cycle operations than equivalent power productions from other renewable energy sources. This is because it involves huge amounts of energy from mining, fuel conversion, fabrication and enrichment. Hence, considering the life cycles emission of CO_2 and spent fuel management perspective, the global efficiency of nuclear energy is significantly less than what is advocated.

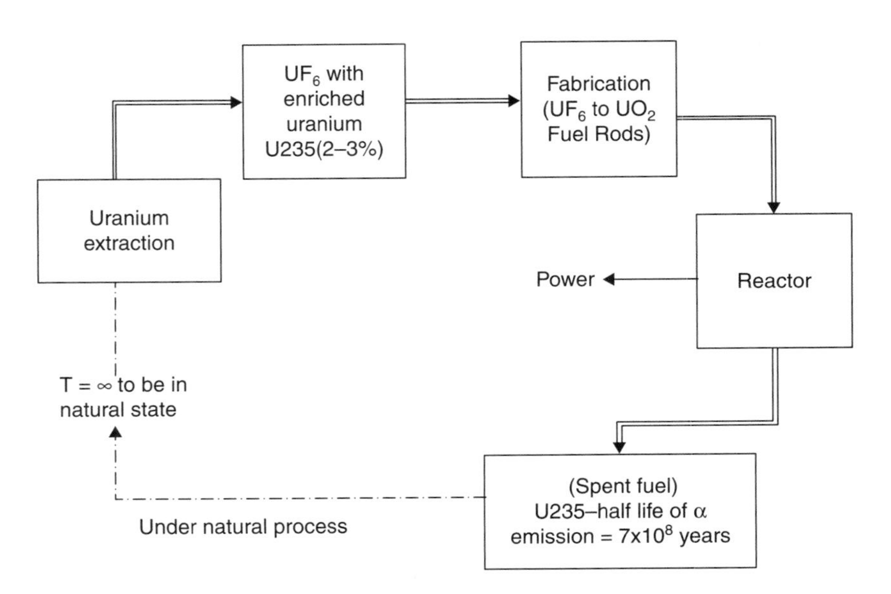

Figure 7.20 Schematic of power generation from nuclear power.

7.5 Carbon Dioxide and Global Warming

Energy production and use are considered as major causes for the greenhouse gas emission. Emission of greenhouse gases, particularly CO_2, is of great concern today. Even though CO_2 is considered as one of the major greenhouse gases, production of natural CO_2 is essential for maintaining lives on Earth. Note that all CO_2 are not same, and plants apparently do not accept all types of CO_2 for photosynthesis. There is a clear difference between the old CO_2 from fossil fuels and the new CO_2 produced from renewable biofuels (Dietze, 1997). The CO_2 generated from fossil fuel burning is an old and contaminated CO_2. As various toxic chemicals and catalysts are used for oil and natural gas refining, the danger of generating CO_2 with higher isotopes cannot be ignored (Islam, 2003; Chhetri *et al.*, 2006a). Hence, it is clear that CO_2 itself is not a culprit for global warming, but the industrial CO_2, which is contaminated with catalysts and chemicals, likely becomes heavier with higher isotopes, and plants cannot accept this CO_2. Plants always accept lighter portions of CO_2 from the atmosphere (Bice, 2001). Thus, CO_2 has to be distinguished between natural and industrial CO_2, based on the source from which it is emitted, and a pathway of the fuel that emits CO_2 follows from source to the combustion. Chhetri and Islam (2006c) showed that even though the total CO_2 is increasing in the atmosphere, the natural CO_2 is decreasing after the industrial revolution (Figure 7.21). They further argued that industrial CO_2, which is not acceptable to plants, are responsible for global warming (Chhetri and Islam, 2006c). Thus, generalizing CO_2 as a precursor for global warming is an absurd concept, and is not valid.

7.6 Renewable vs Non-Renewable: No Boundary as such

The current demand for fossil fuels such as oil, coal, and natural gas will still be significant in the next several decades. Figure 7.22 shows that, as the natural processing time increases, the energy content of the natural fuels increases from wood to natural gas. The average energy value of wood is 18 MJ/kg (Hall, and Overend, 1987) and the energy content of coal, oil, and natural gas are 39.3MJ/kg, 53.6MJ/kg and 51.6MJ.kg, respectively (Website 3). Moreover, this shows that the renewable and non-renewable energy sources

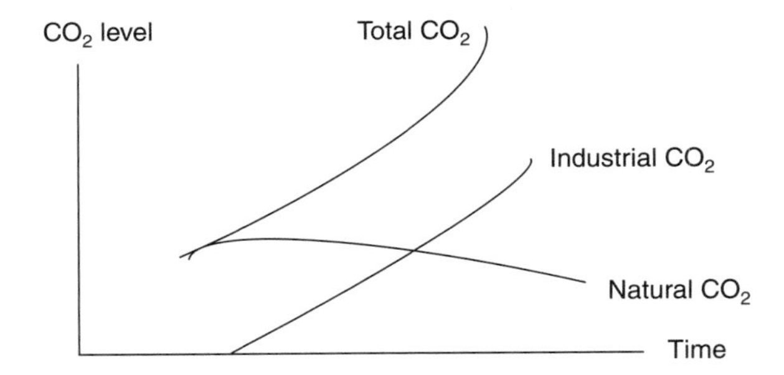

Figure 7.21 Total industrial and natural CO_2 trend (redrawn from Chhetri and Islam, 2006c).

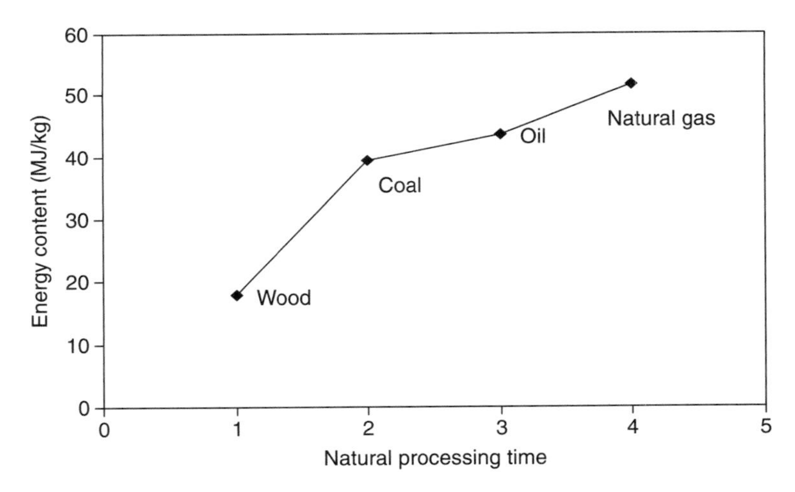

Figure 7.22 Energy content of different fuels (MJ/kg).

have no boundary. It is true that solar, geothermal, hydro, and wind sources are being renewed at every second, based on the global natural cycle. The fossil fuel sources are solar energy stored by trees in the form of carbon and, due to the temperature and pressure, they emerge as coal, oil, or natural gas after millions of years. Biomass is renewed from a few days to a few hundreds years (as a tree can live up to several hundred years). These processes continue forever. There is not a single point where fossil fuel has started or stopped its formation. So, why are these fuels are called non-renewable? The current technology development mode is based on an extremely short-term approach, as our solution of the problems start with the basic assumption of 'Δt tends to = 0'. Only technologies that fulfill the criteria of time approaching infinity are sustainable (Khan and Islam, 2006a). The only problem with fossil fuel technologies is that they are made more toxic after they are refined using high heat, toxic chemicals and catalysts.

From the above discussion, it is clear that fossil fuel can contribute significant amounts of energy by 2050. It is widely considered that fossil fuels will be used up soon. However, there are still huge reserves of fossil fuel. The current estimation on the total reserves is based on the exploration to date. As the number of drillings or exploration activities increases, more recoverable reserves can be found (Figure 7.23). In fact, Figure 7.24 is equally valid if the abscissa is replaced by 'time' and ordinate is replaced by 'exploratory drillings' (Figure 7.24). For every energy source, more exploration will lead to larger fuel reserves. This relationship makes the reserve of any fuel type truly infinity. This relationship alone can be used as a basis for developing technologies that exploit local energy sources.

The U.S. oil and natural gas reserves reported by EIA (2000) and EIA (2002) show that the reserves over the years have increased (Table 7.3). These additional reserves were estimated after the analysis of geological and engineering data. Hence, as the number of exploration increases, the reserves will also increase.

Figure 7.24 shows that the discovery of natural gas reserves increases as exploration activities or drillings are increased. Biogas is naturally formed in swamps, paddy fields, and other places due to the natural degradation of other organic materials. As illustrated in Figure 7.24, there are huge gas reservoirs, including deep gas, tight gas, Devonian shale gas, and gas hydrates, which are not yet exploited. The current exploration level is limited to shallow gas, which is a small fraction of the total natural gas reserve. Hence, by increasing the number of exploration activities, more and more

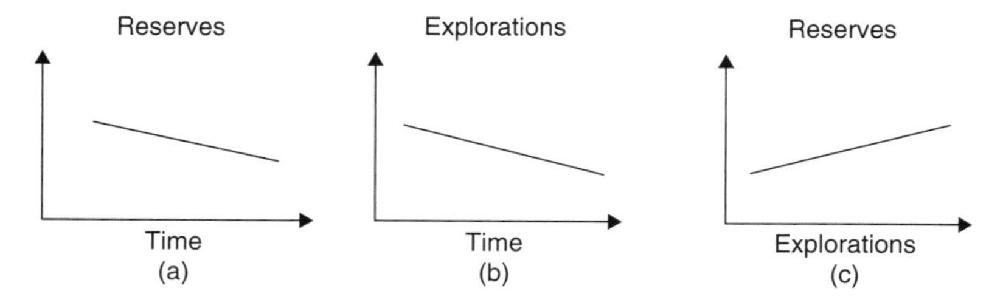

Figure 7.23 Fossil fuel reserves and exploration activities.

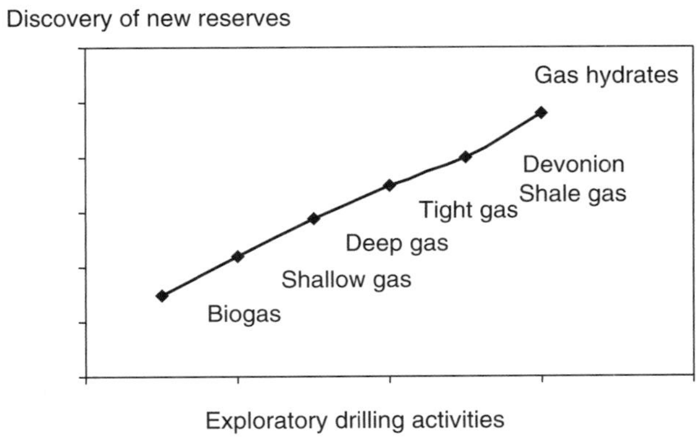

Figure 7.24 Discovery of natural gas reserves with exploration activities.

Table 7.3 U.S. crude oil and natural gas reserve (Million barrels).

	Year	Reserve	% Increment
Crude Oil Reserve	1998	21,034	
	1999	217,65	3.5%
	2000	22,045	1.3%
	2001	22,446	1.8%
Natural Gas	1998	164,041	
	1999	167,406	2.1%
	2000	177,427	6.0%
	2001	183,460	3.4%

reserves can be found, which indicates the availability of unlimited amount of fossil fuels. As the natural processes continue, formation of natural gas also continues forever. This is applicable to other fossil fuel resources, such as coal, light and heavy oil, bitumen, and tar sands.

Figure 7.25 shows the variation of the resource base with time, starting from biomass to natural gas. Biomass is available in huge quantities on earth. Due to natural activities, the biomass undergoes various changes. With heat and pressure on the interior of

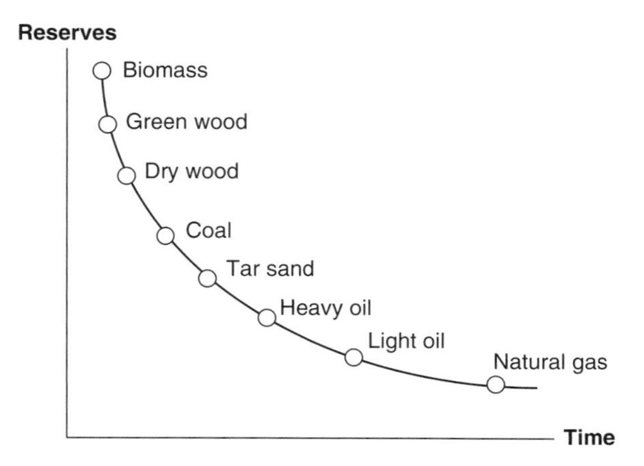

Figure 7.25 Continuity of Resource base.

the earth, formation of fossil fuels starts due to the degradation of organic matters from the microbial activities. The slope of the graph indicates that the volume of reserve decreases as it is further processed. Hence, there is more coal than oil, and more oil than natural gas, meaning unlimited resources. Moreover, the energy content per unit mass of the fuel increases as the natural processing time increases (Figure 7.22). The biomass resource is renewable and if the biological activities continue on the earth, the process of formation of fossil fuel also continues forever. From this discussion, the conventional boundary of renewable and non-renewable is dismantled, and it can be concluded that there is no boundary between the renewable and non-renewable, as all natural processes are renewable. The only problem with fossil fuel arises from the use of toxic chemicals and catalysts during oil refining and gas processing. Provided the fossil fuels are processed using natural and nontoxic catalysts and chemicals, or if we make use of crude oil or gas directly, fossil fuel will still remain as a good supplement in the global energy scenario in the days to come. These resources are totally recyclable.

7.7 Concluding Remarks

Ranking based on scientific characterization is different from the one produced by conventional short-term approaches. In this chapter, various types of energy sources, the status of their development to date, their possible impacts on the environment, and their global efficiencies have been discussed. It is argued that the world will not run out of oil, yet fossil fuel can not be taken as the single source for the global energy supply. It has further been argued that there is no boundary between renewable and non-renewable sources. It is concluded that the current global energy and environmental problems are not due to the use of fossil fuels, but they are due to the current mode of technology development, which is heavily dependent on the use of synthetic chemicals for their refining and processing. A holistic approach in energy development that leads to zero-waste living should be the basis for any technology development. As this cannot be achieved from the current technology development models, a paradigm shift toward knowledge-based technology development and energy management is essential to achieve the true sustainability in energy and environmental management. An integrated use of energy resources such as solar, hydro, biomass, wind, geothermal, and hydrogen could lead to the global energy solution. The concept of direct use of crude oil has been proposed in this research. This chapter has deconstructed current technology development models that are aphenomenal and non-scientific. It has been concluded that only knowledge-based technology and policy options that are technically feasible, environmentally attractive, and socially responsible are the keys to the solution of global energy and environment problems.

8

Good Light and Bad Light

Light is an essential element of life. In addition to being able to view our surroundings, read, and so forth, light provides us with vital energy required to stimulate our brains and nervous systems. The sources of light can be natural or artificial. Natural sources of light, including the sun, are free of charge. Artificial sources of light are the sources made by humans that cost money. They consist of candles, incandescent and fluorescent lamps, and light-emitting diodes (LEDs). Although the light can be provided by both types of light sources, only natural sources ensure the perfect and healthy lighting since it has been an inherent part of the universe. It is commonly understood that the sun is the source of all energy for the Earth. This chapter reviews the heterogeneous sun composition and microstructure that guarantees the perfect lighting. In addition, a light energy model is proposed. This model shows the effect of the size, and therefore the number, of the particles emitting the light on the corresponding spectrum and resulting light quality. This chapter also investigates the light spectra for the sun, incandescent white and red lamps, fluorescent lamp, and red and yellow LEDs. This study indicates that the sun, a natural light source, produces a continuous spectrum, and the best coverage of the visible colors. However, artificial light sources, such as candles, incandescent and fluorescent lamps, and LEDs, consisting of homogeneous materials, are limited in size, light coverage, and service life. For example, their spectra show spikes and troughs, and they do not cover all of the visible lights properly. The use of transparent medical eyeglasses keeps the form of the light spectrum. However, the sunglasses reduce the intensity and brightness of the visible colors, and eliminate some visible colors, which are bad for clear eye vision, but might be good for other applications that are vital to our organism.

8.1 Introduction

Light is a form of energy that produces a sensation of brightness, making vision possible. There are natural and artificial lights. A natural light is a light produced from a natural source, such as the sun or lightning. Any other source of energy can also be natural as long as it does not have unnatural components in it. For instance, a beeswax candle is natural, while paraffin wax is not. For light to be natural, it has to have a natural source as well as natural processes involved. For instance, light emitting from the moon is natural, as is light from an organic olive oil lamp. On the other hand, fluorescent light bulbs, that produce light through passage of electricity into mercury vapor, have artificial sources as well as processes involved. Natural light follows the natural way of lighting up, using only natural materials and processes. Natural sources of light play additional roles in keeping nature in harmony and from chaos. However, an artificial light is the type of light resulting from an artificial source, such as a light bulb. This follows a man-made process, using artificial materials such as glass or plastics. Artificial sources of light are made only to illuminate darkness.

Daylight on earth is a vital part of the light consumed by humans, animals, and plants for a healthy life. Each sunrise, a transition from the restful darkness of night to the energetic brightness of day infuses life into all living things (Liberman, 1991).

Natural light is used by the human body for its normal functioning. The sun emits light that is materialized by particles bombarding the human body, which produce hormones and vitamins amid natural light-body interactions and reactions. So the sun composition is part of the whole process. The human eye and body were created to function in natural light, since the latter is rich in quality and meaning, and it is connected with many profound functions of nature (Hale, 1993). For instance, research has demonstrated that the full spectrum of daylight is important to stimulate the endocrine system properly and that subjects suffer side effects when forced to spend much of their time under artificial light sources that reproduce only a limited portion of the daylight spectrum (Ott, 2000).

Regarding animals, the domestic laying hen, for instance, is a day-active gregarious bird, just as its ancestor the red jungle hen, which evolved in 12 hours of light and 12 hours of darkness in the equatorial jungle (Collias and Collias, 1996; Gunnarsson *et al.*, 2008). Therefore, the lighting environment is crucial for laying hens and their production (Gunnarsson *et al.*, 2008). Birds intended for organic production, but not given access to natural light as pullets, have to adapt to a new lighting environment at an early age, possibly resulting in behavioral problems, e.g., cannibalism (Manser, 1996, Gunnarsson *et al.*, 1999; Gunnarsson *et al.*, 2008). The pullets for organic egg production shall be reared with access to natural light (Gunnarsson *et al.*, 2008).

Concerning plants, they grow as a result of their ability to absorb light energy and convert it into reductive chemical energy, which is used to fix carbon dioxide (Attridge, 1990). The natural light comes mainly from the sun.

This chapter investigates the natural and artificial lights based on their pathways and light spectra. The natural light source considered here is the sun. The artificial light sources studied are candles, incandescent and fluorescent lamps, and LEDs. The sun's composition and microstructure are exposed. Also, a light energy model is developed, showing the outcome of light source composition on light quality. The pathways for both types of lights are described. In addition, the effects of using medical eyeglasses and sunglasses on the eye vision and colors coverage are examined.

It is known now that artificial light is damaging to the eye, brain, and the whole body, due to the stress provoked by artificial light sources and their limited light spectra. Many vision disorders, such as eye fatigue, have been caused by modern light technology products, such as incandescent and fluorescent lamps. In addition, modern technology tools, such as computers and televisions, demand a high and unnatural visual effort. As a result, computer users, for example, experience symptoms related to dry eyes, refractive error, accommodation infacility and hysteresis, exophoria and esophoria, and presbyopia.

8.2 Natural Light Source: Sun

8.2.1 Sun Composition

Figure 8.1 shows a view of the sun, taken at 9:19 a.m. EST on Nov. 10, 2004, by the SOHO (Solar and Heliospheric Observatory) spacecraft. Table 8.1 indicates the sun's composition based on known tangible data. Nevertheless, it is very important to mention here that the sun's elements are infinite, and mostly undiscovered yet. This table of elements is based on the analysis of the solar spectrum, which comes from the photosphere and chromosphere of the sun (Chaisson and McMillan, 1997). About 67 known and tangible elements have been detected in the solar spectrum (Chaisson and McMillan, 1997). Because most of these data are collected from the

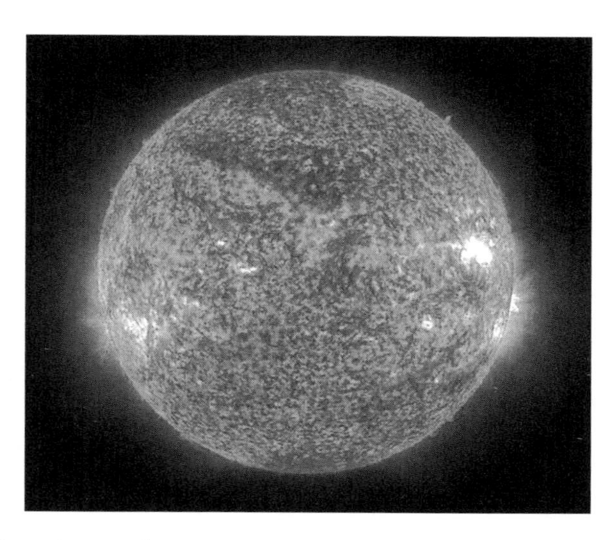

Figure 8.1 Sun picture taken at 9:19 a.m. EST on Nov. 10, 2004, by the SOHO (Solar and Heliospheric Observatory) spacecraft (NASA/European Space Agency, 2004)

Table 8.1 Sun composition (Chaisson and McMillan, 1997).

Element	Abundance (Percentage of Total Number of Atoms)	Abundance (Percentage of Total Mass)
Hydrogen	91.2	71.0
Helium	8.7	27.1
Oxygen	0.078	0.97
Carbon	0.043	0.40
Nitrogen	0.0088	0.096
Silicon	0.0045	0.099
Magnesium	0.0038	0.076
Neon	0.0035	0.058
Iron	0.0030	0.14
Sulfur	0.0015	0.040

outermost layer of the sun, it is likely that the proportion of lighter components is skewed.

8.2.2 Sun Microstructure

The sun matter microstructure consists of molecules, which represent the smallest part of a chemical compound. A molecule is the smallest physical unit of a substance that can exist independently, consisting of one or more atoms held together by chemical forces.

An atom is the smallest part of element, into which it can be divided and still retain its properties, made up of a dense, positively charged nucleus surrounded by a system of electrons. The atom size is around 10^{-10} m (see Table 8.2). Atoms usually do not divide in chemical reactions except for some removal, transfer, or exchange of specific electrons. An atom is composed of proton, neutron, and electron.

The sun emits an infinite number of invisible elements called neutrinos. A neutrino is defined as a stable neutral elementary particle of the lepton group with a zero rest mass and no charge. There are

Table 8.2 Atom structure.

Element	Size (m)
Atom	$\approx 10^{-10}$
Proton	$\approx 10^{-15}$
Neutron	$\approx 10^{-15}$
Electron	$< 10^{-18}$
Nucleus	$\approx 10^{-14}$
Quark	$< 10^{-19}$

Table 8.3 Types of interaction field (Cottingham and Greenwood, 2007).

Interaction Field	Boson	Spin
Gravitational field	'Gravitons' postulated	2
Weak field	W^+, W^-, Z particles	1
Electromagnetic field	Photons	1
Strong field	'Gluons' postulated	1

Table 8.4 Leptons (Cottingham and Greenwood, 2007).

	Mass (MeV/c^2)	Mean Life (s)	Electric Charge
Electron e^-	0.511	∞	$-e$
Electron neutrino ν_e	$< 3 \times 10^{-6}$		0
Muon μ^-	105.658	2.197×10^{-6}	$-e$
Muon neutrino ν_μ			0
Tau τ^-	1777	$(291.0 \pm 1.5) \times 10^{-15}$	$-e$
Tau neutrino ν_τ			0

three types of neutrinos, associated respectively with the electron, muon, and tau particle, and all have a spin of 1/2. Table 8.3 lists the interaction field and Table 8.4 lists the mean life and charges of various particles.

According to the standard model, there are 12 fundamental matter particle types and their corresponding antiparticles. The matter particles are classified into two categories: quarks and leptons. Each category includes six particles and six corresponding antiparticles (Fig. 8.2).

Another type of fundamental particles corresponds to the force carrier particles. They are called gluons, photons, and W and Z, and are responsible for strong, electromagnetic, and weak interactions respectively.

The sun subatomic particles discovered so far are limited in number. However, the sun subatomic particles are infinite, and interact with each other continuously.

Let us consider the sun as the system into consideration for particle size analysis. There is only one sun. However, the sun consists of an infinite number of particles, including just a few of known

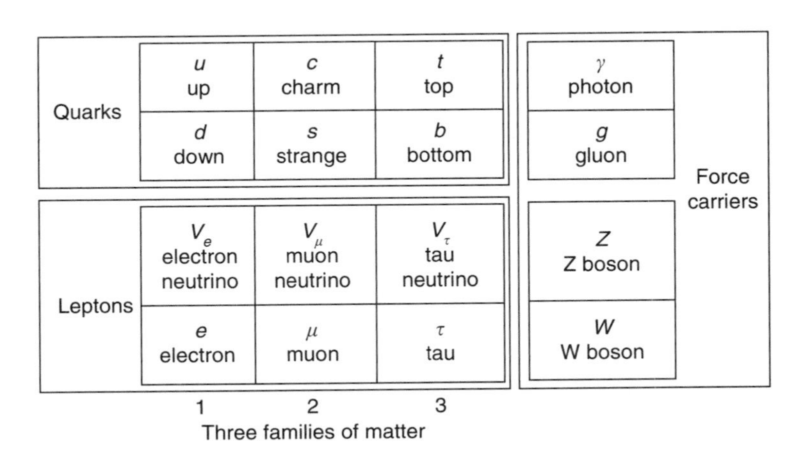

Figure 8.2 Elementary particles.

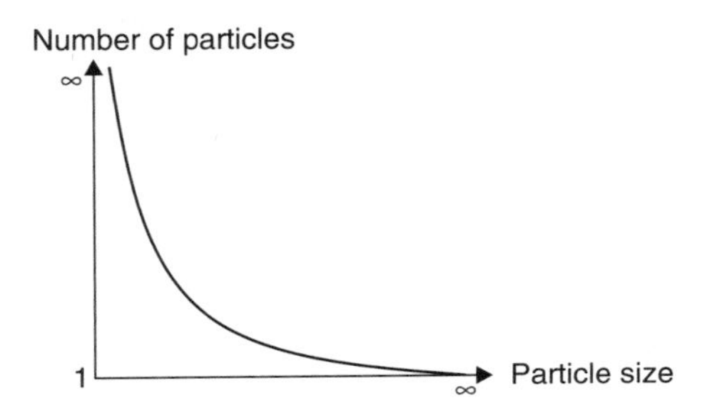

Figure 8.3 Sun particle number as a function of the particle size.

particles. As the size of the subatomic sun element decreases, the number of the considered element in the sun increases, for example. Figure 8.3 indicates that the number of sun particles decreases with their size.

8.3 Artificial Light Sources

This chapter investigates the following artificial light sources (see Figures 8.4 to 8.8) and artificial filters (Figs. 8.9–8.10):

Figure 8.4 Mulled cider candle.

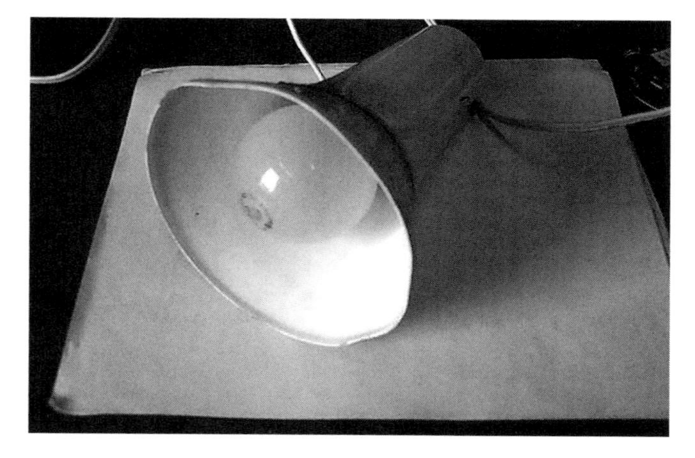

Figure 8.5 Incandescent lamp.

Figure 8.6 Incandescent red lamp.

Figure 8.7 Fluorescent lamp.

Figure 8.8 Light Emitting Diodes (LED).

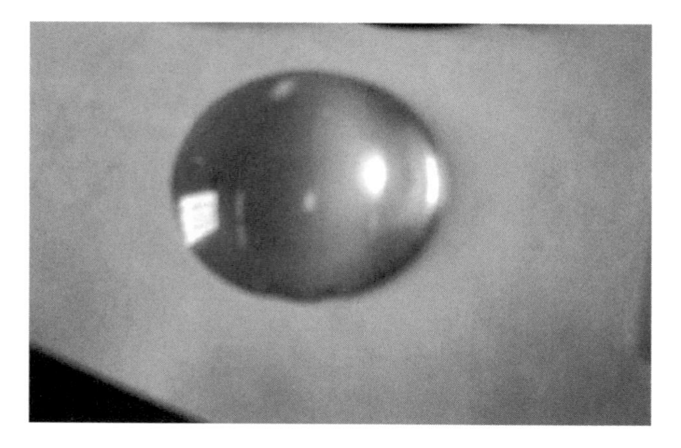

Figure 8.9 Lens.

Figure 8.10 Sunglasses.

1. Mulled cider candle.
2. Incandescent white and red lamps.
3. Fluorescent lamp.
4. Red and yellow light-emitting diodes (LEDs).

8.4 Pathways of Light

Natural light has both a natural source and a natural pathway. For example, the sun that is a natural source of light is an essential element in the universe. One of the benefits of the sun is daylight and night light via the moon. The sun does not produce waste, since all its resulting particles and effects are used by nature. The sunlight service life is infinite. The sun consists of heterogeneous materials and particles. Therefore, this type of light source is natural, heterogeneous, clean, vital, and efficient. Figure 8.11 shows the natural light pathway.

Artificial light sources, such as candles, incandescent lamps, and fluorescent lamps, are made by humans for the unique benefit that is light. Their components and resulting particles and effects are homogeneous, toxic and not accepted by nature, since they generate discomfort, stress, overall light pollution, and various diseases,

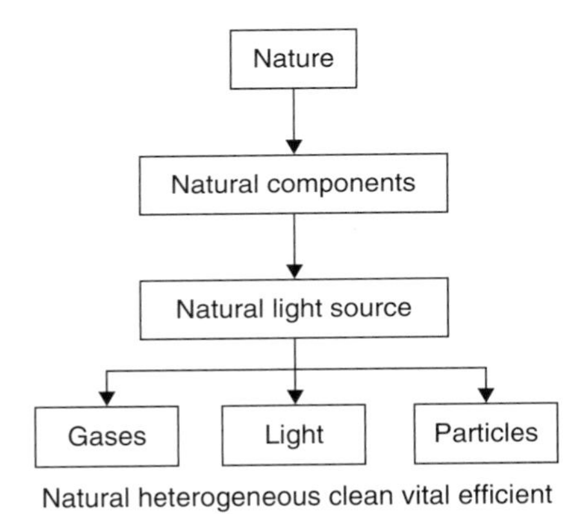

Figure 8.11 Natural light pathway.

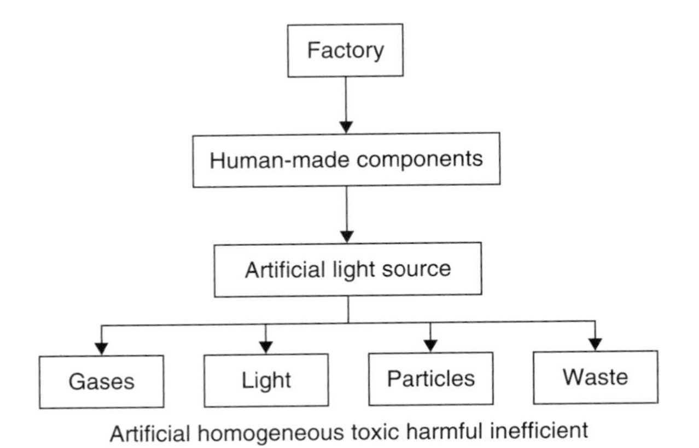

Figure 8.12 Artificial light pathway.

in addition to the waste in the surrounding environment including humans. Moreover, their service lives are limited. So, this type of light sources is artificial, toxic, harmful, and inefficient. Figure 8.12 shows the pathway of artificial light.

Light intensity or energy, efficiency, and quality are functions of the light source composition.

8.5 Natural and Artificial Light Spectra

Table 8.5 shows the perceived color as a function of the wavelength ranges. It is paramount to indicate that all the colors, whether they are visible or not, are needed by the human body.

The artificial lights produced by the candle and the incandescent lamp display continuous spectra, since the light results from material burning. The candle consists of wax, and the incandescent lamp includes a tungsten filament. However, the artificial lights from fluorescent lamps and LEDs show spectra with spikes and troughs. Fluorescent lamps emit light after an electric discharge through their gases, and LEDs are monochromatic. The light spectra for a tungsten filament lamp are shown in Figure 8.13.

The artificial light spectra are not balanced in the visible light region, which proves one of the negative aspects of artificial

Table 8.5 Perceived color based on light wavelength.

Wavelength (nm)	Color
<400	Ultraviolet (invisible)
400–450	Violet
450–490	Blue
490–560	Green
560–590	Yellow
590–630	Orange
630–670	Bright red
670–750	Dark red
>750	Infrared (invisible)

lighting. The sun size is infinite when it is compared to the artificial light sources. As a result, the particles composing the sun produce an infinite number of spectra. However, the particles forming the artificial light sources are limited. Then, their generated spectra are the sums of a limited number of spectra, which explains the incomplete coverage of the visible light region, and the presence of spikes and troughs.

The light spectrum of the light-emitting element is the sum of all the spectra developed by the various particles composing the element. Figures 8.14 through 8.22 show the light spectra for various light sources. It shows that light intensity increases with wavelength for the candle and incandescent lamps. The fluorescent lamp spectrum displays spikes and troughs, confirming the presence of its mercury vapor and gases such as neon, argon, and xenon, in addition to the phosphor coating. The red LED light spectrum

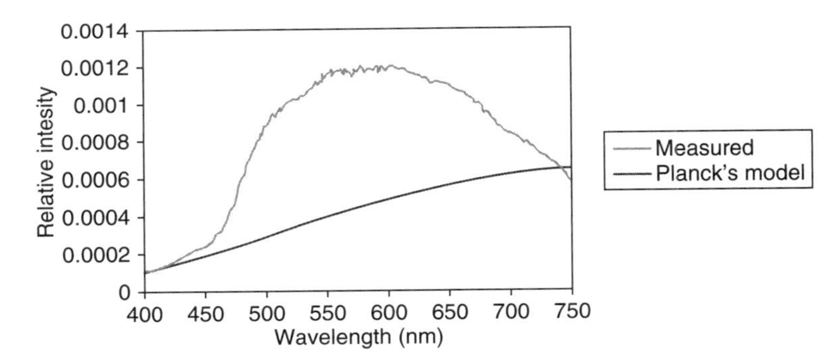

Figure 8.13 Normalized visible light spectra for tungsten filament lamp.

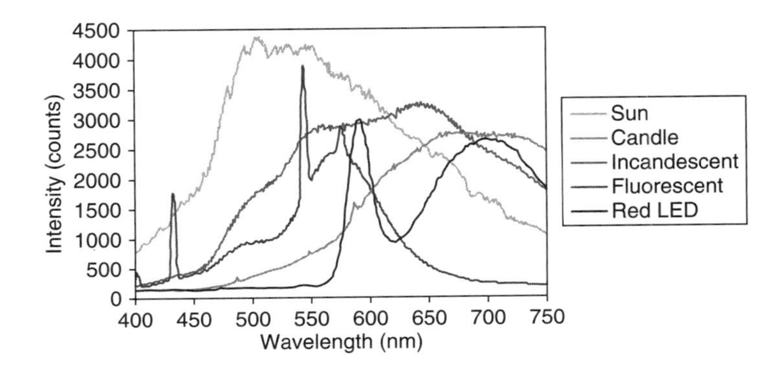

Figure 8.14 Visible light spectra.

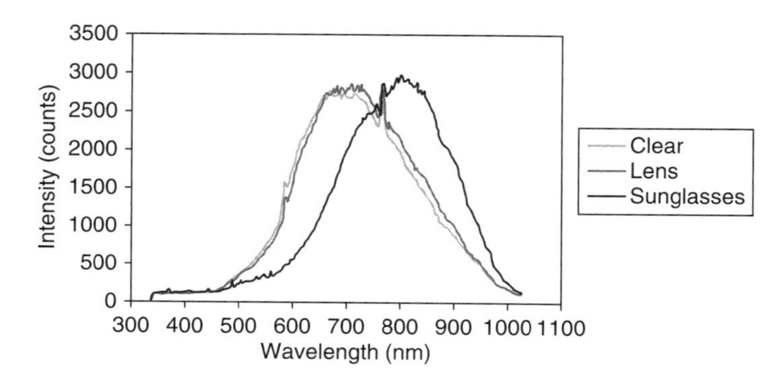

Figure 8.15 Light spectra in presence of artificial lenses.

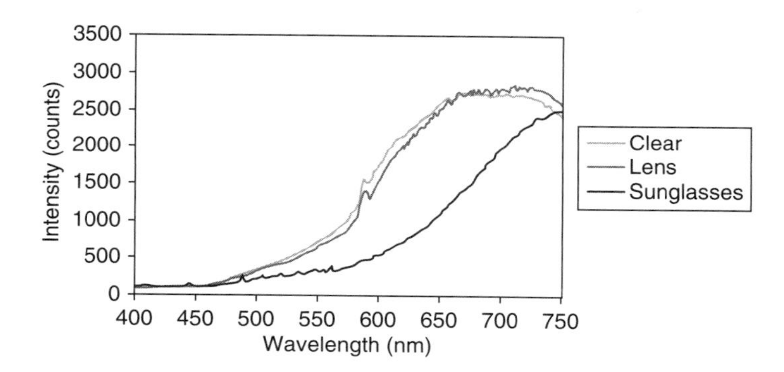

Figure 8.16 Visible light spectra for candle in presence of artificial lenses.

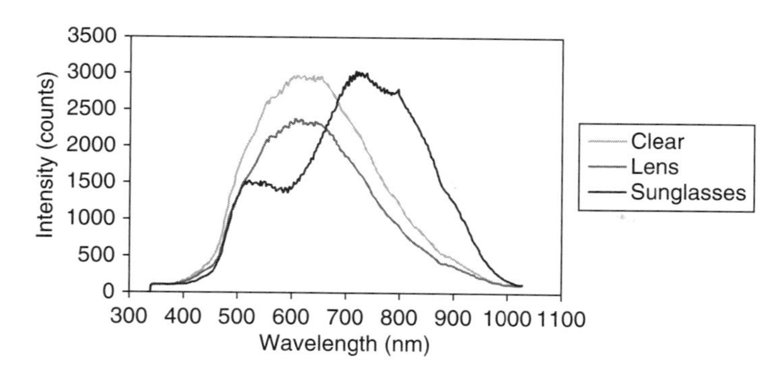

Figure 8.17 Light spectra for incandescent white lamp in presence of artificial lenses.

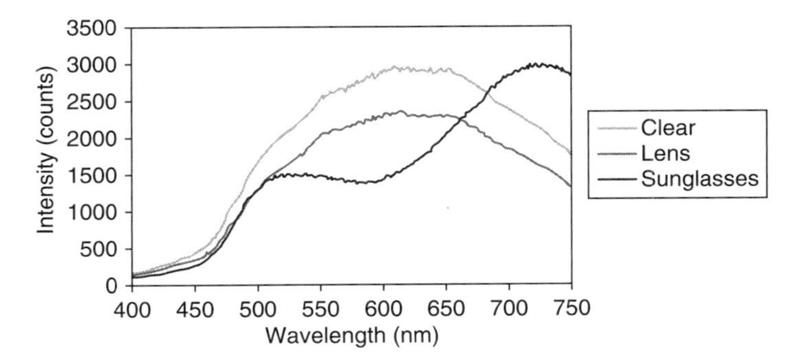

Figure 8.18 Visible light spectra for incandescent white lamp in presence of artificial lenses.

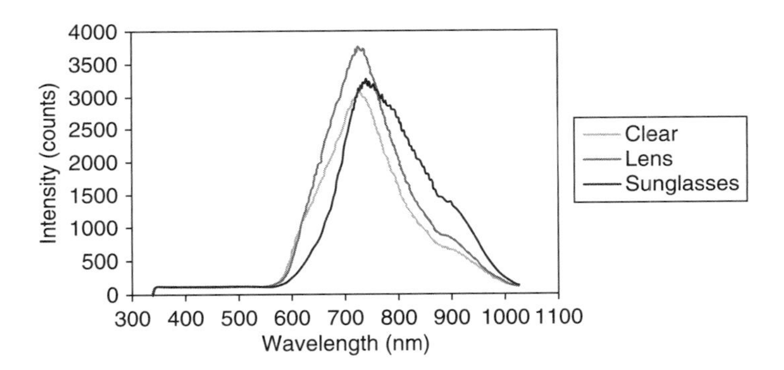

Figure 8.19 Light spectra for incandescent red lamp.

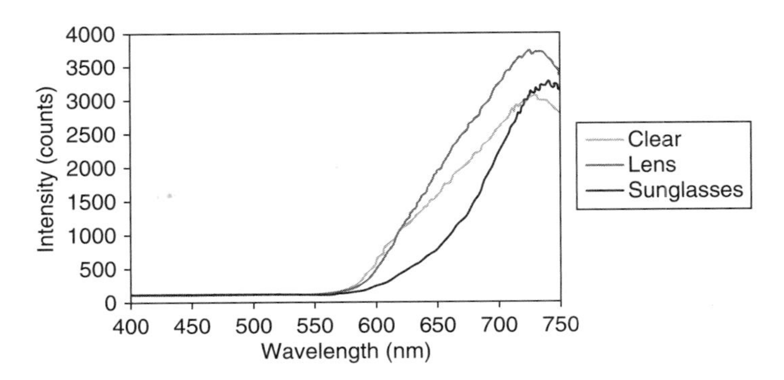

Figure 8.20 Visible light spectra for incandescent red lamp.

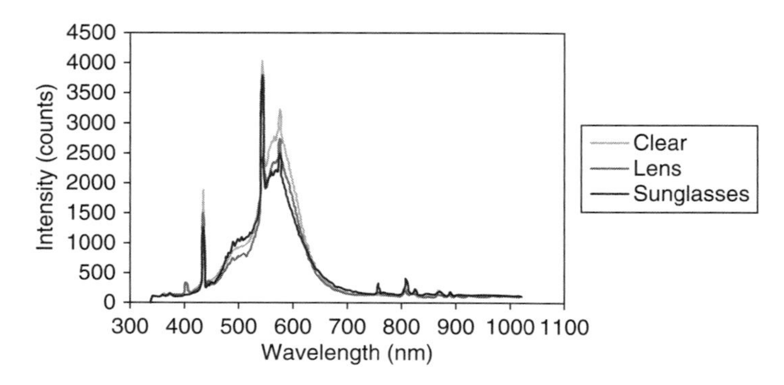

Figure 8.21 Light spectra for fluorescent lamp.

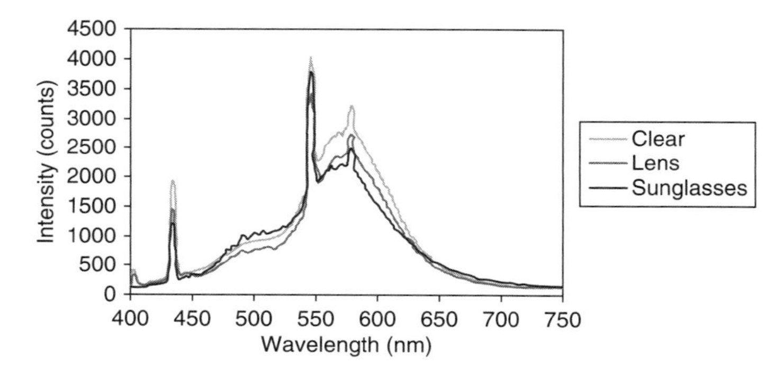

Figure 8.22 Visible light spectra for fluorescent lamp.

exhibits also spikes in the orange and red color areas, due to the phosphor and additional red coating.

The incandescent and fluorescent lamps are considered full-spectrum light sources. A full-spectrum light source is supposed to imitate the spectrum of natural light. However, a natural light source, such as the sun, has a huge amount of material that undergoes spontaneous and continuous combustion to produce light. As a result, an infinite number of spectra are generated, producing the perfect, smooth, and continuous spectrum. Regarding humans, sunlight is the best for perfect visual clarity, color perception, body growth and behavior, mood, mental awareness, performance, and productivity. In addition, the other elements of the human environment, including the flora and fauna, benefit also from sunlight, since it ensures their normal growth and behavior.

Based on light spectra, the wavelength values at the maximum radiation values for the corresponding light sources are:

1. Sun: 504.63 nm.
2. Candle: 644.99 nm.
3. Incandescent lamp: 557.34 nm.
4. Incandescent red lamp: 714.27 nm.
5. Fluorescent lamp: 544.79 nm.
6. Red LED: 591.47 nm.
7. Yellow LED: 675.61 nm.

Then, the equivalent black body temperature values for the various light sources equal:

1. Sun: 5746.79 K.
2. Candle: 4496.19 K.
3. Incandescent lamp: 5203.29 K.
4. Incandescent red lamp: 4060.09 K.
5. Fluorescent lamp: 5323.15 K.
6. Red LED: 4903.04 K.
7. Yellow LED: 4292.42 K.

8.6 Effect of Eyeglasses and Sunglasses on Light Spectra

The use of eyeglasses might be necessary for normal eye vision. However, sunglasses are not essential for normal eye vision. Sunglasses are coated with paints other than white to lower the light intensity and brightness perceived by the eye while driving, for instance. In our study, a lens was used to simulate medical eyeglasses as well as sunglasses. The results are shown in Figure 8.15 through 8.22 for various light sources. This figure shows that the lens reproduces the same form of the spectrum covering the same colors, since it is not coated and, therefore, is transparent.

As expected, the effect of the dark green sunglasses is obvious. The dark green coating affected the final spectrum form and color coverage for all the light sources studied except the incandescent red and fluorescent lamps. A darkening of the colors observed means reduction of the light color intensity and brightness. Concerning the fluorescent lamp, the original spectrum is not continuous and does not cover properly the range of visible lights. Therefore, the sunglasses did not affect the final spectrum for this type of lamps.

8.7 Concluding Remarks

Natural light sources, such as the sun, are free of charge, and produce the perfect light quality, since they cover all the colors needed by living things, including the human body. However, artificial light sources, comprising of candles, incandescent and fluorescent lamps, and LEDs, cost money, and do not ensure a complete color coverage as expected by a natural body. As a result, the light lacks its natural quality. The light energy model developed shows that the light source composition and microstructure play an important role in the final outcome for clear eye vision and normal body functioning, for example. This is noticed by the colors covered by the light. Each color is characteristic of a certain type of particle. So, if a certain color is not well-covered, the human body will not benefit from the corresponding emitted light and bombarded particles. This will definitely cause a weak body and provoke diseases. Then, natural light, such as the sun, is necessary for humans and other living things. However, the artificial light sources can accomplish a limited role when and where it is difficult to get natural light, or for machine use, for instance. But the artificial light sources never achieve the expected quality due to their inefficient service life, the toxicity of their composition, and the deficiency of light color coverage and related essential particles.

9

Do You Believe in Global Warming?

A Hollywood movie director once said, "The ignorance in American culture started when people started to 'believe' (in anything and everything that the Establishment perpetrates)." We know that truth has no chance when the only options one is given are to 'believe' or disbelieve in something. Whenever a topic emerges, the 'believers' and 'disbelievers' start a debate, and the entire polarization takes place between supporting or opposing the dogma, completely shutting down the logical cognition process. Global warming has been a subject of discussion from the late 1970s. It is perpetrated that the building up of carbon dioxide in the atmosphere results in irreversible climate change. Even though carbon dioxide has been blamed as the sole cause for the global warming, there is no scientific evidence that all carbon dioxides are responsible for global warming. A new theory has been developed, which shows that all carbon dioxides do not contribute to global warming. For the first time, carbon dioxide is characterized based on various criteria, such as the origin, the pathway it travels, and the isotope number. In this chapter, the current status of greenhouse gas emissions from various anthropogenic activities is summarized. The role of water in global warming has been discussed. Various energy sources are classified based on their global efficiencies. The assumptions and implementation mechanisms of the Kyoto Protocol have been critically reviewed. It is argued that the Clean Development Mechanism of the Kyoto Protocol has become the "license to pollute," due to its improper implementation mechanism. The conventional climatic models are deconstructed, and guidelines for new models are proposed in order to achieve true sustainability in the long term. A series of sustainable technologies that produce natural CO_2 which do not contribute to global warming has been presented. Various zero-waste technologies that have no negative impact on the environment are keys to reverse the global warming. Because synthetic chemicals, which are inherent to the current technology development mode, are primarily responsible for global warming, there is no hope for reversing global warming without fundamental changes in technology development. The new technology development mode must foster the development of natural products, which are inherently beneficial to the environment.

9.1 Introduction

For some four decades, global warming alarm has been sounded. Indeed the discussion of the possibility that a build-up of carbon dioxide in the atmosphere results in irreversible climate change has been transformed into a "controversy" of the type seen all too often on every other subject: a "pro" *versus* "con" proposition is advanced, dividing people according to their support for one side or the other, all before anything objective and scientific in connection with the originating subject matter is even established.

As far as the science of the question goes: despite the fact that various international and government organizations have set series of standards to reduce the carbon dioxide level in the atmosphere due to anthropogenic activities, the current climatic models show that the global temperature is still increasing. Carbon dioxide has been blamed as the sole cause for the global warming, even though there is no scientific evidence that all carbon dioxides are responsible for global warming. Precisely to address this critical gap, a detailed analysis of greenhouse gas emission, starting from the pre-industrial era, moving to the industrial age to the (for some) "golden era" of petroleum has been carried out in this chapter.

A very large amount of pseudo-science is already afoot on all aspects of this question, much of it used to divide, if not indeed aimed in the first place at dividing, public opinion over whether nature or humanity is the chief culprit. This state of affairs has opened the door to proposing all manner of band-aid solutions that share the trait of enabling peoples of the global North to continue dreaming of two SUVs for every garage, while peoples of the South can continue to fantasize about motorizing their bicycle or oxcart, in other words, the *status quo*, with one foot on the accelerator, while the argument carries on as to when to apply the brake. The crying need for the serious scientific approach taken in the present work has never been greater. On this question, paraphrasing Albert Einstein, it can truly be said that the system that got us into the problem is not going to get us out. Absent a comprehensive characterization of CO_2 in all its possible roles and forms as a starting-point, any attempt to analyze the tangle of symptoms identified with global warming, or to design any solution based on univariate correlations or even correlations of multiple variables, but assuming that the effects of each variable can be superposed linearly and still mean anything, must collapse under the weight of its very incoherence. The absurdity is so well known that one popular graph on the Internet depicts a strictly proportional increase in incidences of piracy in all the world's oceans as a function of increasing global temperature.

The current status of greenhouse gas emissions due to industrial activities, automobile emissions, biogenic and natural sources is systematically presented here. In this chapter, a newly developed theory has been detailed: all carbon dioxides are not same. Thus, not all carbon dioxides may be contributing to global warming. For the first time, carbon dioxide is characterized based on such normally ignored criteria, such as its origin, the pathway it travels, isotope number, and age of the fuel source from which it was emitted. Fossil fuel exploration, production, processing, and consumption are major sources of carbon dioxide emissions; here, various energy sources are characterized based on their efficiency, environmental impact, and quality of energy based on the new criteria. Different energy sources follow different paths, from origin to end use, and contribute the emissions differently.

A detailed analysis has been carried out on potential precursors to global warming. The focus is on supplying a scientific basis, as well as practical solutions identifying the roots of the problem. Similarly, an evaluation of existing models on global warming, based on the scenario of the Kyoto Protocol under satisfactory implementation, as well as under partial implementation, is presented. Shortcomings in the conventional models have been identified based on this evaluation.

The sustainability of conventional global warming models has been argued. Here, these models are deconstructed, and new models are developed based on new sustainability criteria. Conventional energy production and processing uses various toxic chemicals and catalysts that are very harmful to the environment. Moreover, all energy systems are totally dependent on fossil fuel, at least as the primary energy input, or in the form of embodied energy. This chapter offers unique solutions to overcome such problems, based on truly green technologies that satisfy the new sustainability criteria. These green energy technologies are highly efficient technologies producing with zero net waste.

In this chapter, various energy technologies are ranked based on their global efficiency. For the first time, this research offers energy development techniques that produce what might best be described as "good CO_2" which do not contribute to global warming. A thorough discussion of natural transport phenomena, specifically the role of water and its interaction with various energy sources and climate change taking into account the memory of water, is also undertaken in this work. Conventional models are evaluated based on the long-term impact of CO_2 and their contribution to global warming. It is concluded that conventional energy development systems and global warming models are based on ignorance. Only knowledge-based technology development offers solutions to the global warming.

9.2 Historical Development

The history of technological development from the pre-industrial age to the petroleum era has been reviewed. There is a colloquial expression to the effect that exact change plus faith in the Almighty will always get you downtown on the public transit service. On the one hand, with or without faith, all kinds of things could happen with the public transit service, before the matter of exact fare even enters the picture. On the other hand, with or without exact fare, other developments could intervene to alter the availability of the service, and even cancel it. This helps isolate one of the key difficulties in uncovering and elaborating the actual science of increased carbon dioxide concentrations in the atmosphere. All kinds of activities can increase CO_2 output into the atmosphere; but precisely which activities can be held responsible for consequent global warming or other deleterious impacts? Both the activity and its CO_2 output are necessary, but neither by itself is sufficient, for establishing what the impact may be and whether it is deleterious.

Pre-industrial. One commonly encountered argument attempts to frame the historical dimension of the problem more or less as follows: once upon a time, the scale of humanity's efforts at securing a livelihood was insufficient to affect overall atmospheric levels of CO_2. The implication is that with the passage of time and the development of ever more extensive technological intervention in the natural-physical environment by humanity, everything just got worse. In a contemporary world that has systematically removed ever further from the human person any living connection with the gathering or application of meaningful knowledge, such typically linearized evolutionary analyses may pass for meaningful exegesis. However, from prehistoric times onward, there have been important periods of climate change, whose causes could not have had anything to do with human intervention in the environment on anything approaching the scale that is blamed widely today for "global warming." Nevertheless, these had consequences that were extremely significant, and even devastating, for wide swaths of subsequent human life on this planet.

One of the best-known was the period of almost two centuries of cooling in the northern hemisphere during the 13th and 14th centuries CE, in which Greenland is said to have acquired much of its most recent ice cover. This definitively brought to an end any further attempts at colonizing the north and northwest Atlantic by Scandinavian tribes (descended from the Vikings), creating the opening for later commercial fisheries expansion into the northwest Atlantic by Basque, Spanish, Portuguese, and eventually French and British fishermen and fishing enterprises, the starting-point of European colonization of the North American continent.

Industrial Age. Even one-off events like the volcanic eruption in the Indonesian archipelago in 1816, which spewed an enormous volume of dust into the atmosphere traveling around the globe in the jet stream and led to the "year with no summer" in Europe and the northern half of North America, incurred tremendous consequences. In 1817, grain crops on the continent of Europe failed. In industrial Great Britain, where the factory owners and their politicians boasted how that country's relatively (compared to the rest of the world) highly advanced industrial economy had overcome the "capriciousness of nature," hunger and famine actually stalked the English countryside for the first time in more than a century and a half. The famine conditions were blamed on the difficulties attending the import of extra supplies of food from the European continent, and led directly to a tremendous and unprecedented pressure to eliminate the Corn Laws, the system of high tariffs protecting English farmers and landlords from the competition of cheaper foodstuffs from Europe or the Americas. Politically, the industry lobby condemned the Corn Laws as the main obstacle to cheap food, winning broad public sympathy and support. Economically, the Corn Laws actually operated to keep hundreds of thousands employed in the countryside on thousands of small agricultural plots, at a time when the demands of expanding industry required uprooting and forcing this rural population to work as factory laborers. Increasing the industrial reserve army would enable British industry to reduce wages. Capturing command of that new source of ever cheaper labor was in fact the industrialists' underlying aim.

Without the famine of "the year with no summer," it seems unlikely that British industry would have hit upon the political device of targeting the Corn Laws for elimination as the road on which to blast its way into dominating world markets. Even then, because of the still prominent involvement of the anti-industrial lobby of aristocratic landlords who dominated the House of Lords, it would take British industry another nearly 30 years, but between 1846 and 1848 Parliament eliminated the Corn Laws, industry captured access to a desperate workforce fleeing the ruin brought to the countryside, and overall industrial wages were driven sharply downward. On this train of economic development, the greatly increased profitability of British industry took the form of a vastly whetted appetite for new markets at home and abroad, including the export of important industrial infrastructure investments in "British North America," i.e., Canada, Latin America, and India. Extracting minerals and other valuable raw materials for processing into new commodities in this manner brought an unpredictable level of further acceleration to the industrialization of the globe in regions where industrial capital had not accumulated significantly, either because traditional development blocked its role, or because European settlement remained sparse.

Age of Petroleum. The world economy entered the Age of Petroleum mostly since the rise of industrial-financial monopoly in one sector of production after another in Europe and America, before and following the First World War. Corresponding to this has been the widest possible extension of chemical engineering, especially the chemistry of hydrocarbon combination, hydrocarbon catalysis, hydrocarbon manipulation and rebonding, on which the refining and processing of crude oil into fuel and myriad byproducts, such as plastics and other synthetic materials, crucially depend. As a result, there is today no activity, be it production or consumption, in any society that is tied to the production and distribution of such output in which adding to the CO_2 burden in the atmosphere can be avoided or significantly mitigated.

In these developments, carbon and CO_2 are in fact vectors carrying many other actually toxic compounds and byproducts of these chemically-engineered processes. Atmospheric absorption of carbon and CO_2 from human activities or other natural non-industrial activities would normally be continuous. However, what occurs with hydrocarbon complexes combined with inorganic and other substances that occur nowhere in nature is much less predictable, and, on the available evidence, not benign, either. From a certain standpoint, there is a logic in attempting to estimate the effects of these other phenomena by taking carbon and CO_2 levels as vectors. However, there has never been any justification to assume the CO_2 level itself is the malign element.

Such a notion is a non-starter as science in any event, which raises the even sharper question: just what does science have to do with it? There is today no large petrochemical company or syndicate that has not funded some study, group, or studies or groups, interested in CO_2 levels as a global warming index, whether to discredit or to affirm such a connection. It is difficult to avoid the obvious inference that these very large enterprises, fiercely competing to retain their market shares against rivals, do not have a significant stake in engineering a large and permanent split in public opinion based on confusing their intoxication of the atmosphere with rising CO_2 levels. Whether the consideration is refining for automobile fuels, processing synthetic plastics, or concocting synthetic crude, behind a great deal of the propaganda about "global warming" stands a huge battle among oligopolies, cartels and monopolies over market share. The science of "global warming" is precisely the only route on which to separate the key question of what is necessary to produce goods and services that are nature-friendly from the toxification of the environment as a byproducts of the anti-nature bias of chemical engineering in the clutches of the oil barons.

9.3 Current Status of Greenhouse Gas Emission

The current status of greenhouse gas emissions from various anthropogenic activities is summarized. Industrial activities, especially related to the burning of fossil fuels, are major contributors of global greenhouse gas emissions. Climate change due to anthropogenic greenhouse gas (GHG) emissions is a growing concern for global society. In the third assessment report, the Intergovernmental Panel on Climate Change (IPCC) provides the strongest evidence so far that the global warming of the last 50 years is due largely to human activity and the CO_2 emissions that arise when burning fossil fuel (Farahani *et al.*, 2004).

It has been reported that the CO_2 level now is at the highest point in 125,000 years (Service, 2005). Approximately 30 billion tons of CO_2 is released from fossil fuel burning each year (Figure 9.1). The CO_2 concentration level in the atmosphere traced back in 1750 was reported to be 280±10ppm (IPCC, 2001). It has risen continuously since then, and the CO_2 level reported in 1999 was 367 ppm. The

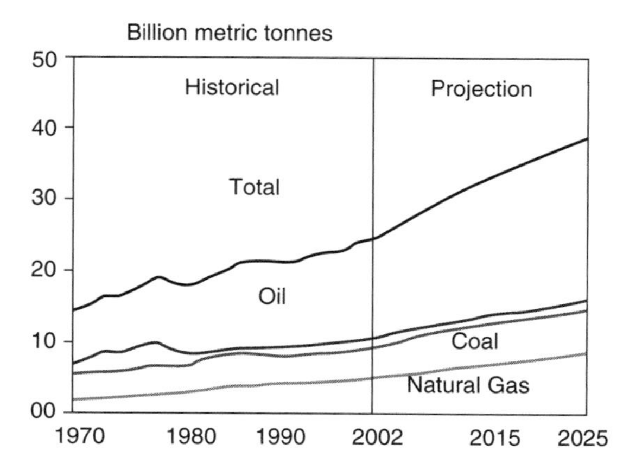

Figure 9.1 World CO_2 Emissions by oil, coal and natural gas, 1970–2025 (adopted from EIA, 2004).

present atmospheric CO_2 concentration level has not been exceeded during the past 420,000 years (IPCC, 2001; Houghton *et al.*, 2001; Houghton, 2004).

The latest 150 years were a period of global warming (Figures 9.2, 9.2a). Global mean surface temperatures have increased 0.5–1.0°F since the late 19th century. The 20th century's 10 warmest years all occurred in the last 15 years of the century. Of these, 1998 was the warmest year on record. Sea level has risen 4–8 inches globally over the past century. Worldwide precipitation over land has increased by about one percent.

The industrial emission of CO_2 consists of process emission and production emission. Coal mining, oil refining, gas processing, petroleum fuel combustion, pulp and paper industry, ammonia, petroleum refining, iron and steel, aluminum, electricity generation, and cement production are the major industries responsible for producing various types of greenhouse gases. Besides these industrial sources, the transportation sector has also a large share of greenhouse gas emission. Greenhouse gas emission from bioresources is also significant. However, National Energy Board of Canada does not consider CO_2 from biomass as a contribution to greenhouse problems (Hughes and Scott, 1997). The justification emerges from the fact that greenhouse gas emission from bioresources such as fuel wood, agricultural waste, and charcoal is carbon neutral, as the plants synthesize this CO_2. However, if various additives are added during the production of fuel, such as pellet making and charcoal

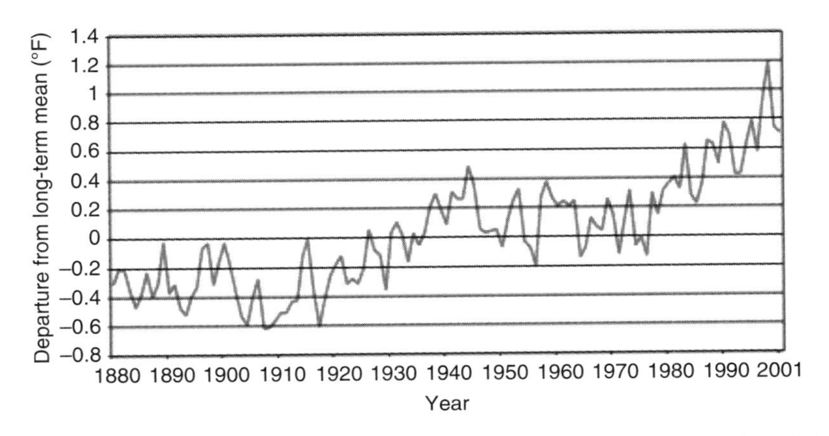

Figure 9.2 Global Temperature Changes from 1880 to 2000 (Modified after EPA Global Warming site: US National Climate Data Center 2001).

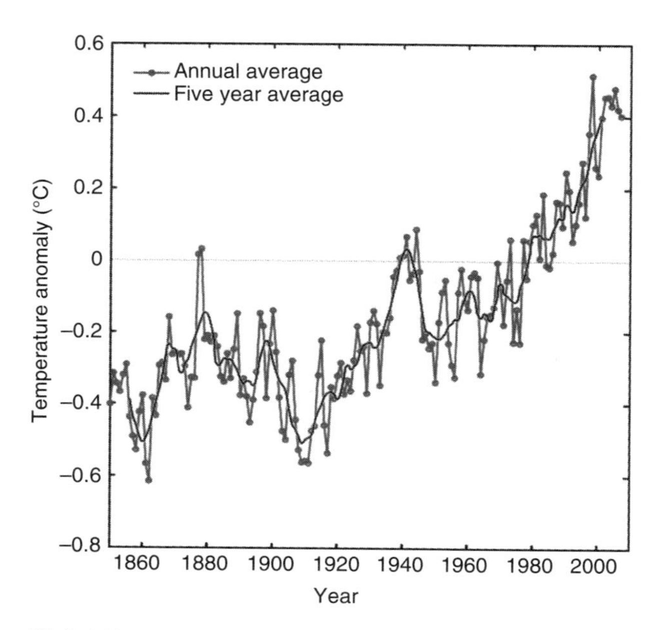

Figure 9.2a Global Temperature Changes reported by Wikimedia (http://commons.wikimedia.org/wiki/Main_Page).

production, the CO_2 produced is no longer carbon neutral. For instance, pellet making involves the addition of binders, such as carbonic additives, coal, and coke breeze, which emit carcinogenic benzene as a major aromatic compound (Chhetri at al., 2006). The CO_2 contaminated with such chemical additives is not favored by plants for photosynthesis, and as a result, CO_2 will be accumulated in the atmosphere. Moreover, deforestation, especially the unsustainable harvesting of biomass due to urbanization and to fulfill the industrial biomass requirement, also results in net CO_2 emission from bioresources.

The worldwide CO_2 emission from the consumption of fossil fuels was 24,409 million metric tons in 2002 and it is projected to reach to 33284 million metric tons in 2015 and 38,790 million tons in 2025 (IEO, 2005). The worldwide CO_2 production from consumption and flaring of fossil fuel in 2003 was 25,162.07 million metric tons. The United States alone had a share of 5802.08 million tons of CO_2 emission in 2003 (IEA, 2005). Current CO_2 emission levels are expected to continue increasing in the future, as fossil fuel consumption is sharply increasing (WEC, 2006). The projection showed that emissions from all sources are emitted to grow by 36% in 2010 (to 18.24

Gt/y) and by 76% in 2020 to 23.31 Gt/y (compared to the 2000 base level). Variation of CO_2 concentration at different time scales is presented in Figure 9.3 This figure shows the increase in CO_2 emission exponentially after 1950. However, present methodology does not classify CO_2 based on its source. Industrial activities during this period also went up exponentially. Because of this industrial growth and extensive use of fossil fuels, the level of 'industrial' CO_2 emission increased sharply (Figure 9.4). The worldwide supply in 1970 was approximately 49 million barrels per day but the supply increased to approximately 84 million barrels per day (EIA, 2006). At the same time, the level of 'natural' CO_2 which comes by burning biomass went down, due to deforestation. However, researchers, industry, and government are focused on the total CO_2, which is not correct in terms of its impacts on global warming. NOAA (2005) defined annual mean growth rate of CO_2 as the sum of all CO_2 added to, and removed from, the atmosphere during the year by human activities

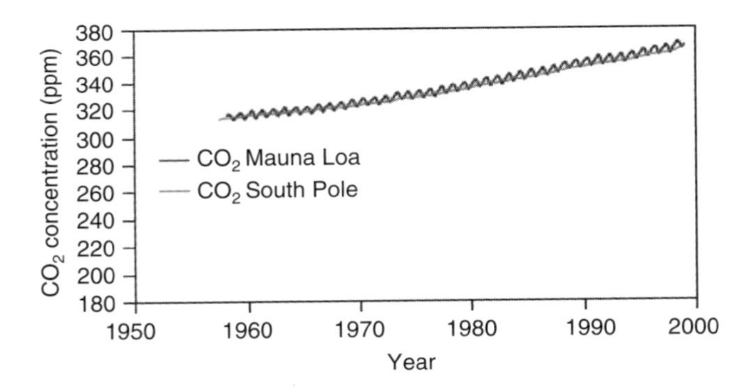

Figure 9.3 Variation in atmospheric CO_2 concentration (IPCC, 2001).

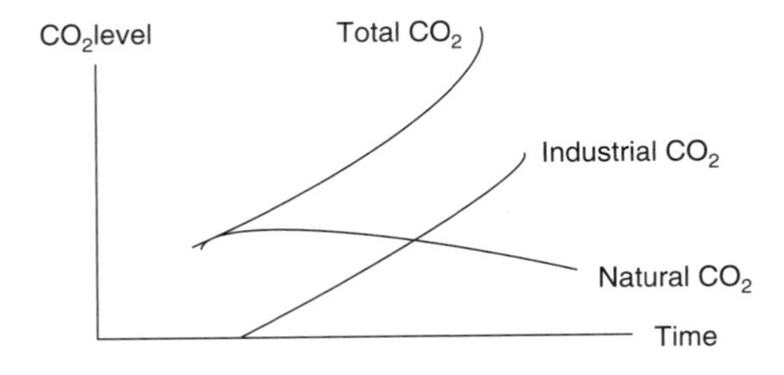

Figure 9.4 Total, industrial, and natural CO_2 trend.

and by natural processes. Natural CO_2 cannot be same as that of industrial CO_2 and should be examined separately.

Some recent studies reported that the human contribution to global warming is negligible (Khilyuk and Chilingar, 2004). The global forces of nature, such as solar radiation, outgassing from the ocean and the atmosphere, and microbial functions are driving the Earth's climate (Khilyuk and Chilingar, 2006). These studies showed that the CO_2 emissions from human-induced activities are far less in quantity than the natural CO_2 emission from ocean and volcanic eruptions. This line of argument is used by others to demonstrate that the cause of global warming is at least a contentious issue (Goldschmidt, 2005). These studies fail to explain the differences between the natural and human induced CO_2 and their impacts on global warming. Moreover, the CO_2 from ocean and natural forest fires were a part of the natural climatic cycle even when no global warming was noticed. All the global forces mentioned by Khilyuk and Chilingar (2006) are also affected by human interventions. For example, more than 70,000 chemicals being used worldwide for various industrial and agricultural activities are exposed in one or the other way to the atmosphere or ocean water bodies that contaminate the CO_2. The CO_2 produced from fossil fuel burning is not acceptable to plants for their photosynthesis, and for this reason, most organic plant matters are depleted in carbon ratio $\delta^{13}C$ (Farquhar *et al.*, 1989; NOAA, 2005). Finally, the notion of 'insignificant' has been used in the past to allow unsustainable practices, such as pollution of harbors, commercial fishing, and massive production of toxic chemicals that were deemed to be "magic solutions" (Khan and Islam, 2006). Today, the banning of chemicals and pharmaceutical products has become almost a daily affair (Globe and Mail, 2006; New York Times, 2006). None of these products were deemed 'significant' or harmful when they were introduced. Khan and Islam (2006) have recently catalogued an array of such ill-fated products that were made available to 'solve' a critical solution (Environment Canada, 2006). In all of these engineering observations, a general misconception is perpetrated, that is: if the harmful effect of a product can be tolerated in the short term, the negative impact of the product is 'insignificant.'

The above argument is bolstered by the notion that short-term temperature fluctuation is natural, and in a long-term scale, it becomes apparent that we are in fact experiencing global cooling. Figure 9.5 shows global reconstructed global temperature change of the last two millennia. Figure 9.6 shows the same on a geological scale. This theory is further supported by carbon dioxide data (Figuure 9.7).

Khilyuk and Chilingar (2006) explained an adiabatic model developed by Khilyuk *et al.* (1994) for the atmosphere together with a sensitivity analysis to evaluate the effects of human-induced CO_2

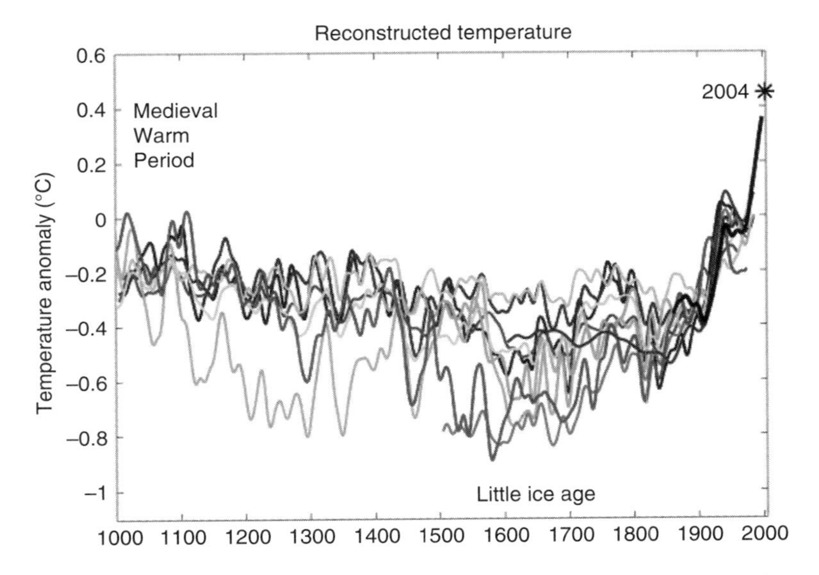

Figure 9.5 Reconstructions of temperatures for the 2nd millennium according to various older articles (bluish lines), newer articles (reddish lines), and instrumental record (black line). (Figure from Wikimedia, http://commons.wikimedia.org/wiki/Main_Page).

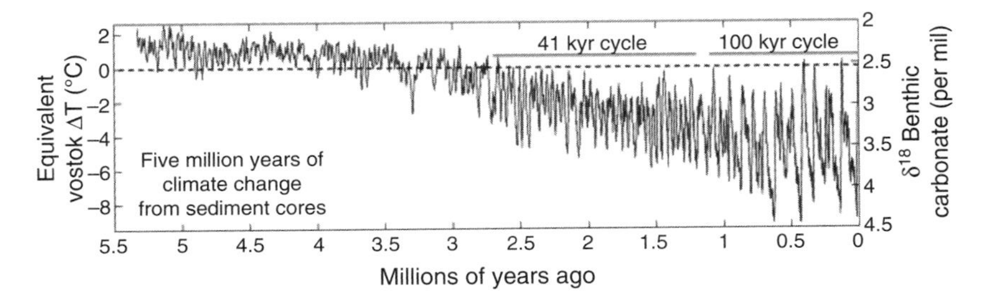

Figure 9.6 Shows that temperature change can easily be explained as natural phenomena, thereby vindicating human intervention and its effect (graph from Wikimedia).

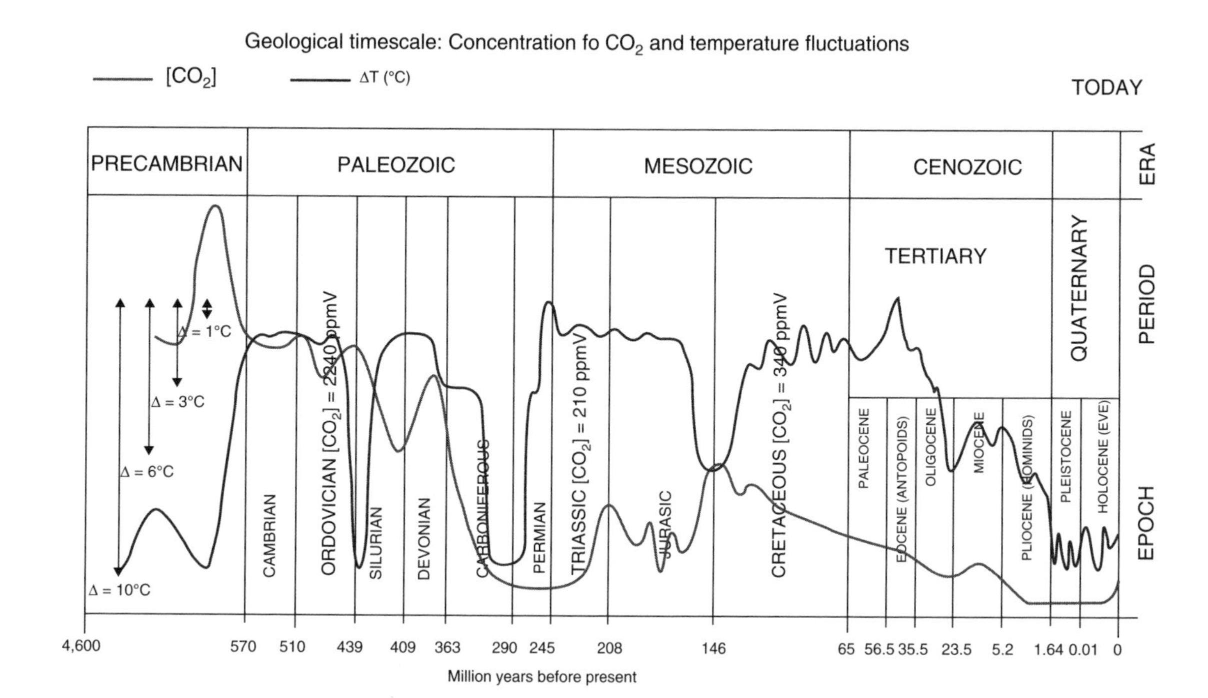

Figure 9.7 Shows that CO_2 concentration today is not alarming if placed in an historical perspective. 1-Analysis of the Temperature Oscillations in Geological Eras by Dr. C.R. Scotese © 2002. 2- Ruddiman, W.F. 2001. Earth's Climate: Past and future. W.H. Freeman & Sons. New York, NY. 3- Mark Pagani et al. Marked Decline in Atmospheric Carbon Dioxide Concentrations During the Paleocene. Science; Vol. 309, No.5735; pp. 600–603. 22 July 2005. Conclusion and interpretation by Nasif Nahle © 2005, 2007. Corrected on 07 July 2008 (CO2: Ordovician Period).

emissions on the global temperature. The model showed that due to the human induced CO_2, the global temperature rise is negligible. However, the adiabatic condition in the atmosphere is one of the most linear thoughts, and the basic assumption of this model is incorrect (Khan *et al.*, 2006). For an adiabatic condition, the following three assumptions are made: perfect vacuum between the system and surrounding area, perfect reflector around the system, like the thermo flux mechanism to resist radiation, and zero heat diffusivity material that isolates the system. None of these conditions can be fulfilled in the atmosphere. Moreover, the study reported that increased emissions of carbon dioxide and water vapor are important for agriculture and biological protection, and the CO_2 from fossil fuel combustion is nontoxic. However, their finding is in contradiction to the fact that plants discriminate against heavier CO_2 and favor CO_2 with lighter carbon isotope ratios. As all chemicals are not the same, all CO_2 is not the same. The CO_2 from power plants is highly toxic, as various toxic chemicals are added during the refining and processing of fossil fuels (Chhetri *et al.*, 2006; Chhetri and Islam, 2006; Khan and Islam, 2006). Since the CO_2 from fossil fuel burning is contaminated with various toxic chemicals, plants do not readily synthesize it. Note that practically all catalysts used are either chemically synthesized or are denatured by concentrating them to a more beneficial state (Khan and Islam, 2006). The CO_2 rejected from plants accumulates in the atmosphere, and is fully responsible for global warming. According to Thomas and Nowak (2006), human activities have already demonstrably changed the global climate, and further, much greater changes are expected throughout this century. The emissions of CO_2 and other greenhouse gases will further accelerate global warming. Some future climatic consequences of human-induced CO_2 emissions, for example, some warming and sea-level rise, cannot be prevented, and human societies will have to adapt to these changes. Other consequences can perhaps be prevented by reducing CO_2 emissions.

Figure 9.8 is the crude oil pathway. The crude oil is refined to convert into various products, including plastics. More than four

Crude oil → Gasolene + Solid residue + diesel + kerosene + volatile HC + numerous petroleum products
Solid residue + hydrogen + metal (and others) → plastic

Figure 9.8 The crude oil pathway (Islam, 2004).

million tons of plastics are produced from 84 million barrels of oil per day. It has been further reported that plastic burning produces more than 4000 toxic chemicals, 80 of which are known carcinogens (Islam, 2004).

In addition to the CO_2, various other greenhouse gases have contributed to global warming. The concentration of other greenhouse gases has increased significantly in the period between 1750–2001. Several classes of halogenated compounds, such as chlorine, bromine, and fluorine, are also greenhouse gases, and are the direct result of industrial activities. None of these compounds was in existence before 1750, but all are found in significant concentration in the atmosphere after that period (Table 9.1). Chlorofluorocarbons (CFCs), hydrohlorofloroca rbons (HCFCs), which contains chlorine, and halocarbons, such as bromoflorocarbons, which contain bromine, are considered potent greenhouse gases. The sulfur hexafluoride (SF_6), which is emitted from various industrial activities, such as the aluminum industry, semi-conductor manufacturing, electric power transmission and distribution, magnesium casting, and from nuclear power generating plants, is also considered a potent greenhouse gas. It is important to note here that these chemicals are totally synthetic in nature and cannot be manufactured under natural conditions. This would explain why the future pathway of these chemicals is so rarely reported.

The transportation sector consumes a quarter of the world's energy, and accounts for some 25% of total CO_2 emissions, 80% of which is attributed to road transport (EIA, 2006). Projections for Annex I countries indicate that, without new CO2 mitigation measures, road transport CO_2 emissions might grow from 2500 million tons in 1990 to 3500 to 5100 million tons in 2020. The fossil fuel consumption by the transportation sector is also sharply increasing in the non-Annex I countries as well. Thus, the total greenhouse gas emission from transportation will rise in the future. It is reported that as much as 90% of global biomass burning is human-initiated, and that such burning is increasing with time (NASA, 1999). Forest products are the major source of biomass, along with agricultural as well as household wastes. The CO_2 from biomass has long been considered to be the source of feedstock during photosynthesis by plants. Therefore, the increase in CO_2 from biomass burning can not be considered to be unsustainable, as long as the biomass is not contaminated through 'processing' before burning. CO_2 from unaltered biomass is distinguished from CO_2 emitted from processed fuels.

Table 9.1 Concentrations, global warming Potentials (GWPs), and atmospheric lifetimes of GHGs.

Gas	Pre-1750 Concentration	Current Topospheric Concentration	GWP (100-yr Time Horizon)	Life Time (years)
carbon dioxide (CO_2)	280 ppm	374.9	1	varies
methane (CH_4)	730ppb	1852ppb	23	12
nitrous oxide (N_2O)	270	319ppb	296	114
CFC-11 (trichlorofluoromethane) (CCl_3F)	0	256 ppt	4600	45
CFC-12 (dichlorodifluoromethane) (CCl_2F_2)	0	546 ppt	10600	100
CFC-113 (trichlorotrifluoroethane) ($C_2Cl_3F_3$)	0	80 ppt	6000	85
carbon tetrachloride (CCl_4)	0	94 ppt	1800	35
methyl chloroform (CH_3CCl_3)	0	28 ppt	140	4.8
HCFC-22 (chlorodifluoromethane) ($CHClF_2$)	0	158 ppt	1700	11.9
HFC-23 (fluoroform) (CHF_3)	0	14 ppt	12000	260
perfluoroethane (C_2F_6)	0	3 ppt	11900	10000
sulfur hexafluoride (SF_6)	0	5.21 ppt	22200	3200
trifluoromethyl sulfur pentafluoride (SF_5CF_3)	0	0.12 ppt	18000	3200

Source: IPCC, 2001

To date, any processing involves the addition of toxic chemicals. Even if the produced gases do not show detectable concentration of toxic products, it is conceivable that the associated CO_2 will be different from CO_2 of organic origin. The CO_2 emission from biomass which is contaminated with various chemical additives during processing has been calculated and deducted from the CO_2 which is good for the photosynthesis that does not contribute to global warming.

9.4 Classification of CO_2

Carbon dioxide is considered to be the major precursor for current global warming problems. Previous theories were based on the "chemicals are chemicals" approach of the two-time Nobel Laureate Linus Pauling's vitamin C and antioxidant experiments. This approach advanced the principle that, whether it is from natural or synthetic sources, and irrespective of the pathways it travels, vitamin C is the same. This approach essentially disconnects a chemical product from its historical pathway. Even though the role of pathways has been understood by many civilizations for centuries, systematic studies questioning their principle is a very recent phenomenon. For instance, only recently (Gale *et al.*, 1995) it was reported that vitamin C did not lower death rates among elderly people, and may actually have increased the risks of dying. Moreover, ß carotene supplementation may do more harm than good in patients with lung cancer (Josefson, 2003). Obviously, such a conclusion cannot be made if subjects were taking vitamin C from natural sources. In fact, the practices of people who live the longest lives indicate clearly that natural products do not have any negative impact on human health (New York Times, 2003) More recently, it has been reported that antioxidant supplements including vitamin C should be avoided by patients being treated for cancer, as the cancer cells gobble up vitamin C faster than normal cells, which might give greater protection for tumors rather than normal cells (Agus *et al.*,1999). Antioxidants that are presently in nature are known to act as anti-aging agents. Obviously, these antioxidants are not the same as those synthetically manufactured. The previously used hypothesis that "chemicals are chemicals" fails to distinguish between the characteristics of synthetic and natural vitamins and antioxidants. The impact of synthetic antioxidants and vitamin C in body metabolism would be different than that of natural sources. Numerous other cases can be cited demonstrating that the pathway involved in producing the final product is of utmost importance. Some examples have recently been investigated by Islam and others (Islam, 2004; Khan *et al.*, 2006; Khan and Islam, 2006; Zatzman and Islam, 2006). If the pathway is considered, it becomes clear that organic produce is not the same as non-organic produce, natural products are not the same as bioengineered products, natural pesticides are not the same as chemical pesticides, natural leather is not same as synthetic plastic, natural fibers are not the same as synthetic fibers, and natural wood is not same as fiber- reinforced plastic (Islam, 2006). In addition to being the only ones that are good for the long term, natural products are also extremely efficient and economically attractive. Numerous examples are given in Khan and Islam (2006). Unlike synthetic hydrocarbons, natural vegetable oils are reported to be easily degraded by bacteria (AlDarbi *et al.*, 2005). Application of wood ash to remove arsenic from aqueous streams is more effective than removal by any synthetic chemicals (Rahman, *et al.*, 2004; Wassiuddin *et al.*, 2002). Using the same analogy, carbon dioxide has also been classified based on the source from where it is emitted, the pathway it traveled, and age of the source from which it came (Khan and Islam, 2006).

Carbon dioxide is classified based on a newly developed theory. It has been reported that plants favor a lighter form of carbon dioxide for photosynthesis and discriminate against heavier isotopes of carbon (Farquhar *et al.*, 1989). Since fossil fuel refining involves the use of various toxic additives, the carbon dioxide emitted from these fuels is contaminated, and is not favored by plants. If the CO_2 comes from wood burning, which has no chemical additives, this CO_2 will be most favored by plants. This is because the pathway the fuel travels from refinery to combustion devices makes the refined product inherently toxic (Chhetri *et al.*, 2006). The CO_2 that the plants do not synthesize accumulates in the atmosphere. The accumulation of this rejected CO_2 must be accounted for in order to assess the impact of human activities on global warming. This analysis provided a basis for discerning natural CO_2 from "manmade $CO_{2,}$" which could be correlated with global warming.

9.5 Role of Water in Global Warming

The flow of water in different forms has a great role in climate change. Water is one of the components of natural transport phenomenon. Natural transport phenomenon is a flow of complex physical processes. The flow process consists of production, storage and transport of fluids, electricity, heat, and momentum (Figure 9.8). The most essential material components of these processes are water and air, which are also the indicators of natural climate. Oceans, rivers, and lakes form both the source and sink of major water transport systems. Because water is the most abundant matter on earth, any impact on the overall mass balance of water is certain to impact the global climate. The interaction between water and air in order to sustain life on this planet is a testimony to the harmony of nature. Water is the most potent solvent, and also has very high heat storage capacity. Any movement of water through the surface and the Earth's crust can act as a vehicle for energy distribution. However, the only source of energy is the sun, and sunlight is the most essential ingredient for sustaining life on earth. The overall process in nature is inherently sustainable, yet truly dynamic. There is not one phenomenon that can be characterized as cyclic. Only recently, scientists have discovered that water has memory. Each phenomenon in nature occurs due to some driving force, such as pressure for fluid flow, electrical potential for the flow of electricity, thermal gradient for heat, and chemical potential for a chemical reaction to take place. Natural transport phenomena cannot be explained by simple mechanistic views of physical processes by a function of one variable. Even though Einstein pointed out the possibility of the existence of a fourth dimension a century ago, the notion of extending this dimensionality to infinite numbers of variables is only now coming to light (Islam, 2006). A simple flow model of natural transport phenomenon is presented in Figure 9.9. This model shows that nature has numerous interconnected processes, such as production of heat, vapor, electricity and light, storage of heat and fluid, and flow of heat as well as fluids. All of these processes continue for infinite time and are inherently sustainable. Any technologies that are based on natural principles are sustainable (Khan and Islam, 2006).

Water plays a crucial role in the natural climatic system. Water is the most essential as well as the most abundant ingredient of life. Just as 70% of the earth's surface is covered with water, 70% of the human body is constituted of water. Even though the value and sanctity of water have been well-known for thousands of years in Eastern cultures, scientists in the West are only now beginning to break out of the "chemicals are chemicals" mode and examine the concept that water has memory, and that numerous intangibles

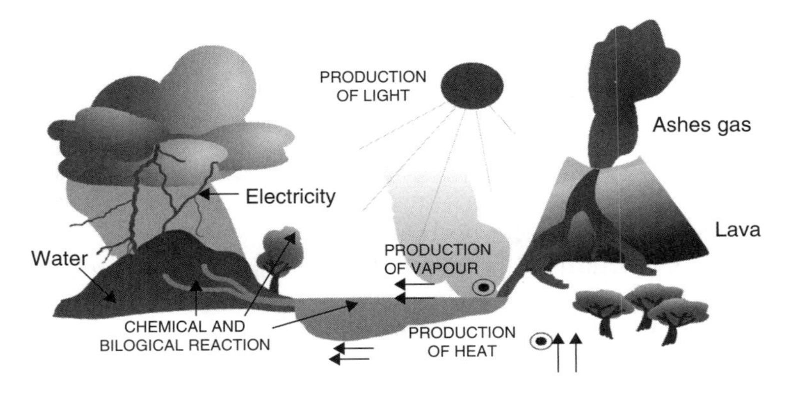

Figure 9.9 Natural transport phenomenon (after Fuchs, 1999).

(most notably the pathway and intention behind human intervention) are important factors in defining the value of water (Islam, 2006).

At the industrial/commercial level, however, preposterous treatment practices, such as the addition of chlorine to 'purify;' the use of toxic chemicals (soap) to get rid of dirt (the most potent natural cleaning agent [Islam, 2006]); the use of glycol (very toxic) for freezing or drying (getting rid of water) a product; the use of chemical CO_2 to render water into a dehydrating agent (opposite to what is promoted as 'refreshing'), then again demineralization followed by the addition of extra oxygen and ozone to 'vitalize;' and the list seems to continue forever. Similar to what happens to food products (we call that the degradation of the following chemical technology chain: Honey → Sugar → Saccharine → Aspartame), the chemical treatment technique promoted as water purification has taken a spiral-down turn (Islam, 2005). Chlorine treatment of water is common in the West and is synonymous with civilization. Similarly, transportation in copper pipe and distribution through stainless steel (reinforced with heavy metal), storage in synthetic plastic containers and metal tanks, and mixing of groundwater with surface water (itself collected from 'purified' sewage water) are common practices in 'developed' countries. More recent 'innovations,' such as Ozone, UV, and even H_2O_2, are proving to be worse than any other technology. Overall, water remains the most abundant resource, yet 'water war' is considered to be the most certain destiny of the 21st century. Modern technology development schemes termed by Robert Curl (a Nobel Laureate in Chemistry) as a 'technological disaster,' seem to have targeted the most abundant resource (Islam, 2006).

Water vapor is considered to be one of the major greenhouse gases in the atmosphere. The greenhouse gas effect is thought to be one of the major mechanisms by which the radiative factors of the atmosphere influence the global climate. Moreover, the radiative regime of the radiative characteristics of the atmosphere is largely determined by some optically active component, such as CO_2 and other gases, water vapor, and aerosols (Kondratyev and Cracknell, 1998). As most of the incoming solar radiation passes through atmosphere and is absorbed by the Earth's surface, the direct heating of the surface water and evaporation of moisture results in heat transfer from the Earth's surface to the atmosphere. The transport of heat by the atmosphere leads to the transient weather system. The latent heat released due to the condensation of water vapors and the clouds play an important role in reflecting incoming short-wave solar radiation, and absorbing and emitting long-wave radiation. Aerosols, such as volcanic dust and the particulates of fossil fuel combustion, are important factors in determining the behavior of the climate system. Kondratyev and Cracknell (1998) reported that the conventional theory of calculating global warming potential only account for CO_2, ignoring the contribution of water vapor and other gases in global warming. Their calculation scheme took into account the other components affecting the absorption of radiation, including CO_2, water vapor, N_2, O_2, CH_4, NOx, CO, SO_2, nitric acid, ethylene, acetylene, ethane, formaldehyde, chlorofluorocarbons, ammonia, and aerosol formation of different chemical composition and various sizes. However, this theory fails to explain the effect of pure water vapor and the water vapor that is contaminated with chemical contaminants.

The impact of water vapor on climate change depends on the quality of water evaporated, its interaction with the atmospheric particulates of different chemical composition, and size of the aerosols. There are at least 70,000 synthetic chemicals being used regularly throughout the world (Icenhower, 2006). It has further been estimated that more than 1000 chemicals are introduced every year. Billions of tons of fossil fuels are consumed each year to produce these chemicals, which are the major sources of water and air contamination. The majority of these chemicals are very toxic and radioactive, and the particulates are being continuously released into the atmosphere. The chemicals also reach water bodies by leakage, transportation loss, and as byproducts of pesticides, herbicide, and water disinfectants. The industrial wastes which are contaminated with these chemicals finally reach the water bodies and contaminate the entire water system. The particulates of these chemicals and aerosols when mixed with water vapor may increase the absorption characteristics in the atmosphere, thereby increasing the possibility of trapping more heat. However, pure water vapor is one of the most essential components of the natural climate system, and will have no impacts on global warming. Moreover, most of the water vapors will end up transforming into rain near the Earth's surface, and will have no effect on the absorption and reflection. The water vapors in the warmer part of the Earth could rise to higher altitudes, as they are more buoyant. As the temperature decreases in the higher altitude, the water vapor gets colder, decreasing its ability to retain water. This reduces the possibility of increasing global warming.

Water is considered to have memory (Tschulakow *et al.*, 2005). Because of this property, the assumption of the impact of water vapor on global warming cannot be explained without the knowledge of memory. The impact will depend on the pathway it traveled before and after the formation of vapor from water. Gilbert and Zhang (2003) reported that nanoparticles change their crystal structure when they are wet. The change of structure taking place in the nanoparticles in the water vapor and aerosols in the atmosphere has profound impact on climate change. This relation has been explained, based on the memory characteristics of water and its pathway analysis. It is reported that water crystals are entirely sensitive to the external environment, and take different shape based on the input (Emoto, 2004). Moreover, the history of water memory can be traced by its pathway analysis. The memory of water might have a significant role to play in technological development (Hossain and Islam, 2006). Recent attempts have been directed toward understanding the role of history on the fundamental properties of water. These models take into account the intangible properties of water. This line of investigation can address the global warming phenomenon. The memory of water not only has impacts on energy and ecosystems, but also has a key role to play in the global climate scenario.

9.6 Characterization of Energy Sources

Various energy sources are classified based on a set of newly developed criteria. Energy is conventionally classified, valued, or measured based on the absolute output from a system. The absolute value represents the steady state of the energy source. However, modern science recognizes that such a state does not exist, and that every form of energy is in a state of flux. This chapter characterizes various energy sources based on their pathways. Each form of energy has a set of characteristics features. Anytime these features are violated through human intervention, the quality of the energy form declines. This analysis enables one to assign greater quality index to a form of energy that is closest to its natural state. Consequently, the heat coming from wood burning and the heat coming from electrical power will have different impacts on the quality. Just as all chemicals are not the same, different forms of heat coming from different energy sources are not the same. The energy sources are based on the global efficiency of each technology, the environmental impact of the technology, and overall value of energy systems (Chhetri *et al.*, 2006). The energy sources are classified based on the age of the fuel source in nature as it is transformed from one form to another (Chhetri *et al.*, 2006).

Various energy sources are also classified according to their global efficiency. Conventionally, energy efficiency is defined for a component or service as the amount of energy required in the production of that component or service, for example, the amount of cement that can be produced with one billion Btu of energy. Energy efficiency is improved when a given level of service is provided with reduced amounts of energy inputs, or services or products are increased for a given amount of energy input. However, the global efficiency of a system is defined as the efficiency calculated based on the energy input, products output, the possibility of multiple use of energy in the system, the use of the system's byproducts, and its impacts to the environment. The global efficiency calculation considers the source of the fuel, the pathways the energy system travels, conversion systems, impacts to human health and the environment, and intermediate as well as byproducts of the energy system. Farzana and Islam (2006) calculated the global efficiency of various energy systems. They showed that global efficiencies of higher quality energy sources are higher than those of lower quality energy sources. With their ranking, solar energy source (when applied directly) is the most efficient, while nuclear energy is least efficient, among many forms of energy studied. They demonstrated that previous findings failed to discover this logical ranking because the focus had been on local efficiency. For instance, nuclear energy is generally considered to be highly efficient, which is a true observation, if one's analysis is limited to one component of the overall process. If global efficiency is considered, of course the fuel enrichment alone involves numerous centrifugation stages. This enrichment alone will render the global efficiency very low. As an example, the global efficiency of a wood combustion process is presented.

Figure 9.6 shows the classification of energy sources based on their global efficiency. Based on the global efficiency, the nuclear energy has the lowest efficiency. Direct solar application has the highest efficiency among the energy sources, because the solar energy source is free and has no negative environmental impacts.

9.7 Problems with the Currently used Models

Current climate models have several problems. Scientists have agreed on the likely rise in the global temperature over the next century. However, the current global climatic models can predict only global average temperatures. Projection of climate change in a particular region is considered to be beyond current human ability. Atmospheric Ocean General Circulation Models (AOGCM) are used by IPCC to model climatic features; however, these models are not accurate enough to provide a reliable forecast on how climate may change. They are linear models and cannot forecast complex climatic features. Some climate models are based on CO_2 doubling and transient scenarios. However, the effect on climate of doubling the concentration of CO_2 in the atmosphere cannot predict the climate in other scenarios. These models are insensitive to the difference between natural and industrial greenhouse gases. There are some simple models in use, which use fewer dimensions than complex models, and do not predict complex systems. The Earth System Models of Intermediate Complexity (EMIC) are used to bridge the gap between the complex and simple models; however, these models are not suitable to assess the regional aspects of climate change (IPCC, 2001).

Unsustainable technologies are the major cause of global climate change. Sustainable technologies can be developed following the principles of nature. In nature, all functions are inherently sustainable, efficient and functional for an unlimited time period. In other words, as far as natural processes are concerned, *'time tends to infinity'*. This can be expressed as t or, for that matter, $\Delta t \rightarrow \infty$. By following the same path as the functions inherent in nature, an inherently sustainable technology can be developed (Khan and Islam, 2006). The 'time criterion' is a defining factor in the sustainability and virtually infinite durability of natural functions. Figure 9.10 shows the direction of a nature-based, inherently sustainable technology, as contrasted with an unsustainable technology. The path of sustainable technology is its long-term durability and environmentally wholesome impact, while unsustainable technology is marked by Δt approaching 0. Presently, the most commonly used theme in technology development is to select technologies that are good for t = 'right now,' or $\Delta t = 0$. In reality, such models are devoid of any real basis (termed "aphenomenal" by Khan *et al.*, (2005)) and should not be applied in technology development if we seek sustainability for economic, social and environmental purposes. While developing the technology for any particular climatic model, this sustainability criterion is truly instrumental. The great flaw of conventional climate models is that they are focused on the extremely short term, t = 'right now,' or $\Delta t = 0$.

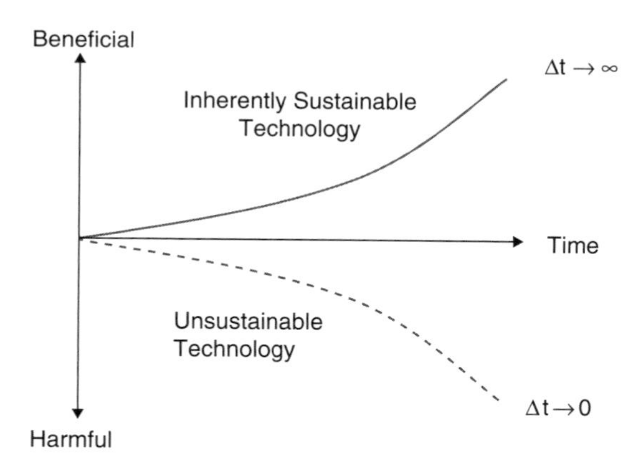

Figure 9.10 Direction of sustainable/green technology (redrawn from Islam, 2005b).

9.8 Sustainabile Energy Development

Different technologies that are sustainable for the long term and do not produce any greenhouse gases are presented. Technology has a vital role to play in modern society. One of the major causes of present-day environmental problems is the use of unsustainable technologies. The use of thousands of toxic chemicals in fossil fuel refining and industrial processes, to the products of personal care, such as body lotion, cosmetics, soaps, and others has polluted much of the world in which we live (Chhetri *et al.*, 2006; The Globe and Mail 2006). Present-day technologies are based on the use of fossil fuel in the form of primary energy supply, production or processing, and the feedstock for products, such as plastic. Every stage of this development involves the generation of toxic waste, rendering products harmful to the environment. According to the criterion presented by Khan and Islam (2005 a; 2005 b), toxicity of products mainly comes from the addition of chemical compounds that are toxic. This leads to continuously degrading quality of the feedstock. Today, it is becoming increasingly clear that the "chemical addition" that once was synonymous with modern civilization is the principal cause of numerous health problems, including cancer and diabetes. A detailed list of these chemicals has been presented by Khan and Islam (2006 b).

Proposing wrong solutions for various problems has become progressively worse. For instance, the United States is the biggest consumer of milk, most of which is 'fortified' with calcium. Yet the U.S. ranks at the top of the list of osteoporosis patients per capita in the world. Similar standards are made about the use of vitamins, antioxidants, sugar-free diet, and so forth. Potato farms on Prince Edward Island in eastern Canada are considered a hotbed for cancer (The Epoch Times, 2006). Chlorothalonil, a fungicide, which is widely used in the potato fields, is considered a carcinogen. The U.S. EPA has classified chlorothalonil as a known carcinogen that can cause a variety of ill effects, including skin and eye irritation, reproductive disorders, kidney damage, and cancer. Environment Canada (2006) published lists of chemicals which were banned at different times. This indicates that all of the toxic chemicals used today are not beneficial, and will be banned from use someday. This trend continues for each and every technological development. However, few studies have integrated these findings to develop a comprehensive cause-and-effect model. This comprehensive scientific model developed by Khan and Islam (2006) is applied for screening unsustainable and harmful technologies right at the onset. Some recently developed technologies that are sustainable for the long term are presented.

One of the sustainable technologies presented in this chapter is the true green biodiesel model (Chhetri and Islam, 2006). As an alternative to petrodiesel, biodiesel is a renewable fuel that is derived from vegetable oils and animal fats. However, the existing biodiesel production process is neither completely 'green' nor renewable because it utilizes fossil fuels, mainly natural gas, as an input for methanol production. It has been reported that up to 35% of the total primary energy requirement for biodiesel production comes from fossil fuel (Carraretto *et al.*, 2004). Methanol makes up about 10% of the feedstock input, and since most methanols are currently produced from natural gas, biodiesel is not completely renewable (Gerpen *et al.*, 2004). The catalysts and chemicals currently in use for biodiesel production are highly caustic and toxic. The synthetic catalysts used for the transesterification process are sulfuric acid, sodium hydroxide, and potassium hydroxide, which are highly toxic and corrosive chemicals. The pathway for conventional biodiesel production and petrodiesel production follows a similar path (Figure 9.11). Both the fuels have similar emission of pollutants, such as benzene, acetaldehyde, toluene, formaldehyde, acrolein, PAHs, and xylene (EPA, 2002). However, the biodiesel has fewer pollutants in quantity than petrodiesel.

Chhetri and Islam (2006) developed a process that rendered the biodiesel production process truly green. This process used waste vegetable oil as biodiesel feedstock. The catalysts and chemicals used in the process were nontoxic, inexpensive, and natural. The catalysts used were sodium hydroxide, obtained from the electrolysis of natural sea salt, and potassium hydroxide, from wood ash. The new process substituted the fossil fuel-based methanol with ethanol produced by grain-based renewable products. Use of natural catalysts and nontoxic chemicals overcame the limitations of the existing process. Fossil fuel was replaced by direct solar energy for heating, making the biodiesel production process independent of fossil fuel consumption.

Khan *et al.* (2006) developed a criterion to test the sustainability of the green biodiesel. According to this criterion, to consider any technology sustainable in the long term, it should be

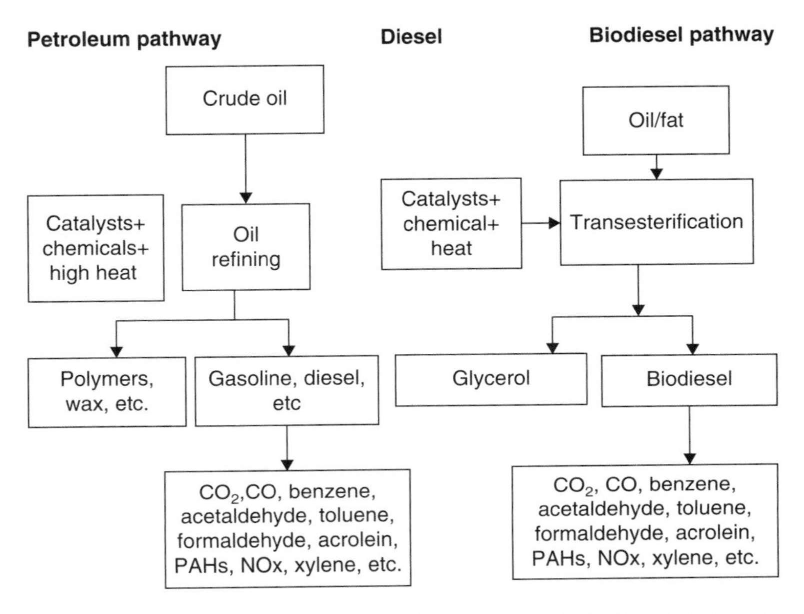

Figure 9.11 Pathway of mineral diesel and conventional biodiesel (Chhetri and Islam, 2006).

environmentally appealing, economically attractive and socially responsible. The technology should continue for infinite time, maintaining the indicators functional for all time horizons. For a green biodiesel, the total environmental benefits, social benefits, and economics benefits are higher than the input for all time horizons. For example, in the case of environmental benefits, green biodiesel burning produces 'natural' CO_2, which can be readily synthesized by plants. The formaldehyde produced during biodiesel burning is also not harmful, as there are no toxic additives involved in the biodiesel production process. The plants and vegetables for biodiesel feedstock production also have positive environmental impacts. Thus switching from petrodiesel to biodiesel fulfils the condition $dCn_t/dt \geq 0$ where Cn is the total environmental capital of life cycle process of biodiesel production. Similarly, the total social benefit (Cs) $dCS_t/dt \geq 0$ and economic benefit (Ce) $dCe_t/dt \geq 0$ by switching from mineral diesel to biodiesel (Khan *et al.*, 2006). Figure 9.12 gives a sustainable regime for an energy system for infinite time and fulfills the environmental, social, and economic indicators. Biodiesel can be used in practically all areas where petrodiesel is being used. This substitution will help

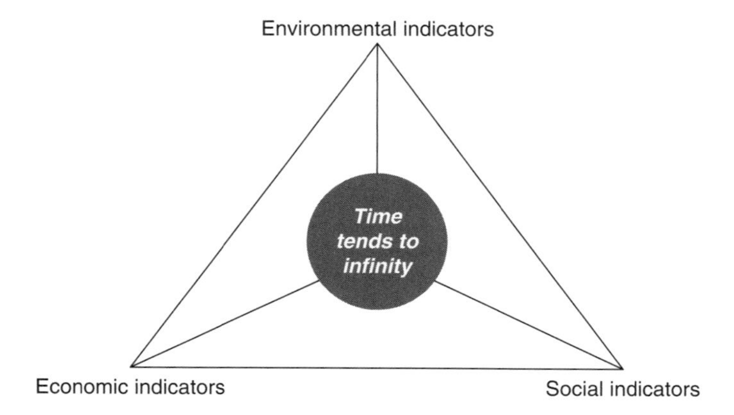

Figure 9.12 Major elements of sustainability in technology development (After Khan *et al.*, 2006).

to significantly reduce the CO_2 responsible for the current global warming problem.

Bioethanol is another sustainable technology that offers a replacement for gasoline engines. The global gasoline consumption is approximately 12 billion liters per year (Martinot, 2005). This is one of the major sources for CO_2 emission. Current gasoline replacement by bioethanol fuel is approximately 32 billion liters worldwide. The conventional bioethanol production from various feedstocks, such as switchgrass and other biomass, involves the use of chemicals for its breakdown in various stages. For example, the ethanol production process from switchgrass involves acid hydrolysis as a major production process. It is reported that the conversion of switchgrass into bioethanol uses concentrated sulfuric acid at 4:1 (acid biomass ratio) which makes the process unsustainable; the produced fuel is a highly toxic fuel and produces fermentation inhibitors, such as 5-hydroxymethylfurfural (5-HMF) and furfural acid, during the hydrolysis process, which reduces the efficiency (Bakker *et al.*, 2004). Moreover, the conventional bioethanol production also consumes huge fossil fuel as a primary energy input, making the ethanol production dependent on fossil fuels.

Development of bioenergy on a large scale requires the deployment of environmentally acceptable, low-cost energy crops, as well as sustainable technologies to harness them with the least environmental impact. Sugarcane, corn, switchgrass, and other ligocellulogic biomass are the major feedstocks for ethanol production.

Chhetri *et al.* (2006) developed a process that makes the bioethanol production process truly green. They proposed the use of nontoxic chemicals and natural catalysts to make the bioethanol process truly environmental friendly. The technology has been tested for long-term sustainability using a set of sustainability criteria. The ethanol produced using the natural and nontoxic catalysts will produce natural CO_2 after combustion, and has no impacts on global warming.

Recently, a jet engine has been designed in order to convert sawdust waste to electricity (Vafaei, 2006). This is one of the most efficient technologies, since it can use a variety of fuels for combustion. This was designed primarily to use sawdust to produce power for the engine (Figure 9.13). In this jet, sawdust is sprayed from the top, where the air blower works to make a jet. Some startup fuel, such as organic alcohol, is used to start up the engine. Once the engine is started, the sawdust and blower will be enough to create power for the engine to run. The main advantage of this jet engine is that it can use a variety of fuels, such as waste vegetable oil and tree leaves. It has been reported that crude oil can be directly burnt in such engines (Vafaei, 2006). The utilization of waste sawdust and waste vegetable oil increases the global efficiency of the systems significantly. The possibility of directly using crude oil can eliminate the various toxic and expensive refining processes, which

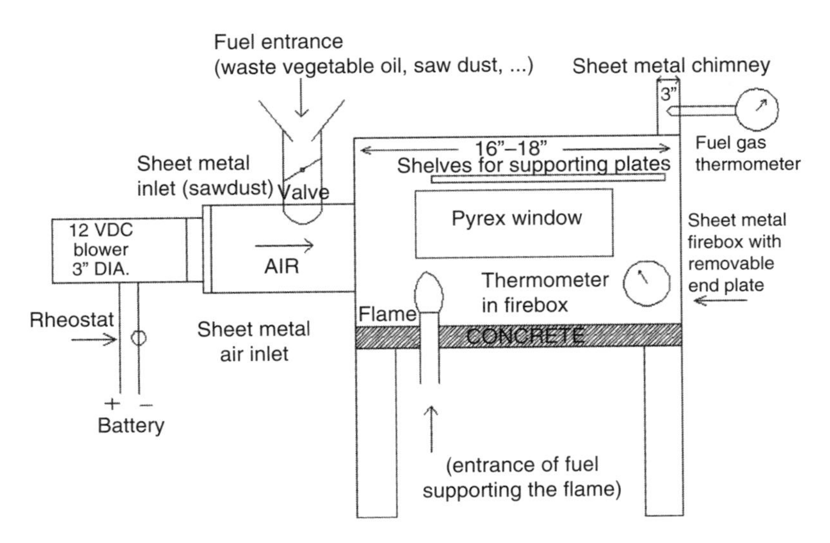

Figure 9.13 Sawdust to electricity model jet engine (Vafaei, 2006).

alone release large amounts of greenhouse gas emission into the atmosphere. This technology is envisaged as one of the most sustainable technologies among the others currently available.

9.9 Zero Waste Energy Systems

Different zero-waste technologies are described that eliminate the production of industrial CO_2. Modern civilization is synonymous with the waste generation (Islam, 2004; Khan and Islam, 2006). This trend has the most profound impact on energy and mass utilization. Conventional energy systems are most inefficient technologies (Khan and Islam, 2012). The more that is wasted, the more inefficient the system is. Almost all industrial and chemical processes produce wastes, and most of them are toxic. The wastes not only reduce the efficiency of a system, but also pose severe impacts on health and the environment, leading to further degradation of global efficiency. The treatment of this toxic waste is also highly expensive. Plastics derivatives from refined oil are more toxic than original feedstocks; the oxidation of a plastic tire at a high temperature produces toxics such as dioxin. The more refined the products are, the more wastes are generated. A series of zero-waste technologies are presented. They are analogous to the 'five zeros' of the Olympic logo which are zero emissions, zero resource waste, zero waste in activities, zero use of toxics, and zero waste in the product life cycle. This model, originally developed by Lakhal and H'Midi (2003) was called the Olympic Green Chain model.

Solar energy is free energy, and is extremely efficient. Direct solar energy is a benign technology. Khan and Islam (2005) developed a direct solar heating unit to heat waste vegetable oil as a heat transfer medium (Figure 9.14). The solar concentrator can heat the oil to more than 300°C, and the heat can be transferred through a heat exchanger for space heating, water heating, or any other purpose. In conventional water heating, the maximum heat that can be stored is 100°C. However, in the case of direct oil heating, the global efficiency of the system is more than 80%. No waste is generated in the system.

Khan *et al.* (2006) developed a heating/cooling and refrigeration system that uses direct solar heat without converting it into electricity. The single-pressure refrigeration cycle is a thermally-driven cycle that uses three fluids. One fluid acts as a refrigerant, the second

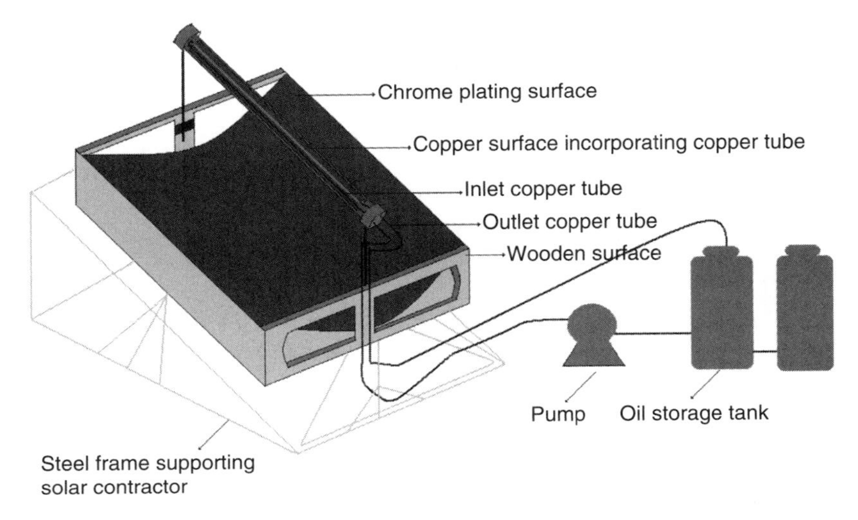

Figure 9.14 Details of solar heating unit (After Khan *et al.*, 2006).

as a pressure-equalizing fluid, and a third as an absorbing fluid. Because the cycle operates at a single pressure, no moving parts, such as a pump or compressor, are required. In order to facilitate fluid motion, the cycle uses a bubble pump that uses heat to ensure the drive. All of the energy input is in the form of heat. Utilization of direct heat for heating, cooling, or refrigeration replaces the use of large amounts of fossil fuel, reducing the CO_2 emission significantly. This type of refrigerator has silent operations, higher heat efficiency, no moving parts, and portability.

Khan *et al.* (2006) developed a novel zero-waste sawdust stove in order to utilize waste sawdust (Figure 9.16). At present sawdust is considered a waste, and management of waste always involves cost. Utilizing the waste sawdust enhances the value addition of the waste material and generates valuable energy as well.

The particulates of the wood burning are collected in an oil-water trap, which is a valuable nano-material for many industrial applications. This soot material can also be used as an ingredient for nontoxic paint.

A high-efficiency jet engine has also been developed (Vafaai and Islam, 2006). This jet engine can use practically any type of solid fuel such as waste sawdust, and liquid fuel such as waste vegetable oil and crude oil to run the engine. A jet is created to increase the surface area in order to increase the burn rate. Direct combustion of

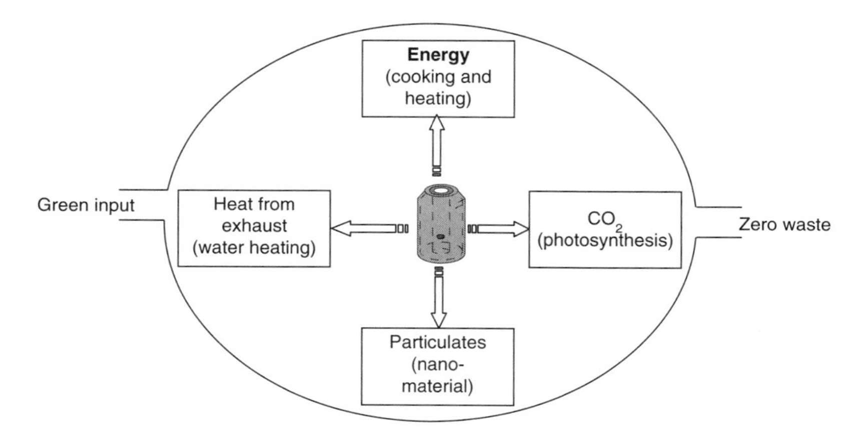

Figure 9.15 Pictorial view of zero-waste models for wood stove (after Chhetri *et al.*, 2006).

crude oil is possible in this engine. This development will eliminate the costly refining and processing of petroleum products. This will have a significant impact in reducing industrial CO_2 as the refining process is what makes the CO_2 a toxic product, due to the use of heavy metals and toxic catalysts.

Khan and Islam (2012) proposed an approach for zero-waste (mass) utilization for a typical urban setting, including processing and regeneration of solid, liquid, and gas. In this process, kitchen waste and sewage waste are utilized for various purposes, including biogas production, desalination, water heating from flue gas, and good fertilizer for agricultural production. The carbon dioxide generated from biogas burning is utilized for the desalination plant. This process achieves zero-waste in mass utilization. The process is shown in Figure 9.16. The technology development in this line has no negative impact on global warming.

9.10 Reversing Global Warming: The Role of Technology Development

A series of techniques are discussed to reduce industrial $CO_{2,}$ which contributes to global warming. Conventional energy sources such as fossil fuel contribute to greenhouse gases emissions. Because various toxic chemicals and catalysts are used for refining/processing

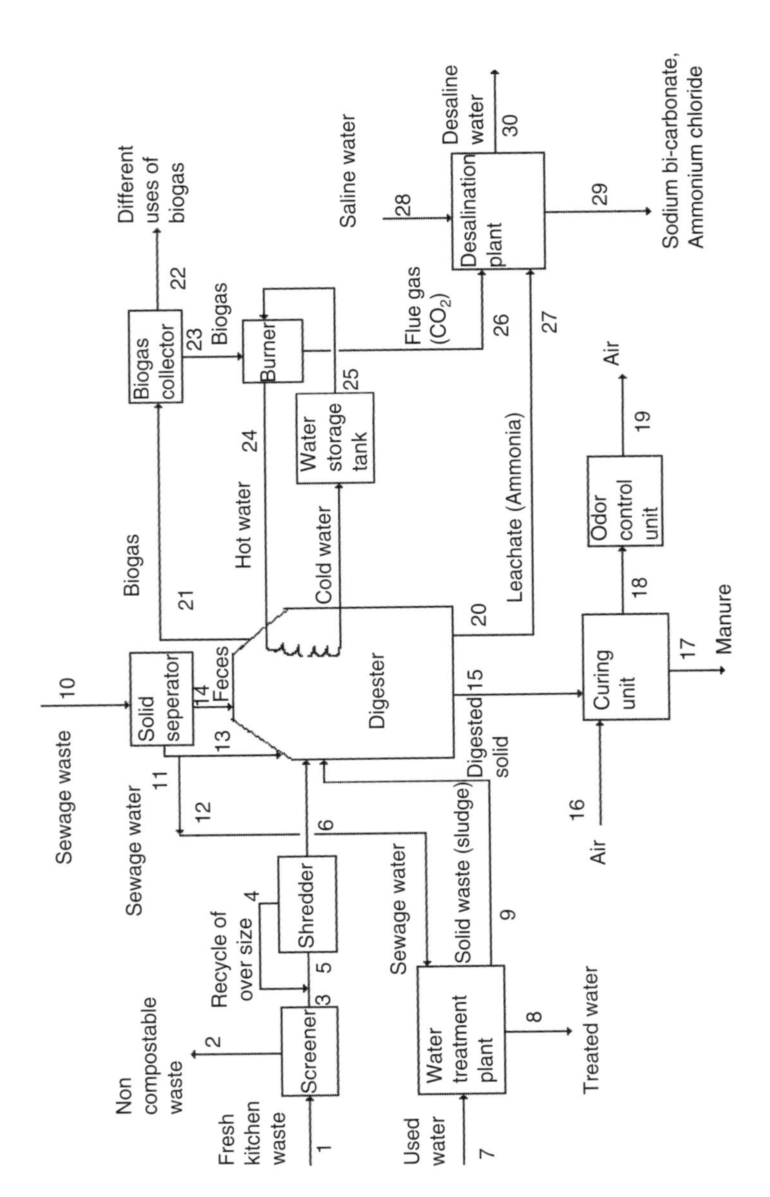

Figure 9.16 Zero-waste mass utilization scheme.

oil and natural gas, the emitted CO_2 is a toxic product. In addition, fossil fuels have greater properties of carbon isotope ^{13}C, making them more likely to be readily absorbed by plants. This leads to the alteration of the characteristic recycle period of carbon dioxide, causing delays that result in an increase in total CO_2 in the atmosphere. Billions of people in the world use traditional stoves fueled by biomass for their cooking and space heating requirements. It is widely held that wood-burning stoves emit more pollution into the atmosphere, compared with oil- and natural gas-burning stoves. However, a small intervention in wood-burning stoves will result in the emission of natural CO_2, which is essential for natural processes. Identifying the limitations of conventional stoves, a new technique has been developed to achieve zero-waste in such technologies. This line of development will have great impact on technological development in the industrial sector and other sectors as well.

9.11 Deconstructing the Myth of Global Warming and Cooling

Global warming and climate change have been hot discussion topics worldwide since the late 1970s. It has been postulated that the buildup of carbon dioxide in the atmosphere results in irreversible climate change. Even though carbon dioxide has been blamed as the sole cause for the global warming, there is no scientific evidence that all carbon dioxides are responsible for global warming. This is not only due to lack of proper scientific research, but also due to the orientation of research based on pre-conceived ideas.

There are two schools of thought regarding global warming. The first is based on the argument that the global warming is caused by the emission of greenhouse gases from various residential, commercial, and industrial activities. This argument has received widespread attention, and several national and international organizations in the world are working to reduce the greenhouse gas emission from different sectors. A recent report by IPCC (2007) declared that most of global warming is contributed by human activities. Global concentrations of greenhouse gases, such as carbon dioxide, methane, and nitrous oxides, among others, are considered major precursors of this warming effect (IPCC, 2001). However, none of the studies differentiate the impacts of such gases according to whether their origins are natural or synthetic.

The available evidence lends no support for any assumption that naturally-sourced concentrations of such gases cause or participate in global warming: on the contrary, they should accordingly be excluded from the models being used to predict climate change (Chhetri and Islam, 2007a; Chhetri and Islam, 2007b; Khan and Islam, 2007; Chhetri and Zatzman, 2008). With the technologies that process or refine by way of synthesizing any part of a concentration of so-called "greenhouse gas" (GHG), however, matters stand rather differently: these should definitely be placed in the dock.

Without taking up a different path from that of current technological development, however, it seems unlikely that any significant abatement of global warming effects can be effected. Indeed: despite spending billions of dollars, no significant results have been achieved so far. Several studies show that current global warming problem is real and is due to the anthropogenic emission of greenhouse gases. However, different models developed to describe this phenomenon are not without controversy. Alexiadis (2007) recently developed a feedback model that was applied to describe the influence of the carbon dioxide emission due to human activity on the global temperature and the atmospheric CO_2 concentration. It was argued that the anthropogenic carbon dioxide is the main driving force in global warming, and even in the case of reduction of the emissions, the temperature will keep increasing for a certain time.

The second school of thought is based on the argument that anthropogenic impacts in global warming are negligible compared to the natural driving forces such as solar radiation, precession of Earth, Earth's outgassing, microbial activities, volcanoes, and ocean currents that cause the global warming (Khilyuk, and Chilingar, 2006). This theory claims that the Earth is actually in a cooling phase rather than warming, based on historical periods beginning from the formation of Earth. This book is based on this second school of thought, and has described in detail the scientific evidences that argue the human impacts on current global warming are negligible. This school of thought is aptly represented by the recent work of Sorochkin *et al.* (2007). In this work, it has been clearly stated in the foreword and preface that current debate in global warming substantially lacks the scientific breadth of the problem and sufficient data to describe this hotly debated subject. The authors argue that computer models are made to forcefully argue human contributions to greenhouse gas emission, but lack historical observation and influence of natural processes that in fact dominate the climate change. It has been stated that temperature evolution is one of the Earth's major dynamic processes, and human impact in this process could be negligible. To the contrary, the authors have also agreed that prior to the industrial revolution, all climate changes were naturally driven, and climate has been changing continuously for last 4.5 billion years in terms of intensity and duration. In this book, the analysis of climate change has been included based on the theory of evolution of Earth, outgassing, precession of Earth, solar systems, and ocean formation from the beginning of the Earth's formation to date.

In Chapter 1 of the book of Sorochkin *et al.* (2007), the authors present a theory of evolution that describes the process of chemical-density differentiation of Earth's matter, which has been considered the main planetary process driving the evolution of Earth. This theory explains the formation and growth of dense iron oxide core, the emergence of chemical density convection in the mantle, and the formation of the lighter silicate crust of Earth. The authors claimed that, although it varies with geological history, the input into the total inner earth's energy is constituted 90% by endogenous energy, 9% by radioactive decay, and the last 1% by tidal deformation generated inside the Earth's body. Moreover, gravitational energy of space matter was the dominant energy at the time of formation of Earth. These natural forces still play a key role in the tectonic and other evolutionary processes on Earth. The authors believe that the impact of these natural forces, which play a significant role in global warming, has been absent in the current scientific debates.

In Chapter 2, the Earth's degassing and stages of formation of the hydrosphere and atmosphere are described. The authors argue that the juvenile Earth was completely deprived of hydrosphere, and the atmosphere consisted almost entirely of nitrogen. From the early Archaen time when the first sea basins were formed, degassed carbon dioxide entered into the atmosphere and formed a carbon dioxide-nitrogen atmosphere. Ocean water then started interacting with the oceanic crust rocks. It is argued that this process resulted in the considerable changes in composition, pressure, and formation of iron-bearing deposits, which suppressed the oxygenation of the atmosphere and generation of abiogenic methane that could have been the basis for life forms on Earth.

In Chapter 3, the authors have described the adiabatic theory of the greenhouse effect as the major theory advanced to account for the global warming phenomenon. As they describe, more than 67% of heat transfer in the atmosphere occurs by convection possessing adiabatic properties. This adiabatic model is used to explain quantitatively the temperature regimes of planetary troposphere and impacts of composition and pressure on climate systems. Based on the adiabatic theory, the authors have compared the relative effects of natural and anthropogenic influences on Earth's climate change, and concluded that the anthropogenic influence is negligible compared with natural factors. Based on this theory, the total temperature rise predicted in the worst case scenario was approximately 0.01°C, attributed to total anthropogenic emission of greenhouse gases, such as carbon dioxide and methane. The authors have also linked the cooling and warming with the Earth's precession angle. It has been described that the earth's climate cools down when the Earth's precession angle is smaller, and vice versa. Hence, they conclude that the traditional explanation of global warming due to man-made sources is no more than a myth.

However, consideration of adiabatic conditions in the atmosphere is rife with linearized and linearizing assumptions, and the model's basic assumption is itself questionable (Khan *et al.*, 2006a). For an adiabatic condition, the following three conditions are generally assumed: the existence of a perfect vacuum between the system and surrounding area, a perfect reflector around the system like the thermo flux mechanism to resist radiation, and the presence of zero heat diffusivity material that isolates the system. Can any of these conditions can be found or maintained anywhere in the atmosphere? In this respect, the predictive value of what the authors have modeled seems little better than the worst of the straight-line predictions of ocean level rise and glacial melt advanced by those marketing scenarios of imminent doom stalking the fundamental natural order of the planet stemming from driving once too often to the local convenience store. No one knows whether the occurrence, either simultaneously or relatively close in time, of similar local effects across various parts of the Earth's surface may be evidence of a larger, more fundamental global phenomenon, so hypothesizing an explanation on the basis of remaining in ignorance about those missing pieces cannot be accepted as scientifically-grounded. On the other hand, however, although the present work indeed addresses phenomena on a scale that seems far more appropriate to account for atmospheric-level climate change (as distinct from phenomena that really address the messing up of human habitat by thoughtlessness), the climate change model it advances based on the adiabatic theory as presented there fails to make its case for how fundamental, atmosphere-wide climate change actually works.

Chapter 4 deals with the evolution of Earth's climate throughout the geological history. For all geological periods, the temperature regimes under the most probable initial and boundary conditions have been computed assuming the adiabatic conditions. This theory explains that the Earth is actually in a cooling geological time. Based on the analysis of geological periods, the Earth's atmospheric temperature is predicted with numerous illustrations. The bacterial nature of Earth's glaciations, which could have occurred due to the removal of nitrogen from atmosphere by nitrogen consuming bacteria that reduced the total pressure of atmosphere, could be one of the main causes of the reduction in temperature. The prediction based on adiabatic theory claims that the earth's temperature 2-3 billion years ago was much higher than the current increase. However, as explained earlier, as there is no adiabatic condition as such in atmosphere to fully support this assumption. A nonlinear model that omits the assumption that the atmospheric systems works adiabatically could fully explain the climate changes to predict the global warming.

In Chapter 5, evolution of climate is used to describe nonuniformity of distribution in time periods of accumulation of mineral deposits. This theory was used to explain the formation of largest iron ore deposits at the end of the Archaean and Early Proterozoic eras. This chapter illustrates the scientific evidence, from prehistoric time to date, that explains how the minerals deposits interacted with the atmospheric gases that had an impact on atmospheric composition and its chemical density.

In Chapter 6, the authors have linked the origin of life on Earth to the formation of reducing environment due to intense generation of abiogenic methane at the beginning of the Archaean period, leading to the formation of the first organic compounds, such as formaldehyde, hydrogen, and cyanide, which served as the building materials for the primitive life forms. The subsequent development of life on Earth unfolded according to biological laws, under strong influence of geochemical and climatic conditions. The authors believe that the main stages of life transformations coincided with the main geotectonic breaks during the development of the Earth. Moreover, the development of life forms significantly influenced the biological processes on Earth. It has also reinforced the statement that nitrogen-consuming bacteria were the main cause to reduce the partial pressure of nitrogen; as a result, the total pressure of atmosphere led to the beginning of cooling phase from the beginning of mid-Proterozoic time.

In Chapter 7, the authors described how solar radiation, Earth's outgassing, and microbial activities operate as the major three forces of nature driving the Earth's climate. The extent of impacts from natural driving forces has been quantified here. Solar luminosity, solar system geometry, and the gaseous composition of atmosphere were considered as the first-order climate drivers. Global distribution of continents and oceans on the Earth's surface were considered the second-order climate drivers. Similarly, orbital and solar variability, large scale oceanic tidal cycles, and variation in the structure of oceanic currents were considered the third-order climate drivers. Volcanoes, natural weathering, regional tectonics, *El Niño*, solar storms and flares, short ocean tidal cycles, meteorite impacts, and human interventions were considered to be the fourth-order climate drivers. It was shown that global forces of nature are at least four to five orders of magnitude greater than those due to human activities. The authors argued that the effects of anthropogenic influence on the global climate are negligible.

The authors are among the most renowned climate scientists in the world. The inclusion of the formation of Earth from prehistoric time to study the global climate has made this book more relevant, as most of the current scientific literature lacks such analysis. Sorokthin, Chilingar and Khilyuk in this book have provided the readers with a new approach that combines both the Earth's geological history as well as climate history highlighting the dynamic nature of Earth's processes. This book is a valuable resource for students, practicing engineers and scientists in the field of geophysics, geology, environment, climate change, and biological sciences. It could be a guideline for policymakers and an interesting resource for those who are interested in global warming debates. Overall, this book can benefit a large section of the scientific and public community which is confronting the lack in explanation of climate theories in the present context.

Nothing in this work, however, should be taken to suggest that the mess being made of human habitat and living environments in the short term cannot be addressed, or should be left to nature to "solve" without anyone lifting a finger here and now. Let us grant that anthropogenic sources of pollution have many local effects throughout the planet, especially on the Earth's surfaces, but with very little consequence for the fundamental mechanisms most responsible for adjusting the earth's global climate at the atmospheric and geological levels over the long term. Does it follow that we cannot or should not be taking action here and now to alleviate and mitigate whatever endangers human habitat and its relationship to the wider environment? On the contrary, action at such levels may very well be the sphere in which much that we do not yet know about the relationship of the local to the global will begin to be sorted out. One simple analysis makes the point clear. It is true that there is no amount of CO_2 that we as humans can produce that would create an imbalance to the global climate. Even if all CO_2s are the same (they are not, see Chhetri and Islam, 2007a for more details), irrespective of their natural or industrial origins, two things happen that can change the fate of human species irreversibly. They are artificial products and their impact on humans (before they reach the global ecosystem). For instance, what would be the impact of Freon or DDT, considering that they did not exist in nature before? In that sense, they are an infinite change from what nature offered for millions of years. So, if the changes due to CO_2 are miniscule, and hence, can be ignored, what happens to the infinite change invoked by, say, Freon, that did not exist in nature before. The same principle applies to every artificial product that surrounds us today. Because this change is infinite, it will impact the global system, no matter how vast the ecosystem is. In the same vein, humans will be the first victims and the most affected, because we live in an environment that is affected by our artificial products more than by nature. This remains an area that neither of the sides in the current global warming debate seems prepared to address.

9.12 Concluding Remarks

It is concluded that the current synthetic chemical-based technological developments are the major causes for global warming and climate change problems. Emission industrial CO_2 that is contaminated by the addition of toxic chemicals during fuel refining, processing, and production activities is responsible for global warming. The natural $CO_{2,}$ which is not only beneficial to the environment, but also an essential ingredient for life and biodiversity in the earth, does not cause global warming. For the first time, natural and industrial CO_2 have been differentiated. Carbon dioxide is characterized based on various criteria, such as the origin, the pathway it travels, and isotope numbers. The current status of greenhouse gas emissions from various anthropogenic activities is discussed. The role of water in global warming has been detailed. Various energy sources are classified based on their global efficiencies. The assumptions and implementation mechanisms of the Kyoto Protocol have been critically reviewed, and it is argued that the Clean Development Mechanism of the Kyoto Protocol has become the 'license to pollute," due to its improper implementation mechanism. The conventional climatic models have been deconstructed, and guidelines for new models have been developed in order to achieve true sustainability in technology development in the long term. A series of sustainable technologies that produce natural CO_2 which do not contribute to global warming were presented. Various zero-waste technologies that have no negative impact on the environment are keys to reverse the global warming. This chapter shows that a complete reversal of the current global warming problem is possible only if pro-nature technologies are developed.

10

Is the 3R's Mantra Sufficient?

The modern age is synonymous with wasting habits, whereas nature does not produce any waste. Ranging from economics models that show positive GDP as more money is wasted, to the 3 R's mantra of Reduce, Recycle, Reuse, all promote wasting. The fundamental notion that mass cannot be created or destroyed dictates that only transformation of materials from one phase to another phase take place. However, the mass balance alone does not guarantee zero waste. In nature, every product created as a waste is useful for someone else, thereby making net waste zero. This postulate necessitates that any product that is the outcome of a natural process must be entirely usable by some other process, which in turn would result in products that are suitable as an input to the process. A perfect system is 100% recyclable, and therefore zero-waste. Such a process will renew zero waste as long as each component of the overall process also operates at with the principle of zero waste.

In a desired zero-waste scheme, the products and byproducts of one process are used for another process. The scientific definition of a zero-waste scheme is followed by an example of zero-waste, with detailed calculations showing how this scheme can be formulated.

Following this, various stages of petroleum engineering are discussed in light of the zero-waste scheme.

10.1 Introduction

Fossil fuel energy sources are predominantly used today. Nearly 90% of today's energy is supplied by oil, gas, and coal (Salameh, 2003). The burning of fossil fuel accounts for more than 100 times greater usage than the energy generated through 'renewable' sources (solar, wind, biomass, and geothermal energy). The panic sets in when it is promoted that fossil fuel is limited, and a switch to 'renewable' is the only sustainable option for the viability of human civilization. The question arises as to how one can begin to make the switch. In this, the science of energy production presents a comprehensive analysis that shows that the panic of running out of energy sources is not in conformance with overall energy and mass balance. In addition, it is shown that the currently used 'renewable' schemes are not truly renewable, and are not even efficient, as compared to conventional petroleum production schemes.

According to present consumption level, known reserves for coal, oil, gas, and nuclear correspond to a duration of the order of 230, 45, 63, and 54 years, respectively (Rubbia, 2006). Note that these numbers correspond to energy production from known reserves with currently established techniques. If petroleum operations can be rendered sustainable, this time limitation will become irrelevant. This is not to say that there should be no effort to make use of other natural energy sources. It is important, however, to remain cognizant about the truly 'natural' status of these energy sources. Crude oil is a natural energy source because it can be used without resorting to unnatural processes. Radioactive ores are also natural, but current technologies are not capable of using them as an energy source without resorting to enrichment processes that are highly unnatural. Solar energy is obviously the most appealing energy source, but the process of turning solar energy into 'usable' energy through a series of inefficient conversions, using toxic photovoltaic, battery materials, and fluorescent light distributors is not sustainable and far more insulting to the environment than flaring natural gas. For instance, the mere fact that the most common usage is the use of photovoltaic that has a maximum efficiency of only 15% (Gupta *et al.*, 2006) can be a reason to reject this particular use of solar energy.

The same comment stands for wind energy, for which direct grinding is sustainable (centuries of practice in the Netherlands), whereas converting to electricity is not. Biofuel in this regard offers an interesting take. Direct burning of wood or vegetation is sustainable, and the resulting CO_2 is beneficial to the environment, with the condition that chemical fertilizer, pesticide, or genetic modification was not used.

The argument made in this book is that all currently used energy solutions are energy-inefficient and mass-wasteful. This chapter establishes that a sustainable technology is based on zero-waste and, therefore, offers the greatest possible global efficiency. Inherent to this is the environmental benefit that is an added bonus to the sustainable technology. Following this, petroleum technologies are discussed, with a focus on current practices and recommendations on how to turn these practices sustainable.

10.2 Petroleum Refining

Crude oil is a mixture of hydrocarbons. These hydrocarbon mixtures are separated into commercial products by numerous refining processes. They have very similar composition to vegetable oils. As a result, many properties of the two sets of fluids are similar, including biodegradability, flashpoint, dead oil viscosity, density, bactericidal properties, and so forth. However, petroleum fluids are rarely used in their original form. Even though it is known that petroleum fluids had been used from ancient times all the way up to the Renaissance in various cultures, in the post-Renaissance culture, petroleum fluids are rarely used directly. One exception was the use of crude oil as mosquito repellent in the former Soviet Union. Even though such use eradicated malaria from much of the Soviet Union, they joined in the production of DDT, ever since the Nobel Prize-winning synthesis of this toxic chemical by Muller, most likely for commercial reasons. As DDT was banned in 1972, the use of crude oil as a pesticide did not return into practice. Today, petroleum fluids are transported to refineries prior to any usage. Oil refineries are enormous complex processes. Figure 10.1 shows an oil refinery complex in Dartmouth, Nova Scotia. Refining involves a series of processes to separate and sometimes alter the hydrocarbons in

Figure 10.1 Dartmouth Refinery, Nova Scotia.

crude oil. The fundamental process of refining involves the breakdown of crude oil into its various components and separating them to sell as a value-added product. Because each component loses its whole properties, chemicals are added to restore original qualities. This is a typical chemical decomposition and re-synthesis process that has been in practice in practically all sectors of the modern age, ranging from the plastic industry to pharmaceutical industries.

Figure 10.2 shows major steps of a conventional refining process. The first step is transportation and storage. In the crude oil refining process, fractional distillation is the main process to separate oil and gas. For this process, a distillation tower is used, which operates at atmospheric pressure, leaving a residue of hydrocarbons with boiling points above 400°C and more than 70 carbon atoms in their chains. Small molecules of hydrocarbons have lower boiling points, while larger molecules have higher boiling points. The fractionating column is cooler at the top than at the bottom, so the vapors cool as they rise. Figure 10.3 shows the pictorial view of the fractional column. It also shows the ranges of hydrocarbons in each fraction. Each fraction is a mix of hydrocarbons, and each fraction has its own range of boiling points, and comes off at a different level in the tower.

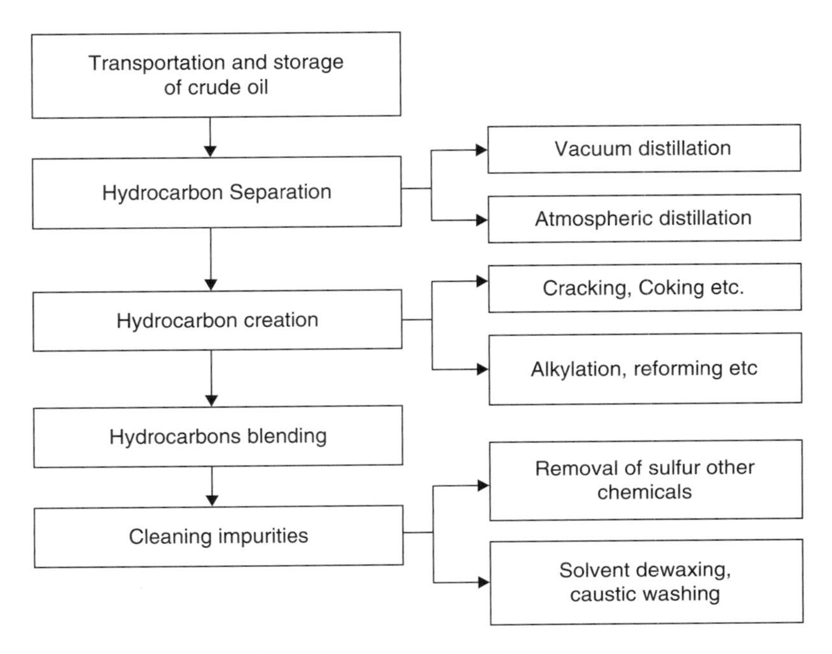

Figure 10.2 Major steps of refining process in Oil Refining.

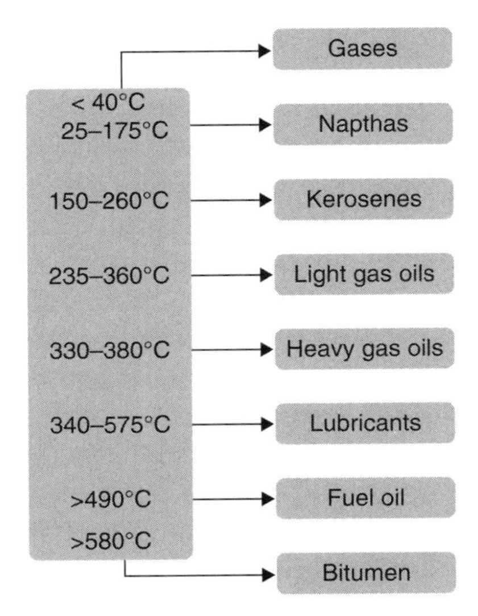

Figure 10.3 Pictorial view of fractional column.

Petroleum refining has evolved continuously in response to changing consumer demand for better and different products, from aviation gasoline and then for jet fuel, each with various degrees of 'refinement' to conform with specific needs of machineries, designed according to certain 'ideal' fluid behavior. A summary of a detailed process flow chart for oil refining steps is presented in Table 10.1. The table also describes the different treatment methods for each of the refining phases.

The third column in the above table shows how the refining process can render natural petroleum fluids into toxic chemicals. If the heat source and catalysts used are themselves products of the unsustainable practices, their contact with petroleum fluids will result into unsustainable products. Unless this is recognized, further refinement of the process, e.g., optimization of catalysts, automatization of heating elements, blending of various additives, and corrosion protection will not solve the sustainability problem.

Catalysts used in processes which remove sulfur are impregnated with cobalt, nickel, or molybdenum. During the separation process, sulfur from crude oil is removed only in exchange of traces of these catalysts. As has been seen in a previous chapter (Chapter 4) of this book, trace elements are not negligible, and must be accounted for in determining long-term impact. These trace elements will accompany the refined oil, and will end up in combustion chambers, eventually polluting the CO_2 emitted from a combustion engine. The inability of current detection techniques to identify these trace elements will not ensure that the pollution of CO_2 does not take place. It is discussed in previous chapters that contaminated CO_2 is not acceptable by plants or trees that reject this strain of CO_2. This process ends up contributing to the overall concentration of CO_2 in the atmosphere, delaying natural consumption and utilization of CO_2 in the ecosystem. If indeed removal of sulfur is the objective, the use of zeolite can solve this problem. It is well known that that naturally occurring zeolite has the composition to act as a powerful agent that would adsorb unwanted matters with high levels of adsorption, ion exchange, and catalytic actions (Shimada *et al.*, 1996). Even before the detailed composition of naturally occurring zeolite is known, the natural state of such a powerful agent should confirm that its usage is not harmful to the environment. Similar properties have been identified in limestone as well as vegetable oils that can be used as a solvent for removing sulfur compounds. The use of zeolite or similar naturally occurring separation

Table 10.1 Details of oil refining process and various types of catalyst used.

Process	Description	Catalyst/Heat/Pressure Used
Distillation Processes	It basically relies on the difference of boiling point of various fluids. Density has also important role to pay in distillation. The lightest hydrocarbon at the top and heaviest residue at the bottom are separated.	Heat
Coking and Thermal process	Coking unit converts heavy feedstocks into solid coke and lower boiling hydrocarbons are produced, which are suitable to offer refinery units to convert to higher value transportation fuel. This is a severe thermal cracking process to form coke. Coke contains high boiling point hydrocarbon and some volatiles, which are removed by calcining at temperatures of 1095–1260^{0}C. Coke is allowed sufficient time to remain in high temperature heaters in insulated surge drums, hence, it is called delayed cooking.	Heat
Thermal Cracking	The crude oil is subjected to both pressure and large molecules are broken into small ones to produce additional gasoline. The naphtha fraction is useful for making many petrochemicals. Heating naphtha in the absence of air makes the molecules split into shorter ones.	Excessive Heat and pressure

(Continued)

Table 10.1 (Cont.) Details of oil refining process and various types of catalyst used.

Process	Description	Catalyst/Heat/Pressure Used
Catalytic Cracking	Converts heavy oils into high gasoline, less heavy oils, and lighter gases. Paraffins are converted to C3 and C4 hydrocarbons. The benzene ring of aromatic hydrocarbons are broken. Rather than distilling more crude oil, an alternative is to crack crude oil fractions with longer hydrocarbons. Larger hydrocarbons split into shorter ones at low temperatures if a catalyst is used. This process is called catalytic cracking. The products include useful short chain hydrocarbons.	Nickels, Zeolites, Acid treated natural alumina silicates, amorphous and crystalline synthetic silica alumina catalyst
Hydro-processing	Hydroprocessing (325°C and 50 atm) includes both hydro cracking (350°C and 200 atm) and hydrotreating. Hydrotreating involves the addition of hydrogen atoms to molecules without actually breaking the molecule into smaller pieces, and improves the quality of various products (e.g., by removing sulfur, nitrogen, oxygen, metals, and waxes and by converting olefins to saturated compounds). Hydrocracking breaks longer molecules into smaller ones. This is a more severe operation using higher heat and longer contact time. Hydrocracking reactors contain fixed, multiple catalyst beds.	Platinum, Tungsten, palladium, nickel, crystalline mixture of silica alumina. Cobalt and Molybdenum oxide on alumina nickel oxide, nickel thiomolybdate tungsten and nickel sulfide and vanadium oxides, nickel thiomolybdate are in most common use for sulfur removal and nickel molybdenum catalyst for nitrogen removal.

Alkylation	Alkylation or "polymerization" - forming longer molecules from smaller ones. Another process is isomerization, where straight chain molecules are made into higher octane branched molecules. The reaction requires an acid catalyst at low temperatures and low pressures. The acid composition is usually kept at about 50%, making the mixture very corrosive.	Sulfuric acid, or hydrofluoric acid, HF (1–40 degrees Celsius, 1–10 atm). Platinum on AlCl3/Al2O3 catalyst uses as new alkylation catalyst
Catalytic Reforming	This uses heat, moderate pressure, and fixed bed catalysts to turn naphtha, short carbon chain molecule fraction, into high-octane gasoline components - mainly aromatics.	Catalyst used is a platinum (Pt) metal on an alumina (Al2O3) base.
Treating Non hydrocarbons	Treating can involve chemical reaction and/or physical separation. Typical examples of treating are chemical sweetening, acid treating, clay contacting, caustic washing, hydrotreating, drying, solvent extraction, and solvent dewaxing. Sweetening compounds and acids desulfurize crude oil before processing and treat products during and after processing.	

materials will be benign to the environment, and would also eliminate additional cost of cobalt, nickel, and molybdenum processing, bringing in double dividends to the petroleum processing industry.

Conventionally synthetic catalysts are used for enhancing the petroleum cracking process. Even when naturally occurring chemicals are used, they are acid-treated. The acid being synthetically produced, the process becomes irreversibly contaminated. More recently, microwave treatment of natural materials is being proposed in order to enhance the reactivity of natural materials (Henda *et al.*, 2006). Microwave heating not being a natural process, this treatment will also render the process unsustainable. However, such treatment is not necessary because natural materials, such as zeolite, clay, and others, do contain properties that would help the cracking process (Lupina and Aliev, 1991). Acid enhancement, if at all needed, can be performed with organic acid or acid derived from natural sources.

Acid-function catalysts impregnated with platinum or other noble metals are used in isomerization and reforming. Research in this topic has focused on the use of refined heavy metal elements and synthetic materials (e.g., Baird, Jr., 1990). These materials are known carcinogens, and have numerous long-term negative effects on the environment. In addition, the resulting products contain aromatic oils, carcinogenic polycyclic aromatic compounds, or other hazardous materials, and may also be pyrophoric. This in itself becomes a difficult short-term problem. When such a problem is addressed, solutions that are no more sustainable are offered. For instance, in order to combat pyrophoricy, a patented technology uses aromatic hydrocarbons, such as alkyl substituted benzenes, including toluene, xylene, and heavy aromatic naphtha. Heavy aromatic naphtha comprises xylene and higher aromatic homologs (Roling and Sintim, 2000). The entire process spirals down further in the path of unsustainability. Table 10.2 also presents the different catalysts used in the different process. Table 10.3 shows the various processes and products during the refining process.

Each of the above functions can also be performed with natural substitutes that are cheaper and are benign to the environment. This list includes zeolites, alumina, silica, various biocatalysts, and enzymes in their natural state. The use of bacteria to decompose large hydrocarbon molecules offers an attractive alternative, as the process is entirely sustainable. Khan and Islam (2007b) also

Table 10.2 Various processes and products in oil refining process.

Process name	Action	Method	Purpose	Feedstock(s)	Product(s)
FRACTIONATION PROCESSES					
atmospheric distillation	separation	thermal	separate fractions	desalted crude oil	gas, gas oil, distillate, residual
vacuum distillation	separation	thermal	Separate without cracking	Atmospheric tower residual	gas, gas oil, lube, residual
CONVERSION PROCESSED – DECOMPOSITION					
catalytic cracking	alteration	catalytic	upgrade gasoline	Gas oil coke, distillate	gasoline, petrochemical feedstock
coking	polymerize	thermal	convert vacuum residuals	Gas oil coke, distillate	gasoline, petrochemical feedstock
hydrocracking	hydrogenate	catalytic	convert to lighter HCs	Gas oil, cracked oil residual	Lighter, higher quality products
hydrogen steam reforming	decompose	catalytic/ thermal	produce hydrogen	Desulfurized gas, O2, steam	hydrogen, CO, CO2
steam cracking	decompose	thermal	crack large molecules	Atm tower, heavy fuel/ distillate	Cracked naphtha, coke, residual
visbreaking	decompose	thermal	reduce viscosity	Atm tower residual	Distillate tar

(Continued)

Table 10.2 (Cont.) Various processes and products in oil refining process.

Process name	Action	Method	Purpose	Feedstock(s)	Product(s)
CONVERSION PROCEESES–UNIFICATION					
alkylation	combining	catalytic	Unit olefins and isoparaffins	Tower isobutane/ cracker olefin	Iso-octane (alkylate)
grease compounding	combining	thermal	Combine soap and oils	Lube oil, fatty acid, alky metal	Lubricating grease
polymerizing	polymerize	catalytic	Unite 2 or more olefins	Cracker olefins	High-octane naphtha, petrochemical stocks
CONVERSION PROCESSES–ALTERATION OR REARRANGEMENT					
Catalytic reforming	Alteration/ dehydration	catalytic	Upgrade low octane naphtha	Coker/ hydro-cracker naphtha	High oct. Reformate/ aromatic
isomerization	rearrange	catalytic	straight chain to branch	Butane, pentane, hexane	Isobutane/ pentane/ hexane
TREATMENT PROCESSES					
amine treating	Treatment	Absorption	Remove acidic contaminants	Sour gas, HCs w/ CO_2 & H_2S	Acid free gases & liquid HCs
desalting	Dehydration	Absorption	Remove contaminants	Crude oil	Desalted crude oil
drying	Treatment	Abspt/ therm	Remove H_2O & sulfur cmpds	Liq Hcs, LPG, alky feedstk	Sweet & dry hydrocarbons

furfural extraction	Solvent extraction	Absorption	Upgrade mid distillate & lubes	Cycle oils & lube feed-stocks	High quality diesel & lube oil
hyfrode-sulfarization	Treatment	Catalytic	Remove sulfur, contaminants	High-sulfur residual/ gas oil	Desulfurized olefins
hydrotreating	Hydrogenation	Catalytic	Remove impurities, saturate HC's	Residuals, cracked HC's	Cracker feed, distillate, lube
phenol extraction	Solvent extraction	Abspt/ therm	Improve visc. index, color	Lube oil base stocks	High quality lube oils
solvent deasphalting	Treatment	Absorption	Remove asphalt	Vac. tower residual, propane	Heavy lube oil, asphalt
solvent dewaxing	Treatment	Cool/ filter	Remove wax from lube stocks	Vac. tower lube oils	Dewaxed lube basestock
solvent extraction	Solvent extr.	Abspt/ precip.	Separate unsat. oils	Gas oil, reformate, distillate	High-octane gasoline
sweetening	Treatment	Catalytic	Remove H_2S, convert mercaptan	Untreated distillate/ gasoline	High-quality distillate/gasoline

Source: OSHA, 2005.

Table 10.3 Emissions from refinery.

Materials transfer and storage
Air releases: VOCs (polluted with catalysts and other toxic additives)
Hazardous/solid wastes: anthracene, benzene, 1,3-butadiene, cumene, cyclohexane, ethylbenzene, ethylene, methanol, naphthalene, phenol, PAHs, propylene, toluene, 1,2,4-trimethylbenzene, xylene (polluted with catalysts and other toxic additives)
Separating hydrocarbons
Air releases: carbon monoxide, nitrogen oxides, particulate matter, sulfur dioxide, VOCs (polluted with catalysts and other toxic additives)
Hazardous/solid wastes: ammonia, anthracene, benzene, 1,3-butadiene, cumene, cyclohexane, ethylbenzene, ethylene, mercury, methanol, naphthalene, phenol, PAHs, propylene, toluene, 1,2,4-trimethylbenzene, xylene (polluted with catalysts and other toxic additives)

suggested the use of gravity segregation to distillate lighter components from heavier ones. The use of solar heating, in conjunction with heating from flares that are available in the oil field, will bring down the heating cost, and make the process sustainable.

10.2.1 Zero-Waste Refining Process

A zero-waste scheme is the only way to sustainability. Recent works of Lakhal and H'Mida (2003) and Lakhal *et al.* (2006) proposed a sustainable refining scheme with the so-called Olympic model. Their work analyzes the structure of the supply chain from production, transportation, and distribution to end users. The specific aspects of the model include: (*i*) The actual contaminants through the supply chain; and (*ii*) Analysis of operations, process, materials design, and selection, according to environmental policy. The research asserts that environmental practices would accrue competitive benefits to petroleum companies and enhance corporate performance (Sharma, 2001). This section defines attributes of a green supply chain for an oil refinery, using the framework to assess greenness efforts of an oil refinery through its supply chain.

The section proceeds to develop the concept of the G
supply chain. The primary features of this model are:

(*i*) Five zeros of waste or emissions (corresponding t the five circles in the Olympic flag):
- Zero emissions (Air, Soil, Water, Solid Waste, Hazardous waste)
- Zero waste of resources (Energy, Materials, Human)
- Zero waste in activities (Administration, Production)
- Zero use of toxics (Processes and Products)
- Zero waste in product life-cycle (Transportation, Use, End-of-Life)

The zero-waste approach is defended by the Zero-waste Organization (Zero Waste, 2005) using a visionary goal of zero waste to represent the endpoint of "closing-the-loop" so that all materials are returned at the end of their life as industrial nutrients, thereby avoiding any degradation of nature. A 100% efficiency of use of all resources (energy, material, and human) is promoted by Zero-waste, working toward a goal of reducing costs, easing demands on scarce resources, and providing greater availability for all. These principles of Zero-waste applied to products reduce impact during manufacture, transportation, during use, and at end of life. For petroleum as a unit of analysis, the concept of the green supply chain is illustrated by Figure 10.4. Such an approach, which is always the norm in nature, is only beginning to be proposed in the petroleum sector (Bjorndalen *et al.*, 2005) or even in the renewable energy sector (Khan *et al.*, 2005b).

The above emissions from the refinery are contaminated with toxic catalysts and other additives, even when they are present in trace amounts. If those toxic agents are not added throughout the supply chain (e.g., well head, separators, pipeline, and refining), these emissions will not be harmful to the environment (similar to organic methane), and only spurious assumptions that organic chemicals are the same as non-organic ones will show a conclusion otherwise.

There are five primary activities in refinery processes. They are: materials transfer and storage, separating hydrocarbons (e.g., distillation), creating hydrocarbons (e.g., cracking/coking, alkylation, and reforming), blending hydrocarbons, removing impurities

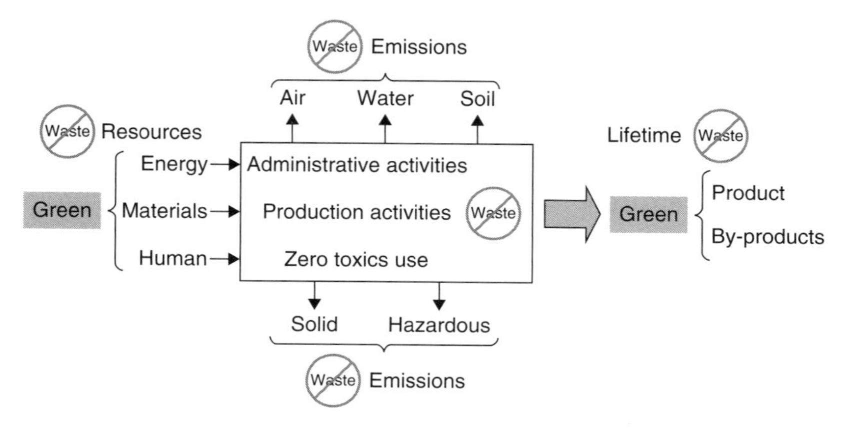

Figure 10.4 The concept of the Olympic Green Supply chain adapted from the Zero-waste approach (adapted Lakhal *et al.*, 2006).

(e.g. sulfur removal), and cooling. Tables 10.3, 10.4, and 10.5 (compiled from Environmental Defense, 2005) enumerate the primary emissions at each activity level. We count seven primary air release emissions and 23 primary hazardous/solid wastes.

Primary Air release: carbon monoxide, nitrogen oxides, particulate matter, particulate matter, sulfur dioxide, and Volatile Organic Compounds VOCs.

Primary Hazardous/solid wastes: 1,2,4-trimethylbenzene, 1,3-butadiene, ammonia, anthracene, benzene, copper, cumene, cyclohexane, diethanolamine, ethylbenzene, ethylene, hydrofluoric acid, mercury, metals, methanol, naphthalene, nickel, PAHs, phenol, propylene, sulfuric acid aerosols or toluene, vanadium (fumes and dust), and xylene.

The most important resource in the refinery process is energy. Unlike the manufacturing industry, labor costs do not constitute a high percentage of expenses in a refinery. In this continuous process, there is no waste of materials in general. The waste of human resources could be measured by ratios of accidents (number of work accidents/number of employers) and absenteeism due to illness (number of days lost for illness/number of work days × number of employees). In the Olympic refinery, ratios of accidents and absenteeism would be near zero.

The refining process uses a lot of energy. Typically, approximately two percent of the energy contained in crude oil is used for distillation. The efficiency of the heating process can be increased

Table 10.4 Primary wasters from oil refinery.

Cracking/Coking	Alkylation and Reforming	Sulfur Re
Air releases: carbon monoxide, nitrogen oxides, particulate matter, sulfur dioxide, VOCs Hazardous/solid wastes, wastewater: ammonia, anthracene, benzene, 1,3-butadiene, copper, cumene, cyclohexane, ethylbenzene, ethylene, methanol, naphthalene, nickel, phenol, PAHs, propylene, toluene, 1,2,4-trimethylbenzene, vanadium (fumes and dust), xylene	Air releases: carbon monoxide, nitrogen oxides, particulate matter, sulfur dioxide, VOCs Hazardous/solid wastes: ammonia, benzene, phenol, propylene, sulfuric acid aerosols or hydrofluoric acid, toluene, xylene Wastewater	Air releases: carbon monoxide, nitrogen oxides, particulate matter, sulfur dioxide, VOCs Hazardous/solid wastes: ammonia, diethanolamine, phenol, metals Wastewater

Table 10.5 Pollution prevention options for different activities in material transfer and storages.

Cracking/ Coking	Alkylation and Reforming	Sulfur Removal	Cooling
Using catalysts with fewer toxic materials reduces the pollution from "spent" catalysts and catalyst manufacturing.	Using catalysts with fewer toxic materials reduces the pollution from "spent" catalysts and catalyst manufacturing.	Use "cleaner" crude oil," containing less sulfur and fewer metals. Use oxygen rather than air in the Claus plant reduces the amount of hydrogen sulfide and nitrogen compounds produced.	Ozone or bleach should replace chlorine to control biological growth in cooling systems Switching from water cooling to air cooling could reduce the use of cooling water by 85 percent.

drastically by combining direct solar heating (with non-engineered thermal fluid) with direct fossil fuel burning. The advantage of this process is a gain in global efficiency, as well as environmental benefit. It is estimated that the total energy requirement for petroleum refining can be reduced to less than 0.5 percent of the energy contained in crude oil by designing the heating systems with zero-waste scheme, as outlined earlier in this chapter.

A number of procedures are used to turn heavier components of crude oil into lighter and more useful hydrocarbons. These processes use catalysts or materials that help chemical reactions without being used up themselves. Table 10.6 shows different toxic catalysts and base metals. Refinery catalysts are generally toxic,

Table 10.6 Catalysts and materials used to produce catalysts base metals and compounds.

Name of Catalysts	Name of Metals Base
Activated alumina, Amine, Ammonia, Anhydrous hydrofluoric acid Anti-foam agents – for example, oleyl alcohol or Vanol, Bauxite, Calcium chloride, Catalytic cracking catalyst, Catalytic reforming catalyst, Caustic soda, Cobalt molybdenum, Concentrated sulphuric acid, Demulsifiers – for example, Vishem 1688, Dewaxing compounds (catalytic) – for example, P4 Red, wax solvents Diethylene glycol, Glycol –Corrosion inhibitors), Hydrogen gas, Litharge, Na MBT (sodium 2-mercaptobenzothiazole) – glycol corrosion inhibitor (also see the taxable list for Oil Refining – Corrosion inhibitors), Na Cap – glycol corrosion inhibitor (also see the taxable list for Oil Refining – Corrosion inhibitors), Nalcolyte 8103, Natural catalysts – being compounds of aluminum, silicon, nickel, manganese, iron and other metals, Oleyl alcohol – anti-foam agent, Triethylene glycol, Wax solvents – dewaxing compounds	Aluminum (Al), Aluminum Alkyls, Bismuth (Bi), Chromium (Cr), Cobalt (Co), Copper (Cu), Hafnium (Hf), Iron (Fe), Lithium (Li), Magnesium (Mg), Manganese (Mn), Mercury (Hg), Molybdenum (Mo), Nickel (Ni), Raney Nickel, Phosphorus (P), Potassium (K), Rhenium (Re), Tin (Sn), Titanium (Ti), Tungsten (W), Vanadium (V), Zinc (Zn), Zirconium (Zr).

Source: CTB (2006).

and must be replaced or regenerated after repeated use, turning used catalysts into a waste source. The refining process uses either sulfuric acid or hydrofluoric acid as catalysts to transform propylene, butylenes, and/or isobutane into alkylation products, or alkylate. Vast quantities of sulfuric acid are required for the process. Hydrofluoric acid or HF, also known as hydrogen fluoride, is extremely toxic, and can be lethal. Using catalysts with fewer toxic materials significantly reduces pollution. Eventually, organic acids and enzymes instead of catalysts must be considered. Thermal degradation and slow reaction rate are often considered to be biggest problems of using organic acids and catalysts. However, recent discoveries have shown that this perception is not justified. There are numerous organic products and enzymes that can withstand high temperatures, and many of them induce fast reactions. More importantly, recent developments in biodiesel indicate that the process itself can be modified in order to eliminate the use of toxic substances (Table 10.7). The same principle applies to other materials, e.g., corrosion inhibitors, bactericides, and so forth. Often, toxic

Table 10.7 Chemicals used in refining.

Chemicals used in Refining	Purpose
Ammonia	Control corrosion by HCL
Tetraethyl lead (TEL) and tetramethyl lead (TML)	Additives to increase the octane rating
Ethyl tertiary butyl ether (ETBE), methyl tertiary butyl ether (MTBE), tertiary amyl methyl ether (TAME),	To increase gasoline octane rating and reduce carbon monoxide
Sulfuric Acid and Hydrofluoric Acid	alkylation processes, some treatment processes.
Ethylene glycol	dewatering
Toluene, methyl ethyl ketone (MEK), methyl isobutyl ketone, methylene chloride ,ethylene dichloride, sulfur dioxide	dewaxing

(Continued)

Table 10.7 (Cont.) Chemicals used in refining.

Chemicals used in Refining	Purpose
zeolite, aluminum hydrosilicate, treated bentonite clay, fuller's earth, bauxite, and silica-alumina	catalytic cracking
nickel	catalytic cracking
Granular phosphoric acid	Polymerization
Aluminum chloride, hydrogen chloride	Isomerization
Imidazolines and Surfactants Amino Ethyl Imidazoline Hydroxy-Ethyl Imidazoline Imidazoline/Amides Amine/Amide/DTA	Oil soluble corrosion inhibitors
Complex Amines Benzyl Pyridine	Water soluble corrosion inhibitors
Diamine Amine Morpholine	Neutralizers
Imidazolines Sulfonates	Emulsifiers
Alkylphenolformaldehyde, polypropeline glycol	Desalting and emulsifier
Cobalt Molybdate, platinum, chromium alumina	
$AlCl_3$-HCl , Copper pyrophosphate	

chemicals lead to very high corrosion vulnerability, and even more toxic corrosion inhibitors are required. The whole process spirals down to a very unstable process, which can be eliminated with the new approach (Al-Darbi *et al.*, 2002).

10.3 Zero Waste in Product Life Cycle (Transportation, Use, and End-of-Life)

The complex array of pipes, valves, pumps, compressors, and storage tanks at refineries are potential sources of leaks into the air, land, and water. If they are not contained, liquids can leak from transfer and storage equipment, and contaminate soil, surface water, and groundwater. This explains why, according to industry data, approximately 85 percent of monitored refineries have confirmed groundwater contamination as a result of leaks and transfer spills (EDF, 2005).

To prevent the risks associated with transportation of sulfuric acid and on-site accidents associated with the use of hydrofluoric acid, refineries can use a solid acid catalyst that has recently proven effective for refinery alkylation. However, the solids are also more toxic than the liquid counterpart. As pointed out earlier, the use of organic acid or organically-prepared acid would render the process inherently sustainable.

A sustainable petroleum process should have storage tanks and pipes aboveground to prevent ground water contamination. There is room for improving the efficiency of these tanks with natural additives. Quite often, the addition of synthetic materials makes an otherwise sustainable process unsustainable. Advances in using natural materials for improving material quality have been made by Saeed *et al.* (2003).

Sulfur is the most dangerous contaminant in a refinery's output products. When fuel oils are combusted, the sulfur in them is emitted into the air as sulfur dioxide (SO_2) and sulfate particles (SO_4). Emissions of SO_2, along with emissions of nitrogen oxides, are a primary cause of acidic deposition (i.e., acid rain), which has a significant effect on the environment, particularly in central and eastern Canada (2002). Fine particulate matter ($PM_{2.5}$), of which sulfate particles are a significant fraction (30–50%), may affect human health adversely. In absence of toxic additives, the produced products will perform equally well, but will not release contaminated natural products to the environment.

10.4 No-Flaring Technique

Flaring is a commonly used technique in oil refinery to burn out low-quality gas. With increasing awareness of the environmental impact, gas flaring is likely to be banned in the near future. This will require significant changes in the current practices of oil and gas production and processing. The low-quality gas that is flared contains many impurities, and during the flaring process, toxic particles are released into the atmosphere. Acid rain, caused by sulfur oxides in the atmosphere, is one of the main environmental hazards resulting from this process. Moreover, flaring of natural gas accounts for approximately a quarter of the petroleum industries emissions (UKOO, 2002). However, the alternative solution that is being offered is not sustainable. Consider the use of synthetic membranes or synthetic solvents to remove impurities, such as CO_2, water, SO_2, and so forth. These impurities are removed and replaced with traces of synthetic materials, either from the synthetic membrane or the synthetic solvent.

Recently, Bjorndalen *et al.* (2005) developed a novel approach to avoid flaring from petroleum operations. Petroleum products contain materials in various phases. Solids in the form of fines, liquid hydrocarbon, carbon dioxide, and hydrogen sulfide are among the many substances found in the products. According to Bjorndalen *et al.* (2005), by separating these components through the following steps, no-flare oil production can be established:

- Effective separation of solid from liquid
- Effective separation of liquid from liquid
- Effective separation of gas from gas

Many separation techniques have been proposed in the past (Basu *et al.*, 2004; Akhtar, 2002). However, few are economically attractive and environmentally appealing. This option requires an innovative approach that is the central theme of this text. Once the components for no-flare have been fulfilled, value-added end products can be developed. For example, the solids can be utilized for minerals, the brine can be purified, and the low-quality gas can be re-injected into the reservoir for enhanced oil recovery. Figure 10.5 outlines the components and value-added end products that will be discussed.

10.4.1 Separation of Solid-Liquid

Even though numerous techniques have been proposed in the past, little improvement has been made in energy efficiency of solid-liquid separation. Most techniques are expensive, especially if a small unit is operated. Recently, a patent has been issued in the United States. This new technique of removing solids from oil is an EVTN system. This system is based on the creation of a strong vortex in the flow to separate sand from oil. The principle is that by allowing the flow to rotate rapidly in a vortex, centrifugal force can be

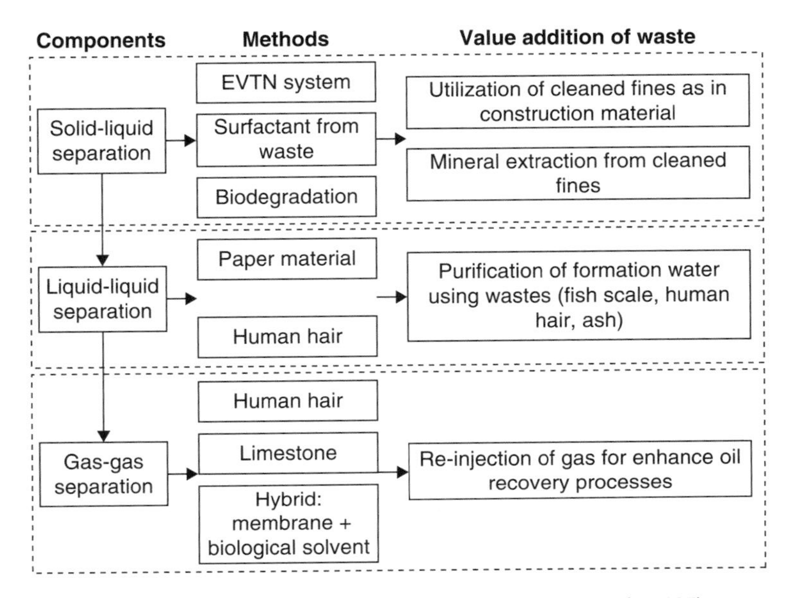

Figure 10.5 Breakdown of No-flaring Method (Bjorndalen *et al.*, 2005).

generated. This force makes use of the density differences between the substances. The conventional filtration technique requires a large filter surface area (in case of high flow rate), replacement of filter material, and back flush. The voraxial technique eliminates this problem. Moreover, it is capable of maintaining high gravity or 'g' force as well as high flow rate, which will be very effective in oil-fines separation (EVTN, 2003). This product shows great potential for the separation of liquid and fines.

Use of surfactants from waste can be another possible technique to separate solids from liquid. The application of the combination of waste materials with water to separate fines from oil is attractive due to the relatively low cost and environmentally sound nature. Waste products, such as cattle manure, slaughterhouse waste, okra, orange peels, pine cones, wood ash, paper mill waste (Lignosulfate), and waste from the forest industry (Ferrous Chloride) are all viable options for the separation of fines and oil. Cattle manure and slaughterhouse waste is plentiful in Alberta, where flaring is very common. Researchers from UAE University have determined that okra is known to act like a soap (Chaalal, 2003). Okra extract can be created through pulverization methods. Orange peel extract should also be examined, since it is a good source of acid. A study conducted at Dalhousie University determined that wood ash can separate arsenic from water, and therefore may be an excellent oil/fines separator. Pine cones, wood ash, and other plant materials may also be viable. Industrial wastes, such as lignosulfate and ferrous chloride, which have been very beneficial in the industrial areas of cellulose production and sewage treatment, respectively, can be potential separators.

Finally, a biodegradation method for stripping solid waste from oily contaminants is currently under study as a collaborative effort between Dalhousie and UAE University. Thermophilic bacteria are found to be particularly suitable for removing low-concentration crude oils from the solid surface (Tango and Islam, 2002). Also, it has been shown that bioremediation of flare pits, which contain many of the same substances that need to be removed for an appealing no-flare design, has been successful (Amatya *et al.*, 2002).

Once the fines are free from liquids, they can be useful for other applications. For example, the fines can be utilized as a substitution of components in construction materials. Drilling wastes have been found to be beneficial in highway construction (Wasiuddin, 2002), and an extension of this work can lead to the usage of fines. Studies have shown that the tailings from oil sands are high in titanium content. With this in mind, the evaluation of valuable minerals in fines will be conducted. To extract the minerals, usually a chemical treatment is used to modify the surface of the minerals. Treatment with a solution derived from natural material has a great potential. Microwave heating has the potential of assisting this process (Haque, 1999; Hua *et al.*, 2002, Henda *et al.*, 2006). It enhances selective floatability of different particles. Temperature can be a major factor in the reaction kinetics of a biological solvent with mineral surfaces. Various metals respond in a different manner under microwave conditions, which can make significant change in floatability. The recovery process will be completed through transferring the microwave-treated fines to a flotation chamber. However, microwave heating might render the enter process unsustainable, because microwave itself is neither efficient (global efficiency-wise) nor natural (the source being unnatural). Similarly effective separation processes can be induced by using naturally produced acids and hydrogen peroxide. More research should be conducted to this effect.

10.4.2 Separation of Liquid-Liquid

Once the oil has been separated from fines via water and waste material, it must be separated from the solution itself, as well as from formation water. Oil-water separation is one of the oldest practices in the oil industry, as it is almost impossible to find an oil reservoir absolutely free of connate water. In fact, the common belief is that all reservoirs were previously saturated with water and, after oil migration, only a part of the water was expelled from the reservoir and replaced by oil. There are two sources of water that cause water to flow to the well-bore. The first source is the connate water that usually exists with oil in the reservoir, saturating the formation below the Oil-Water-Contact (OWC), or in the form of emulsion even above the OWC. The other source is associated with water flooding, that is mainly considered in secondary oil recovery.

In almost all oil reservoirs, connate water coexists with oil as a percentage filling pore spaces. In some reservoirs where water pressure is the main driving mechanism, water saturation may even exceed oil saturation as the production continues. Therefore, when the well-bore is operational, oil production mixed with water is inevitable. As production continues, more water invades to the oil zone and water-cut in the production stream increases consequently.

Before taking an oil well on production, the oil zone is perforated above the OWC to allow oil flow into the well-bore. The fact that part of the formation above the OWC is still saturated with water consequently causes water production. The problem becomes more severe with time, as the continuous drainage of the reservoir causes OWC to move upward, which results in excessive water production. Because of this typical phenomenon occurring in oil reservoirs, it is impossible to avoid water production.

Moreover, water flooding is an obvious practice for enhancing oil recovery after oil production declines. Encountering problems of high water production and early breakthroughs are common obstacles of water flooding practice, which in turn cause high production costs. Water production not only is associated with the high cost considerations (installation and operation), but also is a major contributing factor to the corrosion of production facilities, and to reservoir energy loss. Moreover, the contaminated water can be an environmental contamination source, if it is not properly disposed.

Water production can be tolerated to a certain extent, depending on the economic health of a given reservoir. The traditional practice of the separation of oil from water is applied after simultaneous production of both and then separating them in surface facilities. Single-stage or multi-stage separators are installed where both oil and water can be separated by gravity segregation.

Down-hole water-oil separation has been investigated and discussed since the early days of producing oil from petroleum reservoirs. Hydrocyclone separation has been used (Bowers *et al.*, 2000) to separate oil from water at the surface, but its small size made its application down-hole even more attractive. Hydrocyclone gives the best efficiency, at 25 to 50% water content in the outlet stream, to make the water stream as clean as possible with a few exceptions under ideal conditions. This makes the consideration of this technique limited to a number of circumstances where high water cut is expected and the costs involved are justified.

Stuebinger *et al.* (2000) compared the hydrocyclone to gas-oil-water segregation, and concluded that all down-hole separations are still premature, and suggested more research and investigation for optimizing and enhancing these technologies.

Recently, the use of membrane and ceramic materials has been proposed (Fernandez *et al.*, 2001; Chen *et al.*, 1991). While some of them show promises, these technologies do not fall under the category of economically appealing. The future of liquid-liquid separation lies within the development of inexpensive techniques and preferably down-hole separation technology.

The first stage is the material screening, searching for potential material that can pass oil but not water. The key is to consider the fundamental differences between oil and water in terms of physical properties, molecular size, structure, and composition. These materials must, in essence, be able to adsorb oil and at the same time prevent water from passing through. Having discovered at least one material with this feature, the next stage should be studying the mechanism of separation. Understanding the separation mechanism would assist in identifying more suitable materials and better selection criteria. The third stage is the material improvement. Testing different down-hole conditions using selected materials for separation and possibility of improving materials by mixing, coating, or coupling with others should be investigated. The outcome of this stage would be a membrane sheet material that gives the best results and a suitable technique that optimizes the procedure. Investigating the effect of time on separation process and its remedy should be carried out in this stage. Eventually, a new completion method should be designed. The material with the ability to separate oil and water can be also used in above-ground separation units. The main advantage of this technique will be to reduce the size of the separation units and to increase their efficiency. This is very critical, especially in offshore rigs, where minimizing the size of different units is of the essence, and any save in space would substantially reduce the production cost.

Preliminary studies and initial lab-scale experiments show that the use of a special type of long-fiber paper as the membrane could be a good start (Khan and Islam, 2006). These preliminary experiments have been performed during the material selection stage, and quite encouraging results have been encountered. Lab-scale experimental setup consists of a prototype-pressurized reservoir containing oil-water emulsion, and tubing on which the perforated section is wrapped with the membrane. The oil-water emulsion has been produced, and it was found that the selected material gives 98–99% recovery of oil without producing any water. These findings are subject to the lab conditions in which ΔP is kept about 20 psi. Continuous shaking was employed to maintain the emulsion intact during the whole process of separation. An emulsion made up of varying ratios of oil and water was utilized in different sets of experiments. Almost all of the oil-water ratios gave the same separation efficiency with the material used.

The mentioned paper material is made of long fibrous wood pulp treated with waterproofing material as a filtering medium. This treatment prevents water from flowing through, and at the same time allows oil to pass easily. The waterproofing agent for paper used in these experiments is "rosin soap" (rosin solution treated with caustic soda). This soap is then treated with alum to keep the pH of the solution within a range of 4~5. Cellulose present in the paper reacts reversibly with rosin soap in the presence of alum, and forms a chemical coating around the fibrous structure, which acts as a coating to prevent water from seeping through it. This coating allows long-chain oil molecules to pass, making the paper a good conductor for the oil stream. Because of the reversible reaction, it was also observed that the performance of the filter medium increases with the increased acidity of the emulsion, and vice versa. It was also observed that the filter medium is durable to allow its continuous use for a long time, keeping the cost of replacement and production cut down. Different experiments are being done to further strengthen the findings for longer time periods and for high production pressures. It must be noted that the material used as a filtering medium is environmental friendly, and can easily be modified subject to down-hole conditions. Based on the inherent property of the material used, some other materials can also be selected to give equivalent good results, keeping in mind the effect of surrounding temperature, pressure, and other parameters present in down-hole conditions.

Human hair has great potential in removing oil from the solution. Hair is a natural barrier against water, but it easily absorbs oil. This feature was highlighted during a United States Department of Energy-funded project in 1996 (reported by CNN in February, 1997). A doughnut-shaped fabric container filled with human hair was used to separate oil from a low-concentration oil-in-water emulsion. The Dalhousie petroleum research team has later adapted this technique, with remarkable success in both the separation of oil and water, as well as heavy metals from aqueous streams.

Purification of the formation water after the oil separation will ensure an all-around clean system. Wastes such as fish scales have been known to adsorb lead, strontium, zinc, chromium, cobalt (Mustafiz, 2002; Mustafiz *et al.*, 2002), and arsenic (Rahaman, 2003). Additionally, wood ash can adsorb arsenic (Rahman, 2002). Both of these waste materials are great prospects to be implemented in the no-flare process.

10.4.3 Separation of Gas-Gas

The separation of gas is by far the most important phase of no-flare design. Current technologies indicate that separation may not be needed, and the waste gas as a whole can be utilized as a valuable energy income stream. Capstone Turbine Operation has developed a micro-turbine, which can generate up to 30 kW of power and consume 9000 ft^3/day of gas (Capstone, 2003). Micro-turbines may be especially useful for offshore applications, where space is limited. Another possible use of the waste gas is to re-inject it into the reservoir for pressure maintenance during enhanced oil recovery processes. The low-quality gas can be re-pressurized via a compressor and be injected into the oil reservoir. This system has been tested at two oil fields in Abu Dhabi to some praise (Cosmo Oil, 2003). As well, simple incineration instead of flaring to dispose of solution gas has been proposed (Motyka and Mascarenhas, 2002). This process only results in a reduction of emissions over flaring.

The removal of impurities in solution gas via separation can be achieved both down-hole and at the surface, thus eliminating the need for flare (Bijorndalen *et al.* 2005). Many studies have been conducted on the separation of gases using membranes.

In general, a membrane can be defined as a semi-permeable barrier that allows the passage of select components. An effective membrane system will have high permeability to promote large fluxes. It will also have a high degree of selectivity to ensure that only a mass transfer of the correct component occurs. For all practical purposes, the pore shape, pore size distribution, external void, and surface area will influence the separation efficiencies (Abdel-Ghani and Davies, 1983). Al Marzouqi (1999) determined the pore size distribution of membranes. He compared two models, and validated the models with experimental data. Low concentrations of H_2S in natural gas can be handled well by regenerable adsorbents, such as activated carbon, activated alumina, silica gel, and synthetic zeolite. Non-regenerable adsorbents, i.e., zinc and iron oxides, have been used for natural gas sweetening. Many membrane systems have been developed, including polymer, ceramic, hybrid, liquid, and synthetic.

Polymer membranes have gained popularity in isolating carbon dioxide from other gases (Gramain and Sanchez, 2002). These membranes are elastomers, formed from cross-linked copolymers of high molecular weights. They are prepared as thin films by extrusion or casting. They demonstrate unique permeability properties for carbon dioxide, together with high selectivity towards H_2, O_2, N_2 and CH_4. Ceramic membranes are used to separate hydrogen from gasified coal (Fain *et al.*, 2001). With ceramic materials, very high separation factors have been achieved, based on the ratios of individual gas permeances. Hybrid membranes combine thermodynamically-based partitioning and kinetically-based mobility discrimination in an integrated separation unit. Unfortunately, the permeability of common membranes is inversely proportional to selectivity (Kulkarni *et al.*, 1983). Thus, the development of liquid membranes has led to systems that have both a high permeability and high selectivity. Liquid membranes operate by immobilizing a liquid solvent in a micro porous filter, or between polymer layers.

A synthetic membrane is a thin barrier between two phases through which differential transport can occur under a variety of driving forces, including pressure, concentration, and electrical potential across the membranes. Pressure difference across the membrane can facilitate reverse osmosis, ultra-filtration, micro-filtration, gas separation, and pervaporation. Temperature difference across the membrane can facilitate distillation, whereas concentration difference can be used for dialysis and extraction.

The feasibility of using a novel carbon–multi-wall membrane for separating carbon dioxide from flue gas effluent from a power gas generation plant is being studied (Andrew *et al.*, 2001). This membrane consists of nano-sized tubes with pore sizes that can be controlled. This will enhance the kinetic and diffusion rates, which in turn will yield high fluxes.

The mass transport in nano-materials has not been clearly understood (Wagner *et al.*, 2002). Although previous transient models of separation and adsorption suggest that high selectivity is the origin of selective transport, recent analyses indicate that specific penetrant-matrix interactions actually dominate the effects at the transition stage. The primary difficulties in modeling this transport are that the penetrants are in continuous contact with the membrane matrix material, and the matrix has a common topology with multiple length scales. In order to determine the effectiveness of a membrane for separation of gases, it is important to have an accurate estimate of the pore size distribution (Marzouki, 1999). This is a key parameter in determining the separation factor of gases through nano-porous membranes.

Membrane systems are highly applicable for separation of gases from a mixture. Continuing enhancement in technology development of membrane systems is a natural choice for the future. Proven techniques include materials such as zeolite (Izumi *et al.*, 2002; Romanos *et al.*, 2001; Robertson, 2001; Jeong *et al.*, 2002) when combined with pressure swing adsorption (PSA) techniques. PSA works on the principle that gases tend to be attracted to solids and adsorb under pressure. Zeolite is expensive, and therefore fly ash, along with caustic soda, has been used to develop a less expensive zeolite (Indian Energy Sector, 2002). Waste materials can also be an attractive separation media, due to the relatively low cost and environmentally sound nature. Since caustic soda is a chemical, other materials, such as okra, can be a good alternative. Rahaman *et al.* (2003) have shown that charcoal and wood ash are

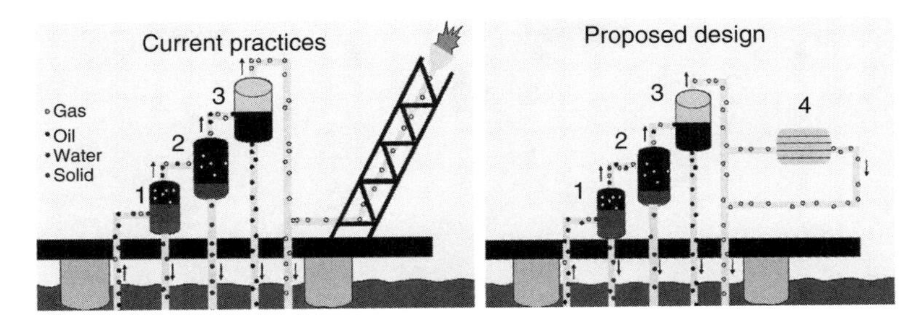

Figure 10.6 Pictorial presentation of faring techniques and proposed no-flaring method (after Bjorndalen *et al.*, 2005).

comparable to zeolite for the removal of arsenic from wastewater. Many of the waste materials discussed in the solid/liquid separation section also have the potential for effective separation materials. Carbon fibers (Fuertes, 2001; Park and Lee, 2003; Gu *et al*, 2002) as well as palladium (Lin and Rei, 2001; Chang *et al.*, 2002; Karnik *et al.*, 2002) fibers have been studied extensively for separation. A low-cost alternative to this technology is human hair, since it is a natural hollow fiber. Wasiuddin et a1. (2002) outlined some of the potential uses of human hair, confirming that it is an effective medium for the removal of arsenic from water.

As well as improving new-age techniques (e.g. polymeric membranes), research will be performed to test new materials, hybrid systems (e.g., solvent/membrane combination), and other techniques that are more suitable for the Atlantic region. For the gas processing research, focus will be given to improve existing liquefaction techniques, addressing corrosion problems (including bio-corrosion), and treatment of regenerated solvents.

10.4.4 Overall Plan

Figure 10.6 compares flare design to the overall plan for a no-flare design. Point 1 on the figure represents solid-liquid separation, Point 2 represents liquid-liquid separation, and Point 3 represents liquid-gas separation. Since current techniques to separate liquids from gas are relatively adequate, liquid-gas separation was not considered in this report. Point 4 represents the new portion of this design. It shows the addition of the gas-gas separator. For the no-flare system to be effective, each one of these points must perform with efficiency.

11

Truly Green Refining and Gas Processing

11.1 Introduction

Many people were surprised at how quickly the infamous BP oil spill in the Gulf of Mexico got cleaned up naturally (Biello, 2010). In fact, the bigger problem in that incident was the abundance of chemical dispersants that were sprayed to 'clean' the oil spill. While crude oil and natural gas are inherently environment-friendly in their natural state, they are seldom used in their natural state because the modern age has seen an influx of engineering design and machine developments that use "ideal fluids." These "ideal fluids" are very different from natural fluids. Consequently, fluids that are available in a natural state are routinely refined, decomposed, or separated from their natural state, making it mandatory for the fluids to undergo refining and gas processing. Because these refining and gas processing techniques contain synthetic chemicals that have undergone unsustainable processes, the processed fluids are rendered unsustainable. With this mode, all emissions contain products from the natural petroleum fluids that are contaminated with synthetic chemicals, making the total emission unacceptable to the ecosystem. In the end, this contamination process plays the

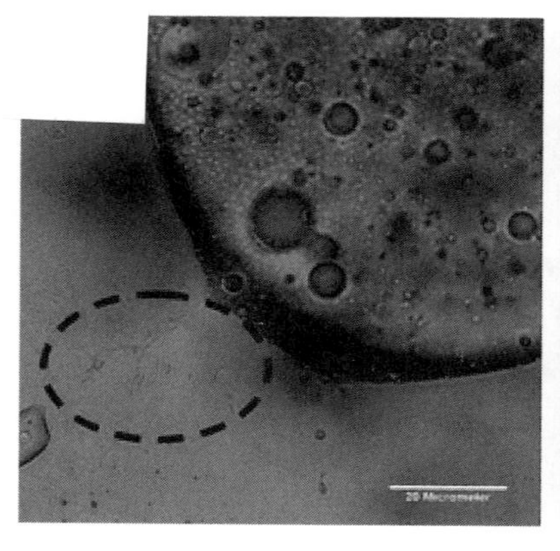

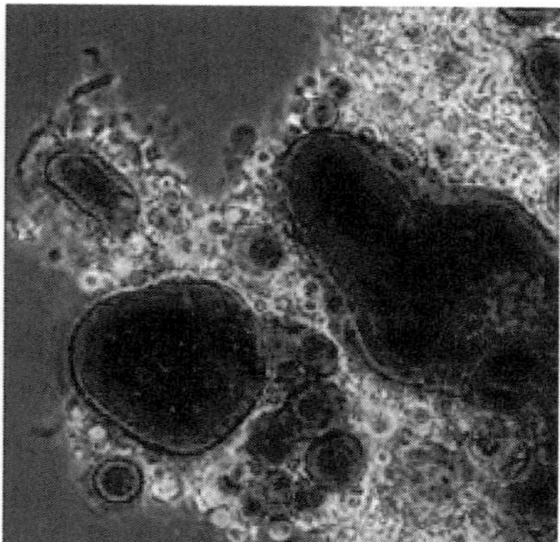

Pictures from Bielo, 2010

most significant role in creating environmental consequences. This chapter shows how conventional refining and gas processing techniques lead to inherent pollution of the environment. This is followed by true green processes that can refine crude oil and purify natural gas, while improving the environmental quality.

11.1.1 Refining

The refining of crude oil and processing of natural gas involve the application of large amount of synthetic chemicals and catalysts such as lead, chromium, glycol, amines, and others. These synthetic chemicals contaminate the end products and are burnt along with the fuels, producing various toxic by-products. The emission of such air pollutants that did not exist in nature before can cause environmental effects that can irreversibly damage the global ecosystem. It is found that refined oils degrade slower and last in the natural environment for a longer duration than crude oil. In this chapter, a pathway analysis of crude oil and fossil fuel refining and their impacts on the natural environment are discussed. Hence, it is clear that a significant improvement in current engineering practices is needed in order to reduce the emissions from refineries. Current engineering practices should follow natural pathways in order to reduce the emission of fluids that are inherently damaging to the environment. Only then can environmental problems, including global warming, be reversed.

Modern transportation systems, ranging from cars to aircrafts, a designed based on the use of oil, gas, and other fossil fuels. Because the use of processed fossil fuels creates several environmental and health problems, the environmental consequences are usually attributed to petroleum production. The total energy consumption in 2004 was equivalent to approximately 200 million barrels of oil per day, which is about 14.5 terawatts, over 85% of which comes from fossil fuels (Service 2005). However, not a drop of this is crude oil, because the machinery used to convert crude oil into "usable" energy uses only "refined" or processed fluids itself. Globally, about 30 billion tons of CO_2 are produced annually from fossil fuels, which includes oil, coal, and natural gas (EIA 2004). Because the contribution of toxic chemicals during the refining process is not accounted for (considered negligible in all engineering calculations), the produced industrial CO_2 is considered solely responsible for the current global warming and climate change problems (Chilingar and Khilyuk 2007). From this point onward, all calculations indicate that burning fossil fuels is not a sustainable option. The subscriber of the alternate theory, however, makes the argument that the total emission of CO_2 from petroleum activities is negligible compared to the total emission of CO_2 from the overall ecosystem. For instance, Chilinger and Khilyuk (2005) argue that the emission of greenhouse gases by burning fossil fuels is not responsible for global warming and, hence, is not unsustainable. In their analysis, the amount of greenhouse gases generated through human activities is scientifically insignificant compared to the vast amount of greenhouse gases generated through natural activities. This scientific investigation is infamously called the "flat earth theory" by the likes of environmental giants such as Al Gore (*60 Minutes* interview, March 23, 2008). It is true that if only the composition of CO_2 is considered, the CO_2 emissions from petroleum activities are negligible. However, this cannot form the basis for stating that global warming is not caused by petroleum activities. Similar to the phenomenal cognition that requires that the first premise be true before arriving to a conclusion, one must realize that the CO_2, contaminated with trace elements from toxic catalysts and other chemicals (during refining, separation, transportation, and processing), plays a very different role than CO_2 that is directly emitted form organic matter. Neglecting this fact would be equivalent to stating that because Freon concentration is negligible, it cannot cause a hole in the ozone layer. Neither side of the global warming debate considers this factor.

Refining crude oil and processing natural gas use large amounts of processed/synthetic chemicals and catalysts, including heavy metals. These heavy metals contaminate the end products and are burnt along with the fuels, producing various toxic byproducts. The pathways of these toxic chemicals and catalysts show that they largely affect the environment and public health. The use of toxic catalysts creates many environmental effects that cause irreversible damages to the global ecosystem. The problem with synthetic additives emerges from the fact that they are not compatible with natural systems and are not assimilated with biomasses in a way that would preserve natural order (Khan *et al.* 2008). This indicates that synthetic products do not have any place in the sustainability cycle. The use of natural catalysts and chemicals should be considered the backbone for the future development of sustainable practices. Crude oil is a truly nontoxic, natural, and biodegradable product, but the way we refine it is responsible for all the problems created by it on earth.

At present, for every barrel of crude oil, approximately 15% additives are added (CEC, 2004). These additives, with current practices, are all synthetic and/or engineered materials that are highly toxic to the environment. With this "volume gain," the following distribution is achieved (Table 11.1).

Each of these products is subject to oxidation, either through combustion or low-temperature oxidation, which is a continuous process. As one goes toward the bottom of the table, the oxidation rate decreases, but the heavy metal content increases, which makes each product equally vulnerable to oxidation. The immediate consequence of this conversion through refining is that one barrel of naturally occurring crude oil (convertible non-biomass) is converted into 1.15 barrels of potentially non-convertible non-biomass that would continue to produce more volumes of toxic components as it oxidizes either though combustion or through slow oxidation (Chhetri *et al.* 2008). As an example, just from the oxidation of the carbon component, 1 kg of carbon, which was convertible non-biomass, would turn into 3.667 kg of carbon dioxide (if completely burnt) that is not acceptable by the ecosystem, due to the presence of the non-natural additives. Of course, when crude oil is converted, each of its numerous components would turn into non-convertible non-biomass. Many of these components are not accounted for or even known, with no scientific estimation of their consequences. Hence, the sustainable option is either to use natural

Table 11.1 Petroleum products yielded from one barrel of crude oil in California.

Product	Percent of Total
Finished Motor Gasoline	51.4%
Distillate Fuel Oil	15.3%
Jet Fuel	12.3%
Still Gas	5.4%
Marketable Coke	5.0%
Residual Fuel Oil	3.3%
Liquefied Refinery Gas	2.8%
Asphalt and Road Oil	1.7%
Other Refined Products	1.5%
Lubricants	0.9%

Source: California Energy Commission (CEC) 2004

catalysts and chemicals during refining, or to design a vehicle that directly runs on crude oil based on its natural properties.

11.1.2 Natural Gas Processing

Natural gas found in natural gas reservoirs is a complex mixture of hundreds of different compounds. A typical natural gas stream consists of a mixture of methane, ethane, propane, butane and other hydrocarbons, water vapors, oil and condensates, hydrogen sulfides, carbon dioxide, nitrogen, other gases, and solid particles (Table 11.2). Even though these compounds are characterized as contaminants, they are not removed because of environmental concerns. It is obvious that water vapor, carbon dioxide, nitrogen, and sulfur components from a natural source are not a threat to the environment. The main reasons for their removal from a gas stream are the following: 1) the heating value of the gas is decreased in the presence of these gases. Consequently, suppliers are required to remove any levels of these gases beyond a desired value, depending on the standard set by the regulatory board. 2) The presence of water vapor, sulfur compounds, and so forth increases the

Table 11.2 Typical composition of natural gas.

Methane	CH_4	70–90%
Ethane	C_2H_6	
Propane	C_3H_8	0–20%
Butane	C_4H_{10}	
Carbon dioxide	CO_2	0–8%
Oxygen	O_2	0–0.2%
Nitrogen	N_2	0–5%
Hydrogen sulfide	H_2S	0–5%
Rare gases	A, He, Ne, Xe	Traces
Water vapor	H_2O (g)	*16 to 32 mg/m^3 (typical)

Source: Natural Gas Organization 2004

*Eldridge Products 2003

possibility of corrosion in the pipeline. This is a maintenance concern that has to do with the type of material used. 3) The presence of water in liquid form or hydrate in solid form can hinder compressors from functioning properly, or even block the entire flow stream. This is a mechanical concern that can affect smooth operations. 4) The presence of H_2S poses immediate safety concerns, in case of accidental leaks. Conventionally, various types of synthetic chemicals, such as glycol, amines, and synthetic membranes including other adsorbents, are used to remove these impurities from natural gas streams. Even though these synthetic-based absorbents and adsorbents work effectively in removing these impurities, there are environmental degradations caused during the life cycle of the production, transportation, and usage of these chemicals. As outlined in Chapter 3, such materials are inherently toxic to the environment, and can render the entire process unsustainable, even when traces of them are left in the gas stream. In this chapter, various gas processing techniques and chemicals used for gas processing are reviewed, and their impacts on the environment are discussed. Some natural and nontoxic substitutes for these chemicals are presented. Also presented is a detailed review of CO_2 and hydrogen sulfide removal methods used during natural gas processing.

There are certain restrictions imposed on major transportation pipelines regarding the make-up of the natural gas that is allowed into the pipeline; the allowable gas is called pipe "line quality" gas. Pipe line quality gas should not contain other elements, such as hydrogen sulfide, carbon dioxide, nitrogen, water vapor, oxygen, particulates, and liquid water that could be detrimental to the pipeline and its operating equipment (EIA 2006). Even though the hydrocarbons, such as ethane, propane, butane, and pentanes, have to be removed from natural gas streams, these products are used for various other applications.

The presence of water in natural gas creates several problems. Liquid water and natural gas can form solid ice-like hydrates that can plug valves and fittings in pipelines (Mallinson 2004). Natural gas containing liquid water is corrosive, especially if it contains carbon dioxide and hydrogen sulfide. It has also been argued that water vapor increases the volume of natural gas, decreasing its heating value, and in turn reducing the capacity of the transportation and storage system (Mallinson 2004). Hence, the removal of free water, water vapors, and condensates is a very important task during gas processing. Other impurities of natural gas, such as carbon dioxide and hydrogen sulfide, are generally considered impurities because acid gases must be removed from the natural gas prior to its transportation (Chakma 1999). Carbon dioxide is a major greenhouse gas that contributes to global warming. It is important to separate the carbon dioxide from the natural gas stream for its meaningful application, such as for enhanced oil recovery. Even though hydrogen sulfide is not a greenhouse gas, it is a source of acid rain deposition. Hydrogen sulfide is a toxic and corrosive gas that is rapidly oxidized to form sulfur dioxide in the atmosphere (Basu *et al.* 2004: Khan and Islam 2007). Oxides of nitrogen found in traces in the natural gas can cause ozone layer depletion and global warming. Figure 11.1 below shows the various stages in the life cycle of natural gas in which emissions are released from the natural gas into the natural environment.

11.2 Pathways of Crude Oil Formation

Crude oil is a naturally occurring liquid, found in formations in the earth, consisting of a complex mixture of hydrocarbons of various lengths. It contains mainly four groups of hydrocarbons: saturated hydrocarbons, which consist of a straight chain of carbon atoms;

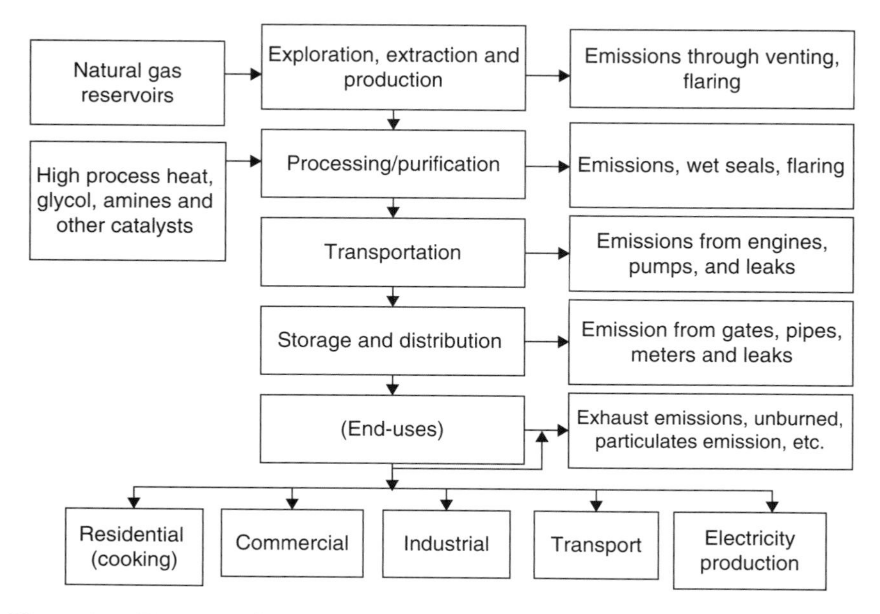

Figure 11.1 Emissions from life cycle of natural gas, from exploration to end use (after Chhetri *et al* 2008).

aromatics, which consist of ring chains; asphaltenes, which consist of complex polycyclic hydrocarbons with complicated carbon rings; and other compounds, mostly consisting of nitrogen, sulfur, and oxygen. Crude oil, natural gas, and coal are formed from the remains of zooplankton, algae, terrestrial plants, and other organic matters after exposure to heavy pressure and the temperature of the earth. These organic materials are chemically changed to kerogen. After more heat, pressure, and bacterial activities, crude oil, natural gas, and coal are formed. Figure 11.2 shows the pathway of the formation of crude oil, natural gas, and coal. These processes are all driven by natural forces.

It is well known that the composition of crude oil is similar to that of plants (Wittwer and Immel 1980). Crude oils represent the ultimate in natural processing, from wood (fresh but inefficient) to coal (older but more efficient) to petroleum fluids (much older but much more efficient). Natural gas has much greater energy efficiency than liquid petroleum. This is not evident in conventional calculations, because these calculations are carried out on the volume basis for gas. If calculations are made on the basis of weight, energy efficiency with natural gas would be higher than crude oil.

The following table summarizes some of the heating values.

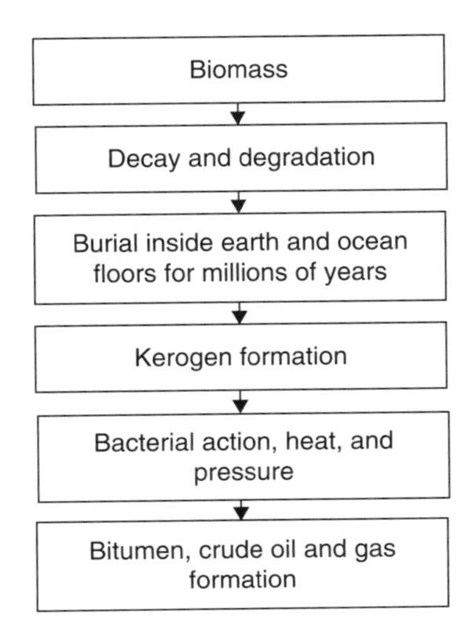

Figure 11.2 Crude oil formation pathway – the natural processes.

Table 11.3 Heating value for various fuels.

Fuel Type	Heating Value
Premium wood pellets	13.6 million Btu/ton
Propane	71,000 Btu/gal
Fuel oil #2	115,000 Btu/gal
Fuel oil #6	114,000 Btu/gal
Seasoned firewood	15.3 million Btu/cord
Ovendried switchgrass	14.4 million Btu/ton
Bituminous coal	26 million Btu/ton
Shelled corn @15% MC	314,000 Btu/bushel
Natural gas	1050 Btu/scf

Source: Some of these data can be obtained from US government sites (website 21)

More detailed heating values are given by J.W. Bartock, as listed in Table 11.4.

Using this knowledge, techniques have been developed to accelerate (through pyrolysis) the natural process of crude oil formation in order to produce synthetic crude from coal starting in the late

Table 11.4 Approximate Heating Value of Common Fuels.

Natural Gas	1,030 Btu/cu ft
Propane	2,500 Btu/cu ft 92,500 Btu/gal
Methane	1,000 Btu/cu ft
Landfill gas	500 Btu/cu ft
Butane	3,200 Btu/cu ft 130,000 Btu/gal
Methanol	57,000 Btu/gal
Ethanol	76,000 Btu/gal
Fuel oil	
Kerosene	135,000 Btu/gal
#2	138,500 Btu/gal
#4	145,000 Btu/gal
#6	153,000 Btu/gal
Waste oil	125,000 Btu/gal
Biodiesel – Waste vegetable oil	120,000 Btu/gal
Gasoline	125,000 Btu/gal
Wood	
Softwood 2–3,000 lb/cord	10–15,000,000 Btu/cord
Hardwood 4–5,000 lb/cord	18–24,000,000 Btu/cord
Sawdust – green 10–13 lb/cu ft	8–10,000,000 Btu/ton
Sawdust – kiln dry 8–10 lb/cu ft	14–18,000,000 Btu/ton
Chips – 45% moisture 10–30 lb/cu ft	7,600,000 Btu/ton
Hogged 10–30 lb/cu ft	16–20,000,000 Btu/ton
Bark 10–20 lb/cu ft	9–10,500,000 Btu/ton
Wood pellets 10% moisture 40–50 lb/cu ft	16,000,000 Btu/ton
Hard Coal (anthracite) 13,000 Btu/lb	26,000,000 Btu/ton

(Continued)

Table 11.4 (Cont.) Approximate Heating Value of Common Fuels.

Soft Coal (bituminous) 12,000 Btu/lb	24,000,000 Btu/ton
Rubber–pelletized 16,000 Btu/lb	32–34,000,000 Btu/ton
Plastic	18–20,000 Btu/lb
Corn–shelled 7,800–8,500 Btu/lb	15–17,000,000 Btu/ton
Cobs	8,000–8,300 Btu/lb 16–17,000,000 Btu/ton

Source: (Website 22)

1970s (Cortex and Ladelfa 1981; Stewart and Klett 1979). However, pyrolysis does not guarantee a natural process. In fact, the use of synthetic catalysts, synthetic acids, and other additives, along with electric heating, will invariably render the process unsustainable and inherently detrimental to the environment, in addition to being inefficient. The same comment stands for numerous processes that have been in place for converting wood into fuel (Guo 2004) and natural gas into synthetic crude (Teel 1994). The premise that if the origin or the process is not sustainable, the final product cannot be sustainable, meaning acceptable to the ecosystem, is consolidated by the following consideration. Consider in Table 11.4 that the heating value of plastic is much higher than that of sawdust (particularly the green type). If the final heating value were the primary consideration, plastic materials would be a far better fuel source than sawdust. However, true sustainability must consider beyond a short-term single criterion. With the sustainability discussion presented in this book, plastic materials would be rejected because the process followed in creating these materials is not sustainable (see Chapter 9).

In general, a typical crude oil has the bulk composition shown in Table 11.5.

In the above list, metals found in crude oil are numerous, including many heavy metals. Some of them are Ni, V, Cu, Cd, Hg, Zn, and Pb (Osujo and Onoiake 2004). In their natural state, these metals are not harmful because they are in a status similar to plants and other organic materials.

Table 11.5 Composition of a typical crude oil.

Elements	Lower Range (Concentration, wt%)	Upper Range (Concentration, wt%)
Carbon	83.9	86.8
Hydrogen	11.0	14.0
Sulfur	0.06	8.00
Nitrogen	0.02	1.70
Oxygen	0.08	1.82
Metals	0.00	0.14

Typical crude oil density ranges from 800 kg/m^3 to 1000 kg/m^3. The following table shows how density can vary with weathering and temperature for the North Slope Alaskan crude oil.

Similarly, viscosity properties of the same crude oil are given in Table 11.7. Note that without weathering (causing low temperature oxidation), the viscosity values remain quite low.

Table 11.8 shows the flash points of the same crude oil. Note how the flash point is low for unweathered crude oil, showing its efficiency for possible direct combustion. The same crude oil, when significantly weathered, becomes incombustible, reaching values comparable to vegetable oils, thus making it safe to handle.

Table 11.9 shows further distribution of various hydrocarbon groups that occur naturally. This table shows how valuable components are all present in crude oil, and if no toxic agents were added in order to refine or process the crude oil, the resulting fluid would remain benign to the environment.

Finally, Table 11.10 shows various volatile organic (VOC) compounds present in a typical crude oil.

11.3 Pathways of Crude Oil Refining

Fossil fuels derived from the petroleum reservoirs are refined in order to suit the various application purposes, from car fuels to airplane fuels and space fuels. Fossils fuels are a complex mixture of hydrocarbons, varying in composition depending on source and

Table 11.6 Density properties of crude oil (Alaskan North Slope).

Weathering (%wt)	Temperature (C)	Density (g/mL)
0	0	0.8777
	15	0.8663
10	0	0.9054
	15	0.894
22.5	0	0.9303
	15	0.9189
30.5	0	0.9457
	15	0.934

Source: (Website 23)

Table 11.7 Viscosity properties of crude oil (Alaskan North Slope).

Weathering (%wt)	Temperature (C)	Viscosity (cP)
0	0	23.2
	15	11.5
10	0	76.7
	15	31.8
22.5	0	614
	15	152
30.5	0	4230
	15	624.7

Table 11.8 Flash points of crude oil (Alaskan North Slope).

Weathering (%wt)	Flash Point (C)
0	<- 8
10	19
22.5	75
30.5	115

Table 11.9 Hydrocarbon groups in a typical crude oil (North Slope Alaska).

Components	0% weathered	10% weathered	22.5% weathered	30.5% weathered
Saturates	75	72.1	69.2	64.8
Aromatics	15	16	16.5	18.5
Resins	6.1	7.4	8.9	10.3
Asphaltenes	4	4.4	5.4	6.4
Waxes	2.6	2.9	3.3	3.6

Table 11.10 Volatile Organic Compounds (VOC) in crude oil (North Slope Alaska).

Component	0% weathered	30.5% weathered
Benzene	2866	0
Toluene	5928	0
Ethylbenzene	1319	0
Xylenes	6187	0
C_3-Benzenes	5620	30
Total BTEX	16300	0
Total BTEX and C_3-Benzenes	21920	30

origin. Depending on the number of carbon atoms the molecules contain and their arrangement, the hydrocarbons in the crude oil have different boiling points. In order to take advantage of the difference in boiling points of different components in the mixture, fractional distillation is used to separate the hydrocarbons from the crude oil. Figure 11.3 shows the fractional distillation column, in which the temperature is lower at the top and increases as it goes down the column.

Figure 11.3 gives a general schematic of the activities from the storage of crude oil to the complete refining process. The stored crude oil is transported into the place where either vacuum

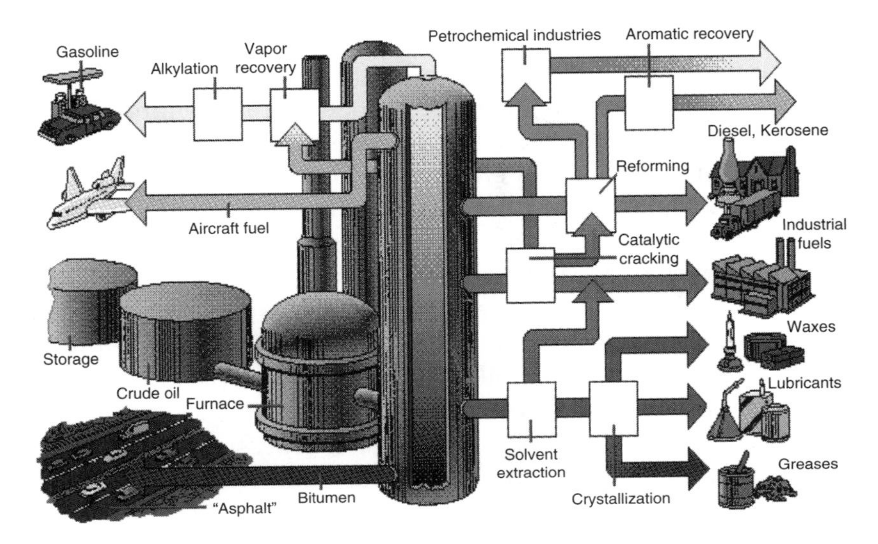

Figure 11.3 Fractional distillation unit for hydrocarbon refining (adapted from Website 1, Encarta).

distillation or atmospheric distillation is used for hydrocarbon separation. Chemical impurities of crude oil, such as sulfur or wax, are separated.

Crude oil is refined through distillation, or fractionation, in order to form several different hydrocarbon groups, such as gasoline, diesel, aircraft fuel, kerosene, asphalt, and waxes. The fractions emerging from crude oil distillation are divided out based on their increasing molecular weight and boiling temperature in the distillation column. The distillation process continues until all the fractions are separated.

Fractional distillation is the process of separating crude oil in atmospheric and vacuum distillation towers into groups of hydrocarbon compounds of different boiling points. The hydrocarbon conversion process consists of alkylation, thermal and catalytic cracking for decomposition, and polymerization for combining the hydrocarbon molecules and rearranging with catalytic reforming. To remove or separate the naphthenes, aromatics, and other undesirable compounds, various treatment processes, such as dissolving, adsorption, and precipitation, are carried out. In addition to this, desalting, drying, hydrodesulfurizing, solvent refining, sweetening, solvent extraction, and solvent dewaxing are also done to remove impurities from the fractions. Other activities, such as formulating and blending, are carried out to produce finished products with desired properties. Refining operations also include the treatment of wastewaters contaminated due to petroleum operations, solid waste management, process water treatment, and cooling and sulfur recovery. Other auxiliary operations include power generation and management for process operations, the flare system, the supply of air, nitrogen, steam, and other necessary system inputs, along with the administrative management of the whole refining systems.

Even though distillation results in separate hydrocarbons, the resulting products of petroleum are directly related to the properties of the processed crude oil. These distillation products are further processed into more conventionally usable products by using cracking, reforming, and other conversion processes.

The pathways of oil refining illustrate that the oil refining process utilizes toxic catalysts and chemicals, and the emission from oil burning also becomes extremely toxic. Figure 11.4 shows the pathway of oil refining. During the cracking of the hydrocarbon molecules, different types of acid catalysts are used, along with high heat and pressure. The process of breaking the hydrocarbon molecules is the thermal cracking. During alkylation, sulfuric acids, hydrogen fluorides, aluminum chlorides, and platinum are used as catalysts. Platinum, nickel, tungsten, palladium, and other catalysts are used during hydro processing. In distillation, high heat and pressure are used as catalysts.

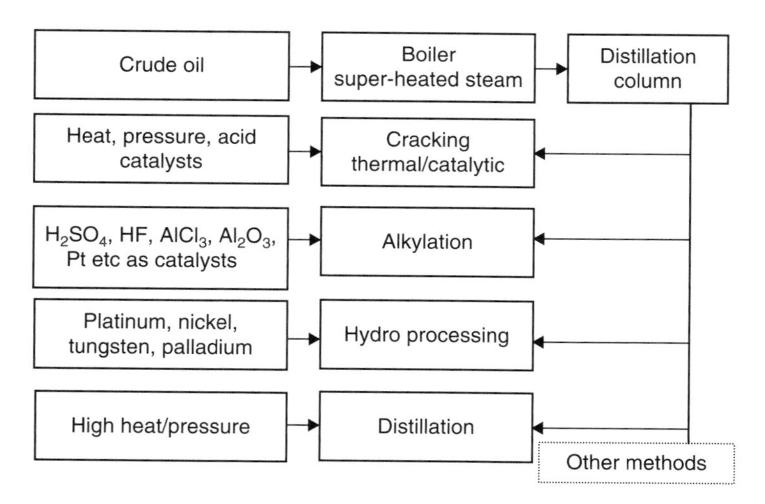

Figure 11.4 Pathway of oil refining process.

11.4 Additives in Oil Refining and their Functions

Oil refining and natural gas processing are very expensive processes, in terms of operation and management. These operations involve the use of several chemicals and catalysts that are very expensive. Moreover, these catalysts and chemicals pose a great threat to the natural environment, including air and water quality. Air and water pollution ultimately have impacts on the health of humans, animals, and plants. For instance, the use of catalysts, such as lead, during crude oil refining to produce gasoline has been a serious environmental problem. Burning gasoline emits toxic gases containing lead particles, and the oxidation of lead in the air forms lead oxide, which is a poisonous compound affecting the lives of every living thing. Heavy metals such as mercury and chromium and the use of these metals in oil refining are major causes of water pollution. In the previous chapter, details of catalysts used in a refining process have been given. Consider the consequences of some of these chemicals.

11.4.1 Platinum

It is well known that platinum salts can induce numerous irreversible changes in human bodies, such as DNA alterations (Jung and Lippard 2007). In fact, an entire branch of medical science evolves around exploiting this deadly property of platinum compounds in order to manufacture pharmaceutical drugs that are used to attack the DNA of cancer cells (Farrell 2004a, 2004b, 2004c, 2005). It is also known that platinum compounds cause many forms of cancer. Once again, this property of platinum is used to develop pharmaceutical drugs that could possibly destroy cancer cells (Volckova *et al.* 2003; Website 1). Also, it is well known that platinum compounds can cause liver damage (Stewart *et al.* 1985). Similar damage to bone marrow is also observed (Evans *et al.* 1984). Platinum is also related to hearing loss (Rybak 1981). Finally, potentiation of the toxicity of other dangerous chemicals in the human body, such as selenium, can lead to many other problems.

The above are immediate concerns to human health and safety. Consider the damage to the environment that might be incurred through vegetation and animals (Kalbitz *et al.* 2008). It is already known that platinum salts accumulate at the root of plants, from which they can easily enter the food chain, perpetually insulting the environment. In addition, microorganisms can play a role to broaden the impact of platinum. This aspect of ecological study has not yet been performed.

11.4.2 Cadmium

Cadmium is considered to be a non-essential and highly toxic element to a wide variety of living organisms, including man, and it is one of the widespread pollutants with a long biological half-life (Plunket 1987; Klaassen 2001; Rahman *et al.* 2004). A provisional, maximum, tolerable daily intake of cadmium from all sources is 1–1.2g/kg body mass (Bortoleto *et al.* 2004) and is recommended by FAO-WHO jointly. This metal enters the environment mainly from industrial processes and phosphate fertilizers, and is transferred to animals and humans through the food chain (Wagner 1993; Taylor 1997; Sattar *et al.* 2004). Cadmium is very hazardous because humans retain it strongly (Friberg *et al.*, 1974), particularly in the liver (half-life of 5 to 10 years) and kidney (half-life of 10 to 40 years). The symptoms of cadmium toxicity produced by enzymatic inhibition include hypertension, respiratory disorders, damage of kidney and liver, osteoporosis, formation of kidney stones, and others (Vivoli *et al.* 1983; Dinesh *et al.* 2002; Davis 2006). Environmental, occupational, or dietary exposure to Cd(II) may lead to renal toxicity, pancreatic cancer (Schwartz 2002), or enhanced tumor growth (Schwartz *et al.* 2000). The safety level of cadmium in drinking water in many countries is 0.01ppm, but many surface waters show higher cadmium levels. Cadmium can kill fish in one day at a concentration of 10 ppm in water, whereas it can kill fish in 10 days at a concentration of 2 ppm. Studies with cadmium have shown harmful effects on some fish at concentrations of 0.2ppm (Landes *et al.* 2004). Plants can accumulate cadmium up to a level as high as 5 to 30 mg/kg, whereas the normal range is 0.005 to 0.02 mg/kg (Cameron 1992). Taken up in excess by plants, Cd directly or indirectly inhibits physiological processes, such as respiration, photosynthesis, cell elongation, plant-water relationships, nitrogen metabolism, and mineral nutrition, all of which result in poor growth and low biomass. It was also reported that cadmium is more toxic than lead in plants (Pahlsson 1989; Sanita´di Toppi and Gabbrielli 1999).

11.4.3 Lead

Lead (II) is a highly toxic element to humans and most other forms of life. Children, infants, and fetuses are at particularly high risk of neurotoxic and developmental effects of lead. Lead can cause accumulative poisoning, cancer, and brain damage, and it can cause mental retardation and semi-permanent brain damage in young children (Friberg *et al.* 1979; Sultana *et al.* 2000). At higher levels, lead can cause coma, convulsion, or even death. Even low levels of lead are harmful and are associated with a decrease in intelligence, stature, and growth. Lead enters the body through drinking water or food, and can accumulate in the bones. Lead has the ability to replace calcium in the bone to form sites for long-term release (King *et al.* 2006). The Royal Society of Canada (1986) reported that human exposure to lead has harmful effects on the kidney, the central nervous system, and the production of blood cells. In children, irritability, appetite loss, vomiting, abdominal pain, and constipation can occur (Yule and Lansdown 1981). Pregnant women are at high risk because lead can cross the placenta and damage the developing fetal nervous system; lead can also induce miscarriage (Wilson 1966). Animals ingest lead via crops and grasses grown in contaminated soi1. Levels in plants usually range from 0.5 to 3 mg/kg, while lichens have been shown to contain up to 2,400 mg/kg of lead (Cameron 1992). Lead ingestion by women of childbearing age may impact both the woman's health (Lustberg and Silbergeld 2002) and that of her fetus, for ingested lead is stored in the bone and released during gestation (Angle *et al.* 1984; Gomaa *et al.* 2002).

11.5 Emissions from Oil Refining Activities

Crude oil refining is one of the major industrial activities that emits CO_2 and many toxic air pollutants and has a high energy consumption. Because of the presence of trace elements, this CO_2 is not readily absorbed by the ecosystem, creating an imbalance into the atmosphere. Szklo and Schaeffer (2002) reported that crude oil refining processes are highly energy-intensive, requiring between 7% and 15% of the crude oil from the refinery processes. The study showed that the energy use in the Brazilian refining industry will further increase by 30% between 2002 and 2009 to reduce the sulfur content of diesel and gasoline as well as to reduce CO_2 emissions.

For example, lube oil production needs about 1500 MJ/barrel, and alkylation with sulfuric acid and hydrofluoric acid requires 360MJ/barrel and 430 MJ/barrel, respectively. The energy consumption would further increase to meet the more stringent environmental quality specifications for oil products worldwide. The full recovery of such chemicals or catalysts is not possible, leading to environmental hazards.

Concawe (1999) reported that CO_2 emissions during the refining process of petroleum products were natural gas (56 kg CO_2/GJ), LPG (64 kg CO_2/GJ), distillate fuel oil (74 kg CO_2/GJ), residual fuel (79 kg CO_2/GJ), and coke (117 kg CO_2/GJ). Cetin *et al.* (2003) carried out a study located around a petrochemical complex and an oil refinery, and reported that the volatile organic compounds (VOCs) concentrations measured were 4–20 times higher than those measured at a suburban site in Izmir, Turkey. Ethylene dichloride, a leaded gasoline additive used in petroleum refining, was the most abundant volatile organic compound, followed by ethyl alcohol and acetone. In addition to the VOCs, other pollutants, such as sulfur dioxide, reduced sulfur compounds, carbon monoxide, nitrogen oxides, and particulate matter, are also emitted from petroleum refineries (Buonicare and Davis 1992). Rao *et al.* (2005) reported that several hazardous air pollutants (HAPs), including hydrocarbons such as Maleic anhydride (pyridines, sulfonates, sulfones, ammonia, carbon disulfide, methylethylamine, arsenic, coppers, beryllium, and others), benzoic acid (benzene, xylene, toluene, formic acid, diethylamine, cobalt, zinc, formaldehyde, cadmium, antimony, and others), and ketones and aldehydes (phenols, cresols, chromates, cyanides, nickel, molybdenum, aromatic amines, barium, radionuclides, chromium, and others), are emitted during refining.

Some studies have shown that oil refineries fail to report millions of pounds of harmful emissions that have substantial negative impacts on health and air quality. A report prepared by the Special Investigation Division, Committee on Government Reform for the United States House of Representatives concluded that oil refineries vastly underreport fugitive emissions to federal and state regulators that exceed 80 million pounds of VOCs and 15 million pounds of toxic pollutants (USHR 1999). The total VOCs emission was reported to be 492 million pounds in the United States. This report confirms that the leaks in the valves are five times higher than the leaks reported to the state and federal regulators. Among other toxic air pollutants, it was also reported that oil refineries were the largest emitter of benzene (over 2.9 million pounds) in the United States during that report period. Other emissions reported were 4.2 million pounds of xylenes, 4.1 million pounds of methyl ethyl ketone, and 7 million pounds of toluene.

Some of these pollutants are byproducts of the catalysts or additives used during the refining process. Other pollutants, such as VOCs, are formed during high temperature and pressure use in the refining process. However, by improving the refining process, it is possible that the emissions can be reduced. For example, mild hydrotreating is conventionally used to remove sulfur and olefins, while severe hydrotreating removes nitrogen compounds, and reduces sulfur content and aromatic rings (Gucchait *et al.* 2005). Hence, a search for processes that release fewer emissions and use more environment-friendly catalysts is very important in reducing the emissions from refineries and energy utilization.

11.6 Degradation of Crude and Refined Oil

The rate of biodegradation of crude oil and refined oil indicates what change has occurred to the refined products compared to its feedstock. Boopathy (2004) showed that under an anaerobic condition, 81% of diesel oil was degraded within 310 days under an electron acceptor condition. However, 54.5% degradation was observed in the same period under sulfate-reducing condition. Aldrett *et al.* (1997) studied the microbial degradation of crude oil in marine environments with thirteen different bioremediation products for petroleum hydrocarbon degradation. This sample was extracted and fractionated in total saturated petroleum hydrocarbons (TsPH) and total aromatic petroleum hydrocarbons (TarPH). The analysis showed that some products reduced the TsPH fraction to 60% of its initial weight and the TarPH fraction to 65% in 28 days. This degradation was reported to be higher than that which degraded by naturally occurring bacteria. Even though the condition of the degradation for diesel and crude oil is different, it is observed that crude oil degrades faster than diesel (refined oil). Al-Darbi *et al.* (2005) reported that natural oils degrade faster in a sea environment due to the presence of a consortia of microorganisms in sea. Livingston and Islam (1999) reported that petroleum hydrocarbons can be degraded by the bacteria present in soil.

11.7 Pathways of Natural Gas Processing

Natural gas is a mixture of methane, ethane, propane, butane and other hydrocarbons, water vapor, oil and condensates, hydrogen sulfides, carbon dioxide, nitrogen, other gases, and solid particles. The free water and water vapors are corrosive to the transportation equipment. Hydrates can plug the gas accessories, creating several flow problems. Other gas mixtures, such as hydrogen sulfide and carbon dioxide, are known to lower the heating value of natural gas by reducing its overall fuel efficiency. This makes mandatory that natural gas is purified before it is sent to transportation pipelines. Gas processing is aimed at preventing corrosion, an environmental and safety hazard associated with the transport of natural gas.

In order to extract the natural gas found in the natural reservoirs, onshore and offshore drilling activities are carried out. Production and processing are carried out after the extraction, which includes natural gas production from the underground reservoir and removing the impurities in order to meet certain regulatory standards before sending for end use. Purified natural gas is transported in different forms, such as liquefied petroleum gas (LPG), liquefied petroleum gas (LNG), or gas hydrates, and distributed to the end users as per the demands. EIA (2006) illustrated a generalized natural gas processing schematic (Figure 11.5). Various chemicals and catalysts are used during the processing of natural gas.

This generalized natural gas processing scheme includes all necessary steps depending on the types of ingredients available in a particular gas. The gas-oil separator unit removes any oil from the gas stream, the condensate separator removes free water and condensates, and the water dehydrator separates moisture from the gas stream. The other contaminants, such as CO2, H2S, nitrogen, and helium, are also separated from the different units. The natural gas liquids, such as ethane, propane, butane, pentane, and gasoline, are separated from methane using cryogenic and absorption methods. The cryogenic process consists of lowering the temperature of the gas stream with the turbo expander and external refrigerants. The sudden drop in temperature in the expander condenses the hydrocarbons in the gas stream, maintaining the methane in the gaseous form. Figure 11.6 illustrates the details of removing the contaminants and chemicals used during natural gas processing. The procedure to remove each contaminant is discussed below.

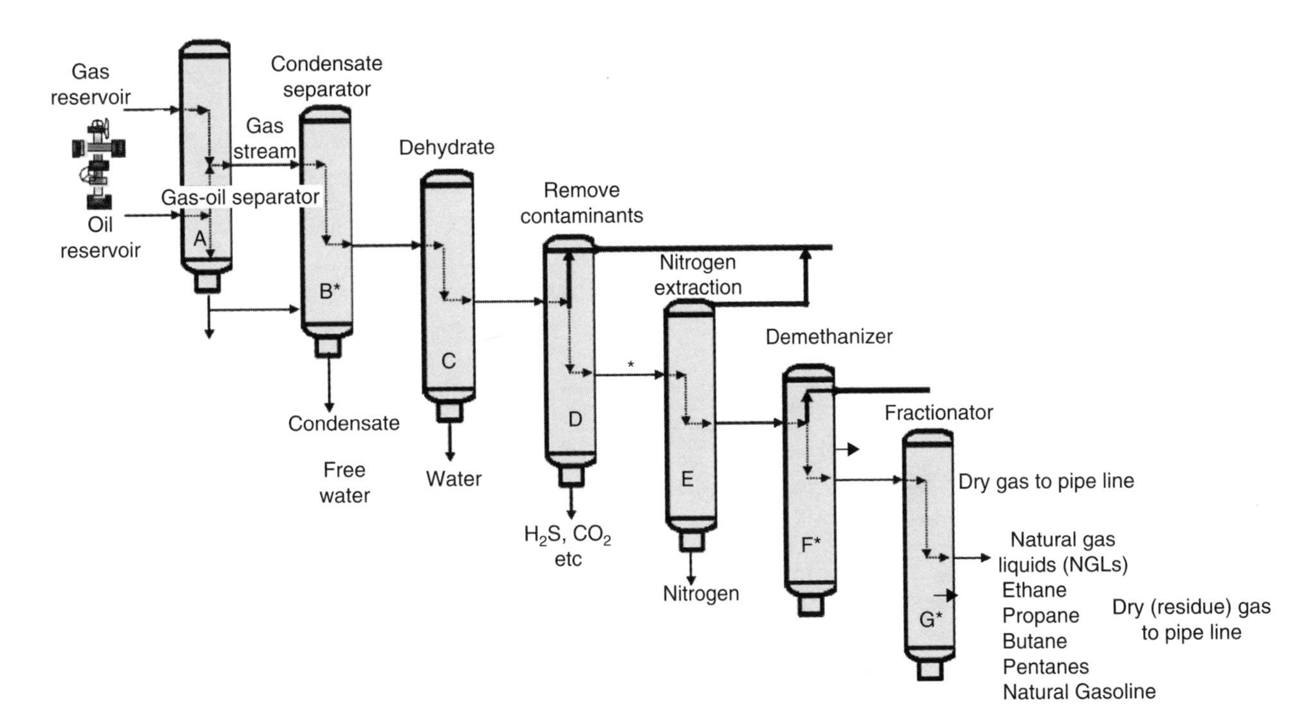

Figure 11.5 Generalized natural gas processing schematic (EIA 2006).

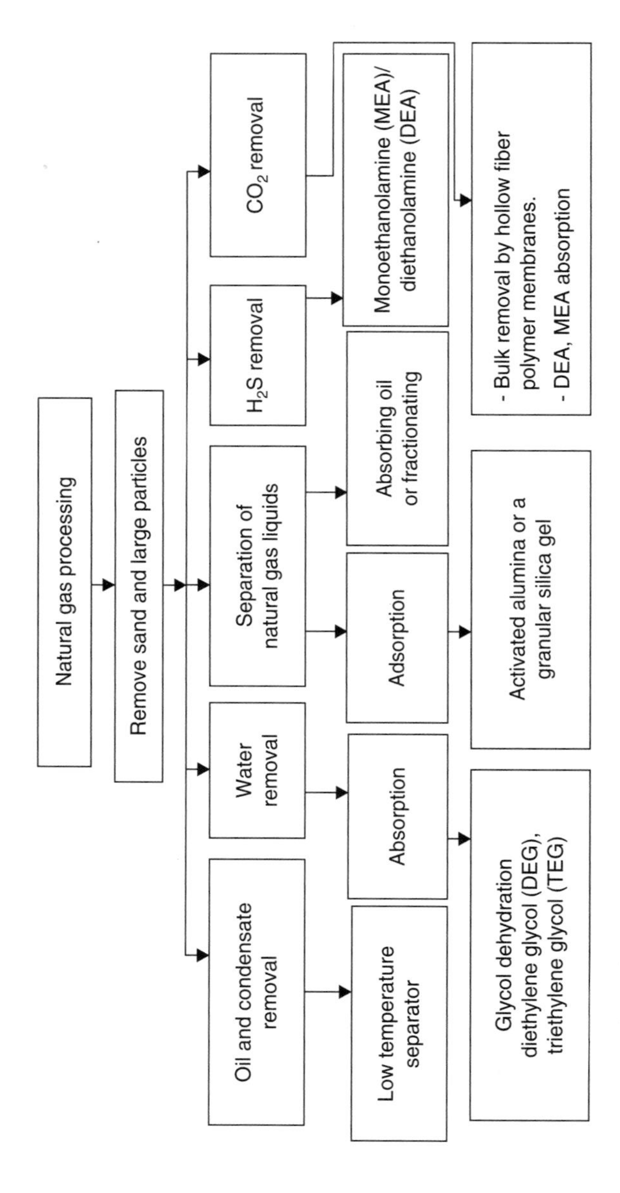

Figure 11.6 Various methods for gas processing.

11.8 Oil and Condensate Removal from Gas Streams

Natural gas is dissolved in oil underground due to the formation pressure. When natural gas and oil are produced, they generally separate simply because of the decrease in pressure. This separator consists of a closed tank where the gravity serves to separate the heavier liquids from lighter gases (EIA 2006). Moreover, specialized equipment, such as the Low-Temperature Separator (LTS), is used to separate oil and condensate from natural gas (Natural Gas Org. 2004). When the wells are producing high pressure gas along with light crude oil or condensate, a heat exchanger is used to cool the wet gas, and the cold gas then travels through a high-pressure liquid knockout, which serves to remove any liquid into a low-temperature separator. The gas flows into the low-temperature separator through a choking mechanism, expanding in volume as it enters the separator. This rapid expansion of the gas lowers the temperature in the separator. After the liquid is removed, the dry gas is sent back through the heat exchanger, followed by warming it by the incoming wet gas. By changing the pressure at different sections of the separator, the temperature varies, causing the oil and water to condense out of the wet gas stream. The gas stream enters the processing plant at high pressure (600 pounds per square inch gauge (psig) or greater) through an inlet slug catcher where free water is removed from the gas, after which it is directed to a condensate separator (EIA 2006).

11.9 Water Removal from Gas Streams

Natural gas may contain water molecules in both vapor and liquid states. Water contained in a natural gas stream may cause the formation of hydrates. Gas hydrates are formed when gas containing water molecules reaches a low temperature (usually $<25°C$) and high pressure ($>1.5MPa$) (Koh *et al.* 2002). Water in natural gas is removed by separation methods at or near the well head. Note that it is impossible to remove all water molecules from a gas stream, and operators have to settle for an economic level of low water content. The water removal process consists of dehydrating natural gas either by absorption, adsorption, gas permeation, or low temperature separation. In absorption, a dehydrating agent, such as glycol, takes out the water vapor (Mallinson 2004). In adsorption, the water vapor is condensed and collected on the surface. Adsorption dehydration can also be used, utilizing dry-bed dehydrating towers, which contain desiccants, such as silica gel and activated alumina, to perform the extraction. Various types of membranes have been investigated to separate the gas from water. However, membranes require large surface areas, and, therefore, compact membranes with high membrane areas are necessary to design an economical gas permeation process (Rojey *et al.* 1997). The most widely used membranes consist of modules with plane membranes wound spirally around a collector tube, or modules with a bundle of hollow fibers.

11.9.1 Glycol Dehydration

It is important to remove water vapor present in a gas stream, because otherwise it may cause hydrate formation at low temperature conditions that may plug the valves and fittings in gas pipelines (Twu 2005). Water vapor may further cause corrosion when it reacts with hydrogen sulfide or carbon dioxide present in gas streams. Glycol is generally used for water dehydration or absorption (Mallinson 2004). Glycol has a chemical affinity for water (Kao *et al.* 2005). When it is in contact with a stream of natural gas containing water, glycol will absorb the water portion out of the gas stream (EIA 2006). Glycol dehydration involves using a glycol solution, either diethylene glycol (DEG) or triethylene glycol (TEG). After absorption, glycol molecules become heavier and sink to the bottom of the contactor, from where they are removed. Boiling then separates glycol and water; however, water boils at 212°F, whereas glycol boils at 400°F (Natural Gas Org. 2004). Glycol is then reused in the dehydration process (Mallinson 2004).

Ethylene glycol is synthetically manufactured from ethylene via ethylene oxide as an intermediate product. Ethylene oxide reacts with water to produce ethylene glycol in the presence of acids or bases or at higher temperatures without chemical catalysts ($C_2H_4O+H_2O \rightarrow HOCH_2CH_2OH$). Diethylene glycol (DEG: chemical formula $C_4H_{10}O_3$) and triethylene glycol (TEG) are obtained as the co-products during the manufacturing of monoethylene glycol (MEG). MSDS (2006) categorizes DEG as hazardous materials for 99–100% concentration.

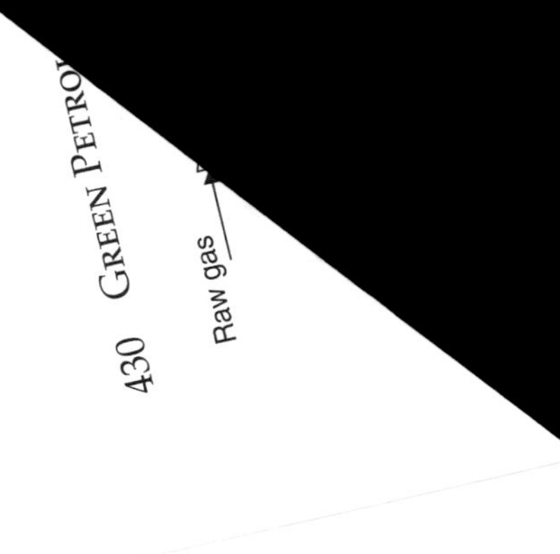

Several incidents of DEG poisoning
for children containing diethylene glyc
killed 80 in Haiti in 1998, 105 people d
an elixir containing diethylene glycol
people died in Nigeria after taking a gl
and 200 died due to a glycol contaminat
1992 (Daza 2006). Hence, it is important to
to glycol, so that such health and environ
avoided.

11.9.2 Solid-Desiccant Dehydration

In this process, solid desiccants, such as activated alumina or granular silica gel materials, are used for adsorption in two or more adsorption tower arrangements (Mallinson 2004; Guo and Ghalambor 2005). Natural gas is passed through these adsorption towers. Water is retained on the surface of these desiccants. The gas rising from the bottom of the adsorption tower will be completely dry gas. These are more effective than glycol dehydrators, and are best suited for large volumes of gas under very high pressure. Two or more towers are generally required because the desiccant in one tower becomes saturated with water and needs to regenerate the desiccant. A solid desiccant system is more expensive than glycol dehydration process.

The solid desiccant adsorption process consists of two beds, with each bed going through successive steps of adsorption and desorption (Rojey *et al.* 1997; Mallinson 2004). During the adsorption step, the gas to be processed is sent through the adsorbent bed and retains the water (Figure 11.7). When the bed is saturated, hot natural gas is sent to regenerate the adsorbent. After regeneration and before the adsorption step, the bed must be cooled. This is achieved by passing through cold natural gas. After heating, the same gas can be used for regeneration.

11.10 Separation of Natural Gas Liquids

Natural gas liquids (NGLs) are saturated with propane, butane, and other hydrocarbons. NGLs have a higher value as separate products. This is one reason why NGLs are separated from the natural gas stream. Moreover, reducing the concentration of

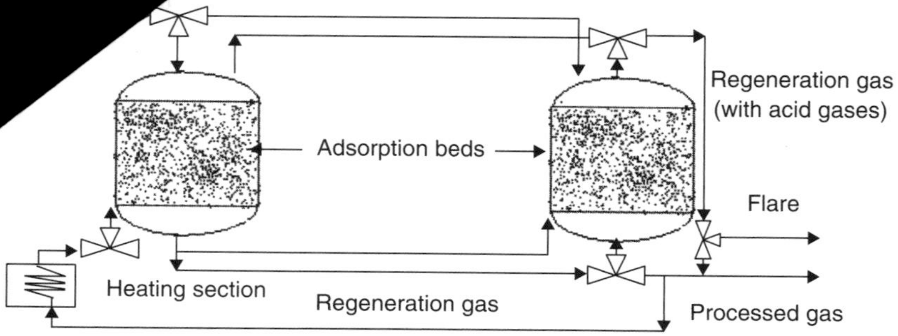

Figure 11.7 Dehydration by adsorption in fixed bed (redrawn from Rojey *et al.* 1997).

higher hydrocarbons and water in the gas is necessary to prevent formation of hydrocarbon liquids and hydrates in the natural gas pipeline. The removal of NGLs is usually done in a centralized processing plant by processes similar to those used to dehydrate natural gas. There are two common techniques for removing NGLs from the natural gas stream: the absorption and cryogenic expander processes.

11.10.1 The Absorption Method

This process is similar to adsorption by dehydration. The natural gas is passed through an absorption tower and brought into contact with the absorption oil that soaks up a large amount of the NGLs (EIA 2006). The oil containing NGLs exits the absorption tower through the bottom. The rich oil is fed into lean oil stills, and the mixture is heated to a temperature above the boiling point of the NGLs and below that of the oil. The oil is recycled, and NGLs are cooled and directed to an absorption tower. This process allows recovery of up to 75% of butanes and 85–90% of pentanes and heavier molecules from the natural gas stream. If the refrigerated oil absorption method is used, propane recovery can reach up to 90%. Extraction of the other, heavier NGLs can reach close to 100% using this process. Alternatively, the fractioning tower can also be used where boiling temperatures vary from the individual hydrocarbons in the natural gas stream. The process occurs in stages as the gas stream rises through several towers, where heating units raise the temperature of the stream, causing the various liquids to separate and exit into specific holding tanks (EIA 2006).

11.10.2 The Membrane Separation

Various types of membranes can be used to remove water and higher hydrocarbons. The conventional membranes can lower the dew point of the gas (Figure. 11.8). The raw natural gas is compressed and the air is cooled, which knocks out some water and NGLs. The gas from the compressor is passed through the membrane, which is permeable to water and higher hydrocarbons. The dry, hydrocarbon-depleted residual gas is sent to the pipeline for use.

11.10.3 The Cryogenic Expansion Process

This process consists of dropping the temperature of the gas stream to a lower level. This can be done by the turbo expander process. Essentially, cryogenic processing consists of lowering the temperature of the gas stream to around –120° Fahrenheit (EIA 2006). In this process, external refrigerants are used to cool the natural gas stream. The expansion turbine is used to rapidly expand the chilled gases, causing the temperature to drop significantly. This rapid temperature drop condenses ethane and other hydrocarbons in the gas stream, maintaining methane in a gaseous form. This process recovers up to 90–95% of the ethane (EIA 2006). The expansion turbine can be utilized to produce energy as the natural gas stream is

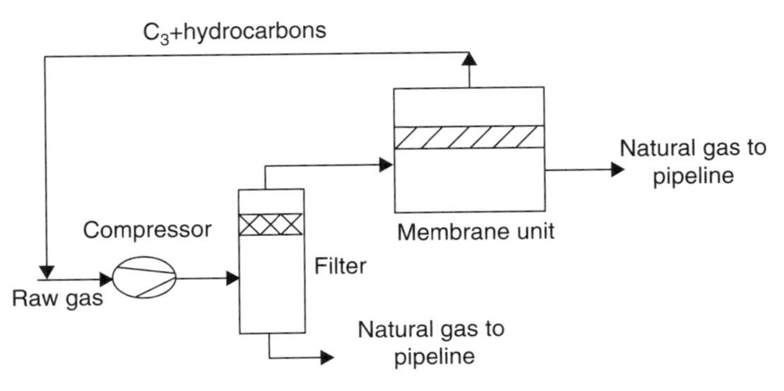

Figure 11.8 Membrane system for NGL recovery and dew point control (redrawn from MTR 2007).

expanded into recompressing the gaseous methane effluent. This helps save energy costs for natural gas processing.

11.11 Sulfur and Carbon Dioxide Removal

CO_2 and H_2S present in the natural gas are considered to have no heating value, and thus they reduce the heating value of natural gas (Mallinson 2004). The solvent in an absorber chemically absorbs acid gases, such as CO_2 and H_2S, and natural gas with reduced acid gas content can be obtained. The chemical solvent containing the absorbed acid gases is regenerated to be used again in the absorption process. The hydrogen sulfide is converted to elemental sulfur, and the CO_2 is released to atmosphere. Since CO_2 is a greenhouse gas, releasing it into the atmosphere will pose environmental threats. With increasing awareness of its environmental impact and the ratification of the Kyoto Protocol by most of the member countries, it is expected that the release of CO_2 into the atmosphere will be limited.

Sulfur exists in natural gas as hydrogen sulfide (H_2S), which is corrosive. H_2S is called a sour gas in the natural gas industry. To remove H_2S and CO_2 from natural gas, amine solutions are generally used (Chakma 1997; EIA 2006). Sulfur removal is generally achieved by using a variant of the Claus process, in which the hydrogen sulfide is partially oxidized. The hydrogen sulfide is absorbed from the natural gas at ambient temperature in a scrubber, or in alkanolamine-glycol solution. The natural gas is run through a tower, which contains the amine solution. This solution has an affinity for sulfur. There are two principal amine solutions used, monoethanolamine (MEA) and diethanolamine (DEA). Both DEA and MEA in the liquid form will absorb sulfur compounds from natural gas as it passes through. After passing through the MEA or DEA, the effluent gas is free of sulfur. The amine solution used can be regenerated (by removing the absorbed sulfur), allowing it to be reused to treat more natural gas. It is also possible to use solid desiccants, such as iron sponges, to remove the sulfide and carbon dioxide. Amines solutions and different types of membrane technologies are used for CO_2 removal from natural gas streams (Wills 2004). However, glycol and amines are toxic chemicals, and have several health and environment impacts (Melnick 1992). Glycols become very corrosive in the presence of oxygen. CO_2 removal is also practiced using molecular gate systems, in which membranes of different sieves, depending on the size of the molecule, are separated (Wills 2004).

11.11.1 Use of Membrane for Gas Processing

The separation of natural gas by membranes is a dynamic and rapidly growing field, and it has been proven to be technically and economically superior to other emerging technologies (Basu *et al.* 2004). This superiority is due to certain advantages that membrane technology benefits from, including low capital investment, low weight, space requirement, and high process flexibility. This technology has higher benefits because higher recovery of desired gases is possible.

Du *et al.* (2006) reported that composite membranes comprised of a thin cationic poly (*N,N*-dimethylaminoethyl methacrylate: PDMAEMA) layer and a microporous polysulfone (PSF) substrate were prepared by coating a layer of PDMAEMA onto the PSF substrate. The membrane showed a high permselectivity to CO_2. The high CO_2/N_2 permselectivity of the membranes make them suitable to use for removing CO_2 from natural gas streams and capturing the flue gas from power plants. By low temperature plasma grafting of DMAEMA onto a polyethylene substrate, a membrane showed a high CO_2 permeance (Matsuyama *et al.* 1996). A study on the performance of microporous polypropylene (PP) and polytetrafluoroethylene (PTFE) hollow fiber membranes in a gas absorption membrane (GAM) system, using aqueous solutions of monoethanolamine (MEA) and 2-amino-2-methyl-1-propanol (AMP) was performed by DeMontigny *et al.* (2006). They reported that the gas absorption systems are an effective technology for absorbing CO_2 from simulated flue gas streams. Markiewicz *et al.* (1988) reported that different types of polymeric membranes were used for the removal of CO_2, H_2S, N, water vapor, and other components. However, the majority of the membranes are synthetically made from polymers, which might cause negative environmental impacts. In order to avoid this problem, nontoxic biopolymers are considered attractive alternatives to the conventional membrane separation systems (Basu *et al.* 2004).

11.11.2 Nitrogen and Helium Removal

A natural gas stream is routed to the nitrogen rejection unit after H_2S and CO_2 are removed, where it is further dehydrated using molecular sieve beds (Figure 11.5). In the nitrogen rejection unit, the gas stream is channeled through a series of passes through a column and a heat exchanger. Here, the nitrogen is cryogenically separated and vented. Absorption systems can also be applied to remove the nitrogen and other hydrocarbons (EIA 2006). Also, helium can be extracted from the gas stream through membrane diffusion.

11.12 Problems in Natural Gas Processing

Conventional natural gas processing consists of applications of various types of synthetic chemicals and polymeric membranes. The common chemicals used to remove water, CO_2, and H_2S are Diethylene glycol (DEG), Triethylene glycol (TEG), Monoethanolamines (MEA), Diethanolamines (DEA), and Triethanolamine (TEA). These synthetic chemicals are considered to have health and environmental impacts during their life cycle from production to end uses. The pathway analysis and their impacts are discussed in the following sections.

11.12.1 Pathways of Glycols and their Toxicity

Matsuoka *et al.* (2005) reported a study on electro-oxidation of methanol and glycol, and found that electro-oxidation of ethylene glycol at 400mV forms glycolate, oxalate, and formate (Figure 11.9). The study further reports that glycolate is obtained by three-electron oxidation of ethylene glycol, and is an electrochemically active product even at 400mV, which leads to further oxidation of glycolate. Oxalate was found to be stable, and no further oxidation was seen or termed as a non-poisoning path. The other product of glycol oxidation is

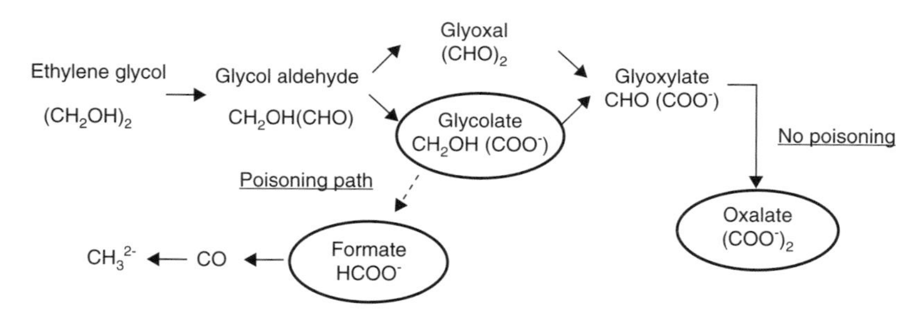

Figure 11.9 Ethylene Glycol Oxidation Pathway in Alkaline Solution (Matsuoka *et al.* 2005).

called formate, which is termed as a poisoning path or CO poisoning path. A drastic difference in ethylene glycol oxidation was noted between 400 and 500mV. The glycolate formation decreased 4–18% and formate increased 15–20%. In case of methanol oxidation, the formate was oxidized to CO_2, but ethylene glycol oxidation produces CO instead of CO_2, and follows the poisoning path over 500 mV. The glycol oxidation produces glycol aldehyde as intermediate products. As the heat increases, the CO poisoning may also increase.

Glycol ethers are known to produce toxic metabolites, such as teratogenic methoxyacetic acid, during biodegradation, and the biological treatment of glycol ethers can be hazardous (Fischer and Hahn 2005). It was reported that abiotic degradation experiments with ethylene glycol showed that the byproducts are monoethylether (EGME) and toxic aldehydes, e.g., methoxy acetaldehyde (MALD).

Glycol passes into the body by inhalation or through the skin. Toxicity of ethylene glycol causes depression and kidney damage (MSDS 2005). As indicated in the MSDS report, ethylene glycol in the form of dinitrate can have harmful effects when breathed in, and by passing through the skin, it can irritate the skin, causing a rash or burning feeling on contact. It can also cause headache, dizziness, nausea, vomiting, abdominal pain, and a fall in blood pressure. High concentration levels can interfere with the ability of the blood to carry oxygen, causing headache, dizziness, a blue color to the skin and lips (*methemoglobinemia*), breathing difficulties, collapse, and even death. This can damage the heart, causing pain in the chest and/or increased heart rate, or it can cause the heart to beat irregularly (arrhythmia), which can be fatal. High exposure may affect the nervous system, and may damage the red blood cells, leading to anemia (low blood count). The recommended airborne exposure limit is 0.31 mg/m3 averaged over an 8-hour work shift. During a study of the carcinogenetic toxicity of propylene glycol on animals, skin tumors were observed (CERHR 2003).

Ingestion of ethylene glycol is a toxicological emergency (Glaser DS 1996). It is commonly found in a variety of commercial products, including automobile antifreeze, and if ingested, it will cause severe acidosis, calcium oxalate crystal formation and deposition, and other fatal organ damage (Davis *et al*. 1997). It is a high volume production (HPV) chemical generally used to synthesize polyethylene terephthalate (PET) resins, unsaturated polyester resins, polyester fibers, and films (SRI 2003). Moreover, ethylene glycols are a constituent in antifreeze, deicing fluids, heat transfer fluids, industrial coolants, and hydraulic fluids. Several studies have consistently demonstrated that the kidney is a primary target organ after acute or chronic exposures of ethylene glycol (NTP 1993; Cruzan *et al.* 2004). It has also been reported that renal toxicity, metabolic acidosis, and central nervous system (CNS) depression are reported in humans in intentional or accidental overdosing (Eder *et al.* 1998). Browning and Curry (1994) reported that because of widespread availability, serious health concerns have been shown for the potential toxicity of ethylene glycol ethers. From the review of these literatures, it is obvious that glycol has health and environmental problems. Hence, searching for alternative materials that have less environmental impacts is very important.

11.12.2 Pathways of Amines and their Toxicity

Amines are considered to have negative environmental impacts. It was reported that occupational asthma was found in a patient handling a cutting fluid containing diethanolamine (DEA). DEA causes asthmatic airway obstruction at concentrations of 0.75 mg/m^3 and 1.0 mg/m^3 (Piipari *et al.* 1998). Toninello (2006) reported that the oxidation of amines appears to be carcinogenic. DEA also reversibly inhibits phosphatidylcholine synthesis by blocking choline uptake (Lehman-McKeeman and Gamsky 1999). Systemic toxicity occurs in many tissue types, including the nervous system, liver, kidney, and blood system. Hartung *et al.* (1970) reported that inhalation by male rats of 6 ppm (25.8 mg/m^3) DEA vapor 8 hours/day, 5 days/week for 13 weeks resulted in depressed growth rates, increased lung and kidney weights, and even some mortality. Rats exposed continuously for 216 hours (nine days) to 25 ppm (108 mg/m^3) DEA showed increased liver and kidney weights and elevated blood urea nitrogen. Barbee and Hartung (1979) reported changes in liver mitochondrial activities in rats following exposure to DEA in drinking water.

Melnick (1992) reported that symptoms associated with diethanolamine intoxication included increased blood pressure, diuresis, salivation, and pupillary dilation (Beard and Noe 1981). Diethanolamine causes mild skin irritation to rabbits at concentrations above 5%, and severe ocular irritation at concentrations above 50% (Beyer *et al.* 1983). Diethanolamine is a respiratory irritant and, thus, might exacerbate asthma, which has a more severe impact on children than on adults (Chronic Toxicity Summary 2001). The summary reports showed that diethanolamine is corrosive to eyes, mucous membranes, and skin. Also, liquid splashed in the eye causes intense pain and corneal damage, and permanent visual impairment may occur. Prolonged or repeated exposure to vapors at concentrations slightly below the irritant level often results in corneal edema, foggy vision, and the appearance of halos, and skin contact with liquid diethylamine causes blistering and necrosis. Exposure to high vapor concentrations may cause severe coughing, chest pain, and pulmonary edema. Ingestion of diethylamine causes severe gastrointestinal pain, vomiting, and diarrhea, and may result in perforation of the stomach. As a large volume of amines are used for natural gas processing and other chemical processes, there are chances that it may have negative environmental and health impacts during their life cycles.

11.12.3 Toxicity of Polymer Membranes

Synthetic polymers are made from the heavier fraction of petroleum derivatives. Hull *et al.* (2002) reported that combustion toxicity of ethylene-vinyl acetate copolymer (EVA) has a higher yield of CO and several volatile compounds, along with CO_2. Due to this reason, biopolymers are being considered as an alternative to synthetic polymers.

11.13 Innovative Solutions for Natural Gas Processing

11.13.1 Clay as a Glycol Substitute for Water Vapor Absorption

Clay is a porous material containing various minerals, such as silica, alumina, and several others. Various types of clays, such as kaolinite and bentonite, are widely used in various industries as sorbents. Abidin (2004) reported that the sorption depends on the available surface area of clay minerals, and is very sensitive to environmental changes. Low *et al.* (2003) reported that the water absorption characteristics of sintered sawdust clay can be modified by the addition of sawdust particles to the clay. The dry clay as a plaster has a water absorption coefficient of 0.067–0.075 ($kg/m^2S^{1/2}$), where the weight of water absorbed is in kg, the surface area in square meters, and the time in seconds (Straube 2000).

11.13.2 Removal of CO_2 Using Brine and Ammonia

A recent patent (Chaalal and Sougueur 2007) showed that carbon dioxide can be removed from exhaust gas reacting with saline water. In such a process, an ammonium solution is combined with CO_2 in two steps. First, ammonium carbonate is formed:

$$NH_3 + CO_2 + H_2O \Rightarrow (NH_4)_2\,CO_3$$

In another step, when $(NH_4)_2\,CO_3$ is supplied with excess CO_2, ammonium hydrogen carbonate is formed as follows:

$$(NH_4)_2\,CO_3 + CO_2 + H_2O \Rightarrow 2NH_4HCO_3\ (aq.)$$

When ammonium hydrogen carbonate reacts with brine, it forms sodium bicarbonate and ammonium chloride:

$$2NH_4HCO_3 + NaCl \Rightarrow NaHCO_3 + NH_4Cl$$

Hence, by using this process, carbon dioxide can be removed from natural gas streams. When sodium bicarbonate is heated between 125–250°C, it is converted to sodium carbonate, driving off water of crystallization, forming anhydrite sodium carbonate or a crude soda ash (Delling *et al.* 1998):

$$NH_4HCO_3 + heat \Rightarrow Na_2CO_3 + NH_4CI + CO_2 + H_2O.$$

This CO_2 can be used for other purposes, such as CO_2 flooding for enhanced gas recovery (Oldenburg *et al.* 2001).

If ammonium chloride is heated, NH_4CI is decomposed into ammonia and hydrochloric acid:

$$NH_4Cl \Rightarrow NH_3 + HCl$$

This ammonia can be reused for this process, and HCl can be used in any other chemical processes.

11.13.3 CO_2 Capture Using Regenerable Dry Sorbents

Capturing CO_2 is also possible by using regenerable sorbents, such as sodium bicarbonate. Green *et al.* (2001) reported that sodium bicarbonate ($NaHCO_3$) can be used as a regenerable sorbent to economically capture CO_2 from dilute flue gas streams.

$$3Na_2CO_3\ (s) + CO_2(g) + H_2O\ (g) = 2NaHCO_{3\ (s)}$$

When $2NaHCO_3$ is heated, CO_2 is released forming $3Na_2CO_3$.

$$2NaHCO_3\ (s) = Na_2CO_3\ (s) + CO_2(g) + H_2O\ (g)$$

In this reaction, CO_2 is absorbed from a flue gas stream by $NaHCO_{3(s)}$. The $NaHCO_{3(s)}$ is regenerated by heat as the CO_2 is released. Thus, this process can be used to capture CO_2 from a low-pressure stream, to form a more concentrated CO_2 to be used for other purposes, such as EOR operations.

11.13.4 CO_2 Capture Using Oxides and Silicates of Magnesium

There are other techniques that can be used to capture CO_2 from exhaust gas streams. Zevenhoven and Kohlmann (2001) studied the use of magnesium silicate to capture CO_2 from exhaust gas streams. The process is called magnesium silicate or magnesium oxide carbonation.

$$MgSiO_3 + CO_2\ (g) \Rightarrow MgCO_2 + SiO_2$$

$$MgO + CO_2\ (g) \Rightarrow MgCO_3$$

The reaction kinetics showed that the carbonation varies depending on the partial pressure of CO_2. The preferable temperature for carbonation is reported to be 200–400°C. The equilibrium constant pressure for the reaction is reported to be much higher than 1 bar (up to 1E+06 bar) to drive the chemical reaction to the right-hand side. Goldberg *et al.* (2001) reported that there was 40–50% conversion after 24 hours at 150–250°C, and 85–125 bar pressure with olivine ($(Mg,Fe)_2SiO_4$) particles of 75–100μm.

Lee *et al.* (2006) reported that potassium-based sorbents are prepared by impregnation of K_2CO_3 on activated carbon as porous support. Table 11.11 is a summary of sorbents prepared by impregnation of potassium carbonate (30% wt.) in the presence of 9% vol. H_2O at 60°C and their corresponding total CO_2 capture capacity. It was

Table 11.11 Sorbents prepared by impregnation of potassium carbonate (30wt. %) in the presence of 9% vol. H_2O at 60°C.

Sorbent	**Total CO_2 capture capacity**
K_2CO_3/AC	86.0
K_2CO_3/Al_2O_3	85.0
K_2CO_3/USY	18.9
K_2CO_3/CsNaX	59.4
K_2CO_3/SiO_2	10.3
K_2CO_3/MgO	119.0
K_2CO_3/CaO	49.0
K_2CO_3/TiO_2	83.0

reported that the CO_2 capture capacity of K_2CO_3/AC, K_2CO_3/TiO_2, K_2CO_3/MgO, and K_2CO_3/Al_2O_3 was 86, 83, 119, and 85mgCO_2/g sorbent, respectively. Moreover, these sorbents could be completely regenerated at 150, 150, 350, and 400°C, respectively. Based on regeneration capacity, TiO_2 was considered to be a potential sorbent for CO_2 capture. Hence, by employing these methods to capture CO_2 from natural gas streams, the conventional chemical-based absorbents, such as DEA and MEA, can be replaced with natural and nontoxic materials.

11.13.5 H_2S Removal Techniques

Hydrogen sulfide (H_2S) is one of the impurities coming through the natural gas streams. A significant amount of hydrogen sulfide is emitted from industrial activities such as petroleum refining (Henshaw *et al.* 1999) and natural gas processing (Kim *et al.* 1992). H_2S is a toxic, odorous (Roth 1993), and corrosive compound that seriously affects internal combustion engines (Tchobanoglous *et al.* 2003). If inhaled, H_2S reacts with enzymes in the bloodstream and inhibits cellular respiration, which can create pulmonary paralysis, sudden collapse, and even death at higher concentrations (Syed *et al.* 2006). Natural gas contains 0–5% of hydrogen sulfide, depending on the reservoir (Natural Gas 2004). Hydrogen sulfide is also present in biogas that is being used as fuel for internal combustion engines and cooking appliances. Lastella *et al.* (2002) reported that hydrogen sulfide in biogas varies from 0.1–2%, depending on the type of feedstock.

The removal of H_2S from chemical processes is very expensive, due to large amounts of chemical, energy, and processing requirements (Buisman *et al.* 1989). Hence, biological treatment for hydrogen sulfide removal is considered a more attractive alternative to chemical treatment (Sercu *et al.* 2005), because biological treatment can overcome the disadvantages of the chemical treatment processes (Elias *et al.* 2002).

Biological removal involves the conversion of H_2S into elemental sulfur (S°) by using bacteria. Among the various bacteria available, Syed *et al.* (2001) showed that *Cholorobium limicola* is a desirable bacterium to use, because it can grow using only inorganic substrates. It also has a high efficiency at converting sulfide into elemental sulfur, as well as the extracellular production of elemental sulfur, which converts CO_2 into carbohydrates (van Niel 1931). This also requires light and CO_2, and the process could be strictly anaerobic. The oxidation product is in the form of elemental sulfur, and can be used in other chemical processes:

$$2nH_2S + NCO_2 \quad \text{Light energy} \quad 2nS^\circ + n\,(CH_2O) + nH_2O$$

A number of chemotrophs, such as *Thiobacillus, Thermothrix, Thiothrix,* and *Beggiato,* can be used for the biodegradation of H_2S. These bacteria use CO_2 as a carbon source, and use chemical energy from oxidation of inorganic compounds such as H_2S, and they produce new cell material. The reaction on the presence of *Thiobacillus thioparas* as reported by Chang *et al.* (1996) and Kim *et al.* (2002) is as follows:

$$2HS^- + O_2 \rightarrow 2S^\circ + 2OH^-$$

$$2S^\circ + 3O_2 + 2OH^- \rightarrow 2SO^{2-}{}_4 + 2H^+$$

$$H_2S + 2O_2 \rightarrow SO^{2-}{}_4 + 2H^+$$

Oyarzun *et al.* (2003) reported that the *Thiobacillus* species are widely used for the conversion of H_2S and other sulfur compounds by biological processes. They have the ability to grow under various environmental stress conditions, such as oxygen deficiency, acid conditions, and low and high pH. Hence, with suitable system design, whereby natural gas stream is passed through such bacteria with certain retention time, H_2S can be removed.

11.14 Concluding Remarks

The crude oil pathway shows that a natural process drives the formation of crude oil without any impacts on other species in the world. However, the pathway analysis of refined oil shows that its processes create several environmental impacts on the globe. Refining crude oil involves the application of large amounts of synthetic chemicals and catalysts, including heavy metals such as lead, chromium, sulfuric acid, hydrofluoric acid, platinum, and others. Moreover, refining the crude oil emits large amounts of VOCs and toxic air pollutants. Refined oils degrade slower and last in the natural environment for a longer duration, affecting the environment in several ways. Because the refining of fossil fuels emits large amounts of CO_2, it has been linked to global warming and climate change. Hence, a paradigm shift in conventional engineering practices is necessary in order to reduce the emissions and impacts on the natural environment.

The review of various natural gas processing techniques and chemicals used during gas processing shows that currently used chemicals are not sustainable. Some natural substitutes for these chemicals have also been experimentally investigated. They offer sustainable alternatives to gas processing.

12

Greening of Flow Operations

12.1 Introduction

Oil and gas have been the primary sources of energy for the past 70 years. All predictions indicate that this trend is likely to continue. In this regard, the role of production technologies cannot be overemphasized. Some 30% of petroleum infrastructure costs relate to production operations, mainly in assuring flow from the wellbore to the processing plants, refineries, and oil tankers. After a petroleum well is drilled, the most important preoccupation of the petroleum industry is to ensure uninterrupted flow through various tubulars, both underground and above-ground. Billions of dollars are spent every year to make sure that access to processing plants, refineries, storage sites, and oil tankers remains open and free of leaks or plugging. This exuberant cost does not include the long-term intangible costs, such as environmental impacts, effects of residuals remaining in petroleum streams, loss of quality due to interference, and others. Any improvement in the current practices can translate into saving billions of dollars in the short term and much more in the long term. A detailed description of conventional remediation schemes is available in the scientific version of this book (Islam *et al.*, 2010). This chapter focuses on green solutions to gas hydrates and corrosion, with some discussion of asphaltenes. For every practice, it is shown that some fundamental adjustments can lead to drastic improvements in performance in the short term as well as in the long term. The short-term benefit is mainly saving material costs, and the long-term benefits are in the true sustainability of the technologies.

12.2 Hydrate Problems

The production and transmission of natural gas is a very complex set of operations. There are virtually hundreds of different compounds in the natural gas stream coming out of the production well. A natural gas stream consists of methane, ethane, propane, butane, gas condensate, liquid petroleum, water vapor, carbon dioxide, hydrogen sulfides, nitrogen, and other gases and solid particles. The overall attraction of natural gas as one of the most environmentally acceptable sources of energy is marred by the presence of some of the most unwanted compounds that are ingredients of the natural gas stream that comes out of the production well. Traces of nitrogen compounds in natural gas are believed to cause ozone layer depletion and contribute to global warming. The H_2S and CO_2 part of the natural gas stream decreases the heating value of natural gas, thereby reducing its overall efficiency as a fuel. These gases are commonly known as acid gases, and they must be removed from the natural gas before it is transported from the production well to the consumer market (Chakma 1999). Hydrogen sulfide, in particular, is a very toxic and corrosive gas that oxidizes instantaneously in the form of sulfur dioxide, and gets dispersed in the atmosphere (Basu *et al.* 2004). These gases render the water content in the gas stream even more corrosive. Hence, the removal of free water, water vapors, and condensates is a very important step during gas processing. The water content in natural gas is exceptionally corrosive, and it has the potential of destroying the gas transmission system. Water content in a natural gas stream can condense and cause sluggishness in the flow. The water content can also initiate the formation of hydrates, which in turn can plug the whole pipeline system (Nallinson 2004). Natural gas hydrates are ice-like crystalline solids that are formed due to the mixing of water and natural gas (typically

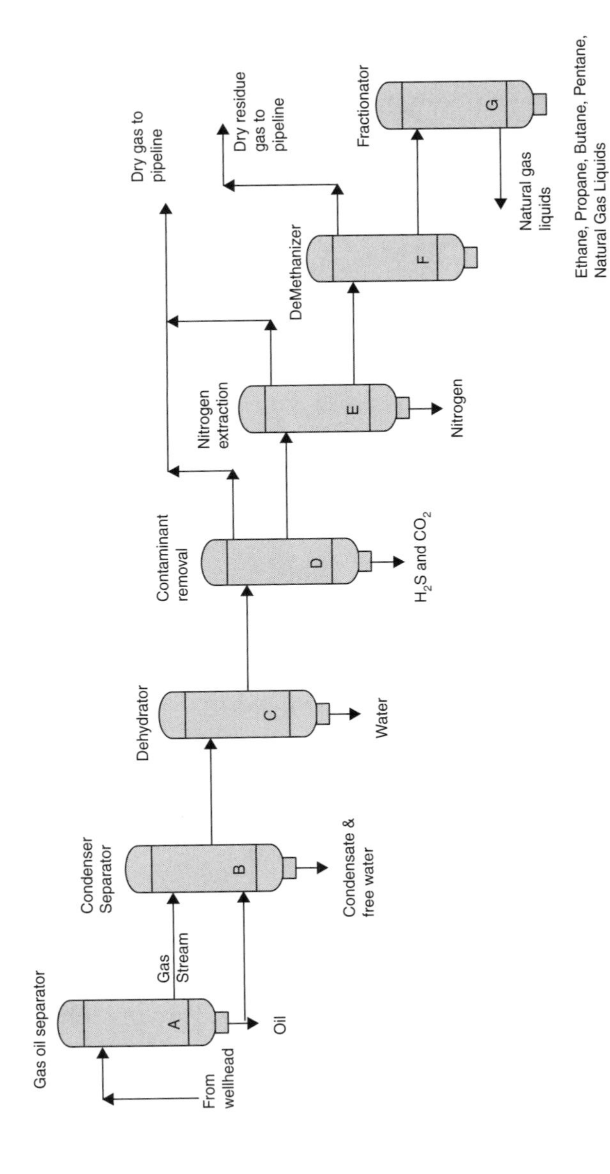

Figure 12.1 Natural gas processing (Tobin *et al*. 2006).

methane). In order to transform the raw natural gas stream into "line quality" gas, certain quality standards have to be maintained, and the natural gas should be rid of these impurities before it can be transported through pipelines. This whole process of purification is known as gas processing, and it guards against corrosion, hydrate formation, and other environmental and safety hazards related to natural gas transportation (Chhetri and Islam 2007).

The above discussion has elaborated the importance of removing water content from the natural gas transmission stream. This would not only provide protection against corrosion problems, but the most important reason behind this task is that it would help prevent the formation of hydrates in the pipeline.

The discovery of hydrates is attributed to Humphrey Davy, who claimed in the early nineteenth century that a solid material can be formed when the aqueous solution of chlorine is cooled below 9°C (Davy 1811). These results were confirmed by Michael Faraday, who proved the existence of these solid compounds, and showed that the composition of the solid is almost 1:10 for chlorine and water, respectively (Faraday *et al.* 1823).

Throughout the remainder of the 19th century, many other scientists experimented with hydrates, e.g., Wroblewski, Cailletet, Woehler, Villard, de Forcrand, Schutzenberger, Cailletet, and Sully Thomas, among others (Schroeder 1926). In particular, Villard was the one who reported the existence of hydrates of methane, ethane, acetylene, and ethylene (Villard 1888).

All the abovementioned research efforts were only of academic interest, and it was not until 1934 that Hammerschmidt discovered that clathrate hydrates were responsible for plugging natural gas pipelines, especially those located in comparatively colder environments, and that hydrate formation was linked to gas transmission in a pipeline (Hammerschmidt 1934). By the turn of the 21st century, Sloan's work on the development of chemical additives and other methods to inhibit hydrate formation led to the construction of the first predictive models of hydrate formation (Sloan 1998).

Natural gas hydrates have generated the interest of petroleum researchers for the past four decades. The role of natural gas hydrates has been evaluated as (1) a future source of abundant energy, (2) a hazard to the marine geo-stability, and (3) a cause of change in the worldwide climate (Kvenvolden 1993).

It has already been projected that natural gas hydrates are one of the most promising future sources of energy. Some estimates put the size of the hydrate reserves at a magnitude that would be enough to last for many decades, if not centuries (Kvenvolden 1988).

Methane hydrates were first found in nature in Siberia in 1964, and it was reported that they were being produced in the Messoyakha Field from 1970 to 1978 (Sapir *et al.* 1973). Another discovery was made in the Mackenzie Delta (Bily 1974) and then on the North Slope of Alaska (Collett 1983).

These and subsequent discoveries of the methane hydrates led many scientists to speculate the universal existence of large reserves of hydrates, because the low temperature-high pressure conditions that are necessary for the formation of hydrates exist all around the globe, especially in the permafrost and deep ocean regions.

Many countries with large energy needs but limited domestic energy resources (e.g., Japan and India) have been carrying out aggressive and well-funded hydrate research and development programs to initiate the production of methane from the hydrates on a commercial basis. These programs led to the recovery of large hydrate nodules, core collections of ocean-bottom hydrate sediments, and the drilling of wells designed specifically to investigate methane hydrate-bearing strata (Max 2000; Park *et al.* 1999). In the global energy outlook, in which rising costs, depleting oil reserves, and future energy needs of emerging economies are constantly extrapolated, methane hydrates are considered the most valuable future energy prospect.

However, it is also hypothesized that these hydrates play a crucial role in nature; they interact with the sea-bottom life forms, help restore the stability of the ocean floor, balance the global carbon cycle, and affect long-term climate change (Dickens *et al.* 1997). These concerns have led to different additives and the examination of the long-term effects of drilling in hydrate reserves for natural gas, corroborating evidences from the cores of different drilling sites (Bains *et al.* 1999; Katz *et al.* 1999; Norris *et al.* 1999). The other concerns related to the technical aspect of the production of methane hydrates include the hazards posed by the hydrate-bearing sediments to conventional oil and gas drilling operations (Max *et al.* 1998)

Even though both the balance between pros and cons of the exploration of gas hydrates for methane production and the credibility of gas hydrates as a future source of cheap and abundant energy may take a long time to be fully established, gas hydrates remain one of the most pressing problems for the natural gas transportation industry. Natural gas hydrates have been one of the potential causes of harm and damage to the natural gas transportation industry by affecting its personnel and infrastructure. Incidents have been reported when the hydrate plugs' projectiles have caused the loss of lives and millions of dollars in material costs. It has also been documented that the natural gas hydrate plugs have adverse effects on drilling activities and threaten the pipelines.

12.3 Corrosion Problems in the Petroleum Industry

The petroleum industry has been the backbone of the world economy for the last 60 years. The United States has been the world leader in petroleum engineering technologies. A large tranche of petroleum infrastructure and maintenance costs relates to production operations, mainly in assuring flow from the wellbore to the processing plants, refineries, and oil tankers. The biggest challenge in assuring petroleum flow through pipelines has been corrosion problems. In one federal study, a total cost of $276 billion was attributed to corrosion in 2002. This presented a rise of more than $100 billion over five years: approximately 3.1% of GDP (Koch *et al* 2002), which is more than the subtotal of its entire agricultural and mining component. Congress was sufficiently alarmed so as to enact a Corrosion Prevention Act (2007), offering a tax incentive of 50% to companies who invest in corrosion abatement and prevention. The petroleum sector carries the biggest burden of this cost, followed by the U.S. Department of Defense. For the petroleum industry, the cost of corrosion and scaling presents anything from 30% (mainland) to 60% (offshore) of the total maintenance expenditure. That is a huge price tag for an industry that has carried the burden of supporting the bulk of the energy needs of this planet, and predictions are that this trend is likely to continue. Yet, few new technologies have emerged to solve this debilitating problem that the petroleum industry faces (Chilingar *et al.* 2008).

Recently Chilingar *et al.* (2008) provided a step-by-step analysis of current practices of flow assurance during petroleum production, focusing on corrosion and scaling problems. They suggest numerous adjustments in practices, and provide a guideline bound to save millions in preparatory research work in a field project. However, little is said about microbial-induced corrosion (MIC).

It is estimated that 30% of the corrosion in the petroleum industry is due to microbial-induced activities (Al-Darbi *et al.* 2002). MIC is extremely harmful to both industry and the environment. It is estimated that 20-30% of all corrosion is microbiologically influenced, with a direct cost of $30-50 billion per year (Javaherdashti 1999). One of the most important types of microbial corrosion is that which is due to the presence of sulfate reducing bacteria (SRB), which is most common in petroleum operations because of the prevailing anaerobic environment (Phelps *et al.* 1991).

Therefore, the protection of structures against MIC has become very critical in many industries, including municipal pipelines, marine, storage vessels, sewage treatment facilities, and so on (Geesey *et al.* 1994). The study of microbiologically-influenced corrosion (MIC) has progressed from phenomenological case histories to a mature interdisciplinary science, including electrochemical, metallurgical, surface analysis, microbiological, biotechnological, and biophysical techniques (Little and Wagner 1994).

Microorganisms, such as bacteria, algae, and fungi, under certain conditions can thrive and accelerate the corrosion of many metals, even in otherwise benign environments. Biological organisms can enhance the corrosion process by their physical presence, metabolic activities, and direct involvement in the corrosion reaction (Hamilton 1985). The occurrence of MIC is often characterized by unexpected severe metal attacks, the presence of excessive deposits, and, in many cases, the rotten-egg odor of hydrogen sulfide (Lee *et al.* 1995).

For a microorganism to grow, environmental conditions must be favorable. Essential nutrients required by most microbes include carbon, nitrogen, phosphorous, oxygen, sulfur, and hydrogen. Other elements required in trace quantities include potassium, magnesium, calcium, iron, copper, zinc, cobalt, and manganese. All organisms require carbon for converting into cell constituents (Tanji 1999).

The main bacteria related to MIC are aerobic slime formers, acetate-producing bacteria, acetate-oxidizing bacteria, iron/manganese oxidizing bacteria, methane producers, organic acid-producing bacteria, sulfur/sulfide-oxidizing bacteria (SOB), and sulfate-reducing bacteria (SRB).

Conventionally, only chemical approaches have been taken in combating MIC. Numerous toxic chemicals are used to destroy the microbes that cause corrosion. Only recently, Al-Darbi (2004) proposed in his PhD work a series of natural alternatives to these toxic agents. These agents are just as efficient as toxic chemicals, but do not contain any toxins.

12.4 Green Solutions for Hydrate Control

The first approach is hypothetical, but it is believed that this can be proven a practical solution with elaborate experimental work. This approach would not alter the present mechanisms and methodology of applying conventional chemicals in processing and the transportation of natural gas, it would only make a change in the pathways of the development of the same chemical.

This approach is based on the assumption that "nature is perfect." It is believed that, if the constituents of the conventional inhibitors were taken from innocuous natural sources without introducing an artificial product or process (Miralai *et al.* 2008), the resulting product would be benign or even beneficial to nature. If the process is sustainable, then the source can be crude oil or natural gas, and the products will be benign to the environment. This approach is equivalent to destroying bacteria with natural chemicals, rather than synthetic ones. It is well known that an olive oil and dead bacteria mixture is not toxic to the environment, whereas conventional pharmaceutical antibiotics are.

12.4.1 Ethylene Glycol

This substance is used for hydrate control. The green solutions involve deriving this chemical from natural sources, using natural processes. The following is the main chemical reaction in the process:

$$\underset{\text{(Ethylene)}}{C_2H_4 + O} \rightarrow \underset{\text{(Ethylene Oxide)}}{C_2H_4O +} \quad \underset{\text{(Water)}}{H_2O} \rightarrow \underset{\text{(Ethylene Glycol)}}{HO - CH_2 - OH}$$

This reaction shows that if ethylene from a source (natural or otherwise) is oxidized, it will convert to ethylene oxide. The introduction of water to ethylene oxide will change it to ethylene glycol. The principal argument put forward here is that if no artificial product (e.g., catalysts that do not exist in natural environment) is added to the left-hand side of the equation, then the resulting ethylene oxide and, eventually, the ethylene glycol will not be detrimental to the environment. This is equivalent to organic farming, in which natural fertilizers and pesticides are used.

There are numerous sources of ethylene in nature; they can be obtained from various fruits and vegetables. A list of fruits and vegetables that can be sources of ethylene is given in the table below:

Table 12.1 Ethylene sensitivity chart.

Perishable Commodities	Temperature C / F	*Ethylene Production
Fruits & Vegetables		
Apple (non-chilled)	1.1 / 30	VH
Apple (chilled)	4.4 / 40	VH
Apricot	–0.5 / 31	H
Artichoke	0 / 32	VL
Asian Pear	1.1 / 34	H
Asparagus	2.2 / 36	VL
Avocado (California)	3.3 / 38	H
Avocado (Tropical)	10.0 / 50	H
Banana	14.4 / 58	M
Beans (Lima)	0 / 32	L
Beans (Snap/Green)	7.2 / 45	L
Belgian Endive	2.2 / 36	VL
Berries (Blackberry)	–0.5 / 31	L
Berries (Blueberry)	–0.5 / 31	L
Berries (Cranberry)	2.2 / 36	L
Berries (Currants)	–0.5 / 31	L
Berries (Dewberry)	–0.5 / 31	L
Berries (Elderberry)	–0.5 / 31	L

(Continued)

Table 12.1 (Cont.) Ethylene sensitivity chart.

Perishable Commodities Temperature C / F *Ethylene Production		
Fruits & Vegetables		
Berries (Gooseberry)	–0.5 / 31	L
Berries (Loganberry)	–0.5 / 31	L
Berries (Raspberry)	–0.5 / 31	L
Berries (Strawberry)	–0.5 / 31	L
Breadfruit	13.3 / 56	M
Broccoli	0 / 32	VL
Brussel Sprouts	0 / 32	VL
Cabbage	0 / 32	VL
Cantaloupe	4.4 / 40	H
Cape Gooseberry	12.2 / 54	L
Carrots (Topped)	0 / 32	VL
Casaba Melon	10.0 / 50	L
Cauliflower	0 / 32	VL
Celery	0 / 32	VL
Chard	0 / 32	VL
Cherimoya	12.8 / 55	VH
Cherry (Sour)	–0.5 / 31	VL
Cherry (Sweet)	–1.1 / 30	VL
Chicory	0 / 32	VL
Chinese Gooseberry	0 / 32	L
Collards	0 / 32	VL
Crenshaw Melon	10.0 / 50	M
Cucumbers	10.0 / 50	L
Eggplant	10.0 / 50	L
Endive (Escarole)	0 / 32	VL

(Continued)

Table 12.1 (Cont.) Ethylene sensitivity chart.

Perishable Commodities	Temperature C / F	*Ethylene Production
Fruits & Vegetables		
Feijoa	5.0 / 41	M
Figs	0 / 32	M
Garlic	0 / 32	VL
Ginger	13.3 / 56	VL
Grapefruit (AZ, CA, FL, TX)	13.3 / 56	VL
Grapes	–1.1 / 30	VL
Greens (Leafy)	0 / 32	VL
Guava	10 / 50	L
Honeydew	10 / 50	M
Horseradish	0 / 32	VL
Jack Fruit	13.3 / 56	M
Kale	0 / 32	VL
Kiwi Fruit	0 / 32	L
Kohlrabi	0 / 32	VL
Leeks	0 / 32	VL
Lemons	12.2 / 54	VL
Lettuce (Butterhead)	0 / 32	L
Lettuce (Head/Iceberg)	0 / 32	VL
Lime	12.2 / 54	VL
Lychee	1.7 /35	M
Mandarine	7.2 / 45	VL
Mango	13.3 / 56	M
Mangosteen	13.3 / 56	M
Mineola	3.3 / 38	L
Mushrooms	0 / 32	L

(Continued)

Table 12.1 (Cont.) Ethylene sensitivity chart.

Perishable Commodities Temperature C / F *Ethylene Production		
Fruits & Vegetables		
Nectarine	–0.5 / 31	H
Okra	10.0 / 50	L
Olive	7.2 / 45	L
Onions (Dry)	0 / 32	VL
Onions (Green)	0 / 32	VL
Orange (CA, AZ)	7.2 / 45	VL
Orange (FL, TX)	2.2 / 36	VL
Papaya	12.2 / 54	H
Paprika	10.0 / 50	L
Parsnip	0 / 32	VL
Parsley	0 / 32	VL
Passion Fruit	12.2 / 54	VH
Peach	–0.5 / 31	H
Pear (Anjou, Bartlett/ Bosc)	1.1 / 30	H
Pear (Prickly)	5.0 / 41	N
Peas	0 / 32	VL
Pepper (Bell)	10.0 / 50	L
Pepper (Chile)	10.0 / 50	L
Persian Melon	10.0 / 50	M
Persimmon (Fuyu)	10.0 / 50	L
Persimmon (Hachiya)	0.5 / 41	L
Pineapple	10.0 / 50	L
Pineapple (Guava)	5.0 / 41	M
Plantain	14.4 / 58	L
Plum/Prune	–0.5 / 31	M

(Continued)

Table 12.1 (Cont.) Ethylene sensitivity chart.

Perishable Commodities	Temperature C / F	*Ethylene Production
Fruits & Vegetables		
Pomegranate	5.0 / 41	L
Potato (Processing)	10.0 / 50	VL
Potato (Seed)	4.4 / 40	VL
Potato (Table)	7.2 / 45	VL
Pumpkin	12.2 / 54	L
Quince	–0.5 / 31	L
Radishes	0 / 32	VL
Red Beet	2.8 / 37	VL
Rambutan	12.2 / 54	H
Rhubard	0 / 32	VL
Rutabaga	0 / 32	VL
Sapota	12.2 / 54	VH
Spinach	0 / 32	VL
Squash (Hard Skin)	12.2 / 54	L
Squash (Soft Skin)	10.0 / 50	L
Squash (Summer)	7.2 / 45	L
Squash (Zucchini)	7.2 / 45	N
Star Fruit	8.9 / 48	L
Swede (Rutabaga)	0 / 32	VL
Sweet Corn	0 / 32	VL
Sweet Potato	13.3 / 56	VL
Tamarillo	0 / 32	L
Tangerine	7.2 / 45	VL
Taro Root	7.2 / 45	N
Tomato (Mature/Green)	13.3 / 56	VL

(Continued)

Table 12.1 (Cont.) Ethylene sensitivity chart.

Perishable Commodities	Temperature C / F	*Ethylene Production
Fruits & Vegetables		
Tomato (Brkr/Lt Pink)	10.0 / 50	M
Tree-Tomato	3.9 / 39	H
Turnip (Roots)	0 / 32	VL
Turnip (Greens)	0 / 32	VL
Watercress	0 / 32	VL
Watermelon	10,0 / 50	L
Yam	13.3 / 56	VL
Live Plants		
Cut Flowers (Carnations)	0 / 32	VL
Cut Flowers (Chrysanthemums)	0 / 32	VL
Cut Flowers (Gladioli)	2.2 / 36	VL
Cut Flowers (Roses)	0 / 32	VL
Potted Plants	–2.8–18.3 / 27–65	VL
Nursery Stock	–1.1–4.4 / 30–40	VL
Christmas Trees	0 / 32	N
Flowers Bulbs (Bulbs/ Corms/		
Rhizomes/Tubers)	7.2–15 / 45–59	VL

Source: Website 18.

**N = None; H = High; L = Low; M = Medium; VH = Very High; VL = Very Low*

12.4.2 Methyl Ethanol Amine (MEA)

The reaction between ammonia and ethylene oxide yields monoethanolamine, the subsequent reaction between monoethanolamine and ethylene oxide produces diethanolamine, and the reaction between diethanolamine and ethylene oxide results in the production of triethanolamine:

$$NH_3 \quad + \quad C_2H_4O \quad \rightarrow \quad (C_2H_4OH)NH_2$$

(Ammonia) (Ethylene Oxide) (Monoethanolamine)

$$(C_2H_4OH)NH_2 \quad + \quad C_2H_4O \quad \rightarrow \quad (C_2H_4OH)_2NH$$

(Monoethanolamine) (Ethylene Oxide) (Diethanolamine)

$$(C_2H_4OH)_2NH \quad + \quad C_2H_4O \quad \rightarrow \quad (C_2H_4OH)_3N$$

(Diethanolamine) (Ethylene Oxide) (Triethanolamine)

In the initial reaction, the sources of ammonia and ethylene oxide can be either synthetic or natural. It is suggested that ethylene oxide from natural sources, as described in the abovementioned processes, be allowed to react with aqueous ammonia (from urine, etc.) in the liquid phase without a catalyst at a temperature range of 50–100°C and 1 to 2 MPa pressure. A reaction would result in the production of monoethanolamine, which, if allowed to proceed further, would produce diethanolamine and triethanolamine.

Ethylene oxide and ammonia from natural sources would render the product non-toxic, the whole process would be environment-friendly, and the byproducts of the reactions would be beneficial as long as the process does not introduce any toxic chemical. Note that even the heat source needs to be sustainable.

12.4.3 Biological Approach

The second approach is based on the hypothesis that natural biological means can be employed by the industry in processing and transporting natural gas. Paez (2001) has isolated cryophilic bacteria from Nova Scotia that can prevent the formation of gas hydrates at pressure ranges of 150 psi. Such actions of bacteria are similar to how LDHIs work.

12.4.3.1 Hydrate Formation Resistance Through Biological Means

The possibilities of completely replacing the present toxic chemicals (used by the gas processing and transportation industry) with substances that are found in nature are immense. The sustainability criteria of these additives are fulfilled only if both the origin and pathway are natural.

The increased activity in natural gas exploration, production, processing, and transportation areas has increased the awareness of the general public regarding the environmental issues. It is believed that, in the future, as the concerns about the toxicity of currently used inhibitors will grow, the environmental consciousness of the consumers would demand major changes to the presently used systems and chemicals.

The industry's approach in this regard has only been to focus on minimizing the waste and increasing recovery and regeneration of the presently used inhibitors. However, it is feared that, if the root cause of the problem means the toxicity issue with the presently used chemicals is not addressed, the current situation is only going to cause further damage to the environment. It is essential that the ones that conform to the first and foremost benchmark, i.e., true sustainability criteria are fulfilled, replace the presently used inhibitors.

It is appropriate to mention here that the use of the microorganisms in the natural gas industry is not new, and it has been used by the industry in certain fields, such as in the bioremediation of the contaminated soil and water and in the enhanced oil recovery. However, the same industry has never used biological means for the inhibition of hydrates. Paez (2001) reported that adequate bacteria can be cultured from sewage water. He hypothesized that the extremophiles that are considered to be ideal are also ubiquitous, and one should be able to isolate them from sewage water. These bacteria can be cultured and inserted into the gas pipeline using a chamber, depicted in Figure 12.2. This approach was previously taken by Al-Maghrabi *et al.* (1999). Extremophiles are the bacteria that live, survive, and grow in extremely harsh conditions. The extremophiles remain active in conditions that are described as inhospitable for other organisms, and the characteristics that allow them to do so are being studied around the developed world. New extremophiles are being discovered, and the already identified ones have been studied. Among the large number of extremophiles, the ones that are needed for the future experiments would be chosen from the category of barophilic and psycrophiles. These barophilic and psycrophilic bacteria thrive under high pressures and low temperatures, and they have been identified as having an optimal growth rate in the range of 60 MPa and 15°C. This pressure range can be higher in some cases and can reach a mark of 80 MPa for the barophilic bacteria, as evident from the discovery of DB21MT-2 and DB21MT-5 by scientists in Japan. Other significant discoveries

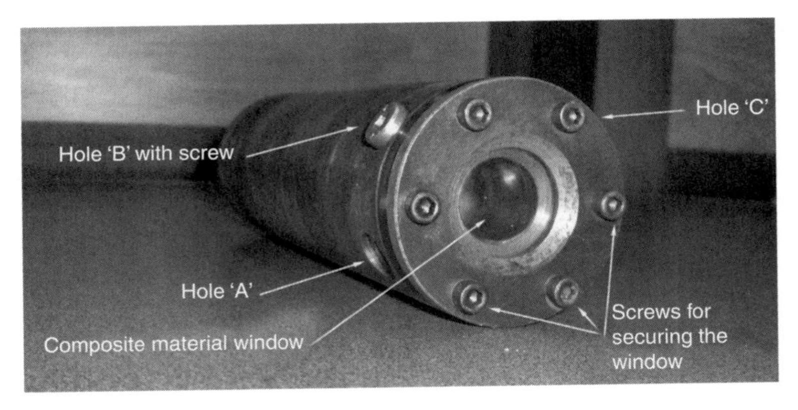

Figure 12.2 The bacteria insertion chamber with a translucent window.

were *Shewanella benthica* and *Moritella* in the barophilic categories (Bohlke *et al.* 2002; Kato *et al.* 1998).

12.4.3.2 *Reaction Mechanisms of Barophiles and Cryophiles*

Researchers have been focusing on the reaction mechanisms of these bacteria under very high pressures conditions. It has been hypothesized that these bacteria regulate the structure of the acids of their membranes to handle these pressures. There are proteins, called ompH, that have the best possible growth environment at high pressures, for which they have an increased ability to take nutrients from the surroundings. The genetic studies of some of these bacteria show that all these barophiles are composed of different DNA-binding factors that vary according to the varying pressure and environmental conditions. These findings led to the culturing of some of the bacteria that exist in the high-pressure zones in the vicinity of 50 MPa (Kato *et al.* 1997).

As mentioned above, the psycrophiles are bacteria that have the property of surviving in very cold temperature conditions. Unfortunately, psycrophiles are the bacteria that researchers have very little knowledge of, as opposed to their cousins, thermopiles, which have a history of research carried out on them (Al-Maghrabi *et al.* 1999). However, it is hypothesized that these organisms regulate the fatty acid arrangement of the phospholipids of the membrane in order to cope with the cold temperature of the surrounding. When the temperature decreases, the composition of the fatty acid in the membrane also changes from a very disordered gel-like material to a very orderly form of a liquid crystal. There are also signs that the flexibility of proteins plays a part in the ability of these organisms to withstand very low temperatures. These activities are the result of the biochemical reactions that involve the enzymatic catalysts (Cummings *et al.* 1999). The efficiency of these reactions is considerably reduced at low temperatures, because the thermodynamic forces also play a certain role in this process. However, the enzymes found in these organisms are more efficient in their manipulations of the metabolic activity. These types of bacteria have been taken from permafrost (temperatures in the range of 5°C) and deep-sea environments, such as 2000m below the sea level (Rossi *et al.* 2003).

12.4.3.3 Bacteria Growth and Survival Requirements

Bacteria, like any other living organisms, need nutrients to survive and grow. The presence of oxygen, the availability of water, temperatures, and pressure are some of the parameters that control bacterial growth.

Any nutrient within a specific cell can be composed of carbon, nitrogen, phosphorous, sulphur, and so forth. These nutrients are available in sugars, carbohydrates, hydrocarbons, carbon dioxide, and some inorganic salts. The main purpose of the nutrient uptake is to generate energy, which is generated by a cell or processed from sunlight. In each of these metabolic reactions, waters play a role of vital importance. This is because 90% of the bacterial body is composed of water molecules.

Different bacteria deal with the presence or absence of oxygen in different ways. There are bacteria that would live in the presence of oxygen only, and there are bacteria that would live in the absence of oxygen only. There are other bacteria that would live in an environment where there is oxygen, though they have the ability to survive in the absence of it. Still there are other types that would live in the absence of oxygen, but they can also survive in the presence of oxygen. These are called obligate aerobes, obligate anaerobes, facultative anaerobes, and facultative aerobes, respectively.

12.4.4 Direct Heating Using a Natural Heat Source

As discussed earlier in this chapter, heating pipelines would eliminate the formation of gas hydrates. However, the heating source should be natural. The advantages of natural heating sources are: 1) they are cheaper than alternatives sources; 2) they are inherently environment-friendly; 3) they have high efficiency; and 4) they are inherently sustainable. The most natural source of such heat is solar. The efficiency of solar heating can be increased drastically if the heating is done directly (not through conversion using photovoltaics). Direct heating can be used in two ways. It could be used with a solar parabolic collector (see Figure 12.3) placed underneath the joints of a pipeline. Because joints are the source of pressure decompression, they are the sites that are usually responsible for the onset of hydrate formation.

The second approach would be to heat a thermal fluid with solar heating. The advantage of this approach is the usage of heat during nighttime when sunlight is not available. Figure 12.4 shows the solar collector, along with a thermal fluid tank and a heat absorption fin.

Figure 12.3 Solar collector for direct heating of a pipeline.

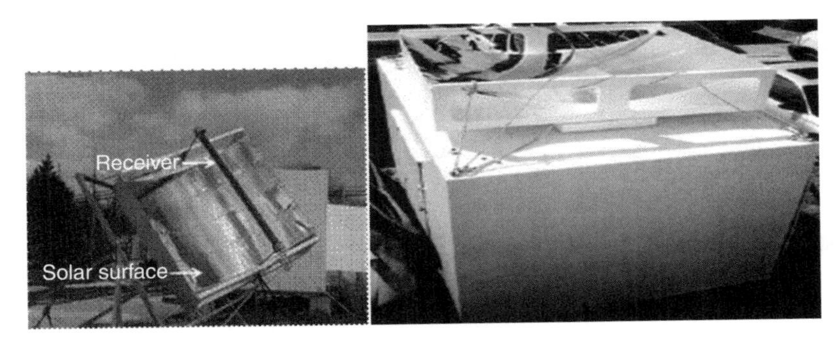

Figure 12.4 Solar parabola, heat absorber (left) and thermal fluid tank enclosurer (right).

12.5 Sustainable Approach to Corrosion Prevention

It is well known that natural materials do not corrode. The reason that corrosion of metal is a great problem is that materials today are processed through unnatural means. The remedy to this problem is not to introduce more unnatural and inherently toxic means to combat corrosion. A sustainable approach to corrosion prevention should mainly focus on using natural materials and, in the absence of natural materials for construction, natural additives to inhibit corrosion.

Many researchers have used different natural materials for corrosion inhibition and control purposes (El-Etre 1998; El-Etre and Abdallah 2000). In a study conducted by Mansour *et al.* (2003), green algae were tested as natural additives for a paint formulation based on vinyl chloride copolymer (VYHH), and its efficiency for protection of steel against corrosion in seawater was evaluated. Both suspended and extracted forms of algae were used to achieve optimum performance of the algae-contained coatings. Poorest performance (protection of steel against corrosion in seawater) was obtained when algae was added in its suspended form, whereas the extracted form exhibited better performance based on impedance measurements.

Instead of adding more toxins, Islam's research group took an entirely different approach. A series of natural oils were found to be adequate for preventing microbial growth, some being particularly suitable for destroying SRB (Al-Darbi *et al.* 2002a, 2002b, 2004a, 2004b, 2005; Saeed *et al.* 2003). Original trials indicated that various natural oils, such as mustard oil, olive oil, and fish oil, have the properties of bactericides, and can destroy SRB effectively. However, if these oils are applied directly, they are not considered to form a base for metallic paints. Typically, it is suggested that the natural oil-based paints suffer from the following shortcomings: 1) slow drying time; 2) high dripping and running; 3) bad sealing with "bleeding" surfaces; 4) heat-resistant properties lower with increased oil content; and 5) resistance to alkali is not predictable, although somehow stable. Considering these shortcomings, natural oils were added in small concentrations to the alkyd paint. Some of these results are shown here. The scanning electron photomicrograph shown in Figure 12.5 is for the surface of a mild steel coupon coated with an alkyd coating mixed with 2% vol. olive oil. No biofilms were detected, except a few small bacterial spots were scattered at different locations on the surface. Blistering with and without rupturing of the coating was observed on some areas of the coated surface, as can be seen from Figure 12.6. This was a clear indication of some local failure in the coating, either as a result of coating disbondment or microbial processes occurring beneath the coating layer.

Figure 12.5 SEM photomicrograph shows some pinholes, spots, and localized attack on the surface of a coated mild steel coupon with alkyd mixed with olive oil.

Figure 12.6 SEM photomicrograph shows blistering on the surface of a coated mild steel coupon with alkyd mixed with olive oil.

It is worth mentioning here that the SRB in the media and in the slim layers (biofilms) converted sulfates in the sample into sulfides, which in turn produced hydrogen sulfide (H_2S). Later, the H_2S and carbon dioxide (CO_2) reacted with water to produce mild acidic products that lower the pH of the substrate (metal) surface to levels favorable for the growth of bacteria, which in the end created a very acidic environment, thereby encouraging the rapid corrosion attack on those metal surfaces (Lee and Characklis 1993; Lee *et al.* 1993).

Figure 12.7 shows the scanning electron photomicrograph of the surface of a mild steel coupon coated with alkyd coating mixed with 2% vol. menhaden fish oil. It was surprising to find very few bacterial spots on this surface, which was shiny and almost clean. No breaches, blistering, or deterioration were later detected on this surface when it was investigated under the microscope.

The above results were attributed to the marked inhibition of bacterial adhesion to the coated surface when one of the natural additives was added to that coating. Also, it is believed that the natural additives increased the modified alkyd coatings' protection efficiency by decreasing the ions and moisture vapor transfer rates through the coating layer.

As a result of those findings, it was concluded that the coated mild steel surfaces with alkyd coating mixed with 2% vol. menhaden fish oil were the most and very well protected surfaces, followed by those coated with alkyd coating mixed with olive oil,

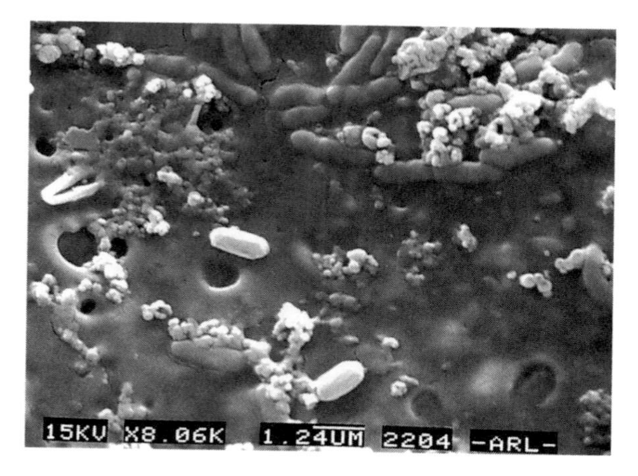

Figure 12.7 SEM photomicrograph shows the surface of a well-protected mild steel coupon coated with alkyd mixed with fish oil.

while the least protected mild steel surfaces were those coated with the original alkyd coating.

Another series of tests were conducted in order to observe the degrading and blistering effects of the acidic environment on the coated surfaces, and later on the corrosion forms and rates on the metallic substrates. Two samples of each coating system were tested in the same environment to be confident of the repeatability of the results. Figure 12.8 shows the blistering effects of the acidic environment on both of the control samples (system A) and the samples coated with the enamel oil-based coating mixed with one of the natural oils (systems B, C, and D). The degree of blistering on each of the samples was evaluated using the ASTM-D714-87

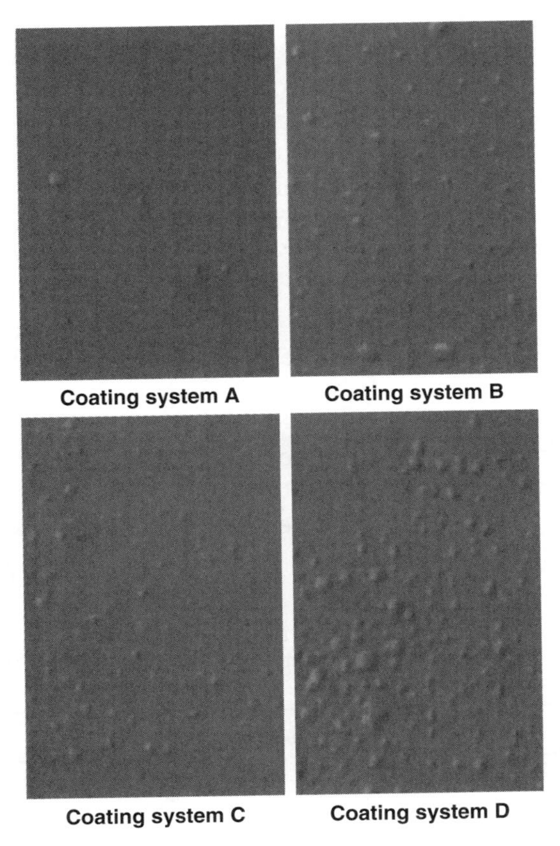

Figure 12.8 Digital photographs show 20 × 30 mm of the surfaces of the coated samples.

(ASTM 2002) photographic reference standard. The results and findings are tabulated in Table 12.2.

The samples coated with the enamel oil-based coating (system A) showed very little or no sign of surface damage, while the samples coated with enamel coating mixed with one of the selected natural oils experienced either a low or high degree of blistering. The highest degree of blistering was observed on the samples coated with the enamel coating mixed with 3% vol. fish oil (system D). This was followed by the samples coated with the enamel coating mixed with 3% vol. mustard oil (system B). A list of various systems is given in Table 12.2.

The samples coated with the enamel coating mixed with 3% vol. olive oil showed anomalous behavior in terms of blister size. Initial surface contamination and a difference in surface preparation could be the reason for the difference in the adhesive strength of the two samples coated with enamel coating mixed with olive oil. From the above observations, it was concluded that the control samples coated with the enamel coating showed better resistance to blistering effects in the acidic environment studied. This also indicates that the presence of natural oils (such as mustard, olive, and fish oils) changes the adhesive properties of the oil-based coatings at low pH environments. These blisters can grow in size and frequency and, hence, can degrade the coating quality and its protection efficiency.

The weight loss is one of the most common methods used to quantify corrosion, mainly when dealing with small panels and coupons (Fontana and Green 1978). In this study, the rate and extent of corrosion on the surfaces of the different samples were estimated using this method. The weight of each sample was measured before and after exposure in the salt fog test corrosion chamber.

Table 12.2 Details of the tested coating systems.

Coating system	Description
A	enamel oil-based coating
B	enamel oil-based coating + 3% vol. mustard oil
C	enamel oil-based coating + 3% vol. olive oil
D	enamel oil-based coating + 3% vol. salmon fish oil

The overall period of exposure for each sample was 3,000 hours. The reported values are the average of duplicate samples of each coating system. From Figure 12.9, it can be seen that the weight loss factor (WLF) was maximum for the samples coated with enamel coating only, closely followed by the samples coated with enamel coating mixed with mustard oil. From that, it was concluded that these two samples experienced high corrosion and erosion rates in the salt fog test corrosion chamber. On the other hand, the samples coated with enamel coating mixed with fish oil showed the lowest weight loss, followed by the samples coated with enamel coating mixed with olive oil. It was obvious that the addition of fish and/or olive oils to the enamel oil-based coating decreased the rate of the coated surface deterioration and, as a result, decreased the associated substrate metal corrosion.

The image analyzer system KS300 was used to monitor and investigate the coated surfaces' deterioration and the growth of the localized corrosion reflected in the form of holes and pits on and beneath the coated surfaces. It was observed that all of the tested

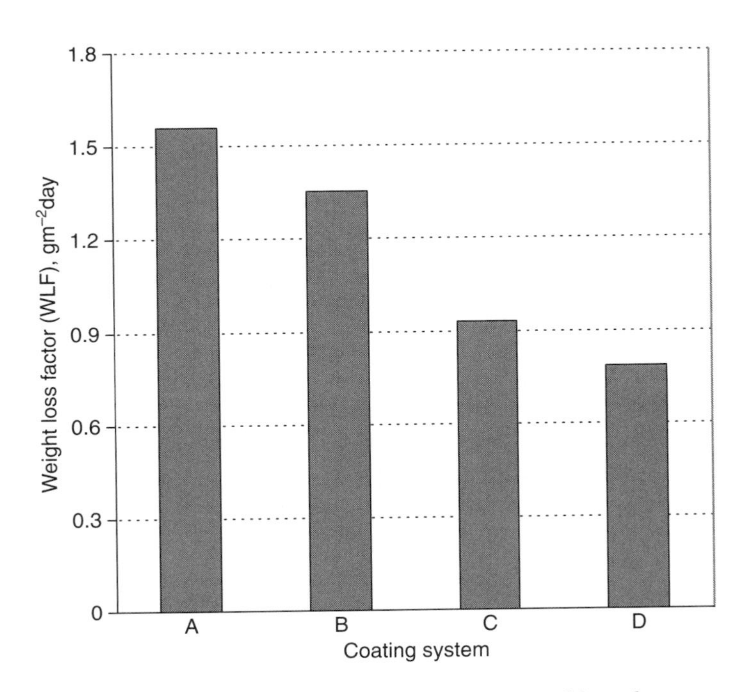

Figure 12.9 The weight loss factor (WLF) values for the mild steel coupons coated with different enamel oil-based coating systems.

coating systems suffered from surface erosion, degradation, and metal corrosion, but with different rates, forms, and extents. The holes and pits on the coated surfaces were photographed using a light microscope using a magnification of 10´. The pictures were then analyzed using the image analyzer technique. These pictures gave an idea about the severity and rates of the coating deterioration and the resulting corrosion. This method gave qualitative as well as quantitative results concerning the extent of corrosion in and around a given pit on the surface of a given coated sample (Muntasser *et al.* 2001). Photographs of some selected pits were taken after 1,000, 2,000, and 3,000 hours of exposure in the salt fog corrosion chamber. The areas of those pits were also measured using the abovementioned image analyzer techniques. The results are graphically represented in Figures 12.10 and 12.11. Figure 12.10 shows the average pits area for the different coating systems at different exposure times. Figure 12.11 shows the growth of pits with time for each coating system.

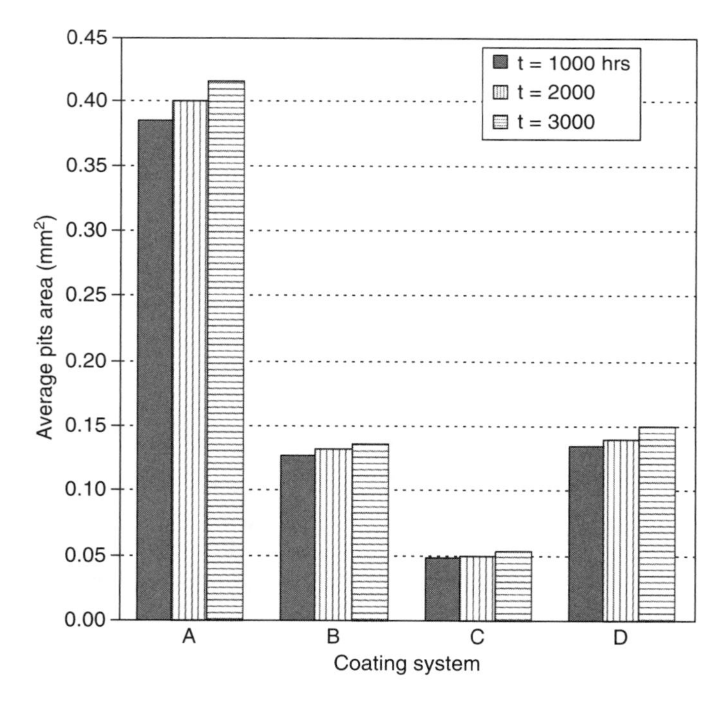

Figure 12.10 The average pits area for the different coating systems at different exposure times inside the salt fog test corrosion chamber.

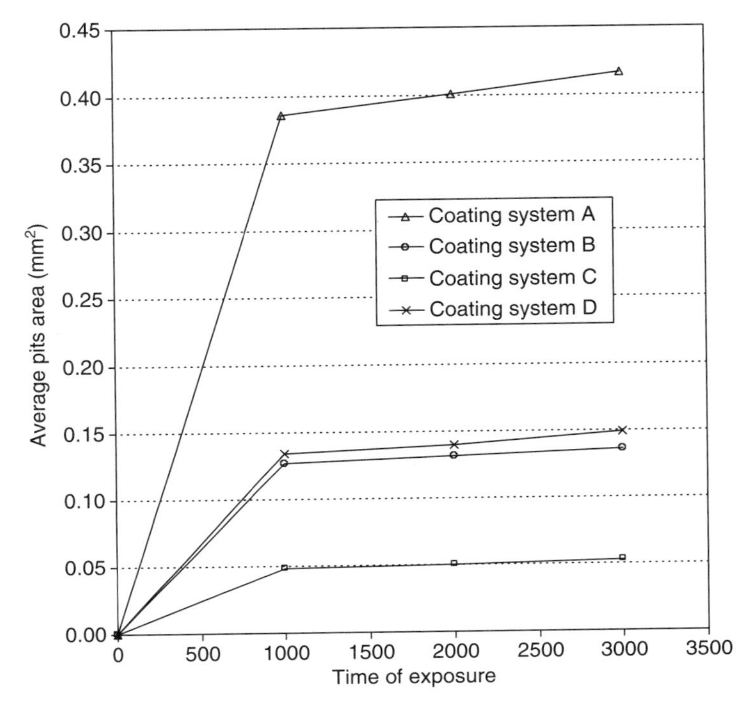

Figure 12.11 Pits average area on the surfaces of the coated mild steel coupons after different exposure times to the simulated marine environment.

Figure 12.12 shows a comparison between the shapes and sizes of the pits on the surfaces of the different coating systems after an exposure time of 3,000 hours inside the salt fog test corrosion chamber. In Figure 12.12, the brownish and reddish colors in and around the pits are the different corrosion products, mainly comprised of ferric and ferrous ions. It is worth mentioning here that several limitations existed regarding the coating application method and the curing process. The size and growth of each pit is influenced by the initial surface contamination or breaks in the coating film.

The surface areas of the pits and the surrounding corrosion products were used to evaluate the performance of each coating system. From Figure 12.12, it was observed that the samples coated with the enamel coating only and enamel coating mixed with mustard oil both showed the highest degree of localized corrosion. The lowest degree of surface degradation and localized corrosion were

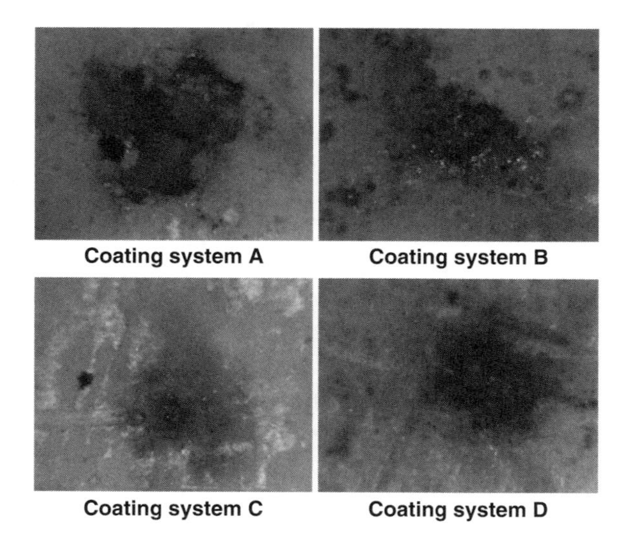

Figure 12.12 Comparison between the shapes and sizes of the pits formed on the surfaces coated with different coating systems after an exposure time of 3,000 hours inside the salt fog test corrosion chamber.

observed in the samples coated with enamel coating mixed with olive oil, where the overall surface damage and rusting on those samples were relatively low. The samples coated with enamel coating mixed with fish oil also suffered from coating degradation and localized corrosion attack, as can be seen from Figure 12.12. The amount of surface damage on these samples was higher compared to those on the surfaces of the samples coated with enamel coating mixed with olive oil.

Both the ESEM and the EDX were used to study and analyze the surfaces of the abovementioned coating systems. The EDX analysis technique is a well-known method used for investigating the surfaces of metals and coatings. Meehan and Walch (Jones-Meehan and Walch 1992) studied coated steel exposed to mixed communities of marine microorganisms using EDX. Their ESEM/EDX analysis detected the breaching of epoxy, nylon, and polyurethane coatings applied to steel coupons.

Figure 12.12 shows the ESEM photomicrograph and the EDX spectra of the surface of a sample coated with enamel coating mixed with mustard oil after an exposure time of 3,000 hours in the salt fog test corrosion chamber. Cracks and pits were observed all over the surface of this sample. The EDX analysis of a particular spot on the surface shown in Figure 12.13a was conducted, and the spectrum is shown in Figure 12.13b. This spectrum revealed a high percentage of Si and Ti, as they do form a major part of the enamel oil-based coating. Iron (Fe) was detected on two different peaks on the spectra, which implies that iron was present in two valence forms (ferrous and ferric). From this observation, it was concluded that the mild steel substrate had corroded and produced both ferric and ferrous oxides as part of the corrosion products. The EDX spectra also showed zinc (Zn) at the energy level of 1.03 KeV. The lower counts of zinc may be justified by the fact that both Zn and the corrosion products in the form of $ZnCl_2$ were leached out and washed away from the surface. This fact makes the coating system much less protective in any aggressive environment.

Figure 12.14 shows the ESEM photomicrograph and the EDX spectrum of the surface of the sample coated with enamel coating mixed with olive oil after 3,000 hours of exposure in the salt fog test corrosion chamber. Figure 12.14a shows that the surface was much less degraded, with fewer pits, holes, and cracks compared

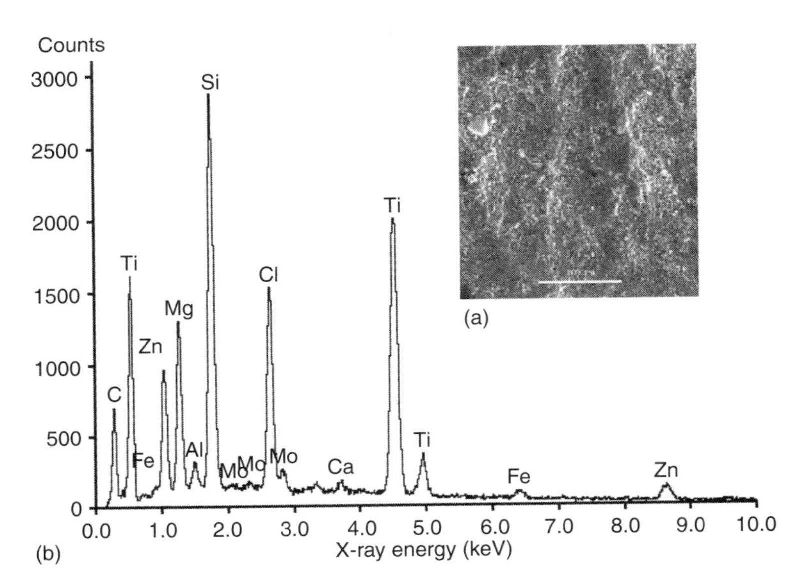

Figure 12.13 (a) ESEM photomicrograph of the mild steel surface coated with coating system B (b) EDX spectra of a spot on the surface shown in (a).

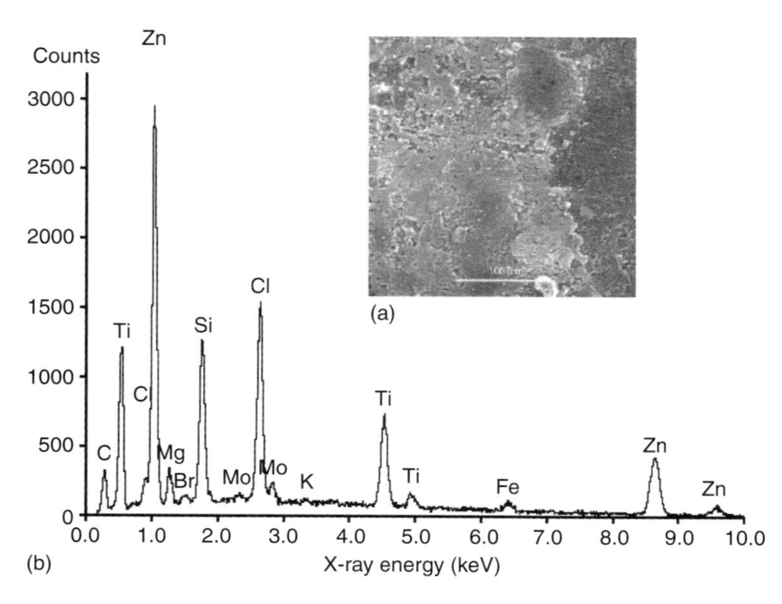

Figure 12.14 (a) ESEM photomicrograph of the mild steel surface coated with coating system C (b) EDX spectra of a spot on the surface shown in (a).

to other coated surfaces. The EDX analysis of a spot on this surface is shown in Figure 12.14b. From the spectra, it can be observed that iron was detected only on one peak. Zinc (Zn), on the other hand, was detected with a very high count as compared to that for coating systems B and D. This can be explained by the fact that the amount of zinc leached out from the coating was quite low. This means that the addition of olive oil formed a homogeneous thin film on the metal surface, which helped make it much more protective.

Figure 12.15 shows the ESEM photomicrograph and the EDX spectrum of the surface of the sample coated with enamel coating mixed with fish oil (coating system D) after 3,000 hours of exposure inside the salt fog test corrosion chamber. Very few localized corrosion attacks were observed on the sample surface, in the form of pits and cracks of almost the same shape and size. The amount of damage on the surface of coating system D was observed to be much lower than that for coating systems A and B. The EDX spectrum of a spot on the surface is shown in Figure 12.15b. From Figure 12.15b, it was observed that Si and Ti had the highest peaks. Iron

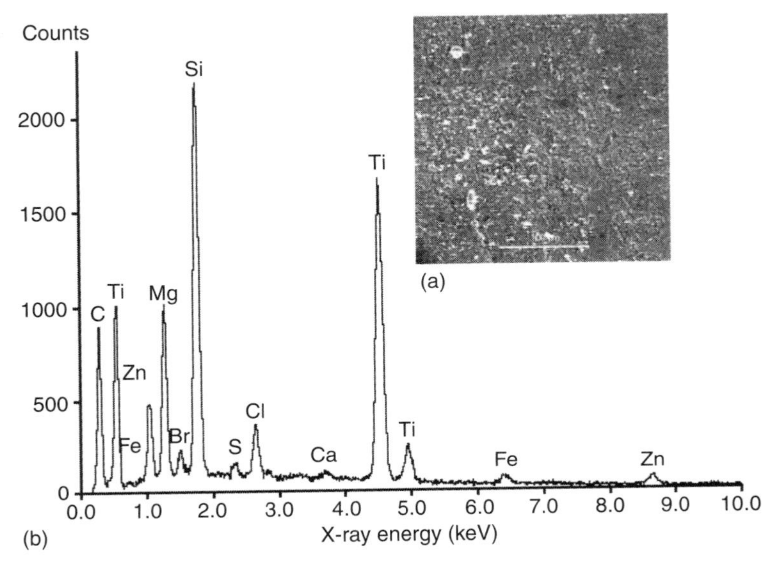

Figure 12.15 (a) ESEM photomicrograph of the mild steel surface coated with coating system D (b) EDX spectra of a spot on the surface shown in (a).

(Fe) was detected on two peaks, implying that it was present in ferrous as well as ferric forms. The amounts of chloride and zinc were detected to be very low. Zinc reacts with chlorides to form $ZnCl_2$, and that could have been washed out because it is a loose product (Munger 1990).

From the above results, it was inferred that coating system C showed the best performance under the simulated marine environment, followed by coating system D, while coating systems A and B experienced the highest surface damage and poorest performance. The leaching of zinc from the coating surface indicates the degradation of the coating system as zinc starts behaving as a sacrificial anode. This phenomenon was observed in coating system B and some in coating system D.

Saeed *et al.* (2003) investigated the antimicrobial effects of garlic and black thorn against *Shewanella putrefaciens*, which is a bacterium implicated in pipeline corrosion. They concluded that both garlic and black thorn possess bacteriostatic effects against *Shewanella putrefaciens* and, therefore, can be used as bactericides to inhibit and prevent biocorrosion in environments containing *Shewanella putrefaciens*.

12.6 Bacterial Solutions for Asphaltene and Wax Damage Prevention

The role of microbes in rapid degradation of petroleum compounds is well known. However, few studies have been done for the possible use of microbes in breaking down asphaltic materials in situ. Often bacteria are considered unfit for growth in harsh salinity and thermal conditions that are prevalent in wellbores of petroleum reservoirs. A few applications of microbial degradation of waxy materials have been reported. However, no systematic study is available in the literature. Progresses have been made, on the other hand, in using microbes for the remediation of petroleum contaminants (Livingston and Islam 1999). However, most previous studies have focussed on bacteria that can survive only at ambient temperatures and in non-saline environments (Baker and Herson 1990; Hills *et al.* 1989). Only recently, Al-Maghrabi *et al.* (1999) have introduced a strain of thermophilic bacteria that are capable of surviving in high-salinity environments. This is an important step, considering that most previous scientists became increasingly frustrated with the slow progress in the areas of bioremediation under harsh conditions.

A great deal of research has been conducted on mesophilic and thermophilic bacteria in the context of leaching and other forms of mineral extraction (Gilbert *et al.* 1988). At least 25% of all copper produced worldwide, for instance, comes from bioprocessing with mesophilic or thermophilic bacteria (Moffet 1994). Metals in insoluble minerals are solubilized either directly by mineral metabolic activities, or indirectly by chemical oxidation brought on by products of metabolic activity, mainly acidic solutions of iron (Hughs 1989). Most of these bacteria, therefore, prefer low pH conditions. One of the best-known mesophiles is the *Thiobacilli* family. These bacteria are capable of catalyzing mineral oxidation reactions (Marsden 1992). *Thiobacillus ferrooxidans* is the most studied organism relevant to the leaching of metal sulphides (Hughs, 1989). This strain of bacteria is most active in the pH range of 1.5–3.5, with an optimum pH of 2.3 and preferred temperatures of 30–35°C. Even though it is generally recognized that these bacteria can survive at temperatures ranging from 30-37°C, there is no data available on their existence in the presence of petroleum contaminants. Also, no effort has been made in trying to identify this strain of bacteria in hot climate areas, even though it is understood that the reaction kinetics will increase at higher temperatures (Le Roux 1987).

Several acidophilic bacteria have been identified that can survive at temperatures higher than the one preferred by mesophiles (Norris *et al.* 1989). In this group, iron- and sulfur-oxidizing eubacteria that grow at 60°C can be considered to be moderately thermophilic (with an optimum temperature of 50°C). At higher temperatures, there are strains of *Sulfolobus* that can readily oxidize mineral sulfides. Other strains, morphologically resembling *Sulfolobus*, belong to the genus *Acidiamus* and are active at temperatures of at least 85°C.

The isolation of thermophilic bacteria does not differ in essence from the isolation of other microorganisms, except in the requirement of high incubation temperatures. This may necessitate measures to prevent media from drying out, or the use of elevated gas pressures in culture vessels to ensure sufficient solubility in the substrate (Lacey 1990).

The use of thermophiles in petroleum and environmental applications has received little attention in the past. However, if thermophiles can survive in the presence of petroleum contaminants, their usefulness can be extended to bioremediation in hot climate areas. On the other hand, if the thermophiles can survive in a saline environment, they can be applied to seawater purification and to microbial enhanced oil recovery. Al-Maghrabi *et al.* (1999) have introduced a thermophilic strain of bacteria that can survive in a saline environment, making them useful for petroleum applications in both enhanced oil recovery and bioremediation.

Figure 12.16 shows bacterial growth curves for a temperature of 45°C. The two compositions reported are 3% and 6% asphalt (crude oil containing heavy petroleum components and 30% asphaltene). Even though the temperature used was not the optimum for this strain of bacteria, the exponential growth is evident. Note that approximately 10 hours of an adaptation period elapsed between the exponential growth phase and the initial stage of the adaptation phase. During this time of adaptation in high temperature and salinity, bacteria concentration remained stable. Also, no appreciable degradation in asphalt was evidenced. During the exponential growth phase, the growth rate was found to be 0.495/hr and 0.424/hr. The growth rates, along with the best-fit exponential equations, are listed in Table 12.3. A confidence of more than 90% for most cases shows relatively high accuracy of the exponential form of the growth curve. Even though the actual growth rate is higher at

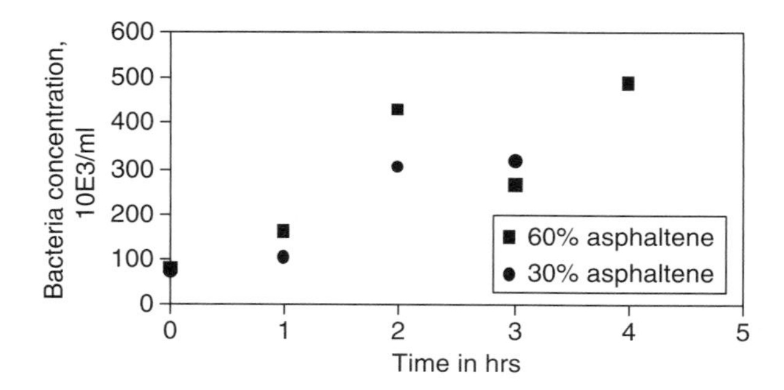

Figure 12.16 Bacterial growth in asphaltene at 45°C.

Table 12.3 Growth rates of bacteria in asphaltene and wax at different concentrations and temperatures.

Concentration	Temperature	Growth rate, μ	Complete equation
30% asphaltene	45C	0.495/hr	$C_t = 75.81e^{(0.495t)}$
60% asphaltene	45C	0.424/hr	$C_t = 99.53e^{(0.424t)}$
30% asphaltene	80C	0.605/hr	$C_t = 56.00e^{(0.605t)}$
60% asphaltene	80C	0.519/hr	$C_t = 77.16e^{(0.519t)}$
30% wax	80C	0.0293/hr	$C_t = 48.038e^{(0.029t)}$

lower asphaltene concentrations, the actual number of bacteria was found to be greater for the high-concentration case. This finding is encouraging, because a greater number of bacteria should correspond to faster degradation of asphaltenes. It also demonstrates that the asphalts (and asphlatenes) form an important component of the metabolic pathways of the bacteria.

Also, for the 6% asphalt case, there appears to be some fluctuation in bacterial growth. This pattern is typical of two or more species of bacteria competing for survival. At a lower temperature (45°C), two types of bacteria are found to survive. At higher temperatures, however, only one type of bacteria (the rod-shaped) continues to grow. Also, the oscillatory nature of bacterial growth can be explained as a process of step-wise breakdown of various components of the petroleum crude. In this process of consumption, the

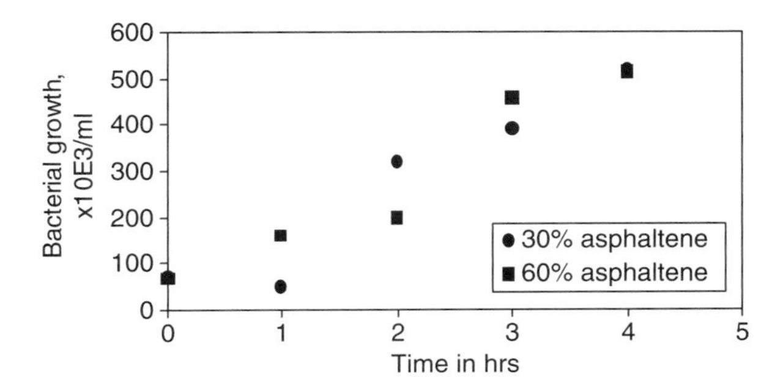

Figure 12.17 Growth of bacteria in asphaltene at 80°C.

asphaltic crude is subject to rapid degradation, along with exponential growth of bacteria following the Monod equation.

Figure 12.17 shows the bacterial growth curve for the two cases of 80°C. The growth rates for these two cases are listed in Table 12.3. Clearly, faster growth was observed for the higher temperature case. Note that all of these cases used 2% salinity. This growth rate shows both viability and enhanced bacterial growth in the presence of higher temperatures. Also, similar to the 45°C case, a larger number of bacteria were found at higher concentrations of asphaltic crudes. Al-Maghrabi *et al.* (1998) made a similar observation when only 3% asphaltene was used. Even though a higher concentration invokes a lower rate of growth, the initial concentration continues to be greater for all temperatures. Also, a much faster increase in bacterial concentration is evidenced at a higher temperature (80°C). This could be due to two factors. The most obvious one is that the bacteria are thermophilic, with an optimum temperature around 80°C. The other explanation is that the crude oil components are easier to break down at higher temperatures. In fact, Al-Maghrabi *et al.* (1999) observed that at 80°C, the interfacial tension between oil and water is lowered significantly, making the oil more vulnerable to microbial degradation. Of course, the viscosity of the crude oil is also reduced at a higher temperature, and this factor cannot be ignored.

Similarly, Figure 12.18 shows bacterial growth in a wax medium. This figure shows the effectiveness of thermophilic bacteria in preventing wax deposition problems. For this particular experiment, 3% wax was added to the bioreactor, while keeping the salinity at 2%. The bacterial growth in the presence of wax is extremely slow,

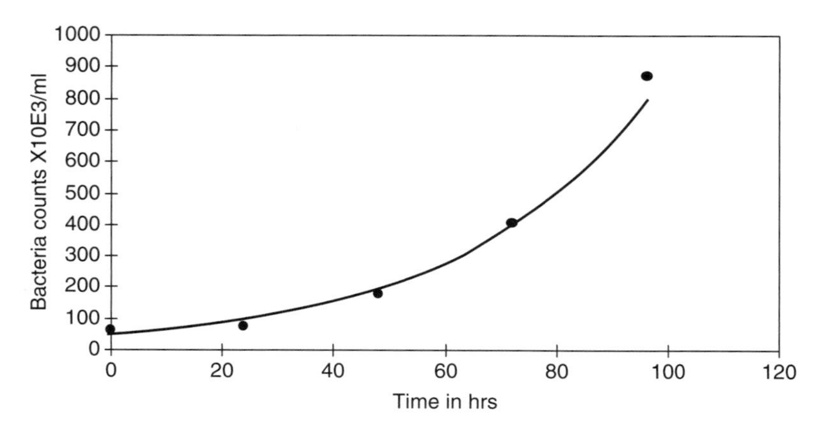

Figure 12.18 Bacterial growth in wax at 80°C.

especially when compared to that in the presence of the asphaltic crude. Table 12.3 shows the bacterial growth at a low value of 0.029/hr, which is an order of magnitude lower than that of the asphaltene case. Also, the actual number of bacteria is lower in this case. As a consequence, the degradation of wax remains very small (less than 5% in 100 hours). However, the presence of bacteria led to the formation of some crystalline structures.

Figure 12.19 shows the effect of salinity on bacterial growth. Both of these two cases represent 10% crude oil (3% asphaltene), but one of them has fresh water in it. The presence of fresh water definitely enhances the bacterial growth. This is conceivable. Note that these two curves were generated at room temperature (22°C). The growth rates (for the exponential phase) are listed in Table 12.3. The growth rate is three times less in the presence of salinity. However, the salinity is very high (10%), and most bacteria would not survive in this environment. Figure 12.18 also shows that the fresh water case reaches maximum bacteria concentration at an earlier stage than does the high-salinity case. This is expected, because the same amount of crude oil was used for both cases, and with the degradation being slower in the presence of high salinity, the bacteria ran out of food faster in the fresh-water case.

Microphotographs of the wax body before and after bacterial action showed the structure change during bacterial actions. The structure of the wax is clearly affected by the presence of bacteria that contribute to the breakdown of the long-chain polymeric

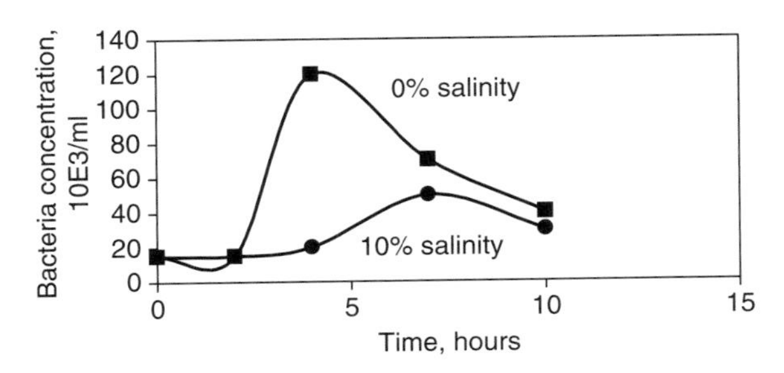

Figure 12.19 Effect of salinity on bacteria growth (10% crude oil).

microstructure of the wax. The emergence of crystalline structures is likely to increase the permeability of the porous medium initially affected by wax deposits.

A series of microphotographs was observed in order to visualize bacterial growth and its consequences. Microphotographs showed the existence of both round-shaped and rod-like bacteria after overnight treatment of the culture medium at room temperature. The round-shaped bacteria are very active in breaking down crude oil at its optimum temperature of 45°C. However, for higher temperatures, the rod-shaped bacterium is the only contributor to the degradation of the asphaltic crude. As the temperature increased, the interfacial tension between oil and water decreased, leading to the formation of water-in-oil emulsions. At the later stage of bacterial growth, more oil breaks down, and the nature of emulsion reverses. Other microphotographs showed the affinity of bacteria to crude oil, in which bacteria are found to gather around the oil droplets. This indicates that, with enhanced agitation, more water will be dispersed, leading to higher rates of biodegradation. Several microphotographs also showed the existence of micro-emulsions at 22°C. This is a water-in-oil emulsion. Such emulsions are an indication of low oil-water interfacial tension. Once emulsions are formed, the bioremediation action of bacteria is likely to be enhanced. Microphotographs were also used to affirm a consistent growth in the size of the bacteria as the temperature increased. These large bacteria contribute to the simultaneous formation of oil-in-water and oil-in-oil emulsions within the same area.

13

Greening of Enhanced Oil Recovery

13.1 Introduction

The predominant theme of this book is that petroleum operations cannot be rendered green unless the industry eliminates the use of artificial chemicals or energy sources. If artificial matter is replaced with natural matter, then sustainability will be assured. Until a century ago, natural materials were used for most engineering applications. This practice dates back to thousands of years of human history.

Natural additives have been used for the longest time, dating back to the regime of the Pharaohs of Egypt and the Hans of China. However, the Renaissance in Europe gave rise to the Industrial Revolution, which became the pivotal point of the emergence of numerous artificial chemicals. Today, thousands of artificial chemicals are being used in everyday products, ranging from health care products to transportation vehicles. With renewed awareness of the environmental consequences and more in-depth knowledge of science, we are discovering that such ubiquitous use of artificial chemicals is not sustainable (Khan 2006). If the pathways of various artificial chemicals are investigated, it becomes clear that such chemicals cannot be assimilated in nature, making an irreparable footprint that can be the source of many other ecological imbalances (Islam 2004; Chhetri *et al.* 2006; Chhetri and Islam 2007). Federal regulators determined that about 4,000 chemicals used for decades in Canada posed enough of a threat to human health and the environment that they subjected the chemicals to safety assessments (The Globe and Mail 2006). These artificial additives are either synthetic themselves, or derived through an extraction process that uses synthetic products.

Crude oil makes a major contribution to the world economy today. Crude oil development and production in oil reservoirs can include up to three distinct phases: primary, secondary, and tertiary (EOR) recovery. During the primary and secondary recovery, only 30% to 50% of a reservoir's original oil in place is typically produced (USDoE 2006). Hence, attention is being paid to enhanced oil recovery (EOR) techniques for recovering more oil from the existing oilfields. The worldwide target for EOR is estimated to be two trillion barrels.

Enhanced oil recovery schemes are broadly in the categories of thermal, chemical, gas injection, and microbial. In all of these applications, the use of artificial or synthetic chemicals is ubiquitous. This chapter presents various ways of using natural chemicals to achieve the same results, in terms of additional oil recovery. With this mode, the products will be environment-friendly (or at least less hostile to the environment) and less expensive than conventional operations.

13.2 Chemical Flooding Agents

Even though the world is facing an energy crisis and, therefore, looking for innovative methods for enhanced oil recovery to produce more oil to meet the current and future energy needs, the EOR schemes have declined recently in the United States and the rest of the world. The major challenge the EOR schemes face today is to produce oil under attractive economic and environmental conditions (Islam 1999; Khan and Islam 2007). Figure 13.1 shows the total EOR production in the U.S. between 1982 and 2006. The EOR production increased from 1982, but started to decrease significantly from 1998. Figure 13.2 shows the decline of U.S. EOR production attributed to chemical flooding at the same period. Despite

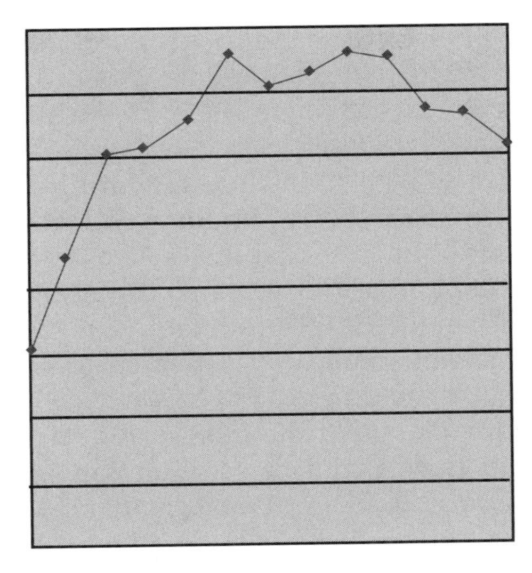

Figure 13.1 Total EOR Production in the U.S. between 1982–2006 (Worldwide EOR Survey 2007).

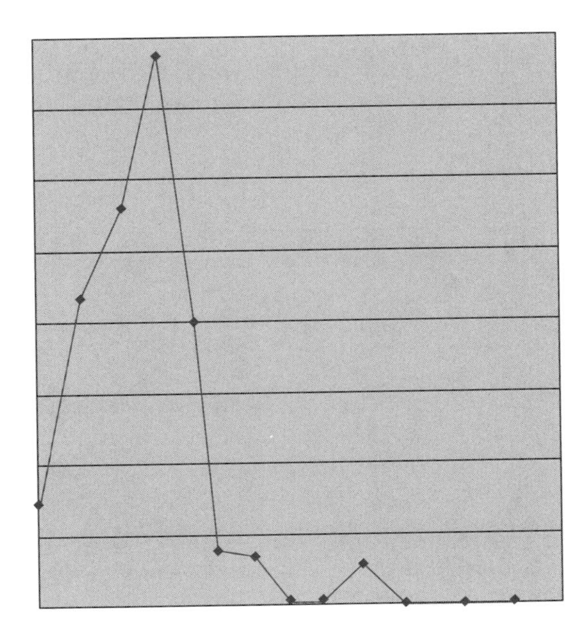

Figure 13.2 Total EOR production in the U.S. by chemical flooding (Moritis 2004).

chemical flooding being one of the widely used EOR techniques, its decline started sharply in the U.S. and in other countries after 1988.

Major reasons for the decline of EOR by chemical flooding are the rising prices of surfactants and their environmental impacts in the long term. The alkalis mostly used during alkaline flooding are sodium hydroxide (NaOH), sodium orthosilicate (Na_4SiO_4), sodium metasilicate (Na_2SiO_3), sodium carbonate (Na_2CO_3), ammonium hydroxide (NH_4OH), and ammonium carbonate ($(NH_4)_2CO_3$) (Burk *et al.* 1987; Taylor *et al.* 1996; Almalik *et al.* 1997). The costs of these chemicals have significantly increased recently. Figure 13.3 shows their cost increment from 1998 to 2006.

Alkaline flooding is one of the EOR recovery processes, and it began in 1925 with the injection of a sodium carbonate, "soda ash," solution in the Bradford area of Pennsylvania (Mayer *et al.* 1983). The alkaline flooding process is simple when compared to other chemical floods, yet it is sufficiently complex to require detailed lab evaluation and careful selection of a reservoir for application. Caustic flooding is an economical option because the cost of caustic chemicals is low compared to other tertiary enhancement systems.

The chemicals most commonly used for alkaline flooding are sodium hydroxide (NaOH), sodium orthosilicate (Na_4SiO_4), sodium metasilicate (Na_2SiO_3), sodium carbonate (Na_2CO_3), ammonium hydroxide (NH_4OH), and ammonium carbonate ($(NH_4)_2CO_3$) (Jennings 1975; Larrondo *et al.* 1985; Rahman 2007). Due to reservoir

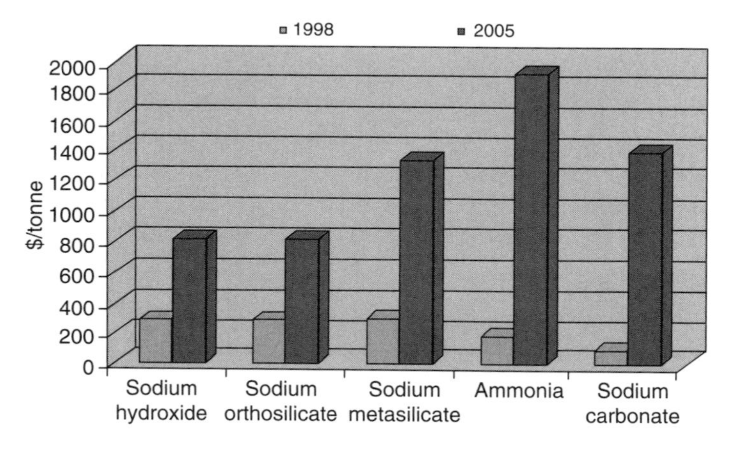

Figure 13.3 Price of common alkali chemicals (Mayer *et al.* 1983; Chemistry Store 2005; ClearTech 2006).

heterogeneity and the mineral compositions of rock and reservoir fluids, the same alkaline solution might induce a different mechanism. A good number of laboratory investigations dealing with the interaction of alkaline solutions with reservoir fluids and reservoir rocks have been reported in the literature (Jennings 1975; Ramakrishnan and Wasan 1983; Trujillo 1983). Due to higher pH value, sodium hydroxide is considered to be the most useful alkaline chemical for oil recovery schemes (Campbell 1977). The price comparison of the most common synthetic alkaline substances between 1982 and 2006 is mentioned in Table 13.1. It shows that alkaline prices have increased by five to twelve times over the last fifteen years. The biggest challenge of any novel recovery technique is to be able to produce under attractive economic and environmental conditions (Islam 1996; Khan and Islam 2007). Due to the high cost of synthetic alkaline substances and the environmental

Table 13.1 Comparison of alkalinity between natural alkaline solution extracted from wood ash and synthetic sodium hydroxide solution at different concentrations.

Synthetic Sodium Hydroxide Solution		
Synthetic sodium hydroxide solution (NaOH)	**Synthetic NaOH Solution (%)**	**pH value**
	2.0% NaOH solution	13.11
	1.5% NaOH solution	13.05
	1.0% NaOH solution	12.74
	0.5% NaOH solution	***12.35***
	0.2% NaOH solution	11.95
Wood ash solutions		
Maple wood ash solution	**Percentage of Wood Ash Solution**	**pH value**
	8% wood ash solution	12.42
	6% wood ash solution	***12.29***
	4% wood ash solution	12.09
	2% wood ash solution	11.83
	1% wood ash solution	11.42

impact, alkaline flooding has lost its popularity. This is reflected in Figures 13.4 and 13.5. These graphs have been generated using data reported by Moritis (2004). However, cost-effective alkali might recover its popularity in the recovery scheme. It has become a research challenge for the petroleum industry to explore the use of low-cost natural alkaline solutions for EOR during chemical flooding. In this paper, wood ash extracted solution is used as a low-cost natural alkaline solution. Several experiments have been conducted to test the feasibility of that natural alkaline solution.

13.2.1 Alkalinity in Wood Ashes

The origin of the word "alkali" is the Arabic word "al qali," which means "from ashes." Wood is the natural source of ashes, and the most important ingredients of natural ashes are sodium and

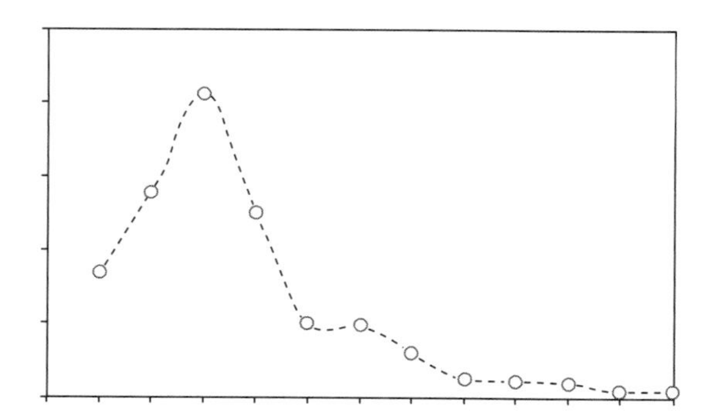

Figure 13.4 Total oil production by chemical flooding projects in the United States.

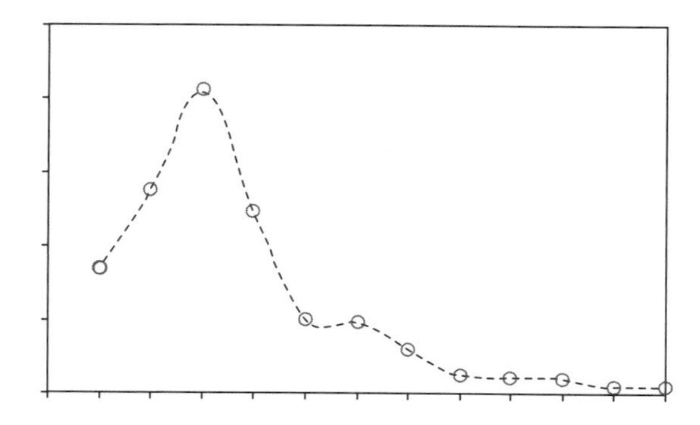

Figure 13.5 Chemical flooding field projects in the United States.

potassium. These metals also form the basis for alkaline solutions used for various applications.

In the modern world, wood ash is a byproduct of combustion in wood-fired power plants, paper mills, and other wood-burning facilities. A huge amount of wood ash is produced every year worldwide, and approximately three million tons of wood ash are produced annually in the United States alone (SAL 2006). Wood ash is a complex heterogeneous mixture of all the non-flammable, non-volatile minerals that remain after the wood and charcoal have burned away. Because of the presence of carbon dioxide in the fire gases, many of these minerals are converted to carbonates (Dunn 2003). The major components of wood ash are potassium carbonate, "potash," and sodium carbonate, "soda ash." From a chemical standpoint, these two compounds are very similar. From the 1700s through the early 1900s, wood was combusted in the United States to produce ash for chemical extraction. Wood ash was mainly used to produce potash for fertilizer and alkali for the industry. On an average, the burning of wood results in about 6–10% ashes. Ash is an alkaline material, with a pH ranging from 9–13 (Rahman *et al.* 2006), and due to its high alkalinity characteristics, wood ash has various applications in different sectors as an environment-friendly alkaline substance.

Rahman (2007) made use of the alkaline properties of wood ash, and formulated a scheme for enhanced oil recovery applications. Figure 13.6 shows the flow chart on his study.

The alkalinity of the ash leachate after repeated batch extraction with deionized distilled water at L/S = 10 is illustrated in Figure 13.7. The pH value is above 10.85 after 10 times repeated extractions, due to the presence of different alkaline components in wood ashes. Therefore, the same wood ashes might be used for alkaline solution extractions several times.

13.2.2 Environmental Sustainability of Wood Ash Usage

In Canada, 4,175,000 km^2 of land out of a total 10,000,000 km^2 of land is covered by forest. Every year, a huge amount of wood ash is produced worldwide, and approximately three million tons of wood ash are produced annually in the United States alone (SAL 2006). Wood ash has great potential to be used as a source of major- and micronutrient elements required for healthy plant growth. Wood ash contains many essential nutrients, mainly calcium, potassium,

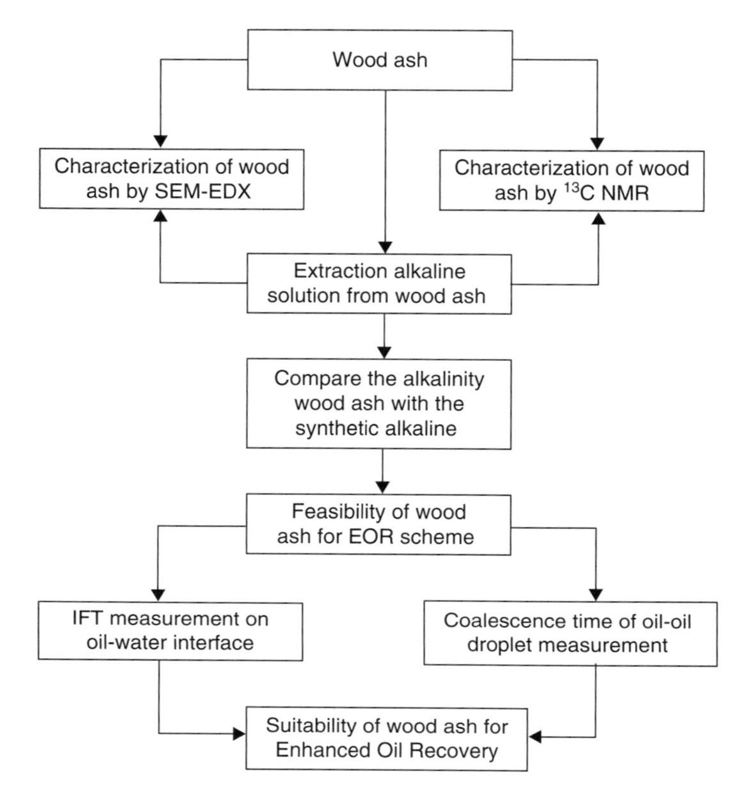

Figure 13.6 Major steps used to study the natural additives for enhanced oil recovery.

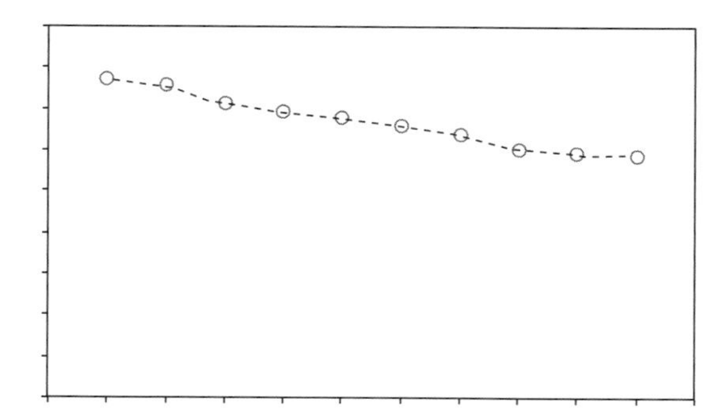

Figure 13.7 pH of the ash leachate as a function of L/S ratio.

magnesium, and phosphorus, for the growth of trees and other plants. Therefore, wood ashes might have potential applications in different sectors as an environment-friendly, sustainable, natural additive.

It has been reported in literature (Anfiteatro, 2007) that wood ashes with neem seed oil, kefir, sea salt, and essential oils might be used to make natural toothpaste, which may help with tooth sensitivity due to the poor condition of tooth enamel, or for the prevention of sensitive teeth. This toothpaste has also shown to eliminate gum bleeding if the paste is used on a daily basis. The toothpaste can remove most stains on teeth, such as those caused by cigarette smoking. It may also strengthen tooth enamel and gums. Apart from the traditional usage of wood ash as a source of alkali to different sectors, it has also been used for a long time to saponify fats in soap making and shampoo producing (Sh 2007). Chhetri *et al.* (2007) developed a process to produce completely natural birth soap using all natural ingredients, such as vegetable oil, coconut oil, olive oil, honey, beeswax, cinnamon powder, neem leaf powder, and natural coloring and flavoring agents instead of synthetic materials. Wood ash extracted alkaline solution was used to saponify the oils to make the natural soap.

This section reveals that the nutritional quality in wood ashes after the alkaline solution extraction and before the extraction from wood ashes is almost the same. However, wood ash could be collected separately from different sources, and it could be disposed in the landfill because the nutrient source is from plants. It could also be industrially utilized for cement manufacture. It can be a glazing agent in the ceramics industry, a road base, puzzolona, an alkaline material for the neutralization of wastes (Liodakis 2005), and it can contribute to the establishment of a sustainable process shown in Figure 13.8.

13.2.3 The Use of Soap Nuts for Alkali Extraction

Recently, Chhetri *et al.* (2008b) reported on the use of ground soap nuts (*Sapindus mukorossi*) to reduce the oil-water interfacial tension. The effect of the surfactant concentrations of 1%, 2%, 4%, 8%, and 12% was investigated. The experimental results showed that the surfactant form can effectively reduce oil-water IFT. The effect of heat on IFT was also studied. It was found that a higher IFT reduction was achieved after heating the system to 50°C. The experimental

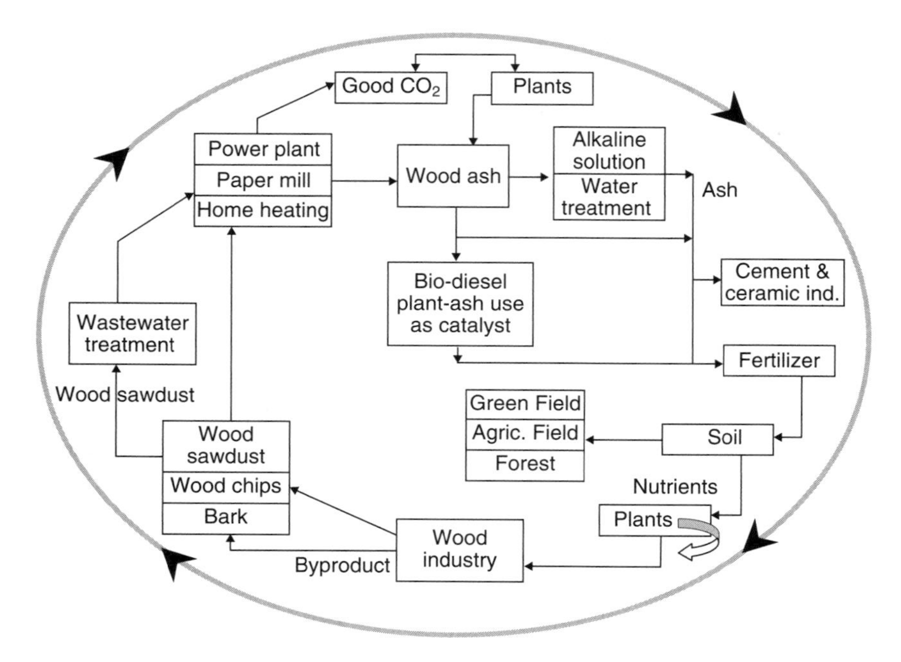

Figure 13.8 Possibilities for a sustainable utlization of wood and wood ash.

results showed that the extract has a great potential to be used as a surfactant for the enhanced oil recovery schemes. This established that surfactants derived from natural sources are economical and environment-friendly options for chemical flooding operations.

13.3 Greening of CO_2 Injection Schemes

The injection of CO_2 is one of the oldest EOR techniques available. Historically, CO_2 has been applied for both miscible and immiscible applications. Because of some very attractive features of CO_2 (e.g., swelling, IFT reduction, and low minimum miscibility pressure), it has received considerable attention in the topic of EOR.

The injection of CO_2 into the hydrocarbon reservoir is known to increase the amount of recoverable oil, yielding economic benefits. The process, however, has not been implemented, out of concerns for the greenhouse effects. International Energy Agency (1995) projects global carbon emissions to grow from about 6 billion metric tons in 1990 to over 8 billion metric tons by 2010, representing an annual growth rate of 1.5%. The participation, signed so far for the Climate Challenges utilities pledging a wide range of greenhouse gas reduction activities, accords for an aggregate of about 44 million metric tons of carbon equivalent (Kane and Klein 1997). However, the scientific protocol to meet this target has not yet been established, and it is becoming increasingly clear that this target is not something that can be achieved.

In the last few years, significant progress has been made in understanding the concepts associated with the greenhouse gas emissions that cause an environmental threat to the planet. It is in this context that CO_2 capture, disposal, and utilization potentials remain an attractive option for the medium to longer term, particularly if the current trend of energy supplies continues. Figure 13.9 shows the global CO_2 utilization potential based on the existing fossil fuel infrastructure and the reliability of associated technologies (Tilley 1997). At present, about 3% of the global oil production comes from EOR. Herzog *et al.* (1993) have mentioned that underground disposal of carbon dioxide has been identified as one of the high priority areas for research related to global climate change. The major obstacle to the disposal of CO_2 underground is the necessity to separate CO_2 from flue/waste gases. Capturing CO_2, its disposal, and utilization have significant challenges to overcome, and the existing technology is considered to be prohibitively expensive. The cost effectiveness for such separations can be achieved by

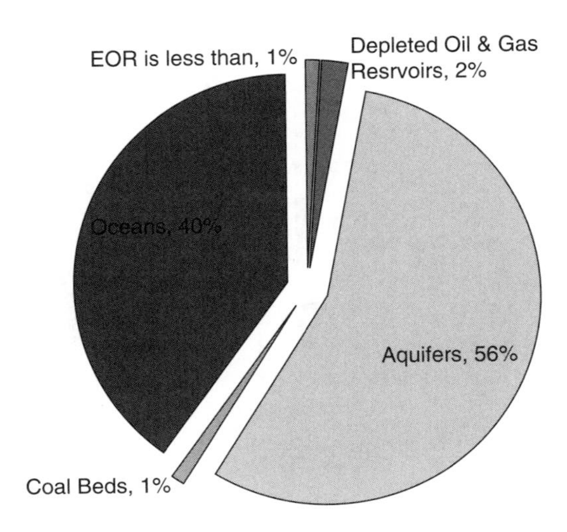

Figure 13.9 Global CO_2 utilization potential (IEA-GHG 1995).

eliminating gas separation, such as through gas re-injection into the sub-surface formations from gas processing plants. Chakma (1996) emphasized that the cost may be acceptable when applied in combination with other measures. It is believed to be too early to predict the implications regarding the disposal of CO_2. Instead, evaluating and discussing the options are the needs of the hour. Chakma (1996) provided examples for the acid gas re-injection projects into depleted oil and gas reservoirs, as well as aquifers that have been used for this purpose.

There have been numerous studies performed on various aspects of storage potential of CO_2 and other greenhouse gases. While many potential solutions have been proposed through these studies, they have also created much confusion, due to conflicting findings, narrow focus, and a lack of interdisciplinary and global approach. A comprehensive review of the latest research developments can dispel some of the misconceptions that have dominated this important aspect of global environment. Turkenberg (1997) reviewed the potential of CO_2 utilization and storage options.

In the overall CO^2 utilization picture, EOR potential is considered to be low, as can be seen from Figure 13.10. The potential to store CO_2 in depleted oil and natural gas fields can be much larger, with estimates ranging from 130 to 500 GT-C, depending on the recoverable amount of oil and gas. Deep aquifers have an estimated storage potential of 90 to 1000 GT-C. This wide range of variation is due to different assumptions made about the necessity of having a

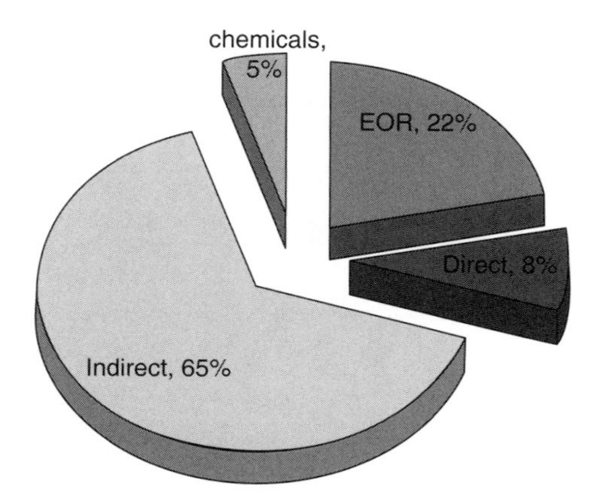

Figure 13.10 CO_2 Global Sinks and Capacities (IEA 1995; Kuuskraa *et al.* 1992; Bamhart *et al.* 1995; Hendriks 1994).

structural trap to assure safe and sustained disposal. Other assumptions for such an estimate are the volume of the aquifers, the percentage of the aquifer to be filled, and the density of CO_2 under reservoir conditions (Hendriks 1994).

Figure 13.11 shows the pictorial view of the CO_2 EOR process. CO_2, when injected into the reservoir, is dissolved in the oil. As a result, the oil viscosity is reduced, and mobility increases. The efficiency of EOR depends on the pressure and, thus, on the reservoir depth. Carbon dioxide is miscible with reservoir oil at high pressures, and greater miscibility has cost benefits associated with increased oil recovery. The threshold pressure, above which miscibility occurs, is called the Minimum Miscibility Pressure (MMP). Miscible CO_2 displacement results in approximately 22% incremental recovery, while immiscible displacement achieves approximately 10% incremental recovery (Taber 1994). Therefore, there is a greater payback for miscible displacement. For this, however, deeper reservoirs are preferred, in which pressures are above the MMP. The minimum miscibility pressure depends on the composition of the oil, higher density, and higher viscosity oils with more multiple aromatic ring structures having a higher MMP (Taber 1994). Historically, CO_2 injection has only been used for recovering oils with an API gravity greater than 22 and a viscosity lower than 10cp, because of greater miscibility and higher recovery efficiencies (Bergman *et al.* 1996).

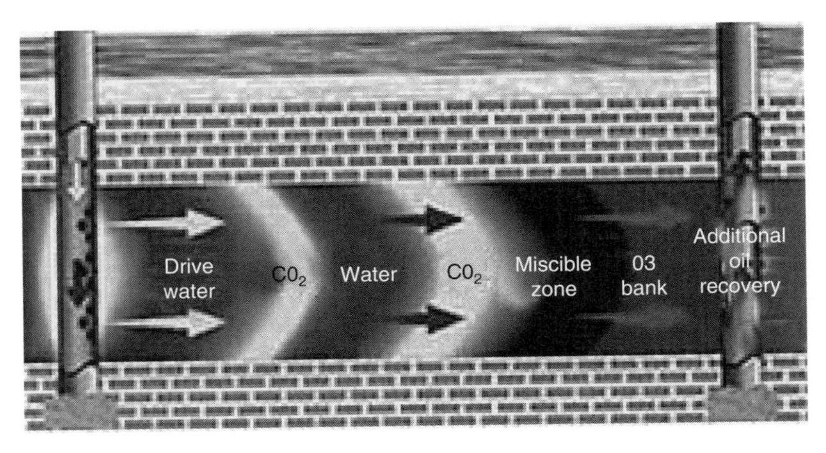

Figure 13.11 Graphic of CO_2 enhanced oil recovery (Courtesy of Occidental Petroleum Corp., DOE, 2004).

Conventionally, it is the norm to use purified CO_2 for gas injection. Often, the cost of purification is prohibitive, and can render a CO_2 injection scheme uneconomical. The argument has been made in the past that CO_2 injection is beneficial to the environment because a gas sequestration scheme and the additional cost of CO_2 purification are justified. This argument is not scientifically correct, and the resulting scheme does not meet the sustainability criterion as proposed by Khan (2007). This becomes clear in the following analysis in this section.

13.3.1 EOR Through Greenhouse Gas Injection

Nitrogen and flue gases are the least expensive among all EOR agents. These gases are also known to be similar (Emmons 1986), and it appears that they can be used interchangeably for oil recovery. However, these gases also have a high MMP. Therefore, miscible displacement is possible only in deep reservoirs with light oils.

Figure 13.12 shows how different gases can lead to similar recovery. This is particularly important because the decision of using a less expensive gas can often change the entire economics of the project.

Moritis (1994) reported that three nitrogen injection projects were operated successfully for years as flue gas injection projects. However, corrosion was a problem, especially for flue gas from

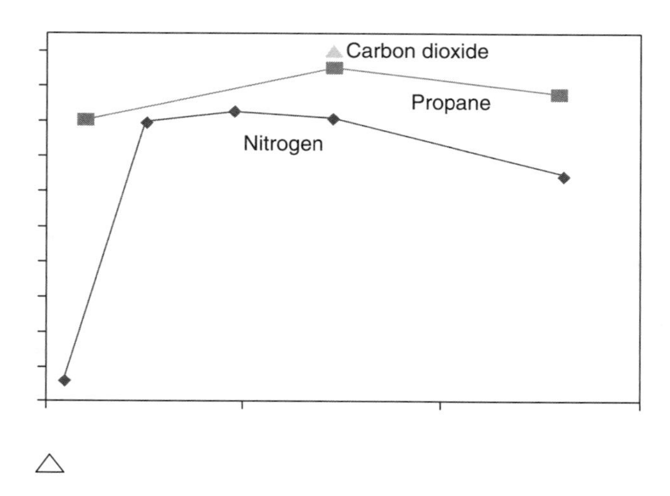

Figure 13.12 Gas injection with horizontal wells (Bansal and Islam 1994).

internal combustion engines. In addition to its low cost and widespread availability, nitrogen is the most inert gas of all injection gases. Therefore, this scheme holds promises for the future.

The suitability of flue gases containing CO_2 for enhanced oil recovery (EOR) was studied by Islam *et al.* (1992) in the context of cold front of in situ combustion. They observed that the flow rate plays a significant role in recovering heavy oil from a reservoir. This flow rate dependence was attributed to viscous fingering, which is a sign of unstable displacement. More recently, Srivastava and Huang (1997) conducted laboratory studies to evaluate various operating strategies for heavy oil recovery with flue gases. They conducted one-dimensional core-flood tests with a 13° API oil and flue gas system. Secondary gas slug flood with live oil was observed to be the most suitable injection strategy for heavy oil recovery from among three different injection strategies investigated (secondary slug, tertiary slug, and tertiary WAG), because this flood had the highest displacement efficiency. The tertiary WAG process recovered more oil than the tertiary slug process, because the former predominantly improved mobility control. Ultimate oil recovery was higher in runs conducted using live oil than in those using dead oil. This was attributed to relatively more favorable mobility of live oil and a slightly higher operating pressure. Srivastava and Huang (1997) reported that the flue gas appeared to be an effective flooding agent, because the oil recoveries were only 2-4 percentage points lower than those obtained with CO_2. They further mentioned that the comparable oil recoveries in flue gas runs are believed to be a combined result of competing mechanisms, namely, the free gas mechanism provided by CO_2 in contrast to the solublization mechanism provided by CO_2, which predominates in CO_2 floods.

The common features of all these estimates lack rigorous scientific analysis. The case in point is the Weyburn project. These estimates were based on MMP studies only. These values were included in a conventional reservoir simulator that had very little scientific merit in terms of modeling complex phenomena, such as viscous fingering, gravity override, reservoir heterogeneity, and others. The production strategy was optimized based on the stable steady state of the displacement front. Besides, no risk analysis was performed for loss of miscibility during the CO_2 displacement phase. Any of these phenomena can make the economic estimates irrelevant. For instance, if viscous fingering occurs during the displacement phase, the time of CO_2 breakthrough will be reduced by more than 50%, resulting in half of the oil recovery and storage capacity of CO_2 (see analysis by Coskuner and Bentsen 1990). It turned out that the CO_2 breakthrough in the Weyburn field took place in a short period of time, approximately around the time it was predicted through experimental studies using models that simulated viscous fingering (Khan and Islam 2007a).

13.3.2 Sour Gas Injection for EOR

The concept of using sour gas to enhance oil recovery is not new. Harvey and Henry (1977) reported one of the first studies on core flood experiments with carbon dioxide and hydrogen sulfide. Even though oil recovery with carbon dioxide (both miscible and immiscible) has been investigated by many researchers, Harvey and Henry (1977) appear to be the first ones to use hydrogen sulfide to recover oil. By using pure hydrogen sulfide, carbon dioxide, and mixtures of the two, they were able to compare recoveries for different cases. They reported that the miscibility pressure was lower for H_2S than for CO_2, consequently making the H_2S displacement process more attractive from the recovery point of view. They observed that miscibility could not be achieved for heavy oil with either H_2S or CO_2. However, even during immiscible displacement, pure H_2S displacement showed a distinct advantage over mixed (H_2S and CO_2) displacement, which in itself performed better than pure CO_2 injection. For both light and heavy oil, more than 15% incremental oil was produced when pure H_2S was used instead of a mixture of H_2S and CO_2. Pure CO_2 showed advantage over water-flood, but consistently showed a lower recovery than that with pure H_2S or a H_2S and CO_2 mixture. In later years, research has been performed on the recovery of both heavy and light oil with CO_2, with more emphasis on immiscible CO_2 displacement for heavy oil. For light oil, on the other hand, focus has been on the impact of H_2S that is generated through bacteriogenic activities (Hill *et al.* 1990; Frazer and Bolling 1991). These studies have identified the nature of the H_2S generation problems, but they did not focus on the recovery aspect of H_2S.

The presence of sour gases poses a difficult problem in terms of oil and gas production, processing, and refining. Unfortunately, little is known in the areas of sour gas behavior in most cases. Advances have been made in other areas of unwanted gas disposal and utilization (see Islam and Chakma 1993). It is important, however, to differentiate between gas injections under stable and unstable flow conditions. If gas is carried out from the top of a reservoir, the flow is likely to be gravity-stabilized. In the presence of an unstable displacement, the gas breakthrough takes place early, leading to a decline in the overall productivity.

In a recent publication by Al-Falahy *et al.* (1998), several solutions were proposed to the sour gas problem supported by numerical simulation and laboratory experiments. The proposed solution deals with both sour gas disposal and oil recovery with sour gas. The problems associated with these techniques were studied in detail in order to depict an accurate picture of the available options. They found that sour gas improves the miscibility behavior of the crude oil, leading to a greater recovery when significant amounts of sour gas are present in the injection stream. They also reported a scheme for separating SO_2 from a gas stream containing CO_2 and other gases. Numerical results indicate that oil recovery as high as 90% can be achieved with pure H_2S injection. Furthermore, they reported that the recoveries changed only slightly when a mixture of gases was used. Mixtures of H_2S and CO_2 and methane followed the high recovery of H_2S injection. The lowest recovery was reported with CO_2. In addition, when an immiscible, unstable gas injection scheme was employed, recovery was significantly lower, limited to a value similar to that with water flooding.

13.4 A Novel Microbial Technique

In an EOR process, the alteration of rock/fluid properties is often sought in order to increase productivity and, ultimately, production. Microbial processes have been presented for improving fluid flow characteristics. In this section, a microbial application for rock characteristic alteration is presented. This technique can be used for sand consolidation as well as fracture remediation to improve wellbore performance.

Because of biological origin and formation, these biomaterials have superior performance over the synthetic or non-natural structure because of both the mechanical strength and the functional properties. The outstanding properties are only visible and attributed to the biomaterials as observed by their well-organized structure and strong interfacial interaction between biomacromolecules and inorganic components.

Yu *et al.* (2004) suggested that every biological system use biomacromolecules as nucleators, cooperative modifiers, and matrixes or molds to exert exquisite control over the processes of biomineralization, which results in unique inorganic–organic composites with various special morphologies and functions. Sondi and Sondi (2005) studied the bioprecipitation of mineralizing organisms that selectively form either intracellular or extracellular metal carbonate polymorphs or unusual morphological properties at ambient pressure and temperature. They found that the nucleation, growth, and morphological properties of the formation of biogenic metal carbonate structures are controlled and regulated by organic macromolecules, mostly peptides and proteins.

One of the most important parameters is particle morphology. To control the precipitation reaction, the nucleation and growth steps must be mastered. The nucleation step is especially important in precipitation reactions. The nucleation rate and the duration of the nucleation process have a direct influence on the final particle size, the size distribution, and the growth mechanism (Donnet *et al.* 2005). The other unique effect is the initial formation of a metastable amorphous phase of calcium carbonate, which rapidly transforms into crystalline entities (Sondi and Matijevic 2001). The rate of the change in the crystal structure strongly depends on the concentration of the enzyme.

To obtain a high space-time yield in industry, the mineralization under high, super saturation conditions in batch or semi-batch processes has become a popular method (Schlomach *et al.* 2006). However, precipitation usually does not naturally take place at low ionic concentrations. Biological processes do not have this limitation. The natural biomineralization process is considered a superior process, not only for its super performance, but also for its sustantaibility in the long term. As a novel and sustainable process, the biomineralization process is now encouraged to replace every non-natural, chemical precipitation process. The application can be widespread, from selective reservoir plugging to teeth feelings.

The wide distribution of microbes in geological environments facilitates the biomineralization process in nature. Natural surface rocks have been observed to have 103 bacteria or fungal cells per gram of stone (Eckhardt 1985). Microbial metabolic activities play an important role in deposition and diagenesis processes in a geological environment. Microbio-mineral-precipitation is not an unusual process in nature. There are numerous examples, such as bacteria and algae precipitate minerals from seawater, and this process plays an important role in the deposition and consolidation of beach-rock formations (Krumbein 1979). Microbes can oxide metals and deposit them in hot spring systems and in deep ocean hydrothermal vents (Ferris *et al.* 1987; Ehrlich 1983). The formation of marine ferromanganese nodules and freshwater ferromanganese deposits has been attributed to bacteria (Ehrlich 1974).

Minerals such as calcite, silicon, oxidized manganese, and oxidized iron usually do not precipitate naturally because of the low ionic concentrations. However, when bacteria interact with ions such as Ca^{+2}, Si^{+}, Fe^{+3} and Mn^{+}, precipitation takes place (Beveridge *et al.* 1985), and plugging or cementing occurs consequently.

The most abundant mineral phase associated with bacteria is a complex (Fe, Al) silicate with a variable composition. The amount of metal sorption and biomineralization largely reflects the availability of dissolved metals in the water. In laboratory studies, Krumbein (1979) found that among 20 bacteria strains, 16 were able to precipitate aragonite from solutions made of seawater and nutrient for bacteria. Some strains yield Mg calcite. In his experiments, up to 350 mg of aragonite were obtained from a liter of the medium.

In sediments, heavy metals, such as alkali and alkaline earth ions, deposit at the surface of membranes of bacterial cells and stain these membranes (Degens and Ittekkot 1982). These naturally stained membranes can further act as templates for mineral deposition and may, in certain environmental conditions, lead to stratabound ore deposits. Microbial-mineral precipitation occurs directly as a result of bacterial metabolic activities, or indirectly as a consequence of regional geochemical condition changes due to bacterial metabolic activities (Kantzas *et al.* 1992).

Similar to most cell surfaces, the bacteria cell walls are anionic (Beveridge *et al.* 1984). These characteristics are independent of whether the bacteria are gram-positive or gram-negative. It is reasonable to assume that bacteria will interact strongly with metallic ions, even in diluted solutions in natural bodies of water. Laboratory experiments had demonstrated that metal accumulation could be substantial within the wall fabric (Hoyle and Beveridge 1983). In a laboratory simulation of a low temperature sediment diagenesis process, Beveridge and Fyfe (1985) found that metal precipitation associated with bacteria was primarily in the wall fabric. Mineral crystals grew with time until all of the wall material had been mineralized, and then crystals developed within the cytoplasm. Eventually, the entire cell became crystalline. In this process, gram-positive walls seem more reactive than their gram-negative counterparts (Beveridge and Fyfe 1985). In a geological formation or an experimental sand pack, bacteria adhere to pore-surfaces with their glycocalyx. They induce mineral crystallization. The dead cells stick together by glycocalyx. These mineral-bio-films eventually plug the pores.

Detailed studies have been carried out on carbonates precipitation induced by environmental condition changes caused by bacterial metabolic activities (Kantzas *et al.* 1992; McCallum and Guhathakurta 1970; Krumbein 1979). In the case of calcium carbonate deposition, bacteria increase the pH of the solution, which in turn reduces the solubility of $CaCO_3$ and induces precipitation.

When fresh medium (mineralization solution) continuously flows through porosity or fracture, continuous mineral precipitation can be maintained to result in plugging. Many experiments have been carried out to study the process of microbial mineral plugging. Bacteria and their population are found to affect this process (Macleod *et al.* 1988; Jack 1988; and Gollapudi *et al.* 1995). Anaerobes and facultative aerobes are preferred in order to enhance oil recovery, because reservoirs are essentially anaerobic, and oxygen injection faces its own constraints. Vegetative cells are more active than starved cells, and can achieve more complete plugging, whereas starved cells have smaller sizes, and they can penetrate to deeper levels and into smaller pores than vegetative cells, which is the advantage of applying starved bacteria to plugging. A higher bacteria concentration (greater population) will induce a quicker and more complete mineral plugging.

Different quantities and qualities of nutrients induce different plugging results (Kantzas *et al.* 1992). Experimental results indicate that the amount of plugging in a core material in a porous medium is roughly proportional to the amount of nutrients passed through the porous medium. There seems to be a critical nutrient injection rate, below which bacteria plugging does not take place. In microbial silicon precipitation, there are certain sugar and amino acid concentrations that are optimum for bacterial silicon uptake.

Microbial-mineral-plugging has been employed by petroleum microbiologists as a method to enhance the production of hydrocarbon resources (Jack 1992). Reservoir heterogeneity has a significant effect on the oil recovery efficiency of a water-flood process. The residual oil saturation (ROS) that remains after water flooding is a potential target for applying reservoir selective plugging techniques using in-situ growth of bacteria (Jenneman *et al.* 1984). In heavy oil fields, where water tends to respond to pumping more readily than viscous oil, wells of primary production commonly water-out at low oil recoveries. This is a serious problem that can occur gradually over several years, or it may be a catastrophic event in which water directly underlies oil in the reservoir. For this situation, control of excess water production may be accomplished by selectively plugging the zones of water encroachment (Jack *et al.* 1991).

Chemically cross-linked polymers may be used as plugging agents. However, they are unsustainable, expensive, and their performance is unpredictable. Microbial plugging is an efficient and less expensive method to these problems. Moreover, this process is sustainable for both the short term and long term. The method involves the introduction of viable bacteria in the aqueous, displacing fluid to be injected into the high-permeability water-swept zones. Once the bacteria are in place, a designed volume of nutrients can be injected into the reservoir to support in-situ metabolism of the bacteria that are capable of initiating physical plugging and reducing original permeability. This will result in a diversion of the displaced fluids from plugged high-permeability zones to unswept zones and, thus, will improve sweep efficiency.

Jack *et al.* (1992) suggested that a significant target for microbio-mineral-plugging might be the plugging of fractures in carbonate reservoirs, which presently thwart late-life strategies for gas and oil recovery by fostering gas and water breakthrough to production wells.

In an experiment conducted by Gollapudi *et al.* (1993), simulated fractures were put into sand packs. They found that increased microbial activities take place in simulated fractures, and the formation of precipitates increases at the fracture surfaces (Gollapudi *et al.* 1995). Islam and Bang (1993) suggested that microbial mineral plugging could be employed to remediate fractures in historic monuments and buildings (Islam and Bang 1993).

Zhong and Islam (1995) reported some experimental studies conducted on the process of microbial fracture remediation. Microbial carbonates precipitation received much attention because it is comparatively a more efficient plugging or consolidation process (McCallum *et al.* 1970; Krumbein 1974; Kmmbein 1979; Kantzas *et al.* 1992). $CaCO_3$ has several polyforms: calcite, aragonite, vaterite, and amorphous calcium carbonate (Seo *et al.* 2005). The first two are the most common, because they are widely found throughout nature, occurring as the main mineral constituents of sedimentary rocks and as inorganic components in the skeletons and tissues of many mineralizing organisms, especially mollusks. Bacteria *Bacillus Pasteurii* can be used efficiently to promote carbonate precipitation and to reduce permeability in an unconsolidated system. These bacteria were employed in this research. Chemical and biochemical reactions and their results, the effects of width and fracture fillings, experimental fracture remediation results, and the effects of chemical and physical factors, medium, and bacteria were studied.

13.5 Humanizing EOR Practices

Especially in EOR operations, thermal EOR, as a reservoir heating method, would require steam generation and bitumen upgrading facilities. Huge amounts of source water are needed for steam generation. The thermal method also generates a large amount of produced water, and recycling of this water would be required in order to reduce the source and disposal volumes to acceptable levels. A hot water extraction process needs to use open pit mines. Sometimes, large-scale tailing ponds are also required. There would be a relatively minor disposal problem for produced sand and fines. Disposal water can be separated into two streams, the most offensive waste being disposed of underground, and the safe stream discharged into a river system. In the underground disposal of wastewater, it is essential to ensure that groundwater sources are not contaminated. Noxious gas emissions into the atmosphere would be an issue of concern. Sulfur dioxide is the main pollutant, and is produced by burning high sulfur crude or oil in the boilers. The injection of flue gases along with steam into the reservoir may have some advantages in reducing atmospheric pollution.

In order to develop EOR schemes that are inherently environmentally friendly, technically effective, and socially responsible, the following steps should be taken (Khan and Islam 2007a):

1. Set up an interdisciplinary team (engineers, scientists, economists, and even social scientists).
2. The problems need to be openly discussed in the presence of top executives and policymakers before solutions can be addressed.

3. Ask each participant to propose his/her own solution to the problem. At least one solution per person is ideal. This should apply to every participant, including those who are from social science or other non-technical disciplines.
4. Document multiple solutions for each problem.
5. Evaluate each solution objectively, irrespective of who is proposing it.
6. Evaluate the cost of the *status quo*.
7. Use the screening criterion of Khan and Islam (2006) to evaluate the long-term benefit/cost of a particular solution.
8. List all of the waste materials naturally generated in a particular project.
9. Select an injection fluid from the waste products (point 8).
10. If a particular solution is not fit for a specific field, investigate the possibility of using that solution in a different field.
11. Develop scaling criteria for each solution.
12. Conduct scaled model experiments using scaling groups that are the most relevant.

14

Deconstruction of Engineering Myths Prevalent in the Energy Sector

14.1 Introduction

In Chapter 2, fundamental misconceptions in science (as in process) were deconstructed. Based on those misconceptions, a series of engineering myths have evolved in the energy sector, with an overwhelming impact on modern civilization. This chapter is dedicated to highlighting those myths and deconstructing them.

14.1.1 How Leeches Fell out of Favor

Over four years ago, the Associated Press reported that the FDA had approved the use of leeches for medicinal purposes (FDA 2004). The practice of drawing blood with leeches, which is thousands of years old, finally redeemed itself by being approved by the FDA. This practice of bloodletting and amputation reached its height of medicinal use in the mid-1800s. How and why did this perfectly natural engineering solution to skin grafting and reattachment surgery fall out of favor? Could it be that leeches couldn't be "engineered?"

In the fall of 2007, as the time for Nobel Prize awards approached, a controversy broke. Dr. James Watson, the European-American who won the 1962 Nobel Prize for his role in discovering the double-helix structure of DNA, created the most widely publicized firestorm in the middle of the Nobel Prize awards month (October 2007). He declared that he personally was "inherently gloomy about the prospect of Africa" because "all our social policies are based on the fact that their intelligence is the same as ours, whereas all the testing says 'not really.'" Here, we see the clash between a first premise and a conclusion based on a different premise. "Their intelligence is the same as ours" stems from the unstated premise that "all humans are created equal," a basic tenet of the "nature is perfect" mantra. "All testing" to which Watson refers, on the other hand, is based on the premise that the theory of molecular genetics/DNA (which is linked with an essentially eugenic outlook) is true.

The entire controversy, however, revolved around whether Dr. Watson is a racist. No one seemed interested in addressing the root cause of this remark, namely, an unshakeable conviction that new science represents incontrovertible truth. This faith has the same fervor as those who once thought and disallowed any other theory regarding the earth being flat.

Consider the apparently magical symmetry of the shapes perpetrated as the "double-helix" structure of DNA. These representations of the "founding blocks" of genes are aphenomenal; they are not consistent with the more detailed descriptions of the different bonding strengths of different amino-acid pairings in the actual molecule. Much as atoms were considered to be the founding block of matter (which was incidentally also rendered with an aphenomenal structure, one that could not exist in nature), these "perfectly" shaped structures are being promoted as founding blocks of a living body. It is only a matter of time before we find out just how distant the reality is from these renderings. The renderings themselves, meanwhile, are aphenomenal, meaning they do not exist in nature (Zatzman and Islam 2007). This is a simple logic that the scientific world, obsessed with tangibles, seems to not understand. Only a week before the Watson controversy unraveled, Mario R. Capecchi, Martin J. Evans, and Oliver Smithies received Nobel Prizes in Medicine for their discovery of "principles for introducing specific gene modifications in mice by the use of embryonic stem cells." What is the first premise of this discovery? Professor Stephen O'Rahilly of the University of Cambridge said, "The development of gene targeting technology in the mouse has had a profound influence on medical research... Thanks to this technology we have a much better understanding of the function of specific genes in pathways in the whole organism and a greater ability to predict whether drugs acting on those pathways are likely to have beneficial effects in disease." (BBC 2007) No one seems to ask why only "beneficial effects" should be anticipated from the introduction of "drugs acting on those pathways." When did intervention in nature, meaning at this level of very real and even profound ignorance about actual pathways, yield any beneficial result? Can one example be cited from the history of the world since the Renaissance?

In 2003, a Canadian professor of medicine with over 30 years of experience was asked, "Is there any medicine that cures any disease?" After thinking for some time, he replied, "Yes. It is penicillin." Then, he was asked, "Why then do doctors tell you these days, 'don't worry about this antibiotic, it has no penicillin?'" "Oh, that's because nowadays we make penicillin artificially (synthetically)," the medicine professor quickly replied.

Do we have a medication today that is not made synthetically? Of course, today, synthetically-manufactured drugs monopolize FDA approvals. The medicinal value of drugs produced other than by an approved factory process are declared of "no medicinal value" or are otherwise extremely negatively qualified. So, when this "miracle drug" becomes a big problem after many decades of widespread application, can we then say that the big problem of penicillin highlights an inherent flaw of the mold, *Penicillium notatum*, which was used some 80 years ago to isolate penicillin?

In 2007, the same question went to a Swedish American doctor. He could not name one medicine that actually cures anything. When he was told about the Canadian doctor's comment about penicillin, he quickly responded, "I was going to use the example of penicillin to say that medicines don't work." It is increasingly becoming clear that synthetic drugs (the only kind that is promoted as "medicinal") do not cure and in fact are harmful. Every week, study after study comes out to this effect. During the week of October 14, 2007, it was publicly recommended that children under six not be given any cough syrup. When will be able to upgrade safe use to, say, 96 years of age?

The truth is that new science has created a wealth of the most tangible kind. This wealth is based on perpetrating artificial products in the shortest-term interests of value-addition. "Artificial" always was and always remains starkly opposite to real, the essence of truth and knowledge. New science generated a vast wealth of techniques for keeping the production of certain desired effects from intervening in the natural environment, but only incidentally has this uncovered knowledge about the actual pathways of nature that people did not possess before.

In this sense, what knowledge has new science created? Before anyone gloats about the success of the 2007 Nobel Prize-winning discovery, think back 60 years. In 1947, another Nobel Prize in the same discipline was announced. The Nobel committee declaration was as follows:

> "Paul Müller went his own way and tried to find insecticides for plant protection. In so doing he arrived at the conclusion that for this purpose a contact insecticide was best suited.
>
> Systematically he tried hundreds of synthesized organic substances on flies in a type of Peet-Grady chamber. An article by the Englishmen Chattaway and Muir gave him the idea of testing combinations with the CCl_3 groups, and this then finally led to the realization that dichloro-diphenyl-trichloro-methylmethane acted as a contact insecticide on Colorado beetles, flies, and many other insect species under test. He determined its extraordinary persistence, and simultaneously developed the various methods of application such as solutions, emulsions, and dusts.
>
> In trials under natural conditions Müller was able to confirm the long persistent contact action on flies, Colorado beetles, and gnats (Culex).
>
> Recognition of the intense contact activity of dichloro-diphenyl-trichloromethylmethane opened further prospects. Indeed, the preparation might be successfully used in the fight against bloodsucking and disease-carrying insects such as lice, gnats, and fleas – carriers incapable of being reached by oral poisons. In the further trials now conducted, DDT showed a very large number of good properties. At requisite insecticidal dosages, it is practically non-toxic to humans and acts in very small dosages on a large number of various species of insects. Furthermore, it is cheap, easily manufactured, and exceedingly stable. A surface treated with DDT maintains its insecticidal properties for a long time, up to several months." (Nobel Prize presentation)

Each of Paul Müller's premises, as stated above, was false. Today, Professor Yen wrote, "Every meal that we take today has DDT in it."

Had Dr. Müller acted on knowledge rather than a short-term instinct of making money, he would have realized that "practically non-toxic to humans" was a scientific fraud. How does this premise differ from that of James Watson or the trio who received the Nobel Prize sixty years after Paul Müller?

There is a widespread belief that the Nobel awards systematically acknowledge transformations of humanity to higher levels of civilization, which are affected as the result of unique contributions of individuals in the sciences and/or at the level of global public consensus-building (e.g., the peace prize). While the Nobel awards process has indeed always been about selecting unique individuals for distinction, the transformations supposedly affected have been another matter entirely. Enormous quantities and layers of disinformation surround the claims made for the transformative powers of these individuals' work. Far from being culturally or socially transformative, prize-winning work, whether it was work underpinning new life-saving medical technologies or work in the physical sciences underpinning advances in various engineering fields, the awards have been intimately linked with upholding and/or otherwise extending critically valuable bits of the political and/or economic *status quo* for the powers-that-be. The DDT example cited above is particularly rich not only in its upholding of an aphenomenal model of "science," but also, and certainly not least, because of the context in which it was found to have been proven so valuable. According to the Nobel Committee's presentation in 1948, Dr Müller's DDT work possessed the following:

> "...a short and crowded history which, from the medical point of view, is closely connected with the fight against typhus during the last World War. In order to give my presentation the correct medical background, I will first mention one or two points concerning this disease.
>
> "Typhus has always occurred as a result of war or disaster and, hence, has been named "Typhus bellicus," "war-typhus," or "hunger-typhus." During the Thirty Years' War, this disease was rampant, and it destroyed the remains of Napoleon's Grand Army on its retreat from Russia. During the First World War, it again claimed numerous victims. At that period more than ten million cases were known in Russia alone, and the death rate was great. Admittedly, the famous Frenchman Nicolle had already, in 1909, shown that the disease was practically solely transmitted by lice, for which discovery he received

the Nobel Prize and, thus, paved the way for effective control. But really successful methods for destroying lice in large quantities, thus removing them as carriers, were not yet at hand.

"Towards the end of the Second World War, typhus suddenly appeared anew. All over the world research workers applied their energies to trying to discover an effective delousing method. Results, however, were not very encouraging. In this situation, so critical for all of us, deliverance came. Unexpectedly, dramatically, practically out of the blue, DDT appeared as a deus ex machina...

"A number of Swiss research workers such as Domenjoz and Wiesmann ... concerned themselves with further trials of the substance. Mooser's researches aimed directly at a prophylaxis of typhus. On the 18th of September 1942, he gave a significant lecture to the physicians of the Swiss First Army Corps, on the possibilities of protection against typhus by means of DDT.

"At that time, the Allied Armies of the West were struggling with severe medical problems. A series of diseases transmittable by insects, diseases such as typhus, malaria and sandfly fever, claimed a large number of victims and interfered with the conduct of the War. The Swiss, who had recognized the great importance of DDT, secretly shipped a small quantity of the material to the United States. In December of 1942 the American Research Council for Insectology in Orlando (Florida) undertook a large series of trials which fully confirmed the Swiss findings. The war situation demanded speedy action. DDT was manufactured on a vast scale whilst a series of experiments determined methods of application. Particularly energetic was General Fox, Physician-in-Chief to the American forces.

"In October of 1943 a heavy outbreak of typhus occurred in Naples and the customary relief measures proved totally inadequate. General Fox thereupon introduced DDT treatment with total exclusion of the old, slow methods of treatment. As a result, 1,300,000 people were treated in January 1944 and in a period of three weeks the typhus epidemic was completely mastered. Thus, for the first time in history a typhus outbreak was brought under control in winter. DDT had passed its ordeal by fire with flying colors." (Quote from Nobel Prize presentation speech)

In 1942, it was becoming crucial for the Americans and the British to find ways to preserve such limited troops because they then had deployed against the Third Reich, but why? The Red Army had not yet broken the back of von Paulus' Sixth Army at Stalingrad, but it was already apparent that the Soviet Union was not going to be defeated by the Nazi invader. A growing concern among U.S. and British strategic planners at the time was that the Red Army could eventually break into eastern Europe before the Americans, and the British were in a position to extract a separate peace from the Hitler regime on the western front. By capturing control of the Italian Peninsula before the Red Army reached the Balkan Peninsula, the Anglo-Americans would be ready to confront the Red Army's arrival on the eastern side of the Adriatic (Cave Brown 1975).

From 1933 to early 1939, the leading western powers bent over backwards to accommodate Italian fascist and German fascist aggression and subversion in Europe. Their hope was to strike a mutual deal that would maintain a threatening bulwark against their common enemy of that time, the Soviet Union (Taylor 1961). Awarding the 1938 Nobel Peace Prize to the Nansen Office, a body that had worked extensively with Germans refugeed by the misfortunes of the First World War (1914–18) in central Europe, was clearly designed to serve that purpose. At this point, the Third Reich had swallowed the German-speaking region of Czechoslovakia, the Sudetenland, and incorporated it into the Reich. The *Anschlüss* annexing Austria to the Third Reich had been set up with the ruling circles in Vienna in the name of protecting German nationality and national rights from "Slavic race pollution." Actions to "protect" German minorities in Poland were publicly placed on the agenda by the Hitler regime (Finkel and Leibovitz 1997).

In brief, the track record of the Nobel awards, in general, has been to uphold the aphenomenal approach to the knowledge of nature and the truth in the sciences and to uphold only those "transformations" that bolster the status quo. This is precisely how and why the *imprimatur* of the Nobel Committee ends up playing such a mischievous role. Consider what happened within days of the Watson imbroglio. Many may have thought it ended with Watson's suspension from the Cold Spring Harbor Laboratory for his remarks. Others may have thought it ended when he retired permanently a week following his suspension. Less than 48 hours following Dr. Watson's retirement announcement, however, the weekly science column of *The Wall Street Journal* demonstrated the matter was far from over (Hotz 2007):

> "...Whatever our ethnic identity, we share 99% of our DNA with each other.

Yet, in that other one percent, researchers are finding so many individual differences they promise to transform the practice of medicine, enabling treatments targeted to our own unique DNA code. Viewed through the prism of our genes, we each have a spectrum of variation in which a single molecular misstep can alter the risk of disease or the effectiveness of therapy…

Scientists and doctors struggle for ways to translate the nuances of genetic identity into racially defined medical treatments without reviving misconceptions about the significance, for example, of skin color.

The problem arises because it may be decades before anyone can afford their own genetic medical profile. Meanwhile, doctors expect to rely on racial profiling as a diagnostic tool to identify those at genetic risk of chronic diseases or adverse reactions to prescription drugs.

Researchers at Brown University and the University of London, and the editors of the journal *PLoS Medicine*, last month warned about inaccurate racial labels in clinical research. In the absence of meaningful population categories, researchers may single out an inherited racial linkage where none exists, or overlook the medical effects of our environment…

It's not the first time that medical authorities have raised a red flag about racial labels. In 1995, the American College of Physicians urged its 85,000 members to drop racial labels in patient case studies because 'race has little or no utility in careful medical thinking.' In 2002, the editors of the *New England Journal of Medicine* concluded that '"race" is biologically meaningless.' And in 2004, the editors of *Nature Genetics* warned that "it's bad medicine and it's bad science."

No one denies the social reality of race, as reinforced by history, or the role of heredity. At its most extreme, however, the concept of race encompasses the idea that test scores, athletic ability, or character is rooted in the genetic chemistry of people who can be grouped by skin color. That's simply wrong, research shows. Indeed, such outmoded beliefs led to the resignation Thursday of Nobel laureate James Watson from the Cold Spring Harbor Laboratory in New York, for disparaging comments he made about Africans.

Diseases commonly considered bounded by race, such as sickle cell anemia, are not…

Researchers studying physical appearance and genetic ancestry in Brazil, for example, discovered that people with white skin owed almost a third of their genes, on average, to

African ancestry, while those with dark skin could trace almost half of their genes to Europe, they reported in the Proceedings of the National Academy of Sciences. 'It's clear that the categories we use don't work very well,' said Stanford University biomedical ethicist Mildred Cho.

Government reporting requirements in the U.S. heighten the difficulty. Since 2001, clinical researchers must use groupings identified by the U.S. Census that don't recognize the underlying complexities of individual variation, migration and family ancestry. In addition, medical reports in the PubMed, Medline and the U.S. National Library of Medicine databases were cataloged until 2003 by discredited 19th-century racial terms.

Today, there are no rigorous, standardized scientific categories. A recent study of 120 genetics and heredity journals found that only two had guidelines for race and ethnic categories, though half of them had published articles that used such labels to analyze findings.

Eventually, genomics may eliminate any medical need for the infectious shorthand of race. 'We need to find the underlying causes of disease,' said David Goldstein, at Duke University's Center for Population Genomics & Pharmacogenetics. 'Once we do, nobody will care about race and ethnicity anymore.'

No one dares stop the merry-go-round to point out that this is all molecular eugenics, *not* molecular genetics. There is something fundamentally intriguing and deficient about connecting everything humans can become entirely to genetic/genomic inheritance. Unfortunately, and contrary to its false promise to figure out the science of the building blocks of life itself, this pseudoscience has developed by using the genome as tangible evidence for all of the assumptions of eugenics regarding inferiority and superiority. These may be building blocks, but then there is the mortar, not to mention the pathway and the time factor involved by which a living organism emerges seemingly from "lifeless matter."

These are matters of grave consequence on which men and women of science should take a clear stand.

14.1.2 When did Carbon Become the Enemy?

An aphenomenal first premise does not increase knowledge, no matter through how many steps we take the argument. We can cite numerous examples to validate this statement. In this, it is neither necessary nor sufficient to argue about the outcome or final conclusion. For instance, we may not be able to discern between two identical twins, but that is our shortcoming (because we spuriously consider DNA tests the measure of identity), and that does not make them non-unique. Two molecules of water may appear to be identical to us, but that is because we wrongly assumed molecules are made of spherical, rigid, non-breakable particles, called atoms. In addition, the aphenomenal features inherent to atom (e.g., rigid, cylindrical, uniform, symmetric, and so forth) and the presumed fundamental unit of mass became the norm of all subsequent atomic theories. Not only that, even light and energy units became associated with such aphenomenal features and were the predominant concepts used in electromagnetic theory as well as in the theory of light (Zatzman *et al.* 2008a, 2008b). The false notion of atoms as fundamental building blocks is at the core of modern atomic theory and is responsible for much of the confusion regarding many issues, including blaming carbon for global warming, blaming natural fat for obesity, sunlight for cancer, and numerous others. Changing the fundamental building block from real to aphenomenal makes all subsequent logic go haywire. Removing this fundamentally incorrect first premise would explain many natural phenomena. It would explain why microwave cooking destroys the goodness of food while cooking in woodstoves improves it, why sunlight is the essence of light and fluorescent light is the essence of death, why moonlight is soothing and can improve vision while dim light causes myopia, why paraffin wax candles cause cancer while beeswax candles improve lungs, why carbon dioxide from "refined oil" destroys greenery (global warming) and that from nature produces greenery, and the list truly goes on. However, this is not the only focus of new science that is motivated by money and employs only the short-term or myopic approach.

In 2003, Nobel Chemistry Laureate Robert Curl characterized the current civilization as a "technological disaster." We do not have to look very far to discover how correct this Nobel Prize winner's analysis is, or why. Every day, headlines appear that show, as a human race, we have never had to deal with more unsustainable technologies. How can this technology development path be changed? We do not expect to change the outcome without changing the origin and pathway of actions. The origin of actions is intention. So, for those who are busy dealing with "unintended consequences," we must say that there is no such thing as "unintended." Every action has an intention and every human being has an option of acting on conscience or "desire." Theoretically, the former symbolizes long-term (hence, true) intentions, and the latter symbolizes short-term (hence, aphenomenal) intentions. Zatzman and Islam (2007) explicitly pointed out the role of intentions on social change and technology. Arguing that the role of intentions, i.e., of the direction in which the scientific research effort was undertaken in the first place, has not been sufficiently recognized, they proposed explicitly to include the role of intentions in the line of social change. This theory was further incorporated by Chhetri and Islam (2008) with technology development. They argued that no amount of disinformation can render a technology sustainable that was not well-intended or was not directed at improving the overall human welfare, or what many today call "humanizing the environment." It is a matter of recognizing the role of intentions and being honest at the starting point of a technology's development. For science and knowledge, honesty is not just a buzzword. It is the starting point. Chhetri and Islam (2008) undertook the investigation that gave rise to their work because they could find no contemporary work clarifying how the lack of honesty could launch research in the direction of ignorance. From there they discussed how, given such a starting point, it is necessary to emulate nature in order to develop sustainable technologies for many applications, ranging from energy to pharmacy and health care.

All aphenomenal logic promoted in the name of science and engineering has an aphenomenal motive behind it.

Recently, Chilingar and associates presented a scientific discourse on the topic of global warming (Sorokhtin *et al.* 2007). Because the book gives the impression that global warming cannot be caused by human activities, it has the potential of alienating readers that are interested in preserving the environment. However, the book is scientifically accurate and the conclusions are the most scientific, based on new science. Chilingar and his associates are scientifically accurate, yet they made a serious mistake in ignoring some facts. As it turns out, this aspect is not considered by anyone, including those who consider themselves the pioneers of the pro-environment movement. When these facts are considered, the theory of global warming becomes truly coherent, devoid of doctrinal slogans.

14.2 The Sustainable Biofuel Fantasy

Biofuels are considered to be inherently sustainable. Irrespective of what source was used and what process was followed to extract high-value fuels, they are promoted as clean fuels. In this section, this myth is deconstructed.

14.2.1 Current Myths Regarding Biofuel

Increasing uncertainty in global energy production and supply, environmental concerns due to the use of fossil fuels, and high prices of petroleum products are considered to be the major reasons to search for alternatives to petrodiesel. For instance, Lean (2007) claimed that the global supply of oil and natural gas from the conventional sources is unlikely to meet the growth in energy demand over the next 25 years. As a result of this cognition, biofuels are considered to be sustainable alternatives to petroleum products. Because few are accustomed to questioning the first premise of any of these conclusions, even the ardent supporters of the petroleum industry find merit in this conclusion. Considerable funds have been spent in developing biofuel technology, and even the mention of negative impacts of food (e.g., corn) being converted into fuel was considered to be anti-civilization. It is assumed that biodiesel fuels are environmentally beneficial (Demirbas 2003). The argument put forward is that plant and vegetable oils and animal fats are renewable biomass sources. This argument follows other supporting assertions, such as the idea that biodiesel represents a closed carbon dioxide cycle because it is derived from renewable biomass sources. Biodiesel has a lower emission of pollutants compared to petroleum diesel. Plus, it is biodegradable, and its lubricity extends engine life (Kurki *et al.* 2006) and contributes to sustainability (Khan *et al.* 2006; Kurki *et al.* 2006). Biodiesel has a higher cetane number than diesel fuel, no aromatics, no sulfur, and contains 10–11% oxygen by weight (Canakci 2007).

Of course, negative aspects of biofuels are also discussed. For instance, it is known that the use of vegetable oils in the compression ignition engines can cause several problems due to its high viscosity (Roger and Jaiduk 1985). It is also accepted that the use of land for the production of edible oil for biodiesel feedstock competes with the use of land for food production. Moreover, the price of edible plant and vegetable oils is considered to be higher than petrodiesel. Based on this argument, alarms were sounded when oil prices dropped in fall 2008, as though a drop in petroleum fuels would kill the "environmentally friendly" biofuel projects, thereby killing the prospect of a clean environment. As a remedy to this unsubstantiated and aphenomenal conclusion, waste cooking oils and non-edible oils are promoted to take care of the economic concerns. It is known that the use of waste cooking oil as biodiesel feedstock reduces the cost of biodiesel production (Canakci 2007), since the feedstock costs constitutes approximately 70–95% of the overall cost of biodiesel production (Connemann and Fischer 1998).

14.2.2 Problems with Biodiesel Sources

The main feedstocks of biodiesel are vegetable oils, animal fats, and waste cooking oil. These are the mono alkyl esters of fatty acids derived from vegetable oil or animal fat. The fuels derived may be alcohols, ethers, esters, and other chemicals made from cellulosic biomass and waste products, such as agricultural and forestry residues, aquatic plants (microalgae), fast growing trees and grasses, and municipal and industrial wastes.

Subramanyam *et al.* (2005) reported that there are more than 300 oil-bearing crops identified that can be utilized to make biodiesel. Beef and sheep tallow, rapeseed oil, sunflower oil, canola oil, coconut oil, olive oil, soybean oil, cottonseed oil, mustard oil, hemp oil, linseed oil, microalgae oil, peanut oil, and waste cooking oil are considered potential alternative feedstocks for biodiesel production (Demirba 2003). However, the main sources of biodiesel are rapeseed oil, soybean oil, and, to a certain extent, animal fat, with rapeseed accounting for nearly 84% of the total production (Demirba 2003). Henning (2004) reported that *Jatropha curcus* also has a great potential to yield biodiesel. The UK alone produces about 200,000 tons of waste cooking oil each year (Carter *et al.* 2005). This provides a good opportunity to utilize waste into energy.

Various types of algae, some of which have an oil content of more than 60% of their body weight in the form of tryacylglycerols, are the potential sources for biodiesel production (Sheehan *et al.* 1998). Many species of algae can be successfully grown in wastewater ponds and saline water ponds, utilizing CO_2 from power plants as their food. Utilizing CO_2 from power plants to grow algae helps to sequester CO_2 for productive use, and at the same time reduces the build up of CO_2 in the atmosphere. Waste cooking oil is also considered a viable option for biodiesel feedstock. Even though the conversion of waste cooking oil into usable fuel has not been in practice at a commercial level, the potential use of such oil can solve two problems: 1) environmental problems caused by its disposal to water courses; and 2) problems related to competition with food sources.

Because the pathway is not considered in conventional analysis, the role of the source or the processes involved is not evident. If the pathway were to be considered, it would become evident that biodiesel derived from genetically modified crops cannot be considered equivalent to biodiesel derived from organically grown crops. Recently, Zatzman *et al.* (2009) outlined the problems, much of which are not detectable with conventional means associated to genetic engineering. While genetic engineering has increased tangible gains in terms of crop yield and the external appeal of the crop (symmetry, gloss, and other external features), it has also added potential fatal, unavoidable side effects. In the context of honey bees, the most important impact of GE is through direct contact of genetically altered crops (including pollen) and through the plant-produced matters (including even organic pesticide and fertilizers). A series of scholarly publications have studied the effects of GE products on honey bees. Malone and Pham-Delègue (2001) studied the effects of transgenic products on honey bees and bumblebees. Obrycki *et al.* (2001) studied genetically engineered insecticidal corn that might have severe impacts on the ecosystem. Pham-Delègue *et al.* (2002) produced a comprehensive report in which they attempted to quantify the impacts of genetically modified plants on honey bees. Similarly, Picard-Nioi *et al.* (1997) reported the impacts of proteins used in genetically engineered plants on honey bees. The need for including non-target living objects was highlighted by Losey *et al.* (2004).

It is true that genetic engineering activities have been carried out at a pace unprecedented for any other technology. This subject has also been hailed to have made the most significant breakthroughs. Unfortunately, these "breakthroughs" only bear fruit in the very short term, within which period the impacts of these technologies do not manifest in measurable (tangible expression) fashion. Even though there is a general recognition that there are "unintended consequences," the science behind this engineering has never been challenged. Often these "unintended consequences" are incorrectly attributed to the lack of precision, particularly in placing the location of the DNA in the new chromosome site. The correct recognition would be that it is impossible to engineer the new location of the gene, and at the same time it is impossible to predict the consequences of the DNA transfer without knowing all possible sites that the DNA will travel to throughout the time domain. Khan (2006) made this simple observation and contended that, unless the consequences are known for the time duration of infinity, an engineering practice cannot be considered sustainable. Similar, but not as bold, statements were previously made by Schubert (2005), who questioned the validity of our understanding of genetic engineering technology and recognized the unpredictability of the artificial gene. Zatzman and Islam (2007a) recognized that an "artificial" object, even though it comes to reality by its mere presence, behaves differently than the object it was supposedly emulating. This explains why vitamin C acts differently depending on its origin (e.g., organic or synthetic), and so does every other artificial product including antibiotics (Chhetri *et al.*, 2007; Chhetri and Islam, 2007).

Similar statements can be made about chemical fertilizers and pesticides that are used to boost crop yield as well as hormones and other chemicals that are used on animals. Therefore, biodiesel derived from organic crops and biodiesel derived from genetically modified crops infested with chemical fertilizer and pesticides would have quite different outputs to the environment, thereby affecting the sustainability picture. Similarly, if the source contains beef tallow from a cow that is injected with hormones and fed artificial feeds, the resulting biodiesel will be harmful to the environment, and could not be compared to petrodiesel that is derived from fossil fuel. Note that the fossil fuel was derived from organic matters, with the exception that nature processed the organic matter to pack it with a very high energy content. If the first premise is that nature is sustainable, then fossil fuel offers much greater hope of sustainability than contemporary organic sources that are infested with chemicals that were not present even 100 years ago.

14.2.3 The Current Process of Biodiesel Production

Recently (Chhetri and Islam 2008b) detailed the process involved in biodiesel production. Conventionally, biodiesel is produced either in a single-stage or a double-stage batch process, or by a continuous flow type transesterification process. These are either acid catalyzed or base catalyzed processes. The acids generally used are sulfonic acid and sulfuric acid. These acids give very high yields in alkyl esters, but these reactions are slow, requiring high temperatures above 100°C and more than three hours to complete the conversion (Schuchardt 1998). Alkali catalyzed transesterification is much faster than acid catalyzed transesterification, and all commercial biodiesel producers prefer to use this process (Ma and Hanna 1999). The alkalis generally used in the process include NaOH, KOH, and sodium methoxide. For an alkali-catalyzed transesterification, the glycerides and alcohol must be anhydrous because water changes the reaction, causing saponification. The soap lowers the yield of esters and makes the separation of biodiesel and glycerin complicated.

No catalysts are used in the case of supercritical methanol methods, where a methanol and oil mixture is superheated to more than 350°C, and the reaction completes in 3–5 minutes to form esters and glycerol. Saka and Kudsiana (2001) carried out a series of experiments to study the effects of the reaction temperature, pressure, and molar ratio of methanol to glycosides in methyl ester formation. Their results revealed that supercritical treatment of 350°C, 30 MPa, and 240 seconds with a molar ratio of 42 in methanol is the best condition for transesterification of rapeseed oil for biodiesel production. However, the use of methanol from fossil fuel still makes the biodiesel production process unsustainable and produces toxic byproducts. Since this process uses high heat as catalysts, then the use of electricity to heat the reactants will increase the fossil fuel input in the system. The direct heating of waste cooking oil by solar energy using concentrators will help to reduce the fossil fuel consumption, and the process will be a sustainable option.

The major objectives of transesterification are to break down the long chain of fatty acid molecules into simple molecules and reduce the viscosity considerably in order to increase the lubricity of the fuel. The transesterification process is the reaction of a triglyceride (fat/oil) with an alcohol, using a catalyst to form esters and glycerol. A triglyceride has a glycerin molecule as its base with three long chain fatty acids attached. During the transesterification process, the triglyceride is broken down with alcohol in the presence of a catalyst, which is usually a strong alkaline like sodium hydroxide. The alcohol reacts with the fatty acids to form the mono-alkyl ester, or biodiesel and crude glycerol. In most production, methanol or ethanol is the alcohol used, and it is base catalyzed by either potassium or sodium hydroxide. After the completion of the transesterification reaction, the glycerin and biodiesel are separated (gravity separation). The glycerin is either reused as feedstock for methane production, or refined and used in pharmaceutical products. A typical transesterification process is given in Figure 14.1.

Interestingly, the current biodiesel production uses fossil fuel at various stages, such as agriculture, crushing, transportation, and the process itself (Carraretto *et al.* 2004). Figure 14.2 shows the share of energy use at different stages from farming to biodiesel production. Approximately 35% of the primary energy is consumed during the life cycle from agriculture farming to biodiesel production. This energy basically comes from fossil fuel. To make the biodiesel completely green, this portion of energy also has to be derived from renewable sources. For energy conversion and crushing, direct solar energy can be effectively used, while renewable biofuels can be used for transportation and agriculture.

14.2.3.1 *Pathways of Petrodiesel and Biodiesel*

Figure 14.3 below shows the pathways of additives for the production of conventional petrodiesel and biodiesel. Highly toxic

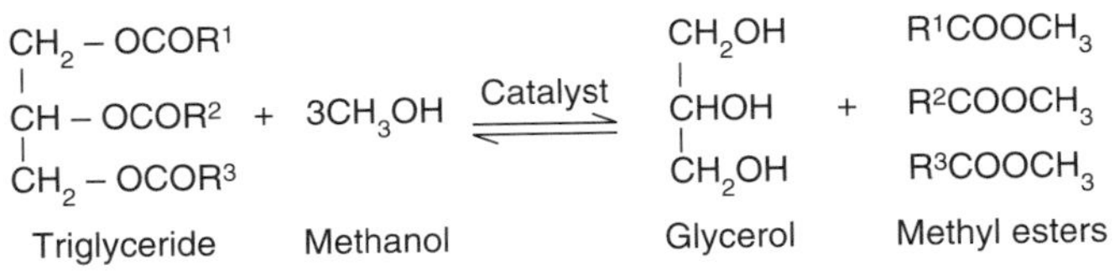

Triglyceride		Methanol		Glycerol		Methyl esters
CH_2-OCOR^1				CH_2OH		R^1COOCH_3
$CH-OCOR^2$	+	$3CH_3OH$	$\xrightleftharpoons{\text{Catalyst}}$	$CHOH$	+	R^2COOCH_3
CH_2-OCOR^3				CH_2OH		R^3COOCH_3

Figure 14.1 Typical equation for transesterification, where R1, R2, and R3 are the different hydrocarbon chains.

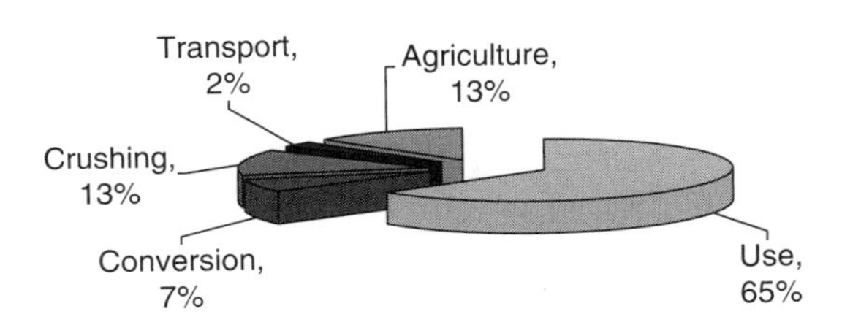

Figure 14.2 Share of energy at different stages of biodiesel production.

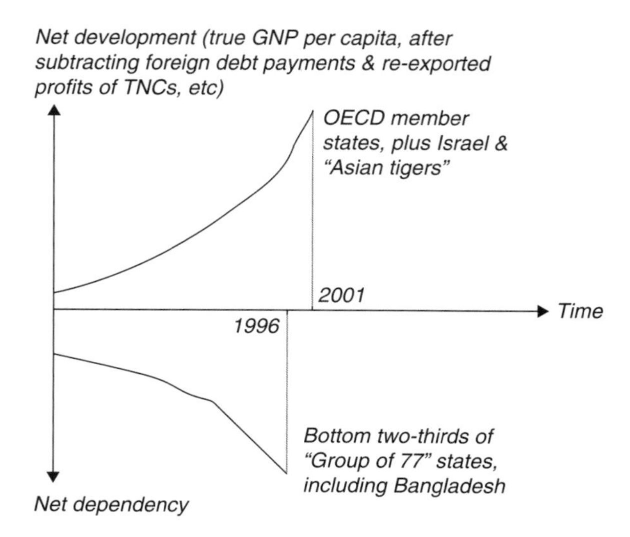

Figure 14.3 Pathways of petrodiesel and biodiesel production.

catalysts, chemicals, and excessive heat are subjected to crude oil in the oil refineries. Chemicals such sulfuric acid (H_2SO_4), hydrofluoric acid (HF), aluminium chloride ($AlCl_3$), and aluminium oxide (Al_2O_3), and catalysts such as platinum and high heat are applied for oil refining. These all are highly toxic chemicals and catalysts. The crude oil that is originally nontoxic yields a more toxic product after refining. Conventional biodiesel production follows a similar pathway. Methanol (CH_3OH) made from natural gas, a highly toxic chemical that kills receptor nerves even without feeling pain, is used for the alcoholysis of vegetable oil. Sulfuric acid or hydroxides of potassium or sodium, which are highly toxic and caustic, are added to the natural vegetable oils.

14.2.3.2 Biodiesel Toxicity

The toxicity of biodiesel is measured by the fuel toxicity to the human body and by the health and environmental impacts due to exhaust emission. Tests conducted for acute oral toxicity of a pure biodiesel fuel and a 20% blend (B20) in a single-dose study on rats reported that the LD50 of pure biodiesel, as well as B20, was found to be greater than 5000 mg/kg (Chhetri *et al.*, 2008). Hair loss was found on one of the test samples in the B20. The acute dermal toxicity of neat biodiesel tested for LD50 was greater than 2000 mg/kg.

The United States Environmental Protection Agency (2002) studied the biodiesel effects on gaseous toxics and listed 21 Mobile Source Air Toxics (MSATs) based on that study. MSATs are significant contributors to the toxic emissions and are known or suspected to cause cancer or other serious health effects. Of the 21 MSATs listed, six are metals. Of the remaining 14 MSATs, the emission measurements were performed for the eleven components, namely, acetaldehyde, acrolein, benzene, 1,3-butadiene, ethylbenzene, formaldehyde, n-hexane, naphthalene, styrene, toluene, and xylene. However, the trend in benzene, 1–3-butadiene, and styrene was inconsistent.

It is obvious that the biodiesel produced from the current methods is highly toxic. This is because of the highly toxic chemicals and catalysts used. The current vegetable oil extraction method utilizes n-hexane, which also causes toxicity. This research proposes a new concept that aims at reducing the toxicity by employing nontoxic chemicals, natural catalysts, and the extraction process itself.

Toxicity can emerge from the source as well as the process that is used to produce biodiesel. The amount of catalysts had an impact in the conversion of esters during the transesterification process. The titration indicated that the optimum amount of catalysts for the particular waste cooking oil was 8 grams per liter of oil (Chhetri *et al.* 2008). They carried the reaction using 4, 6, 8, 10, and 12 grams per liter of sodium hydroxide catalyst. With 4 grams per liter, no reaction was observed, because there was no separated layer of ester and glycerin. With the concentrations of 6, 8, and 10 grams per liter, approximately 50%, 94%, and 40% ester yield, respectively, were obtained (Figure 14.4). It was observed that the production of ester decreased with the increase in sodium hydroxide. With 12 grams per liter of catalysts, a complete soap formation was observed. This is because the higher amount of catalysts caused soap formation (Attanatho *et al.* 2004). The rise in soap formation made the ester dissolve into the glycerol layer. Triplicate samples were used, and the maximum standard deviation from the mean was found to be approximately 4%.

The transesterification reaction usually takes place in ethanol and oil in two phases, taking more time and energy to complete the reaction. However, Boocock *et al.* (1996, 1998) added tetrahydrofuran as a co-solvent, which transformed two oil-methanol phases into one-phase systems, helping the reaction to occur in normal temperatures at a much faster rate. Because the tetrahydrofuran is

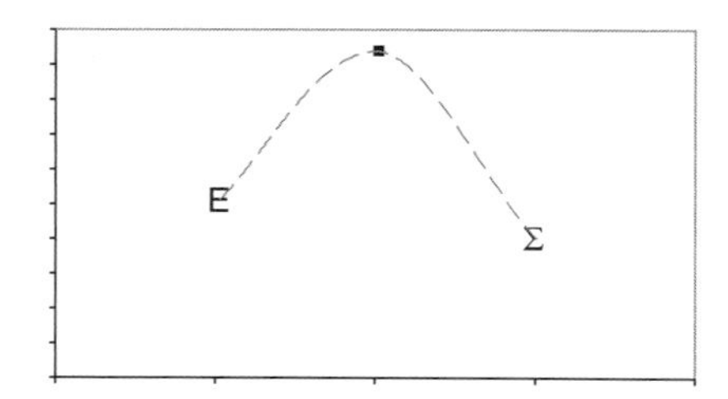

Figure 14.4 Conversion efficiency under different catalysts concentration (Chhetri *et al.* 2008).

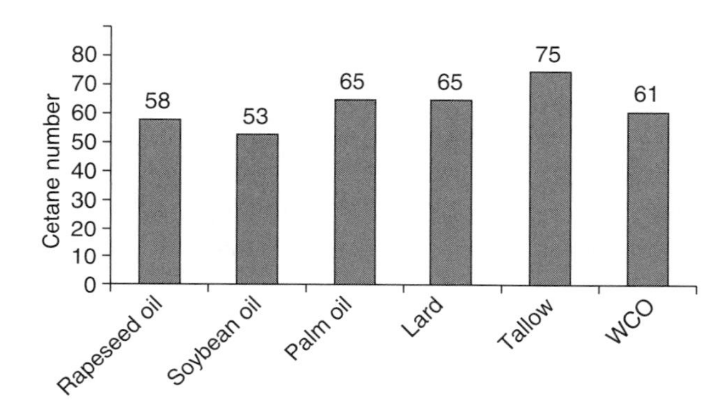

Figure 14.5 Cetane numbers of biodiesel derived from different feedstocks.

a toxic chemical, they suggested that the tetrahydrofuran should be derived from biological sources.

Typically, toxicity can also come from higher cetane number sources. Cetane numbers are the indicators of ignition properties of the diesel fuel. The higher the cetane number, the more efficient the fuel is. Because of higher oxygen content, biodiesel has a higher cetane number compared to petroleum diesel. Rapeseed oil, canola oil, linseed oil, sunflower oil, soybean oil, beef tallow, and lard are being used as feedstock for biodiesel production (Peterson *et al.* 1997; Ma and Hanna 1999).

The cetane number of the diesel fuel is the indicator of the ignition quality. Figure 14.5 shows the cetane numbers of biodiesel fuels derived from different feedstocks. The cetane index of waste cooking oil from the experiment was found to be 61. The cetane number is not much different than the cetane index (cetane number = cetane index-1.5+2.6) (Issariyakul *et al.* 2007). Hilber *et al.* (2006) reported the cetane numbers of rapeseed oil, soybean oil, palm oil, lard, and beef tallow to be 58, 53, 65, 65, and 75, respectively. Among these biodiesel feedstocks, beef tallow has the highest cetane number. The higher cetane number indicates a higher engine performance of beef tallow compared to other fuels, resulting in lower emissions of pollutants. Because beef tallow has the higher amount of saturated fatty acids, the increase in the saturated fatty acids content positively enhances the cetane number of biodiesel. The oxidative stability of biodiesel fuels also increases due to the presence of higher amounts of saturated fatty acids. However, the drawback of higher amounts of saturated fatty acid content in biodiesel fuel is that the cold filter plugging point occurs at a higher temperature.

Kemp (2006) reported the distribution of biodiesel production costs as shown in Figure 14.6. Oil feedstock is the major cost of biodiesel production, accounting for over 70% of the total costs. Hence, if the waste vegetable oil is used as biodiesel feedstock, the economics of biodiesel can be significantly improved. Moreover, the use of waste oil also reduces the waste treatment costs. Over 12% of biodiesel costs are the chemicals that are inherently toxic to the environment. For each gallon of diesel, this cost is higher for biodiesel than for petrodiesel. In addition, because most chemicals are derived from fossil fuel, the cost incurred in biodiesel is more acute to the biodiesel industry. In terms of toxicity, these chemicals offer the same degree of toxicity (if not more) as the petrodiesel, even if the petrodiesel is produced by conventional means. In addition, it is actually easier to produce petrodiesel than biodiesel, because the crude oil is much more amenable to refining than conventional

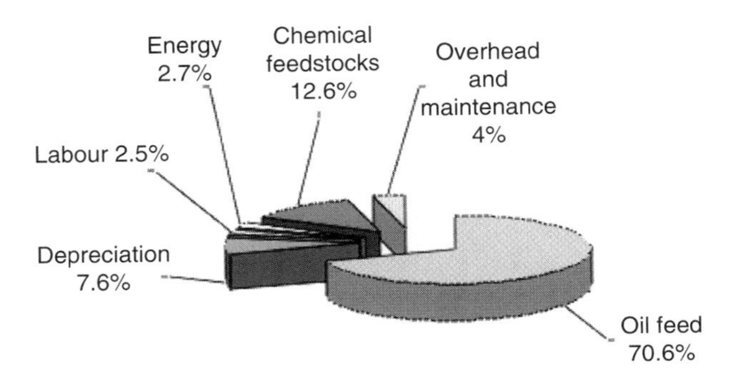

Figure 14.6 Distribution of biodiesel production costs, % (Kemp 2006).

biodiesel sources. The disparity increases if alternative sources are used, such as waste oil, non-edible oil, and so forth.

14.3 "Clean" Nuclear Energy

14.3.1 Energy Demand in Emerging Economies and Nuclear Power

The increasing global energy demand will put great pressure on fossil fuel resources. In order to meet this challenging energy demand, India and China have been attracted toward building nuclear power plants. Recent agreement between India and the United States, in order to develop nuclear power for civil purposes, has opened up an opportunity for India to become a nuclear power intensive country in the region (BBC 2006). As a matter of fact, India already has several nuclear power facilities producing 2550 MWe and 3622 MWe under construction. China has also developed nuclear energy for power generation, and has 5977 MWe as of December 31, 2003. By the end of 2007, 9GWe was attributed to nuclear energy in 11 nuclear power plants (WNA, 2010). Additional reactors are planned, including some of the world's most advanced, to give a sixfold increase in nuclear capacity to at least 60 GWe or possibly more by 2020, and then a further substantial increase to 160 GWe by 2030.

14.3.2 Nuclear Research Reactors

There was a significant research focus over the last 50 years in nuclear energy technologies. According to IAEA (2004), there were 672 research reactors worldwide as of June 2004, all aimed at investigating different aspects of nuclear energy. The research reactors were used for various purposes, including basic nuclear science, material development, and radioisotope management and their application in other fields such as medicine, food industries, and training. Figure 14.7 shows the total number of research reactors commissioned and shut down as of June 2004 (IAEA 2004). Of the 672 research reactors, 274 are operational in 56 countries, 214 are shut down, 168 have been decommissioned, and 16 are planned or under construction. Figure 14.7 shows the decline in the number of research reactors that were operational earlier, which shows an increase in the number

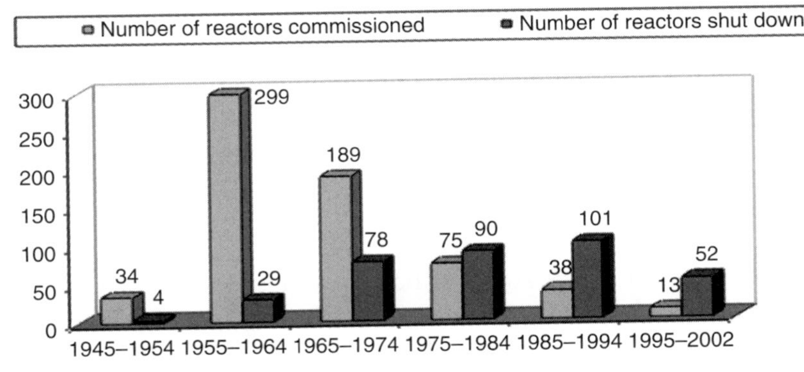

Figure 14.7 Number of research reactors shut down and commissioned (IAEA 2004).

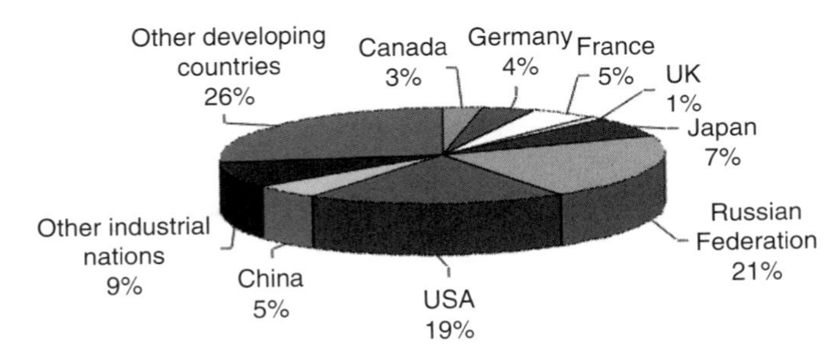

Figure 14.8 Operational research reactors in IAEA Member States (IAEA 2004).

being shut down. This is due to the fairly stable research regime in this technology. There are few research reactors being built, showing that the research opportunity is still available.

Figure 14.8 shows that altogether, 274 research reactors are operational in IAEA member states. Of these reactors, about 70% are in industrialized countries, with the Russian Federation and the United States having the largest number. Most of this research was carried out for fission reaction. Despite the significant amount of research carried out for fusion reaction, there was no positive outcome. Due to the severe environmental consequences created by fission reaction, fusion reaction has been considered to be comparatively less environmentally hazardous. However, scientists are struggling to find ways to carry out fusion reaction in lower temperatures, because currently, it needs huge amounts of energy input, making this technology far from economic realization.

14.3.3 Global Estimated Uranium Resources

The total amount of conventional uranium that can be mined economically is estimated to be approximately 4.7 million tons (IAEA 2005). This amount is considered sufficient for the next 85 years to meet the nuclear electricity generation rate based on 2004. However, if the conventional technology were converted to fast reactor technology, the current resources would be enough for the next hundreds of years (Sokolov 2006). It has further been reported that, if the uranium in phosphate is considered, total uranium reserves will reach up to 35 million tons. The world's nuclear energy capacity is expected to increase from the present capacity of 370 GWe to somewhere between 450 GWe (+22%) and 530 GWe (+44%). To supply the increased requirement of the uranium feedstock, the annual uranium requirement will rise by about 80,000 tons or 100,000 tons (Sokolov 2006). The common belief is that nuclear energy sources would outlast fossil fuel resources. With the currently used sustainability criteria, which ignore the scientific features of natural resources, this argument appears to be valid. However, with the scientific argument put forward in recent years (see previous chapters), it becomes clear that there is an inherent continuity in natural resources, and there is no risk of running out of natural resources. At the same time, it is preposterous to suggest that nuclear technology, which has such a low global efficiency and such negative long-term impacts on the environment, is sustainable.

Figure 14.9 shows the global approximate uranium reserves. As of today, Australia has the highest reserves (24%) in the world, followed by Kazakhstan (17%) and Canada (9%). Australia is the largest exporter of uranium oxide in the world, and averaged almost 10,000 t/year of uranium oxide, which is about 22% of the world's uranium supply (NEA 2005; IAEA 2005). Australia has 38% of the world's lowest-cost uranium resources (under US$ 40/kg).

14.3.4 Sustainability of Nuclear Energy

Sustainability of any technology is evaluated based on its positive and negative impacts on humans and society. In the case of nuclear energy, the sustainable development implies assessing its characteristics related to its economic, environmental, and social impacts. Both positive and negative impacts are to be considered for its analysis, under which nuclear energy can contribute to, or create problems for, sustainable development.

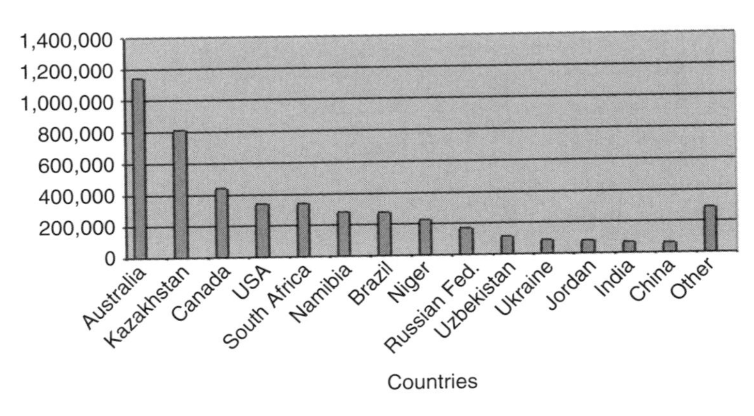

Figure 14.9 Global uranium reserves (NEA 2005; IAEA 2005).

The availability of natural resources, which are necessary as inputs, is one of the indicators to determine whether any technology is sustainable or not. The intensity of energy use and material flow, including those to the environmental emissions such as carbon dioxide, during the project life cycle are also the issues to be considered in evaluating sustainability. The impact of technology on public health, the environment, land use, and the natural habitat of living beings, including potential for causing major and irreversible environmental damages, is to be evaluated in order to assess the sustainability of nuclear technology. Based on these facts, the sustainability of nuclear energy has been evaluated.

14.3.4.1 *Environmental Sustainability of Nuclear Energy*

The impacts and the pathway analysis of radioactive release from a nuclear energy chain are shown in Figure 14.10 (NEA-OECD, 2003). The impacts of a nuclear energy chain are both radiological and non-radiological, which can be caused due to routine and accidental release of radioactive wastes into the natural environment. The sources of these impacts are the releases of materials through atmospheric, liquid, and solid waste pathways. Gaseous release directly reaches into the atmosphere, and the air becomes contaminated. The contaminated air is either inhaled by humans or deposited into the soil. Once the soil is contaminated through the nuclear waste, surface and ground water bodies are affected, and the water will eventually be ingested through agricultural products or seafood. In this way, nuclear wastes will increase the cost of human health, with the actual cost continuously increasing with time.

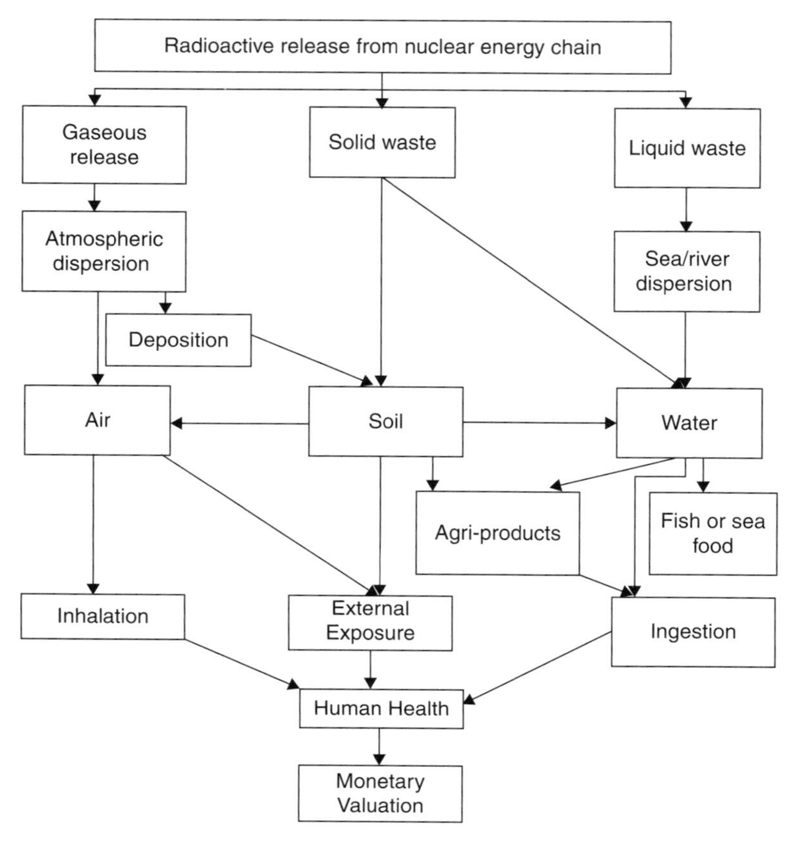

Figure 14.10 Radioactive release from nuclear energy chain (Redrawn from NEA-OECD 2003).

Nuclear power generates spent fuels of roughly the same mass and volume as the fuel that the reactor takes in, because there is only fission but not oxidation. Yet there are different types of waste produced from overall nuclear power systems, primarily a solid waste, spent fuel, some process chemicals, steam, and heated cooling water. Non-nuclear counterparts of nuclear power, such as fossil fuels, also produce various emissions, including solid, liquid, and gaseous wastes on a pound per pound basis. Therefore, the potential environmental cost of waste produced by a nuclear plant is much higher than the environmental cost of most wastes from fossil fuel plants (EIA 2007).

14.3.4.2 *Nuclear Wastes*

Figure 14.11 is the total amount of cumulative, worldwide spent fuel reprocessing and storage for the past and also projected for 2020. The amount is increasing because of the limited processing facilities available and the delays created for disposal due to the unavailability of waste repository. According to IAEA (2004), there has been some progress for the development of repositories in Finland, Sweden, and the United States for disposing high level wastes. The repository in Finland is considered to be in operating condition by 2020 if all legal, licensing, and construction operations go as planned by the Finnish government. The United States has decided to develop a spent fuel repository at the Yucca Mountain

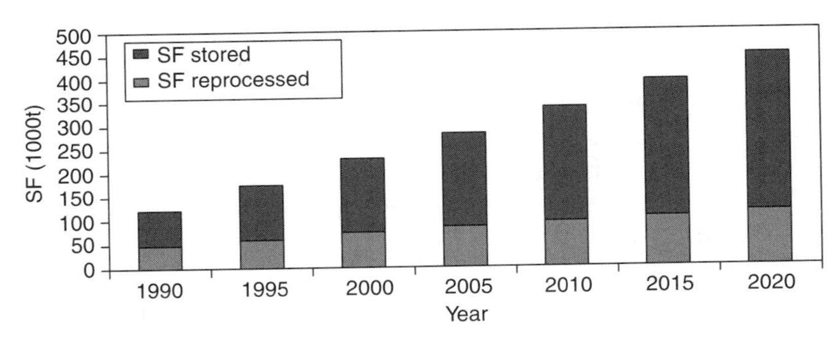

Figure 14.11a Cumulative worldwide spent fuel reprocessing and storage, 1990–2020 (IAEA 2004).

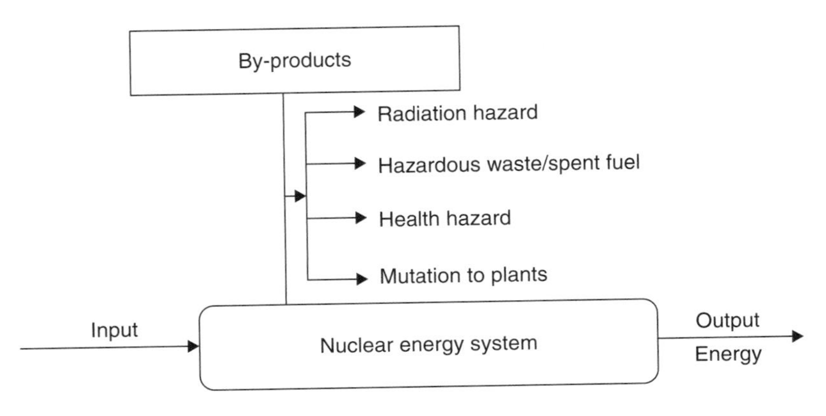

Figure 14.11b Input and output in a nuclear energy system.

disposal site, ready to be operated by 2020. Recently, interests are growing to develop international or regional, rather than national and local, repositories (AP 2006).

Kessler (2000) argued that the IAEA proposed an increase in nuclear power (32 new nuclear plants annually for the next forty years) as a substitute for coal-fired power as one mean toward reducing carbon dioxide emissions. However, the challenge of waste generation is yet to be resolved. Therefore, the production of power by nuclear fission may not be a good choice. Based on the present study, each 1000 MW reaction would produce 30 tons of waste annually, which would need to be sequestered at least for 100 thousand years, whereas the life of a nuclear reactor is assumed to be 50 years. Moreover, it has also been argued that pursuing nuclear energy as an energy solution would be a dangerous move.

The sustainability of any system is evaluated based on the total input into the system and the intended output. In any system, there are byproducts of the systems, which are either beneficial or can be reused, or are harmful and need huge investments to treat before they are safely discharged. In a nuclear energy system, the difference between the input and output environmental resources (d_e) before and after the system is operated should be at least equal to the system before it is implemented. However, in the case of nuclear energy, several byproducts that are highly problematic to human health and the natural environment make the total benefit less than the advantages due to the systems implementation, which is expressed as $d_e/d_t \geq 0$. This system does not fulfill the sustainability criterion developed by Khan *et al.* (2007). Hence, the nuclear energy technology is not acceptable, due to its environmental problems.

14.3.4.3 Social Sustainability of Nuclear Energy

Nuclear energy is one of the most debatable energy sources among others due to its environmental concerns. It is obvious that the potential contribution of even the most novel or sophisticated nuclear technique is compared with and judged against its non-nuclear competitors. However, the health and environmental concerns of nuclear energy are not comparable with any other sources of energy. Yet the cost, reliability, safety, simplicity, and sustainability are important elements for the basis of decisions by governments, private companies, universities, and citizens or consumers.

Despite the previously held notion that nuclear technology is a miracle technology, several nuclear accidents and strategic explosions have proven that nuclear technology can never be a socially responsible technology. The atomic explosions in Hiroshima and Nagasaki of Japan in 1945 left hundreds of thousands of people dead, and many people are still affected even today due to the release of radioactivity. The radiation contaminated everything it came into contact with, and people are still paying the price of the bombing in 1945.

The Chernobyl accident was one of the most dangerous accidents that occurred due to the loss of the coolant in the system (Ragheb 2007). According to Wareham (2006), major radioisotopes present in a nuclear reactor core that have biological significance include caesium-137 (which mimics potassium and is deposited in muscles), strontium-90 (which mimics calcium and is, therefore, deposited mostly in bone), plutonium, and iodine-131. Iodine contamination contributes to the acceleration of thyroid cancers after the idodine is ingested through the air or radiation-contaminated milk.

The health effect of the Chernobyl accident was short-term (deterministic) and long-term (stochastic). Out of 499 people that were admitted for observation right after the Chernobyl accident, 237 people suffered from acute radiation effects (Metivier 2007). Out of the 237, 28 died immediately. Ulcer, leukemia, and thyroid tumors were identified in some of them. The long-term effects were different types of cancers, including thyroid cancer in children. The soil, water, vegetables, and other foods including milk were contaminated, which showed long-term implications for the health systems. Several patients were reported to have developed mental and physiological problems after the accidents. For a period of 12 years after the Chernobyl accident, thyroid carcinoma increased by 4,057 cases in Belarus, as compared to the same period of time before (Metivier 2007). Some studies linked several types of cancers, including leukemia in children, in Germany, Sweden, Greece, and other European countries. Moreover, it is expected that there are many incidences still to be expected from the Chernobyl nuclear accident. Non-radiological symptoms such as headaches, depression, sleep disturbance, the inability to concentrate, and emotional imbalance have been reported, and are considered to be related to the difficult conditions and stressful events that followed the accident (Lee 1996).

Radioactive effects of nuclear energy are not only compelling for the loss of billions of dollars, but also for the irreversible damage to the living beings on Earth. The emission of radioactivity has also reportedly contributed to affecting the DNA of the living cells in several people. The problems start right from the uranium mining, where several chemicals such as sulfuric acids are used to leach the uranium from the rocks. People gain exposure from the mining, milling, enrichment, fuel fabrication, and the hazardous waste it generates after the energy generation. Nuclear energy has created a severe fear among citizens globally. This has also created conflicts among countries. The natural uranium as an input creates energy, hazardous radiation, and hazardous waste as an output of the systems. It is obvious that the benefits of the product output are negative if we consider all of the environmental and health impacts compared to other energy sources, in which similar amounts of energy can be generated without having such adverse environmental and health impacts. Hence, the change in social capital (d_s) with the introduction of nuclear energy with time (d_t) is negative ($d_s/d_t \geq 0$). Thus, nuclear technology is socially unsustainable, and hence, unacceptable.

14.3.4.4 Economic Sustainability of Nuclear Energy

Even though the economics of any energy technology are a function of local conditions and the way energy is processed, the economics of nuclear energy are one of the most contentious issues globally. Tester *et al.* (2005) reported that the cost of electric energy production from light water reactors (LWRs) is typically 57% capital, 30% operation, and 13% maintenance and fuel cost. Because the nuclear power plants have very high capital costs, any factors that affect capital costs, such as inflation rates of the investment, interest rates, gestation period, and power plant efficiencies, will affect the overall economics of the nuclear energy. It was observed that the cost of existing nuclear power plants is competitive with that of conventional coal or gas fueled power plants, but it is not competitive with the most recently developed gas-fired combined cycle power plants. Hence, nuclear energy is in general not competitive with the conventional energy sources. Despite several attempts, including modularized designs, automation, and optimization in construction, the capital costs of the nuclear plants have sufficiently increased the overall cost of the nuclear power system.

Longer gestation periods were one of the major causes to make nuclear energy the most expensive. According to EIA (2001), more than half of the nuclear power plants ordered were not completed. Tester *et al.* (2005) cited that the per kilowatt cost of electricity generated from a Seabrook, New Hampshire nuclear power station reached $2200/kW, even though it was estimated to cost $520/kW. In addition to this, changes in the regulatory requirement and their uncertainties and high interest rates have been other sensitive factors that have impacts on the energy cost. Delay in the licensing process due to the nuclear opponents is also a measurable factor that is making nuclear energy more expensive than anticipated earlier.

Many proponents of nuclear energy consider it one of the cheapest forms of energy. However, there are debates on the issue of the cost of electricity produced from nuclear systems. Comparing the cost of electricity produced with other energy sources, such as coal, natural gas, and wind energy, gives a fairly good idea how the economics of the nuclear energy works. Table 14.1 shows a comparison of the cost per kilowatts of electricity generation from different energy sources. It is understandable that nuclear energy cannot compete with its non-nuclear counterparts.

Nivola (2007) cited the study carried out at the Massachusetts Institute of Technology (MIT), which offered the cost of nuclear energy, incorporating the cost of construction, licensing, and operating a newly commissioned light water reactor, compared to coal or natural gas. The estimate for electricity from nuclear power was 6.7 cents per kilowatt, which was far higher than that of a pulverized coal-fired power plant (4.2 cents/kWh) and a combined cycle

Table 14.1 Cost in US$ per kW for different types of energy generation.

Generating Methods	Generating Cost (US$/kW)	References
Coal	0.01–0.04	Service 2005
Nuclear	0.037	Uranium Information Center 2006
Natural gas	0.025–0.05	Service 2005
Wind	0.05–0.07	Service 2001

natural gas power plant (5.6 cents/kWh), if the gas price was considered to be $6.72 per thousand cubic feet. Moreover, the gestation period for a nuclear power plant is exceptionally long, which contributes to a higher cost. The other factors that contribute to the higher cost are from the nuclear power plants having to fulfill regulatory requirements for health, safety, social, and environmental reasons.

Bertel and Merrison (Merriman and Burchard 1996) reported the average cost of electricity generation from nuclear systems in different parts of the world (Table 14.2). It is obvious that nuclear power is not the cheapest energy option. Because it has to compete with oil and natural gas, and the price of natural gas has been fairly stable recently, it emerges as one of the more expensive forms of electricity. Moreover, it is understood that nuclear energy has to be compared with the price of new forms of renewable energy sources, such as solar, wind, and combined cycle biomass systems. However, the comparison is impossible, because they have different impacts on the environment, with nuclear energy being the most devastating form.

Table 14.2 Nuclear electricity generating costs (Bertel and Morrison 2001).

Country	Discount Rate %	Investment %	O& M %	Fuel %	Total Cost US Cents/kWh
Canada	10	79	14	6	4.0
Finland	10	73	14	13	5.6
France	10	70	14	16	4.9
Japan	10	60	21	19	8
Republic of Korea	10	71	20	9	4.8
Spain	10	70	13	17	6.4
Turkey	10	75	17	9	5.2
United States	10	68	19	13	4.6

Table 14.3 indicates the price of nuclear energy development at various stages compared to other energy sources, such as coal and natural gas. Even though the per kilowatt cost for nuclear energy is reported to be lower than coal and natural gas, this analysis does not take into account the cost for waste management. In terms of waste management, nuclear energy performs worse than other sources. Hence, if the life cycle is considered in the economic evaluation, including waste disposal, social, and environmental costs, nuclear energy will have the highest per kilowatt cost among the energy sources compared.

Nuclear power plants have one of the lowest global efficiencies, meaning that considering the whole life cycle of the nuclear energy chain, from exploration to electricity production, the overall process efficiency is low. Moreover, large amounts of fossil fuel resources are used during exploration, mining, milling, fuel processing, enrichment, and so forth. Hence, nuclear energy sources contribute emissions of greenhouse gases. The cost that is necessary to offset the greenhouse gas emissions has never been taken into consideration, which shows the exaggeration of the cost that it has more benefits compared to its non-nuclear counterparts.

Nuclear energy has more future liabilities, such as decommissioning, the storage of spent fuel, the disposal of tools and equipment being used in the plants, clothes, globes, and other materials that are contaminated and cannot be disposed in the ordinary sewage systems. These issues should be considered during the inception period of a nuclear power project, so as to make sure that these burdens are not passed on to future generations. This is true only theoretically. Knowing the characteristics of nuclear wastes, which

Table 14.3 Price and structure of price of the main sources of power in Europe in 2005 (Fiore 2006).

Cost structure	**Nuclear (%)**	**Coal (%)**	**Gas (%)**
Investment	54	39	20
Exploitation	20	14	11
Fuel	25	46	69
R&D	1		
Average Price €/MWh	28.4	33.7	35

have half-lives of millions and billions of years, it is impossible to contain the necessary arrangement of nuclear waste management in a short time. The economic benefit in the short term cannot be justified because its impacts remain for billions of years in the form of nuclear radiation.

The proliferation of dangerous materials from exploration, mining, fuel processing, and waste disposal and its relation with energy security are also determining factors in the long-term cost. The safe disposal of waste has not been limited to a technical problem; it is a complex environmental, social, and political problem. The resolution of this issue, considering the abovementioned problems, needs clear-cut technical, economical, and socio-environmental responsibility towards society. The resolution needs amicable alternatives and justifiable reasons in order to avoid possible conflicts in the areas where nuclear plants are planned, which clearly involves a huge cost that needs to be taken into account during the project evaluation.

Nuclear energy is considered to have the highest external costs, such as health and environmental costs, due to electricity production and use. The impacts created by radiation hazards are rarely curable, and irreversible damages are caused to the plant and human metabolisms. Economic evaluation of all radiation hazards to health and the environment will make nuclear power unattractive. (See Chapter 5, section 5.14 for further details.)

14.3.5 Global Efficiency of Nuclear Energy

Milling uranium consists of grinding particles in uniform, which yields a dry powder-form of material consisting of natural uranium as U_3O_8. Milling uranium has up to 90% efficiency (η_1) (Mudd 2000). The local efficiency of nuclear energy conversion has been reported to reach up to 50% (Ion 1997). However, considering the life cycle process from extraction to electric conversion and the end uses, the global efficiency of the system is one of the lowest because of the expensive leaching process during mining and series of gas diffusion or centrifugation required for the uranium enrichment. Conventional mining has an efficiency of about 80%, η_2 (Gupta and Mukherjee, 1990). In a milling process, which is a chemical plant and usually uses sulfuric acid for leaching, about 90% of the uranium is extracted (η_3) (Mudd 2000). There is also a significant conversion loss in the conversion of uranium to

UF6, the efficiency (η_4) of which is considered approximately 70%. Enrichment efficiency (η_5) is less than 20%. Considering 50% thermal to net electric conversion (η_6) and 90% efficiency in transmission and distribution (η_7), the global efficiency (η_g) of the nuclear processing technology ($\eta_1 \times \eta_2 \times \eta_3 \times \eta_4 \times \eta_5 \times \eta_6 \times \eta_7$) is estimated to be 4.95%. Because global efficiency considers the environmental and social aspects also, and if we consider the environmental impact due to the radioactive hazards and the cost of the overall system, the global efficiency is even lower than that which is mentioned above. Moreover, the utilization of wastes into valuable products would increase the overall efficiency of the system, in principle. However, there is no way that nuclear waste can be reused. Instead, it will have more devastating impacts on humans and the environment. Nuclear energy has a negative efficiency for waste management.

14.3.6 Energy from Nuclear Fusion

Nuclear fusion power is generated by the fusion reaction of two atomic nuclei. Two light atomic nuclei fuse together to form a heavier nucleus and release energy. The sun in the solar system is the perfect example of fusion reaction occurring every moment in the universe. Fusion can occur only at very high temperatures where materials are at a plasma state. At a very high temperature, the electrons separate from the nuclei, creating a cloud of charged particles, or ions. This cloud of equal amounts of positively charged nuclei and negatively charged electrons is believed to be in a plasma state. Hydrogen fusion can produce a million times more energy than burning hydrogen with oxygen. Fusion will have less radioactive hazards than fission processes. The only problem with fusion is the amount of heat necessary for the reaction to take place. Hence, nuclear fusion is still not a dependable energy source for the foreseeable future (Tornquist 1997).

It is generally considered that fusion needs external energy to trigger the reaction, and may not be feasible until a breakthrough technology is developed. Some scientists have claimed that they have successfully achieved cold fusion at room temperature on a small scale (Kruglinksi 2006). However, other scientists think that this could not be replicated. Goodstein (1994, 2000) made an earlier, similar claim. However, this claim was not confirmed, and could not be replicated (Merriman and Burchard 1996).

Unlike fission reactors, fusion reactors are considered less problematic for the environment. Fusion reactors are considered effective at minimizing radioactive wastes. This is done by developing a waste management strategy, including the maximum possible recycling of materials within the nuclear industry classifying the radioactive and non-radioactive materials (Zucchetti 2005). Fusion reactors were promoted as a zero-waste option by recycling the radioactive materials and disposing the non-radioactive materials. However, it is impossible to have non-radioactive materials in the reaction. Islam and Chhetri (2008) reported that recycling plastic could result in more environmental hazards, due to the emissions of new byproducts such as bisphenol-A. Similarly, all radioactive materials are not recyclable, and there are several activities related to nuclear industries, including metallurgical activities that are linked to negative impacts on the environment and large amounts of energy consumption. However, if scientifically proven, fusion at room temperature would cause a major shift in the current nuclear technology.

15

Conclusions

15.1 Introduction

This book offers a true paradigm shift in energy management, starting with a scientific discourse on what a true paradigm shift is. With a scientific discussion of change, the book shows how previous civilizations handled their energy needs, both in practice and theory. This delinearized history helped determine the root causes of the crises encountered in the information age. Addressing the root cause would invoke changes in the long term, avoiding cosmetic changes that have dominated the modern world. Scientifically, this is equivalent to re-examining the first premise of all theories and laws. If the first premise does not conform to natural laws, then the model is considered unreal (not just unrealistic), dubbed an "aphenomenal model." With these aphenomenal models, all subsequent decisions lead to outcomes that conflict with the stated "intended" outcomes. At present, such conflicts are explained either with doctrinal philosophies or with a declaration of paradox. Our analysis shows that doctrinal philosophy is aphenomenal science and is the main reason for the current crisis that we are experiencing. The statement of a paradox helps us procrastinate in solving the problem, but it does nothing to solve the problem. Both of these states keep us squarely in what we call the Einstein box. (Albert Einstein famously said, "The thinking that got you into the problem, is not going to get you out.") Instead, if the first premises are replaced with a phenomenal premise, the subsequent cognition encounters no contradictions with the intended outcomes. The end results show how the current crises can not only be arrested, but also reversed. As a result, the entire process would be reverted from unsustainable to sustainable.

With the above recasting of the problem, the science and engineering of energy management are worked out in this book, with clear directions of how to change the current practices so that they become sustainable, and remain that way. Using these sustainable practices would cause a change in direction. If the previous practices were taking us from bad to worse, the proposed practices would take us from good to better. This conclusion holds true for all petroleum engineering operations and the economic analysis that justifies such operations at a decision-making stage. This closes the vicious loop (unsustainable engineering → technological disaster → environmental calamity → financial collapse) that we have become familiar with at the dawn of the information age.

The introduction of this book was written to motivate the reader to be a believer in nature and the ability of its "best creation." The conclusion is written to motivate the reader to practice this belief.

15.2 The HSS®A® (Honey → Sugar → Saccharin®→Aspartame®) Pathway

The HSS®A® pathway is a kind of metaphor for many other things that originate from a natural form and become subsequently engineered through many intermediate stages into "new" products. The following discussion lays out how it works. Once it is understood how disinformation works, one can figure out a way to reverse the process by avoiding aphenomenal practices.

Over the years it has become a common idea among engineers and the public to associate an increase in the quality, and/or qualities, of a final product with the insertion of additional intermediate stages of refining the product. If honey, taken more or less directly from a natural source, without further processing, was fine, surely the sweetness that can be attained by refining sugar must be better. If the individual wants to reduce their risk of diabetes, then surely further refining of the chemistry of "sweetness" into such products as Saccharin® must be better still. And why not use even more sophisticated chemical engineering to further convert the chemical essence of this refined sweetness into forms that are stable in liquid phase, such as Aspartame®?

In this sequence, each additional stage is defended and promoted as having overcome some limitation of the last stage. But at the end of this chain, what is left in, say, Aspartame® of the antibacterial qualities of honey? Looking from the end of this chain back to its start, how many lab rats ever contracted cancer from any amount of intake of honey? Honey is known to be the only food that has all of the nutrients, including water, to sustain life. How many true nutrients does Aspartame® have? From the narrowest engineering standpoint, the kinds and number of qualities in the final product at the end of this Honey → Sugar → Saccharin® → Aspartame® chain have been transformed, but from the human consumer's standpoint of the use-value of "sweet-tasting," has there been a net qualitative gain going from honey all the way to Aspartame®?

From the scientific standpoint, honey fulfils both conditions of phenomenality, namely, origin and process. That is, the source of honey (nectar) is real (even if it means flowers were grown with chemical fertilizers, pesticides, or even genetic alteration) and the process is real (honeybees cannot make false intentions, therefore, they are perfectly natural), even if the bees were subjected to air pollution or a sugary diet. The first engineering intervention creates 'sugar'. Here the source is real (organic or non-organic) but the process of making sugar is artificial. Nature never bleaches sweeteners to make them white, 'free flowing', or others. The second level of intervention is invoked through creation of Saccharin® It has another real origin, but this time the original source (crude oil) is a very old food source compared to the source of sugar. With steady-state analysis, they both will appear to be of the same quality! As the chemical engineering continues, we resort to the final transition to Aspartame®. Indeed, nothing is real or natural about Aspartame®, as both the origin and the process are artificial. So, the overall transition from honey to Aspartame® has been from 100% phenomenal to 100% aphenomenal. In order to justify this transition from real to artificial, Economics of tangibles that introduces wasting as a means to 'economize' is invoked. This remains the standard of neoclassical economics.

There is an entire economics of scale that is developed and applied to determine how far this is taken in each case. For example, honey is perceptibly "sugar" to taste. We want the sugar, but honey is also antibacterial, and cannot rot. Therefore, the rate at which customers will have to return for the next supply is much lower and slower than the rate at which customers would have to return to resupply themselves with, say, refined sugar. Or even worse, to extend the amount of honey available in the market (in many developing countries, for example), sugar is added. The content of this "economic" logic then takes over, and drives what happens to honey and sugar as commodities. There are natural limits to how far honey as a natural product can actually be commodified, whereas, for example, refined sugar is refined to become addictive, so that the consumer becomes hooked, and the producer's profit is secured.

The matter of intention is never considered in the economics of scale. As a result, however, certain questions go unasked. No one asks whether any degree of external processing of what began as a natural sugar source can or will improve its quality as a sweetener. Exactly what that process, or those processes, would be is also unasked. No sugar refiner is worried about how the marketing of his product in excess is contributing to a diabetes epidemic. The advertising that is crucial to marketing this product certainly will not raise this question. Guided by the "logic" of the economies of scale, and the marketing effort that must accompany it, greater processing is assumed to be and accepted as being *ipso facto* good, or better. As a consequence of the selectivity inherent in such "logic," any other possibility within the overall picture, such as the possibility that as we go from honey to sugar to saccharin to aspartame, we go from something entirely safe for human consumption to something cancerously toxic, does not even enter the frame.

Such a consideration would prove to be very threatening to the health of a group's big business in the short term. All of this is especially devastatingly clear when it comes to crude oil. Widely and falsely believed to be toxic before a refiner touches it, refined petroleum products are utterly toxic, but they are not to be questioned, since they provide the economy's lifeblood.

Edible, natural products in their natural state are already good enough for humans to consume at some safe level and process further internally in ways useful to the organism. We are not likely to overconsume any unrefined natural food source. However, the refining that accompanies the transformation of natural food sources into processed-food commodities also introduces components that interfere with the normal ability we have to push a natural food source aside after some definite point. Additionally, with externally processed "refinements" of natural sources, the chances increase that the form in which the product is eventually consumed must include compounds that are not characteristic anywhere in nature, and that the human organism cannot usefully process without excessively stressing the digestive system. After a cancer epidemic, there is great scurrying to fix the problem. The cautionary tale within this tragedy is that if the HSS®A® principle were considered *before* a new stage of external processing were added, much unnecessary tragedy could be avoided.

There are two especially crucial premises of the economics-of-scale that lie hidden within the notion of "upgrading by refining": (a) unit costs of production can be lowered (and unit profit therefore expanded) by increasing output **Q** per unit time t, i.e., by driving $\partial\mathbf{Q}/\partial t$ unconditionally in a positive direction; and (b) only the desired portion of the **Q** end-product is considered to have tangible economics and, therefore, also intangible social "value," while any unwanted consequences, e.g., degradation of, or risks to, public health, damage(s) to the environment, and so forth, are discounted and dismissed as false costs of production.

Note that, if relatively free competition still prevailed, premise (a) would not arise even as a passing consideration. In an economy lacking monopolies, oligopolies, and/or cartels dictating effective demand by manipulating supply, unit costs of production remain mainly a function of some given level of technology. Once a certain proportion of investment in fixed-capital (equipment and ground-rent for the production facility) becomes the norm generally among the various producers competing for customers in the same market, the unit costs of production cannot fall or be driven arbitrarily below a certain floor level without risking business loss. The unit cost thus becomes downwardly inelastic.

The unit cost of production can become downwardly elastic, i.e., capable of falling readily below any asserted floor price, under two conditions: (1) during moments of technological transformation of the industry, in which producers who are first to lower their unit costs by using more advanced machinery will gain market shares, temporarily, at the expense of competitors; or (2) in conditions where financially stronger producers absorb financially weakened competitors.

In neoclassical models, which assume competitiveness in the economy, this second circumstance is associated with the temporary cyclical crisis. This is the crisis that breaks out from time to time in periods of extended oversupply or weakened demand. In reality, contrary to the assumptions of the neoclassical economic models, the impacts of monopolies, oligopolies, and cartels have entirely displaced those of free competition, and have become normal rather than the exception. Under such conditions, lowering unit costs of production (and thereby expansion of unit profit) by increasing output $\mathbf{Q}$ per unit time t, i.e., by driving $\partial \mathbf{Q}/\partial t$ unconditionally in a positive direction, is no longer an occasional and exceptional tactical opportunity. It is a permanent policy option: monopolies, oligopolies, and cartels manipulate supply and demand because they can.

Note that premise (b) points to how, where, and why consciousness of the unsustainability of the present order can emerge. Continuing indefinitely to refine nature out by substituting ever more elaborate chemical "equivalents," hitherto unknown in the natural environment, has started to take its toll. The narrow concerns of the owners and managers of production are at odds with the needs of society. Irrespective of the private character of their appropriation of the fruits of production, based on concentrating so much power in so few hands, production has become far more social. The industrial-scale production of all goods and services as commodities has spread everywhere, from the metropolises of Europe and North America to the remotest Asian countryside, the deserts of Africa, and the jungle regions of South America. This economy is not only global in scope, but also social in its essential character. Regardless of the readiness of the owners and managers to dismiss and abdicate responsibility for the environmental and human health costs of their unsustainable approach, these costs have become an increasingly urgent concern to societies in general. In this regard, the HSS®A® principle becomes a key and most useful guideline for sorting what is truly sustainable for the long term from what is undoubtedly unsustainable.

The human being that is transformed further into a mere consumer of products is a being that is marginalized from most of the possibilities and potentialities of the fact of his/her existence. This marginalization is an important feature of the HSS®A® principle. There are numerous things that individuals can do to modulate, or otherwise affect, the intake of honey and its impacts, but there is little, indeed, nothing, that one can do about Aspartame® except drink it. With some minor modification, the HSS®A® principle helps illustrate how the marginalization of the individual's participation is happening in other areas.

What has been identified here as the HSS®A® principle, or syndrome, continues to unfold attacks against both the increasing global striving toward true sustainability on the one hand, and the humanization of the environment in all aspects, societal and natural, on the other. Its silent partner is the aphenomenal model, which invents justifications for the unjustifiable and for "phenomena" that have been picked out of thin air. As with the aphenomenal model, repeated and continual detection and exposure of the operation of the HSS®A® principle is crucial for future progress in developing nature-science, the science of intangibles and true sustainability.

Table 15.1 summarizes the outcome of the HSS®A® pathway. While this pathway is less than a century old, the same pathway was implemented some millennia ago, and has influenced the modern day thinking over the last millennium. This is valid for all aspects of life, ranging from education to economics.

Table 15.1 The HSS®A® pathway and its outcome in various disciplines.

Natural State	2nd Stage of Intervention	Second Stage of Intervention	3rd Stage of Intervention
Honey	Sugar	Saccharin®	Aspartame®
Education	Doctrinal teaching	Formal education	Computer-based learning
Science	Religion	Fundamentalism	Cult
Science and nature-based technology	New Science	Engineering	Computer-based design
value-based (e.g., gold, silver) economy	Coins (non gold or silver)	Paper money (disconnected from gold reserve)	Promissory note (electronic)

15.3 HSS®A® Pathway in Energy Management

If the first premise of "nature needs human intervention to be fixed" is changed to "nature is perfect," then engineering should conform to natural laws. This book presents detailed discussion on how this change in the first premise helps answer all questions that remained unanswered regarding the impacts of petroleum operations. It also helps demonstrate the false, but deeply rooted, perception that nuclear, electrical, photovoltaic, and "renewable" energy sources are "clean" and carbon-based energy sources are "dirty." This book establishes that crude oil, being the finest form of nature-processed energy source, has the greatest potential for environmental good. The only difference between solar energy (used directly) and crude oil is that crude oil is concentrated, and can be stored, transported, and reutilized without resorting to HSS®A® degradation. Of course, the conversion of solar energy through photovoltaics creates technological (low efficiency) and environmental (toxicity of synthetic silicon and battery components) disasters. Similar degradation takes place for other energy sources as well. Unfortunately, crude oil, an energy-equivalent of honey, has been promoted as the root of the environmental disaster (global warming and consequences, e.g., CNN 2008). Ignoring the HSS®A® pathway that crude oil has suffered has created the paradoxes, such as "carbon is the essence of life and also the agent of death" and "enriched uranium is the agent of death and also the essence of clean energy." These paradoxes are removed if the pathway of HSS®A® is understood. Table 15.2 shows the HSS®A® pathway that is followed for some of the energy management schemes. One important feature of these technologies is that nuclear energy is the only one that does not have a known alternative to the HSS®A® pathway. However, nuclear energy is also being promoted as the wave of the future for energy solutions, showing once again that every time we encounter a crisis, we come up with a worse solution than what caused the crisis in the first place.

It is important to note that the HSS®A® pathway has been a lucrative business because most of the profit is made using this mode. This profit also comes with disastrous consequences to the environment. Modern day economics do not account for such long-term consequences, making it impossible to pin down the real cost of this degradation. In this book, the intangibles that caused the technological and environmental disasters are explicitly pointed out, both in engineering and economics. As an outcome of this

Table 15.2 The HSS®A® pathway in energy management schemes.

Natural State	1st Stage of Intervention	2nd Stage of Intervention	3rd Stage of Intervention
Honey	Sugar	Saccharin®	Aspartame®
Crude oil	Refined oil	High-octane refining	Chemical additives for combating bacteria, thermal degradation, weather conditions, etc.
Solar	Photovoltaics	Storage in batteries	Re-use in artificial light form
Organic vegetable oil	Chemical fertilizer, pesticide	Refining, thermal extraction	Genetically modified crop
Organic saturated fat	Hormone, antibiotic	Artificial fat (transfat)	No-transfat artificial fat
Wind	Conversion into electricity	Storage in batteries	Re-usage in artificial energy forms
Water and hydro-energy	Conversion into electricity	Dissociation utilizing toxic processes	Recombination through fuel cells
Uranium ore	Enrichment	Conversion into electrical energy	Re-usage in artificial energy forms

analysis, the entire problem is re-cast in developing true science and economics of nature that would bring back the old principle of value proportional to price. This is demonstrated in Figure 15.1. This figure can be related to Table 15.2 in the following way:

- Natural state of economics = economizing (waste minimization, meaning "minimization" and "ongoing intention" in the Arabic term)

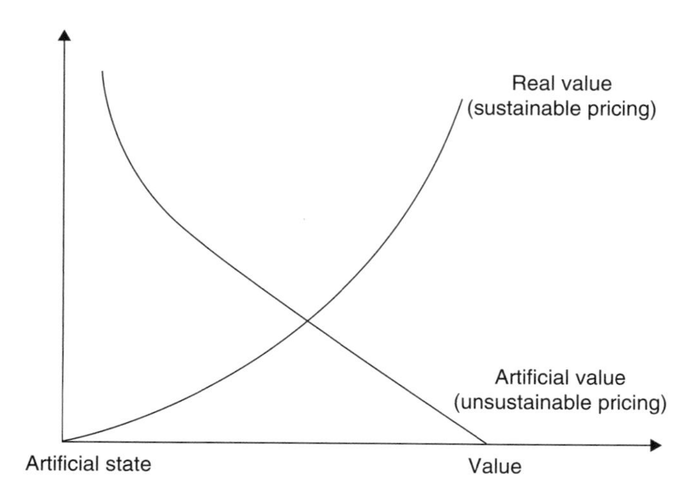

Figure 15.1 Economics and accounting systems have to be reformulated in order to make stated value proportional to real value.

- First stage of intervention = move from intention-based to interest-based
- Second stage of intervention = make wasting the basis of economic growth
- Third stage of intervention = borrow more from future to promote the second stage of intervention

The above model is instrumental in turning a natural supply and demand economic model into an unnatural perception-based model. This economic model then becomes the driver of the engineering model, closing the loop of the unsustainable mode of technology development.

15.4 The Conclusions

Based on the analyses presented in this book, the following conclusions can be drawn:

1. Crude oil and natural gas are the best sources of energy available to mankind. They have the potential of becoming the agent of positive change to the environment.

2. The environmental consequences that have been observed ever since the golden era of petroleum production are not inherent to natural crude oil, but are due in part to the processes involved in petroleum refining, gas processing, and material manufacturing (e.g., plastic).
3. Environmental consequences of all other alternatives to fossil fuel are likely to be worse in the long run. This is not obvious because engineering calculations as well as economical analysis do not account for the entire history (the continuous time function or intangibles) of the process. If intangibles were included, other alternatives would clearly become less efficient and more harmful than petroleum products, even if the current mode of petroleum operations continues.
4. By changing the current mode of petroleum operations (particularly in treating petroleum products with artificial chemicals) to sustainable processes, the current trend of environmental disasters can be effectively reversed.
5. By including the intangibles, new models of energy pricing would increase profitability as the real value of the final product is increased, as opposed to the current model that sees the highest profit margin for the most toxically processed products.
6. The phenomenal, knowledge-based model presented in this book has the potential of revolutionizing the current mode of technology development and economic calculations.

References

Ackermann, R., 1961, "Inductive Simplicity," *Philosophy of Science* 28 (2), pp. 152–61.

Adelaide, 2006, Waite Solid-state NMR Facility, University of Adelaide, Australia <http://www.waite.adelaide.edu.au/NMR/nmrlist.html> accessed: December 30, 2006.

Adeoti, O., Ilori, M.O., Oyebisi, T.O., and Adekoya, L.O., 2000, "Engineering design and economic evaluation of a family-sized biogas project in Nigeria," *Technovation* 20, pp. 103–108.

Ahn, J.S. and Lee, K.H., 1986, "Studies on the volatile aroma components of edible mushroom (*Tricholoma matsutake*) of Korea," *Journal of the Korean Society of Food and Nutrition*, 15, pp. 253–257 [cited in BUA, 1994].

Ahnell, A. and O'Leary, H., 1997, "Drilling and production discharges in the marine environment," in S.T. Orszulik, ed., *Environmental technology in the oil industry*, Blackie Academic & Professional, London, U.K., pp. 181–208.

Akhtar, J., 2002, *Numerical and Experimental Modelling of Contaminant Transport in Ground Water and of Sour Gas Removal from Natural Gas Streams*, Chalmers University of Technology, Goteborg, Sweden.

AlAdhab, H., Kocabas, I., and Islam, M.R., 1998, "Field-Scale Modeling of Asphaltene Transport in Porous Media", paper SPE paper no 49557, Proc. SPE ADIPEC 98, Abu Dhabi, UAE.

Al-Darbi, M., Muntasser, Z., Tango, M., and Islam, M.R., 2002, "Control of microbial corrosion using coatings and natural additives," *Energy Sources*, 24(11), pp. 1009–18.

Al-Darbi, M.M., Saeed, N.O., and Islam, M.R., 2002a, "Biocorrosion and Microbial Growth Inhibition using Natural Additives," The 52rd Canadian Chemical Engineering Conference (CSChE 2003). Vancouver, BC, Canada.

Al-Darbi, M.M., Saeed, N.O., Ackman, R.G., Lee, K., and Islam, M.R., 2002b, "Vegetable And Animal Oils Degradation In Marine Environments,"

Proc. Oil and Gas Symposium, CSCE Annual Conference, refereed proceeding, Moncton, June.

Al-Darbi, M.M., Agha, K.R., Hoda, R., and Islam, M.R., 2004a, "A Novel Method for Preventing Corrosion," Int. Eng. Conf., Jordan, April.

Al-Darbi, M.M., Agha, K.R., and Islam, M.R., 2004b, "Natural Additives for Corrosion Prevention," Int. Eng. Conf., Jordan, April.

Al-Darbi, M.M., Saeed, N.O., Islam, M.R., and Lee, K., 2005, "Biodegradation of Natural Oils in Sea Water," *Energy Sources*, vol. 27, no. 1–2, pp. 19–34.

Aliev, R.R. and Lupina, M.I., 1995, "Formulation of cracking catalyst based on zeolite and natural clays," *Chemistry and Technology of Fuels and Oils*, Volume 31, Number 4, April, pp. 150–154.

AlFalahy, M.A., Abou-Kassem, J.H., Chakma, A., and Islam, M.R., 1998, "Sour Gas Processing, Disposal and Utilization as Applied in UAE Reservoirs," SPE paper 49504 presented at the 8th Abu Dhabi International Petroleum Exhibition and Conference, Abu Dhabi, UAE, October 11–14.

Alguacil, D.M., Fischer, P., and Windhab, E.J., 2006, "Determination of the Interfacial Tension of Low Density Difference Liquid–Liquid Systems Containing Surfactants by Droplet Deformation Methods," *Chemical Engineering Science*, Vol. 61, pp. 1386–1394.

Ali, M. and Islam, M.R., 1998, "The Effect of Asphaltene Precipitation on Carbonate Rock Permeability: An Experimental and Numerical Approach," SPE 38856, *SPE Production & Facilities*, Aug., pp. 78–83.

Allen, C.A.W., 1998, *Prediction of Biodiesel Fuel Atomization Characteristics Based on Measured Properties*, Ph.D. Thesis, Faculty of Engineering, Dalhousie University, pp 200.

Al-Maghrabi, I., Bin-Aqil, A.O., Chaalal, O., and Islam, M.R., 1999, "Use of Thermophilic Bacteria for Bioremediation of Petroleum Contaminants," *Energy Sources*, vol. 21 (1/2), pp. 17–30.

Al-Sulaiman, F. A. and Z. Ahmed, 1995, "The Assessment of Corrosion Damage to Automobiles in the Eastern Coast Area of Saudi Arabia," *Proceedings of the Institution of Mechanical Engineers -D- Jnl of Automobile Engineering* 209(1), p. 3.

Anfiteatro, D.N., 2007, "Dom's Tooth-Saving Paste," <http://users.chariot.net.au/~dna/toothpaste/toothpaste.htm> accessed March 22, 2007.

Annual Book of ASTM Standards, 2002, *Paints, Related Coatings, and Aromatics* (vol. 06.01).

Anon, 1996, "Greenhouse Issues," Newsletter of the IEA Greenhouse Gas R & D Program, Number 22, January 1996.

Anonymous, 2008, http://www.mc.vanderbilt.edu/biolib/hc/journeys/book7.html, accessed Nov., 2008.

AP, 2004, "FDA approves leeches as medical devices," available on website: http://www.msnbc.msn.com/id/5319129/, last accessed on December 12, 2008.

AP, 2006, "Nuclear Tower Crumbles on Purpose Sunday," May 21, (www.cbsnews.com/Stories/2006/05/21/ National/Main1638433.html).

Appleman, B.R., 1992, "Predicting Exterior Marine Performance of Coatings from Salt Fog: Two Types of Errors," *Journal of Protective Coatings and Linings*, October, p. 134.

Ariew, R. and Barker, P., 1986, "Duhem on Maxwell: Case-Study in Interrelations of History of Science & Philosophy of Science," *Proceedings of the Biennial Meetings of the Philosophy of Science Association* I "Contributed Papers," pp. 145–156.

Argonne National Laboratory, 2005, "Natural Decay Series: Uranium, Radium, and Thorium," *Human Health Fact Sheet*, August, 2005.

Askey, A., Lyon, S.B., Thompson, G.E., Johnson, J.B., Wood, G.C., Cooke, M., and Sage., P., 1993, "The Corrosion of Iron and Zinc by Atmospheric Hydrogen Chloride," *Corrosion Science* 34, p. 233.

ASM International Handbook, 1987, *Metals*, ASM (vol. 13).

Astrita, G., Savage, D., and Bisio, A., 1983, *Gas Treating with Chemical Solvents*, John Wiley and Sons, New York.

ATSDR Agency for Toxic Substances and Disease Registry, 2006, "Medical Management Guide Line for Sodium Hydroxide," <http://www.atsdr.cdc.gov/MHMI/mmg178.html> accessed June 07, 2006.

Attanatho, L., Magmee, S., and Jenvanipanjakil, P., 2004, The Joint International Conference on 'Sustainable Energy and Environment.' 1–3 December 2004, Huan Hin, Thailand.

Ayala, J., Blanco, F., Garcia, P., Rodríguez, P., and Sancho, J., 1998, "Asturian Fly Ash as a Heavy Metals Removal Material," *Fuel*, Vol. 77 (11), pp. 1147–1154.

Bachu, S., Gunter, W.D., and Perkins, E.H., 1994, "Aquifer Disposal of CO_2: Hydrodynamical and Mineral trapping," *Energy Conversion and Management*, volume 35, pp. 264–279.

Bailey, R.T. and McDonald, M.M., 1993, "CO_2Capture and Use for EOR in Western Canada, General Overview," *Energy Conversion and Management*, Volume 34, no. 9–11, pp. 1145–1150.

Baird, Jr., W.C., 1990, Novel Platinum-Iridium Refining Catalysts, US Patent no. 4966879, Oct. 30.

Baker, K.H. and Herson, D.S., 1990, "In situ biodegradation of contaminated aquifers and subsurface soils," *Geomicrobiology J.* vol. 8, pp. 133–146.

Bamhart, W.D. and Coulthard, C., 1995, "Weyburn CO_2 Miscible Flood Conceptual Design and Risk Assessment," 6th Petroleum Conference of the South Saskatchewan Section, Petroleum Society of CIM, Regina, Oct. 16–18.

Bansal, A. and Islam, M.R., 1994, "Scaled Model Studies of Heavy Oil Recovery from an Alaskan Reservoir Using Gravity-assisted Gas Injection," *J. Can. Pet. Tech.* Vol. 33, no. 6, pp. 52–62.

Barnwal, B.K. and Sharma, M.P., 2005, "Prospects of biodiesel production from vegetable oils in India," *Renewable and Sustainable Energy Reviews*, Vol.9, pp. 363–378.

Basu, A., Akhter, J., Rahman, M.H., and Islam, M.R., 2004, "A review of separation of gases using membrane systems," *J Pet Sci Tech*, vol. 22, no. 9–10, pp. 1343–1368.

Basu, A., White, R.L., Lumsden, M.D., Bisop, P., Butt, S., Mustafiz, S., and Islam, M.R., 2007, "Surface Chemistry of Atlantic Cod Scale," *J. Nature Science and Sustainable Technology*, Vol. 1, no. 1, pp. 69–78.

BBC, 2006, "US and India seal nuclear accord," March 02, 2006, Thursday. http://news.bbc.co.uk/2/hi/south_asia/4764826.stm (Accessed on January 09, 2007).

BBC, 2007, "Key gene work scoops Nobel Prize," Oct. 8, available on http://news.bbc.co.uk/2/low/health/733491.stm, last accessed Dec. 12, 2008.

BBC, 2008, "Oil sets fresh record above $109," http://news.bbc.co.uk/2/hi/business/ 7289070.stm.

Bellassai, S.J., 1972, "Coating Fundamentals," *Materials Performance*, no. 12, p. 33.

Bentsen, R.G., 1985, "A New Approach to Instability Theory in Porous Media," *Soc. Pet. Eng. J.*, Oct, pp. 765–779.

Bergman, P.D., Drummond, C.J., Winter, E.M., and Chen, Z-Y., 1996, "Disposal of Power Plant CO_2 in Depleted Oil and Gas Reservoirs in Texas," *Proceedings of the third International Conference on Carbon Dioxide Removal*, Massachusetts Institute of Technology, Cambridge, MA, USA, 9–11 Sept.

Bertel, E. and Morrison, R., "Nuclear Energy Economics in a sustainable Development Perspective," NEA News 2001– No. 19.1, pp 14–17. Washington DC: US Department of Energy.

Beveridge, T.J. and Fyfe, W.S., 1985, "Metal Fixation by Bacteria Cell Walls," *Canada Journal of Earth Science*, vol. 22, pp. 1893–1898.

Beyer, K.H., Jr., Bergfeld, W.F., Berndt, W.O., Boutwell, R.K., Carlton, W.W., Hoffmann, D.K., and Schroeter, A.L., 1983, "Final report on the safety assessment of triethanolamine, diethanolamine, and monoethanolamine," *J. Am. Coll. Toxicol.* 2, pp. 183–235.

Bezdek, R.H., 1993, "The environmental, health, and safety implications of solar energy in central station power production," *Energy*, Volume 18, Issue 6, June, pp. 681–685.

Biello, D., 2010, "How Fast Can Microbes Clean Up the Gulf Oil Spill?" *Scientific American*, Aug. 24, 12 pp.

Bjorndalen, N., Mustafiz, S., and Islam, M.R., 2005, "No-flare design: converting waste to value addition," *Energy Sources*, 27(4), pp. 371–80.

Blank, L.T. and Tarquin, A.J., 1983, *Engineering economy*, New York, McGraw-Hill, Inc.

Blomstrom, D.C. and Beyer, E.M. Jr., 1980, "Plants metabolise ethylene to ethylene glycol," *Nature*, 283, pp. 66–68.

Boehman, A.L., 2005, "Biodiesel production and processing," *Fuel Processing Technology 86*, pp. 1057–1058.

Bone III, L., 1989, "Accelerated Testing of Atmospheric Coatings for Offshore Structures," *Material Performance*, November, p. 31.

Boocock, D.G.B., Konar, S.K., Mao, V., and Sidi, H., 1996, Fast one-phase oil-rich processes for the preparation of vegetable oil methyl starch, *Biomass Bioenergy*, vol. 11, 43–50.

Boocock, D.G.B., Konar, S.K., Mao, V., Lee, C., and Buligan, S., 1998, "Fast formation of high purity methyl esters from vegetable oils," *JAOCS* 75 (9), pp. 1167–1172.

Bork, A.M., 1963, "Maxwell, Displacement Current, and Symmetry," *American Journal of Physics* 31, pp. 854–9.

Bork, A.M., 1967, "Maxwell and the Vector Potential", *Isis* 58 [2], pp. 210–22.

Boyle, G., Everett, B., and Ramage, J., eds., 2003, *Energy Systems and Sustainability: Power for a Sustainable Future*, Oxford University Press Inc., New York, 2003.

Brecht, B., 1947, *Selected Poems*, New York, Harcourt-Brace, Translations by H.R. Hays.

Brenner, D.J. and Hall, E.J., 2007, "Computer Tomography – An Increasing Source of Radiation Exposure," *New England Journal of Medicine*, issue 357, pp. 2277–2284.

Budge, S.M., Iverson, S.J., and Koopman, H.N., 2006, "Studying trophic ecology in marine ecosystems using fatty acids: A primer on analysis and interpretation," *Marine Mammal Science*, In press.

Buisman, C.J.N., Post, R., Ijspreet, P., Geraats, S., and Lettinga, G., 1989, "Biotechnological process for sulphide removal with sulphur reclamation," *Acta Biotechnologica* 9, pp. 271–283.

Bunge, Mario, 1962, "The Complexity of Simplicity," *Journal of Philosophy* 59 [5], pp. 113–35.

Burk, J.H., 1987, "Comparison of Sodium Carbonate, Sodium Hydroxide, and Sodium Orthosilicate for EOR," *SPE Reservoir Engineering*, pp. 9–16.

Butler, N., 2005, The Global Energy Challenge, Council on Foreign Relations, the Corporate Conference, New York, N.Y., March 11, 2005.

Calle, S., Klaba, L., Thomas, D., Perrin, L., and Dufaud, O., 2005, "Influence of the size distribution and concentration on wood dust explosion: Experiments and reaction modeling," *Powder Technology*, Vol. 157 (1–3), September, pp. 144–148.

Cameco, U., 2007, "101- Nuclear Energy," May, 2006, www.cameco.com/ uranium_101/ nuclear_electricity (Acessed on June 30, 2007).

Campbell, T.C., 1977, "A Comparison of Sodium Orthosilicate and Sodium Hydroxide for Alkaline Waterflooding," *Journal for Petroleum Technology*, SPE 6514, pp. 1–8.

Canadian Gas Potential Committee 1997, "Natural Gas Potential in Canada."

Canakci, M., 2007, "The Potential of Restaurant Waste Lipids as Biodiesel Feedstocks," *Bioresource Technology* 98, pp. 183–190.

Carraretto, C., Macor, A., Mirandola, A., Stoppato, A., and Tonon, S., 2004, "Biodiesel as alternative fuel: Experimental analysis and energetic evaluations," *Energy* 29, pp. 2195–2211.

Carter, D., Darby, D., Halle, J., and Hunt, P., 2005, *How To Make Biodiesel*, Low-Impact Living Initiative, Redfield Community, Winslow, Bucks, UK. ISSN 0-9649171-0-3.

Cave Brown, Anthony, 1975, *Bodyguard of Lies*, New York, Harper & Row.

Caveman Chemistry, 2006, <http://cavemanchemistry.com/browse.html> July 22, 2006.

CEF Consultants Ltd, 1998, "Exploring for Offshore Oil and Gas (Nov)," No. 2 of Paper Series on *Energy and the Offshore*, Halifax, NS, <http://www.cefconsultants.ns.ca/2explore.pdf> [Accessed on May 18, 2005].

CERHR, 2003, NTP-CERHR, expert panel report on reproductive and developmental toxicity of propylene glycol. National Toxicology Program U.S. Department of Health and Human Services. NTP-CERHR-PG-03.

Chakma, A., 1996, "Acid Gas Re-injection a practical way to eliminate CO_2 emissions from gas processing plants," *Proceedings of the third International Conference on CO_2 Removal*, Massachusetts Institute of Technology, Cambridge, MA, USA, 9–11 September, 1996.

Chakma, A., 1997, "CO_2 Capture Processes Opportunities for Improved Energy Efficiencies," *Energy Conversion and Management*, volume 38, pp. 51–56.

Chakma, A., 1999, "Formulated solvents: new opportunities for energy efficient separation of acid gases," *Energy Sources 21*(1–2).

Chalmers, A.F., 1973a, "The Limitations of Maxwell's Electromagnetic Theory," *Isis* 64 [4], pp. 469–483.

Chalmers, A.F., 1973b, "On Learning from Our Mistakes," *British Journal for the Philosophy of Science* 24 [2], pp. 164–173.

Chang, F.F. and Civan, F., 1997, "Practical Model for Chemically Induced Formation Damage," *J. Pet. Sci. Eng.*, vol. 17, 123–137.

Chemistry Store, 2005, <http://www.chemistrystore.com/index.html> June 07, 2006.

Chengde, Z., 1995, "Corrosion Protection of Oil Gas Seawater Submarine Pipeline for Bohai Offshore Oilfield," *The International Meeting on Petroleum Engineering*, Beijing, China, SPE 29972.

Chhetri, A.B., 1997, *An Experimental Study of Emissions Factors from Domestic Biomass Cookstoves*, A Thesis Submitted for the Partial Fulfilment of the Requirements for the Degree of Master of Engineering, Asian Institute of Technology, Bangkok, Thailand, AIT Thesis no. ET-97–34, 147 pp.

Chhetri, A.B., 2007, "Scientific Characterization of Global Energy Sources," *J. Nat. Sci. and Sust. Tech.*, 2007 1(3), pp. 359–395.

Chhetri, A.B. and Islam, M.R., 2007a, "Reversing Global Warming," *J. Nat. Sci. and Sust. Tech.* 1(1), pp. 79–114.

Chhetri, A.B. and Islam, M.R., 2007b, "Pathway Analysis of Crude and Refined Oil and Gas," *Int. Journal of Environmental Pollution*. Submitted.

Chhetri, A.B. and Islam, M.R., 2008, *Inherently Sustainable Technology Development*, Nova Science Publishers, New York, 452 pp.

Chhetri, A.B. and Islam, M.R., 2009, "Greening of petroleum operations," *Advances in Sustainable Petroleum Engineering and Science*, vol. 1, no. 1, pp. 1–35.

Chhetri, A.B., Khan, M.I., and Islam, M.R., 2008, "A Novel Sustainably Developed Cooking Stove," *J. Nat. Sci. and Sust. Tech.*, 1(4), pp. 589–602.

Chhetri, A.B., M.S. Tango, S.M. Budge, K.C. Watts and M.R. Islam, 2008, "Non-Edible as New Sources of Biodiesel Production", *Int. J. Mol. Sci.* 2008, vol. 9, 169–180.

Chhetri, A.B., Rahman, M.S., and Islam, M.R., 2006, "Production of Truly 'Healthy' Health Products," 2nd Int. Conference on Appropriate Technology, July 12–14, Zimbabwe.

Chhetri, A.B., Rahman, M.S., and Islam, M.R., 2007, "Characterization of Truly 'Healthy' Health Products," *J Characterization and Development of Novel Materials*, submitted.

Chhetri, A.B., Watts, K.C., and Islam, M.R., 2008c, "Soapnut extraction as a natural surfactant for Enhanced Oil Recovery," in press.

Chhetri, A.K., Zatzman, G.M., and Islam, M.R., 2008, "Book review: O.G. Sorokhtin, G.V. Chilingar, and L.F. Khilyuk, 2007, *Global Warming and Global Cooling, Evolution of Climate on Earth*," *J Nat Sci & Sust Tech.* Vol. 1, No. 4, pp. 693–698.

Civan, F. and Engler, T., 1994, "Drilling Mud Filtrate Invasion – Improved Model and Solution," *J. Pet. Sci. Eng.*, vol. 11, pp. 183–193.

Clayton, M.A. and Moffat, J.W., 1999, "Dynamical Mechanism for Varying Light Velocity as a Solution to Cosmological Problems," *Physics Letters* B, vol. 460, No.3–4, pp. 263–270.

ClearTech, 2006, "Industrial Chemicals, North Corman Industrial Park, Saskatoon S7L 5Z3, Canada," <http://www.cleartech.ca/products.html> accessed: May 08, 2006.

CMAI, 2005, "Chemical Market Associates Incorporated," <www.kasteelchemical.com/slide.cfm> accessed: May 20, 2006.

Cockburn, Alex, 2007, "Al Gore's Peace Prize," *Counterpunch* (13–14 October), at http://www.counterpunch.org/cockburn10132007,html.

Cohen, D.E., 2008, *What matters*, Sterling Publishing Co Inc (United States), 2008, Hardback, 336 pages.

Cohen, I.B., 1995, "Newton's method and Newton's style," in Cohen and Westfall, R.S., eds., *Newton: Texts, Background, Commentaries*, New York, WW Norton, pp. 126–143.

Coltrain, D., 2002, "Biodiesel: Is It Worth Considering?," Risk and Profit Conference Kansas State University Holiday Inn, Manhattan, Kansas August 15–16.

Connaughton, S., Collins, G., and O'Flaherty, V., 2006, "Psychrophilic and mesophilic anaerobic digestion of brewery effluent: a comparative study," *Water Research* 40(13), pp. 2503–2510.

Connemann, J. and Fischer, J., 1998, "Biodiesel in Europe 1998: biodiesel processing technologies," Paper presented at the International Liquid Biofuels Congress, Brazil, 15 pp.

Cooke, C.E. Jr., Williams, R.E. and Kolodzie, P.H., 1974, "Oil Recovery by Alkaline Water Flooding," *Jour. Pet. Tech.*, pp. 1356–1374.

Correra, S., 2004, "Stepwise Construction of An Asphaltene Precipitation Model," vol. 22, issue 7&8, pp. 943–959.

Coskuner, G. and Bentsen, R.G.J., 1990, "A Scaling Criterion for Miscible Displacements," *Can. Pet. Tech.*, volume 29, no. 1, pp. 86–88.

Crump, K.S., "Numerical Inversion of Laplace Transforms Using Fourier Series Approximations," *J. Assoc. Comput. Mach.*, vol. 23 (1), pp. 89–96, 1976.

Currie, D.R. and Isaacs, L.R., 2005, "Impact of exploratory offshore drilling on benthic communities in the Minerva gas field, Port Campbell, Australia," *Marine Environmental Research*, Vol. 59, pp. 217–233.

Daly, H.E., 1992, "Allocation, distribution. and scale: towards an economics that is efficient. just and sustainable," *Ecological Economics*, Vol. 6, pp. 185–193.

Davis, R.A., Thomson, D.H., Malme, C.I., and Malme, C.I., 1998, Environmental Assessment of Seismic Explorations, Canada/Nova Offshore Petroleum Board, Halifax, NS, Canada.

De Esteban, F., 2002, "The Future of Nuclear Energy in the European Union," Background Paper for a Speech Made to a Group of Senior Representatives from Nuclear Utilities in the Context of a "European Strategic Exchange," Brussels, 23rd May 2002.

De Groot, S.J.D., 1996, "Quantitative assessment of the development of the offshore oil and gas industry in the North Sea," *ICES Journal of Marine Science*, vol. 53, pp. 1045–1050.

Deakin, S. and Konzelmann, S.J., 2004, "Learning from Enron," *Corporate Governance*, vol 12, No. 2, pp. 134–142.

Degens, E. T. and Ittekkot, V., 1982, "In Situ Metal-Staining of Biological Membranes in Sediments," *Nature*, vol. 298, pp. 262–264.

Demirba, A., 2003, "Biodiesel fuels from vegetable oils via catalytic and non-catalytic supercritical alcohol transesterifications and other methods, a survey," *Energy Conversion and Management*, Volume 44 (13), pp. 2093–2109.

Department of Environment and Heritage, 2005, Australian Government, Greenhouse Office, "Fuel consumption and the environment," <www.greenhouse.gov.au/fuellabel/ environment.html> accessed on February 18, 2006.

Deydier, E., Guilet, R., Sarda, S., and Sharrock, P., 2005, "Physical and Chemical Characterisation of Crude Meat and Bone Meal Combustion Residue: 'Waste or Raw Material?'" *Journal of Hazardous Materials*, Vol. 121 1–3, pp. 141–148.

deZabala, E.F. and Radke, C.J., 1982, "The Role of Interfacial Resistances in Alkaline Water Flooding of Acid Oils," paper *SPE* 11213 presented at the 1982 SPE Annual Conference and Exhibition, New Orleans, pp. 26–29.

Dincer, I. and Rosen, M.A., 2005, "Thermodynamic aspects of renewable and sustainable development," *Renewable & Sustainable Energy Reviews*, 9, pp. 169–89.

Dingle, H., 1950, "A Theory of Measurement," *The British Journal for the Philosophy of Science*, Vol. 1, No. 1 (May), pp. 5–26.

Diviacco, P., 2005, "An open source, web-based, simple solution for seismic data dissemination and collaborative research," *Commuters & Geosciences*, 31, pp. 599–605.

Donaldson, E.C. and Chernoglazov, V., "Drilling Mud Fluid Invasion Model," *J. Pet. Sci. Eng.*, vol. 1(1), pp. 3–13, 1987.

Donnet, M., Bowen, P., Jongen, N., Lemaitre, J. and Hofmann, H., 2005, "Use of Seeds to Control Precipitation of Calcium Carbonate and Determination of Seed Nature," *Langmuir*, vol. 21, pp. 100–108.

Drake, S., 1970, "Renaissance Music and Experimental Science," *Journal of the History of Ideas*, Vol. 31, No. 4. (October – December), pp. 483–500.

Drake, S., 1973, "Galileo's Discovery of the Law of Free Fall," *Scientific American* v. 228, #5, pp. 84–92.

Drake, S., 1977, "Galileo and the Career of Philosophy," *Journal of the History of Ideas*, Vol. 38, No. 1 (January – March), pp. 19–32.

Du, W., Xu, Y., Liu, D., and Zeng, J., 2004, "Comparative study on lipase-catalyzed transformation of soybean oil for biodiesel production with different acyl acceptors," *Journal of Molecular Catalysis B: Enzymatic 30*, pp. 125–129.

Duhem, P., 1914, *The Aim and Structure of Physical Theory*, Princeton, Princeton U P, translation by P. Wiener for an English-language ed. published 1954.

Dunn, K., 2003, *Caveman Chemistry*, Chapter-8, Universal Publishers, USA.

Dyer, S.B., Huang, S., Farouq Ali, S.M., and Jha, K.N., 1994, "Phase Behavior and Scaled Model Studies of Prototype Saskatchewan Heavy Oils with Carbon Dioxide," *J. of Can. Pet. Tech.*, pp. 42–48.

Eckhardt, F.E.W., 1985, "Solulization, transport and deposition of mineral cations by microorganisms—Efficient rock weathering agents," in Drever, J.I., ed., *The Chemistry of weathering,* Dordresht, Netherlands, D. Reidd, pp. 161–173.

Ehrlich, H.L., 1974, "The Formation of Ores in the Sedimentary Environment of the Sea with Microbial Participation: The Case Of Ferro-Manganese Concretions," *Soil Science,* vol. 119, pp. 36–41.

Ehrlich, H.L., 1983, "Manganese Oxidizing Bacteria from a Hydrothermally Active Region on the Galapogos Rift," *Ecol. Bull. Stockholm,* vol. 35, pp. 357–366.

EIA (Energy information Administration), 2003, International Energy Annual 2003 Report, Washington DC [U.S. Department of Energy Report, 2005].

EIA (Energy Information Administration), 2001, Annual Energy Review, 2001.

EIA, 2006, Energy Information Administration/International Energy Outlook, 2006b.

EIA, 2006a, Energy Information Administration, System for the Analysis of Global Energy Markets (2006). International Energy Outlook 2006, Office of Integrated Analysis and Forecasting U.S. Department of Energy Washington, DC 20585, 2006a, www.eia.doe.gov/oiaf/ieo/index.html.

EIA, 2006b, Energy Information Administration, International Energy Annual, 2003 (May-July, 2005).

EIA, 2008, Short-Term Energy Outlook, http://www.eia.doe.gov/emeu/steo/pub/ contents.html (Accessed on March 11, 2008).

EIA, Annual Energy Outlook 2005, Market Trends- Energy Demand, Energy Information Administration, Environmental Issues and World Energy Use. EI 30, 1000 Independence Avenue, SW, Washington, DC 20585, 2005.

EIA, Nuclear Issues Paper, 2006, Energy Information Administration, Official energy Statistics from the U.S. government, 2006c, www.eia.doe.gov/cneaf/nuclear/page/ nuclearenvissues.html (accessed on January 11, 2007).

El-Etre, A.Y., 1998, *Corrosion Science* 39(11), p. 1845.

El-Etre, A.Y. and Abdallah, M., 2000, *Corrosion Science* 42(4), p. 731.

Elkamel, A., Al-Sahhaf, T., and Ahmed, A.S., 2002, "Studying the Interactions Between an Arabian Heavy Crude Oil and Alkaline Solutions," *Journal Petroleum Science and Technology,* Vol. 20 7), pp. 789–807.

Ellwood, C.A., 1931, "Scientific Method in Sociology," *Social Forces* Vol. 10, No. 1 (October), pp. 15–21.

Emmons, F.R., 1986, "Nitrogen Management at the East Binger Unit Using an Integrated Cryogenic Process," SPE paper 15591 presented at the SPE Annual Technical Conference and Exhibition, New Orleans, LA, Oct. 5–8.

Energy Information Administration, 2005, EIA's International Energy Outlook 2005 <www.eia.doe.gov/neic/experts/expertanswers. html> accessed on February 18, 2006.

Environment Canada, 2003, Transportation and environment, Environment Canada <www.ec.gc.ca/transport/publications/biodiesel/bio diesel12.html>Accessed November 25, 2005.

Environment Canada, 2007, Canadian Climate Normals 1971–2000 [online] Available: (http://www.climate.weatheroffice.ec.gc.ca/climate_nor mals/stnselect_e.html) [February 10, 2007].

Environmental Defense, 2004, <http://www.environmentaldefense. org/> (Accessed on June 2, 2006).

EPA, 2000, Development Document for Final Effluent Limitations Guidelines and Standards for Synthetic-Based Drilling Fluids and other Non-Aqueous Drilling Fluids in the Oil and Gas Extraction Point Source Category. EPA- 821-B-00-013, U.S. Environmental Protection Agency, Office of Water, Washington, DC 20460, December, <http:// www.epa.gov/waterscience/guide/sbf/fi nal/eng.html>.

EPA, 2002, A Comprehensive analysis of biodiesel impacts on exhaust emissions. Air and radiation. Draft technical report. EPA420-P-02-001.

Erol, M., Kucukbayrak, S., and Ersoy-Mericboyu, A., 2007, "Characterization of Coal Fly Ash for Possible Utilization in Glass Production," *Fuel*, Vol. 86, pp. 706–714.

Farouq Ali, S.M., Redford, D.A., and Islam, M.R., 1987, "Scaling Laws for Enhanced Oil Recovery Experiments," *Proc. of the China-Canada Heavy Oil Symposium*, Zhou Zhou City, China.

Fernandes, M.B. and Brooks, P., 2003, "Characterization of carbonaceous combustion residues: II. Nonpolar organic compounds," *Chemosphere*, Volume 53, Issue 5, November, pp. 447–458.

Ferris, F.G., Fyfe, W.S., and Beveridge, T.J., 1987, "Manganese Oxide Deposition in a Hot Spring Microbial Mat," *Geomicrobiologv Journal*, vol. 5, No. 1, pp. 33–42.

Ferris, F.G., Fyfe, W.S., and Beveridge, T.J., 1988, "Metallic Ion Binding by Bacillus Subtilis: Implications for the Fossilization of Microorganisms," *Geology*, vol. 16, pp. 149–152.

Feuer, L.S., 1957, "The Principle of Simplicity," *Philosophy of Science* 24 [2], pp. 109–22.

Feuer, L.S., 1959, "Rejoinder on the Principle of Simplicity," *Philosophy of Science* 26 [1], pp. 43–5.

Fink, F.W. and Boyd, W.K., 1970, *The Corrosion of Metals in Marine Environments*, Bayer & Company Inc., USA.

Finkel, Alvin and Leibovitz, Clement, 1997, *The Chamberlain-Hitler Collusion*, London, Merlin.

Fiore, K., 2006, "Nuclear energy and sustainability: Understanding ITER," *Energy Policy*, 34, pp. 3334–3341.

Fischer, A. and Hahn, C., 2005, "Biotic and abiotic degradation behaviour of ethylene glycol monomethyl ether (EGME)," *Water Research*, Vol. 39, pp. 2002–2007.

Fontana, M. and Green, N., 1978, *Corrosion Engineering*, McGraw Hill International, New York.

Fraas, L.M., Partain, L.D., McLeod, P.S., and Cape, J.A., 1986, "Near-Term Higher Efficiencies with Mechanically Stacked Two-Color Solar Batteries," *Solar Cells* 19 (1), pp. 73–83, November.

Frank, J., 2006, "Inconvenient Truths About the Ozone Man: Al Gore the Environmental Titan?" *Counterpunch*, 31 May, at http://www.counterpunch.org/frank05312006,html

Frazer, L.C. and Bolling, J.D., 1991, "Hydrogen Sulfide Forecasting Techniques for the Kuparuk River Field," SPE paper 22105 presented at the International Arctic Technology Conference, Anchorage, Alaska.

Freud, P. and Ormerod, W., 1996, "Progress towards storage of CO_2," *Proceedings of the Third International Conference on CO_2 Removal, Massachusetts Institute of Technology*, Cambridge, MA, USA, 9–11 September.

Geesey, G., Lewandewski, Z., and Flemming, H., 1994, *Biofouling and Biocorrosion in Industrial Water Systems*, Lewis Publishers, MI, USA.

Geilikman, M.B. and Dusseault, M.B., "Fluid Rate Enhancement from Massive Sand Production in Heavy-Oil Reservoirs," *J. Pet. Sci. Eng.*, vol. 17, pp. 5–18, 1997.

Gerpen, J.V., Pruszko, R., Shanks, B., Clements, D., and Knothe, G., 2004, "Biodiesel Analytical methods," National Renewable Energy Laboratory, Operated for the U.S. Department of Energy.

GESAMP (I MO/FAO/UNESCO/WMO/WHO/IAEA/UNEP), 1993, "Joint Group of Experts on the Scientific Aspects of marine Pollution," *Impacts of Oil and Related Chemicals and Wastes on the Marine Environment*, GESAMP Reports and Studies No. 50. London, International Maritime Organization.

Gesser, H.D., 2002, *Applied Chemistry*, Kluwer Academic, Plenum Publishers, NY, USA.

Gessinger, G., 1997, "Lower CO_2 emissions through better technology," *Energy Conversion and Management*, volume 38, pp. 25–30.

Gilbert, S.R., Bounds, C.O., and Ice, R.R., "Comparative Economics Of Bacterial Oxidation And Roasting As A Pre-Treatment Step For Gold Recovery From An Auriferous Pyrite Concentrate," *CIM* 81(910), 1988.

Giridhar, M., Kolluru, C., Kumar, R., 2004, "Synthesis of Biodiesel in Supercritical Fluids," *Fuel* 83, pp. 2029–2033.

Gleick, J., 1987, *Chaos – making a new science*, New York, Penguin Books, 352 pp.

Godoy, J., 2006, "Environment: Heat Wave Shows Limits of Nuclear Energy," Inter Press Service News Agency. July 27, 2006.

Goldberg, N.N., and Hudock, J.S., 1986, Oil and Dirt Repellent Alkyd Paint, United States Patent 4600441.

Gollapudi, U.K., Knutson, C.L., Bang, S.S., and Islam, M.R., 1995, "A New Method for Controlling Leaching Through Permeable Channels," *Chemosphere*, vol. 46, pp. 749–752.

Gonzalez, G. and Moreira, M.B.C., "The Adsorption of Asphaltenes and Resins on Various Minerals," in Yen and Chilingar, eds., *Asphaltenes and Asphalts*, Elsevier Science B.V., Amsterdam, pp. 249–298 (1994).

Goodman, Nelson, 1961, "Safety, Strength, Simplicity," *Philosophy of Science* 28 [2], pp. 150–1.

Goodstein, D., 2000, "Whatever Happened to Cold Fusion?," *Accountability in Research*, Vol. 8, p. 59.

Goodstein, D., 2004, "Whatever Happened to Cold Fusion?," *The American Scholar*, 527 pp.

Goodwin, L., 1962, "The Historical-Philosophical Basis for Uniting Social Science with Social Problem-Solving," *Philosophy of Science*, Vol. 29, No. 4. (October), pp. 377–392.

Gore, A., 1992, *Earth in the Balance: Ecology and the Human Spirit*, Houghton Mifflin Company, Boston, New York, London, 407 pp.

Gore, A., 2006, *An Inconvenient Truth*, New York, Rodale. Also a DVD, starring Al Gore presenting the book's content as a public lecture, plus additional personal reflections. Produced by Davis Guggenheim.

Government of Saskatchewan, Energy and Mines News Release, 1997, "$1.1 Billion Oil Project Announced in Southeast Saskatchewan," June 26.

Grigg, R.B. and Schechter, D.S., 1997, "State of the Industry in CO_2 Floods," SPE paper 38849 presented at the Annual Technical Conference and Exhibition, San Antonio, Texas, October 5–8.

Gruesbeck, C. and Collins, R.E., 1982, "Entrainment and Deposition of Fine Particles in Porous Media," *Soc. Pet. Eng. J.*, Dec., pp. 847–856.

Guenther, W.B., 1982, "Wood Ash Analysis: An Experiment for Introductory Courses," *J. Chem. Educ.*, Vol. 59, pp. 1047–1048.

Gunal, G.O. and Islam, M.R., 2000, "Alteration of asphaltic crude rheology with electromagnetic and ultrasonic irradiation," *J. Pet. Sci. Eng.*, vol. 26 (1–4), pp. 263–272.

Gunter, W.D., Gentzis, T., Rottenfuser, B.A., and Richardson, R.J.H., 1996, "Deep Coal-bed Methane in Alberta, Canada: A Fossil Fuel Resource with the Potential of Zero Greenhouse Gas Emissions," *Proceedings of the Third International Conference on CO_2 Removal*, Massachusetts Institute of Technology, Cambridge, MA, USA, 9–11 September.

Gunter, W.D., Bachu, S., Law, D.H.-S., Marwaha, V., Drysdale, D.L., Macdonald, D.E., and McCann, T.J., 1996, "Technical and Economic Feasibility of CO_2 disposal in aquifers within the Alberta sedimentary basin," *Energy Conversion and Management*, volume 37, nos. 6–8, pp. 1135–1142.

Gunter, W.D., Gentzis, T., Rottenfusser, B.A., and Richardson, R.J.H., 1997, "Deep coal-bed Methane in Alberta, Canada: A Fossil Fuel Resource with the Potential of Zero Greenhouse Gas Emissions," *Energy Convers. Mgmt.*, vol. 38 suppl., S217-S222.

Gupta, C.K. and Mukherjee, T.K., 1990, *Hydrometallurgy in Extraction Processes*, vol. 1, CRC Press, New York, 248 pp.

Haldane, J.B.S., 1957, Karl Pearson 1857–1957, *Biometrika*, Vol. 44, Nos. 3/4., (December), pp. 303–313.

Hamming, R.W., 1973, *Numerical Methods for Scientists and Engineers*, New York, McGraw-Hill, 2nd Edition. Ix, 719 pp.

Hanson, R.S. and Hanson, T.E., 1996, "Methanotrophic bacteria," *Microbial. Rev.* 60, pp. 439–471.

Haque, K.E., 1999, Microwave Energy for Mineral Treatment Processes – A Brief Review," *International Journal of Mineral Processing*, 57(1), pp. 1–24.

Harris, G.M. and Lorenz, A., 1993, "New Coatings for the Corrosion Protection of Steel Pipelines and Pilings in Severely Aggressive Environments," *Corrosion Science* 35(5), p. 1417.

Harvey, A.H. and Henry, R.L., 1977, "A Laboratory Investigation of Oil Recovery by Displacement with Carbon dioxide and Hydrogen sulfide," SPE 6983, unsolicited manuscript.

Hau, L.V., Harris, S.E., Dutton, Z., and Behroozi, C.H., 1999, "Light Speed Reduction to 17 Meters Per Second in an Ultra Cold Atomic Gas," *Nature*, vol. 397, pp. 594–598.

Haynes, H.J., Thrasher, L.W., Katz, M.L., and Eck, T.R., 1976, Enhanced Oil Recovery, National Petroleum Council, An Analysis of the Potential for Enhanced Oil Recovery from Known Fields in the United States.

Henda, R., Herman, A., Gedye, R., and Islam, M.R., 2005, "Microwave enhanced recovery of nickel-copper ore: communication and floatability aspects," *Journal of Microwave Power & Electromagnetic Energy*, 40(1), pp. 7–16.

Hendriks, C.A., 1994, *Carbon Dioxide Removal from Coal-Fired Power Plants*, Ph.D. Thesis, Department of Science, Technology and Society, Utrecht University, Utrecht, The Netherlands.

Henning, R.K., 2004, "Integrated Rural Development by Utilization of Jatropha Curcas L. (JCL) as Raw Material" and as "Renewable Energy: Presentation of The Jatropha System" at the international Conference, Renewables 2004 in Bonn, Germany.

Herzog, H.E., Drake, Tester, J., and Rosenthal, R., 1993, A Research Needs Assessment for the Capture, Utilization and Disposal of Carbon Dioxide from Fossil Fuel-Fired Power Plants DOE/ER-30194, US Department of Energy, Washington, D.C.

Hesse, M., Meier, H. and Zeeh, B., 1979, *Spektroscopische Methoden in Der Organischen Chemie*, Thieme VerlagStuttgart.

Hilber, T., Mittelbach, M., and Schmidt, E., 2006, "Animal fats perform well in biodiesel," *Render,* February 2006, www.rendermagazine.com (accessed on Aug 30, 2006).

Hill, D.E., Bross, S.V., and Goldman, E.R., 1990, "The Impacts of Microbial Souring of a North Slope Oil Reservoir," paper presented at the International Congress on Microbially Influenced Corrosion, Knoxville, TN, Oct.

Hills, R.G., Porro, I., Hudson, D.B., and Wierenga, P.J., 1989, "Modeling one-dimensional infiltration into very dry soils 1. Model development and evaluation," *Water Resources Res.* vol. 25, pp. 1259–1269.

Himpsel, F., 2007, *Condensed-Matter and Material Physics: The Science of the World Around Us*, The National Academies Press, ISBN-13: 978-0-309-10965-9, pp. 224.

Hitchon, B., 1996, *Aquifer Disposal of Carbon Dioxide – hydrodynamics and Mineral Trapping: Proof of Concept*, Geo-science Publ. Ltd., Sherwood Park, Alberta, Canada.

Holbein, B.E., Stephen, J.D., and Layzell, D.B., 2004, Canadian Biodiesel Initiative, Final Report; Biocap Canada, Kingston, Ontario, Canada.

Holdway, D.A., 2002, "The Acute and Chronic Effects of Wastes Associated with Offshore Oil and Gas Production on Temperature and Tropical Marine Ecological Process," *Marine Pollution Bulletin*, 44, pp. 185–203.

Holloway, S. and van der Straaten, R., 1995, "The joule II project, 'The Underground Disposal of Carbon Dioxide,'" *Energy Conversion and Management*, volume 36, no. 6–9, pp. 519–522.

Holloway, S., 1996, "An Overview of the Joule II Project, 'The Underground Disposal of Carbon Dioxide,'" *Proceedings of the Third International Conference on CO_2 Removal*, Massachusetts Institute of Technology, Cambridge, MA, USA, 9–11 September.

Holmberg, S.L., Claesson, T., Abul-Milh, M., and Steenari, B.M., 2003, "Drying of Granulated Wood Ash by Flue Gas from Saw Dust and Natural Gas Combustion," *Resources, Conservation and Recycling*, Vol. 38, pp. 301–316.

Holt, T., Jensen, J.I., and Lindeberg, E., 1995, "Underground Storage of CO_2 in Aquifers and Oil Reservoirs," *Energy Conversion and Management*, volume 36, no. 6–9, pp. 535–538.

Holton, G., 1969, "Einstein, Michelson, and the 'Crucial' Experiment," *Isis* 60 [2], pp. 132–197.

Hotz, Robert Lee, 2007, "Scientists Using Maps Of Genes for Therapies Are Wary of Profiling," *The Wall Street Journal*, Friday 26 October, Page B1, (http://online.wsj.com/article/SB119334828528572037.html)

Hoyle, B. and Beveridge, 1983, "The Binding of Metallic Ions to the Outer Membrane of *Escherichia Coli*," *Applied and Environment Microbiology*, vol. 46, pp. 749–752.

Hu, P.Y., Hsieh, Y.H., Chen, J.C., and Chang, C.Y., 2004, "Characteristics of Manganese-Coated Sand Using SEM and EDAX Analysis," *J. Colloid and Interface Sc.*, Vol. 272, pp. 308–313.

Huang, S., de Wit, P., Shatilla, N., Dyer, S., Verkoczy, B., and Knorr, K., 1989, Miscible Displacement of Saskatchewan's Light and Medium Oil Reservoirs, Confidential Technical Report by Saskatchewan Research Council, Petroleum Research.

Huang, S.S., de Wit, P., Srivastava, R.K., and Jha, K.N., 1994, "A Laboratory Miscible Displacement Study for the Recovery of Saskatchewan's Crude Oil," *J. Can. Pet. Tech.*, pp. 43–51, April.

Hughs, M.N. and Poole, R.K., 1989, *Metals and Microorganisms*, London, Chapman and Hall, pp. 303–357.

Hull, T.R., Quinn, R.E., Areri, I.G., and Purser, D.A., 2002, "Combustion toxicity of fire retarded EVA," *Polymer Degradation and Stability* 77, pp. 235–242.

IAEA, 2004, Nuclear Technology Review, International Atomic Energy Agency, P.O. Box 100, Wagramer Strasse 5, A-1400 Vienna, Austria.

Imberger, J., 2007, Interview on debate on climate change, CNN, Aug. 19.

International, Energy Agency GHG, 1994a, "Carbon Dioxide capture from Power Stations," IEA Greenhouse Gas R&D Program, Cheltenham, UK.

International, Energy Agency GHG, 1995, R & D program report, "Carbon Dioxide Utilization," IEA Greenhouse Gas R&D Program, Cheltenham, UK.

International, Energy Agency GHG, 1996b, "Ocean Storage of CO_2, environmental impact," IEA Greenhouse Gas R&D Program, Cheltenham, UK.

Ion, S.E., "Optimising Our Resources," The Uranium Institute, Twenty Second Annual Symposium 3.5 September, London, 1997.

IPCC, 2001, *Climate Change 2001: The Scientific Basis*, J.T. Houghton, Y. Ding, D.J. Griggs, M. Noguer, P.J. Van der Linden, X. Dai, K. Maskell, and C.A. Johnson, eds, Cambridge University Press, Cambridge, UK, 881 pp.

IPCC, 2007, Climate Change 2007: The Physical Science Basis, Summary for Policymakers, Contribution of Working Group I to the Fourth Assessment Report of the Intergovernmental Panel on Climate Change, February, 2007.

Islam, M.R., 1990, "New Scaling Criteria for Chemical Flooding Experiments," *Journal of Canadian Petroleum Technology*, Vol. 29 1, pp. 30–36.

Islam, M.R., 1994, "Role of Asphaltenes on Oil Recovery and Mathematical Modeling of Asphaltene Properties," in Yen and Chilingar, eds., *Asphaltenes and Asphalts*, Elsevier Science B.V., Amsterdam, pp. 249–298.

Islam, M.R., 1994, "Role of Asphaltenes in Heavy Oil Recovery," in Yen and Chilingarian, eds., *Asphaltenes and Asphalts 1*, Elsevier Scientific Publishers, Amsterdam-New York, pp. 249–295.

Islam, M.R., 1995, "Potential of Ultrasonic Generators for Use in Oil Wells and Heavy Crude Oil/Bitumen Transportation Facilities," in Sheu and Mullins, eds.,,*Asphaltenes – Fundamentals and Applications*, Plenum Press, New York (1995), pp. 191–218.

Islam, M.R., 1996, "Emerging Technologies in Enhanced Oil Recovery," *Energy Sources*, Vol. 21, pp. 97–111.

Islam, M.R., 1999, "Emerging Technologies in Enhanced Oil Recovery," *Energy Resources* volume 21, no. 1–2, pp. 97–112.

Islam, M.R., 2003, "Adding Value to Atlantic Canada's Offshore Industry," presented at the 3rd Atlantic Canada Deepwater Workshop, Halifax, N.S., April 9–10, 2003.

Islam, M.R., 2004, "Unraveling the mysteries of chaos and change: knowledge-based technology development," *EEC Innovation*, vol. 2, no. 2 and 3, pp. 45–87.

Islam, M.R., 2004, "Unraveling the Mysteries of Chaos and Change: The Knowledge-Based Technology Development," *EEC Innovation*, Vol. 2 2, pp. 45–87.

Islam, M.R., 2006, "A Knowledge-Based Water and Waste-Water Management Model," International Conference on Management of Water, Wastewater and Environment: Challenges for the Developing Countries, September 13–15, Nepal.

Islam, R., 2008a, "If Nature is perfect, what does 'denaturing' mean?," *Perspective on Sustainable Technology*, M.R. Islam (editor), Nova Science Publishers, New York, 191 pp.

Islam, R., 2008b, "Editorial: How much longer can humanity afford to confuse 'theory' that supports only those practices that make the most money in the shortest time with knowledge of the truth?," *Journal of Nature Science and Sustainable Technology*, vol. 1, no. 4, pp. 510–519.

Islam, M.R. and Bang, S.S., 1993, "Use of Silicate and Carbonate Producing Bacteria in Restoration of Historic Monuments and Buildings," *Patent Disclosure*, South Dakota School of Mines and Technology.

Islam, M.R. and Chakma, A., 1992, "A New Recovery Technique for Heavy Oil Reservoirs with Bottom-Waters," *SPE Res. Eng.*, vol. 7, no. 2, pp. 180–186.

Islam, M.R. and Chakma, A., 1993, "Storage and utilization of CO_2 in Petroleum Reservoirs – A Simulation Study," *Energy Conversion and Management*, volume 34 9, pp. 1205–1212.

Islam, M.R., Chakma, A., and Jha, K., 1994, "Heavy Oil Recovery by Inert Gas Injection with Horizontal Wells," *J. Pet. Sci. Eng.* 11 3, pp. 213–226.

Islam, M.R., Erno, B.P., and Davis, D., 1992, "Hot Water and Gas Flood Equivalence of In Situ Combustion," *J. Can. Pet. Tech.*, 31 8, pp. 44–52.

Islam, M.R. and Farouq Ali, S.M., 1989, "Numerical Simulation of Alkaline/Cosurfactant/Polymer Flooding," *Proceedings of the UNITAR/UNDP Fourth Int. Conf. on Heavy Crude and Tar Sand.*

Islam, M.R. and Farouq Ali, S.M., 1990, "New Scaling Criteria for Chemical Flooding Experiments," *J. Can. Pet. Tech.*, volume 29 1, pp. 29–36.

Islam, M.R., Shapiro, R., and Zatzman, G.M., 2006, "Energy Crunch: What more lies ahead?" The Dialogue: Global Dialogue on Natural Resources, Center for International and Strategic Studies, Washington DC, April 3–4, 2006.

Issariyakul, T., Kulkarni, M.G., Dalai, A.K., and Bakhshi, N.N., 2007, "Production of biodiesel form waste fryer grease using mixed methanol/ethanol system," *Fuel Processing Technology* 88, pp. 429–436.

Jack, T.R., Ferris, F.G., Stehmeier, L.G., Kantzas,A., and Marentette, D.F., 1992, "Bug Rock: Bactriogenic Mineral Precipitation System for Oil Patch Use," in Remuzic, E. and Woodhead, A., eds., *Microbial Enhancement of Oil- Recovery- Recent Advances,* pp. 27–36, Elsevier Publishing Co., New York

Jack, T.R., Stehmeier, L.G., Ferris, F.G., and Islam, M.R., 1991, "Microbial Selective Plugging to Control Water Channeling," in *Microbial Enhancement of Oil-Recovery --Recent Advances,* Elsevier Publishing Co., New York

Jack, T.R., 1988, "Microbially Enhance Oil Recovery," *Biorecovery,* vol. 1, pp. 59–73.

Jememan, G.E., Knapp, R.M., Mclnerey, M.J., and Menzie, E.O., 1984, "Experimental Studies of In-Situ Microbial Enhanced Oil Recovery," *Society of Petroleum Engineers Journal,* vol. 24, pp. 33–37.

Jennings, H.Y. Jr., 1975, "A Study of Caustic Solution-Crude Oil Interfacial Tensions," *Society of Petroleum Engineers Journal*, SPE-5049, pp. 197–202.

Jepson, P.D., Arbelo, M., Deaville, R., Patterson, I.A.P., Castro, P., Baker, J.R., Degollada, E., Ross, H.M., Herráez, P., Pocknell, A.M., Rodríguez, F., Howie, F.E., Espinosa, A., Reid, R.J., Jaber, J.R., Martin, J., Cunningham, A.A., and Fernández, A., 2003, "Gas-bubble lesions in stranded cetaceans. Was sonar responsible for a spate of whale deaths after an Atlantic military exercise?," *Nature*, 425, pp. 575–6.

Jones-Meehan, J. and Walch, M., 1992, *ASME, International Power Generation Conference.* Atlanta, GA, USA, Publ. ASME, NY, USA, p.1.

Kalia, A.K., and Singh, S.P., 1999, "Case study of 85 m^3 floating drum biogas plant under hilly conditions," *Energy Conversion & Management*, 40, pp. 693–702.

Kamath, V.A., Yang, J., and Sharma, G.D., 1993, "Effect of Asphaltene Deposition on Dynamic Displacement of oil by Water," SPE paper 26046 presented at SPE Western Regional Meeting, Anchorage, May 26–28.

Kane, Robert L. and Klein, Daniel E., 1997, "United States Strategy for Mitigating Global Climate Change," *Energy Conversion and Management*, volume 38, S13–S18.

Kantzas, A., Ferns, F.G., Stehmeier, L., Marentette, D.F., Jha, K.N., and Mourits, F.M., 1992, "A New Method of Sand Consolidation through Bacteriogenic Mineral Plugging," *Petroleum Society of CIM*, CIM paper No. 92–4.

Kao, M.J., Tien, D.C., Jwo, C.S., and Tsung, T.T., 2005, "The study of hydrophilic characteristics of ethylene glycol," *Journal of Physics: Conference Series* 13, pp. 442–445.

Kaoma, J. and Kasali, G.B., 1994, *Efficiency and Emissions of Charcoal use in the Improved Mbuala Cookstoves*, Published by the Stockholm Environment Institute in Collaboration with SIDA, ISBN:91 88116: 94 8.

Kashyap, D.R., Dadhich, K.S., and Sharma, S.K., 2003, "Biomethanation under psychrophilic conditions: a review," *Bioresource Technology* 87(2), pp. 147–153.

Katz, D.A., 1981, "Polymers," Available online at: http://www.chymist.com/Polymers.pdf Accessed February 15th, 2006.

Kelly, N., 2006, "The Role of Energy Efficiency In Reducing Scottish and UK CO_2 Emissions," *Energy Policy* 34, pp. 3505–3515.

Kemp, W.H., 2006, *Biodiesel basics and beyond. A comprehensive guide to production and use for the home and farm*, Aztext Press, 300 pp.

Kessler, E., 2000, "Energies and Policies. Editorial," *Energies*, 1, pp. 38–40, DOI: 10.3390/en1010038

Khan, M.I., 2006, *Towards Sustainability in Offshore Oil and Gas Operations*, Ph.D. Dissertation, Department of Civil and Resource Engineering, Dalhousie University, Canada, 442 pp.

Khan, M.I., 2006, "Development and Application of Criteria for True Sustainability," *Journal of Nature Science and Sustainable Technology*, Vol. 1, No. 1, pp. 1–37.

Khan, M.I., Chhetri, A.B., and Islam, M.R., 2008, "Analyzing Sustainability of Community-Based Energy Development Technologies," *Energy Sources – Part B*, vol. 2, pp. 403–419.

Khan, M.I. and Islam, M.R., 2003a, "Ecosystem-Based Approaches to Offshore Oil and Gas Operation: An Alternative Environmental Management Technique," SPE Annual Technical Conference and Exhibition, Denver, USA. October 6–8.

Khan, M.I. and Islam, M.R., 2003b, "Wastes Management in Offshore Oil and Gas: A Major Challenge in Integrated Coastal Zone Management," in L.G. Luna, ed., *CARICOSTA 2003 – 1st International Conference on*

Integrated Coastal Zone Management ICZM, University of Oriente, Santiago du Cuba, May 5–7, 2003.

Khan, M.I., Zatzman, G.M., and Islam, M.R., 2005b, "A Novel Sustainability Criterion as Applied in Developing Technologies and Management Tools," in *Second International Conference on Sustainable Planning and Development*, Bologna, Italy.

Khan, M.I. and Islam, M.R., 2005b, "Assessing Sustainability of Technological Developments: An Alternative Approach of Selecting Indicators in the Case of Offshore Operations," *Proceedings of ASME International, Mechanical Engineering Congress and Exposition*, Orlando, Florida, November 5–11.

Khan, M.I. and Islam, M.R., 2005b, "Sustainable marine resources management: framework for environmental sustainability in offshore oil and gas operations," Fifth International Conference on Ecosystems and Sustainable Development. Cadiz, Spain, May 03–05, 2005.

Khan, M.I. and Islam, M.R., 2005c, "Achieving True technological sustainability: pathway analysis of a sustainable and an unsustainable product," International Congress of Chemistry and Environment, Indore, India, 24–26 December 2005.

Khan, M.I. and Islam, M.R., 2007, *The Petroleum Engineering Handbook: Sustainable Operations*, Gulf Publishing Company, Houston, TX, 461 pp.

Khan, M.I. and Islam, M.R., 2007, *True Sustainability in Technological Development and Natural Resources Management*, Nova Science Publishers, New York, USA, 381 pp.

Khan, M.I, Zatzman, G., and Islam, M.R., 2005, "New Sustainability Criterion Development of Single Sustainability Criterion as Applied in Developing Technologies," *Jordan International Chemical Engineering Conference* V, Paper No.: JICEC05-BMC-3–12, Amman, Jordan, 12 – 14 September 2005.

Khan, M.M. and Islam, M.R., 2006c, "A new downhole water-oil separation technique," *J. Pet. Sci. Tech.*, vol. 24, no. 7, pp. 789–805.

Khan, M.M., Mills, A., Chaalal, O., and Islam, M.R., 2006c, "Novel Bioabsorbents for the Removal of Heavy Metals from Aqueous Streams," *Proc. 36th Conference on Computer and Industries*, June, Taiwan.

Khan, M.M., Prior, D., and Islam, M.R., 2005, "Direct- usage solar refrigeration: from irreversible thermodynamics to sustainable engineering," Jordan International Chemical Engineering Conference V, 12–14 September, Amman, Jordan.

Khan, M.M., Prior, D., and Islam, M.R., 2007a, "A novel combined heating, cooling, refrigeration system," *J. Nat. Sci. and Sust. Tech.*, 1(1), pp. 133–162.

Khan, M.M., Prior, D., and Islam, M.R., 2007, "Zero-Waste Living with Inherently Sustainable Technologies," *J. Nature Science and Sustainable Technology*, vol. 1, no. 2, pp. 263–270.

Khan, M.M., Zatzman, G.M., and Islam, M.R., "The Formulation of a Comprehensive Mass and Energy Balance Equation," *Proc. ASME International Mechanical Engineering Congress and Exposition*, Boston, MA, Nov. 2–6, 2008.

Khilyuk, L.F., Katz, S.A., Chilingarian, G.V., and Aminzadeh, F., 2003, "Global Warming: Are We Confusing Cause and Effect?," *Energy Sources* 25, pp. 357–370.

Khilyuk, L.F. and Chilingar, G.V., 2004, "Global Warming and Long Term Climatic Changes: A Progress Report," *Environmental Geology* 46(6–7), pp. 970–979.

Khilyuk, L.F. and Chilingar, G.V., 2006, "On Global Forces of Nature Driving the Earth's Climate. Are Humans Involved?," *Environmental Geology* 50 (6), pp. 899–910.

Kim, B.W., Chang, H.N., Kim, I.K., and Lee, K.S., 1992, "Growth kinetics of the photosynthetic bacterium *Chlorobium thiosulfatophilum* in a fed-batch reactor," *Biotechnology Bioengineering* 40, pp. 583–592.

Klass, L.D., 1998, *Biomass for Renewable Energy, Fuels and Chemicals*, Academic Press: New York, pp. 1–2.

Kline, M., 1972, Mathematical Thought from Ancient to Modern Times, Oxford Univ. Press, New York.

Kline, R., 1995, "Construing 'Technology' as 'Applied Science'- Public Rhetoric of Scientists and Engineers in the United States, 1880–1945," *Isis* Vol. 86 No. 2 (June), pp. 194–221.

Klins, M.A., 1984, *Carbon Dioxide Flooding: Basic Mechanisms and Project Design*, HRDC, Boston, MA.

Knipe, P. and Jennings, P., 2007, "Electromagnetic radiation emissions from raps equipment," Available:(http://wwwphys.murdoch.edu.au/Solar2004/Proceedings/Systems/Knipe_Paper_EM.pdf) [February 10, 2007].

Koama, J., Kasali, G.B., and Ellegard, A., 1994, "Efficiency and Emissions of Coal Combustions in Two Unvented Cookstoves," *Energy, Environment and Development Series* no 4. Published by Stockholm Environment Institute, ISBN :9188714 020.

Kocabas, I. and Islam, M.R., 1998, "A Wellbore Model for Predicting Asphaltene Plugging," SPE paper 49199, *Proc. of the SPE Annual Technical Conference and Exhibition*, New Orleans.

Kocabas, I. and Islam, M.R., 2000, "Field-Scale Modeling of Asphaltene Transport in Porous Media," *J. Pet. Sci. Eng.*, vol. 26(1–4), pp. 19–30.

Koch, G.H., Brongers, M.P.H., Thompson, N.G., Virmani, Y.P., and Payer, J.H., 2002, *A Supplement to Materials Performance* 41(7), p. 2.

Koertge, N., 1977, "Galileo and the Problem of Accidents," *Journal of the History of Ideas*, Vol. 38, No. 3 (July – September), pp. 389–408.

Koh, C.A., Westacott, R.E., Zhang, W., Hirachand, K., Creek, J.L., and Soper, A.K., 2002, "Mechanisms of gas hydrate formation and inhibition," *Fluid Phase Equilibria, 194–197*, pp. 143–151.

Koide, H., Takahashi, M., Shindo, Y., Noguchi, Y., Nakayama, S., Iijima, M., Ito, K., and Tazaki, Y., 1994, "Subterranean Disposal of Carbon Dioxide at Cool Formation Temperature," *Conference Proceedings, CLEAN AIR 94*, pp. 63–72.

Koide, H., Tazaki, Y., Noguchi, Y., Nakayama, S., Iijima, M., Ito, K., and Shindo, Y., 1992, "Subterranean Containment and long-term Storage of CO_2 in Unused Aquifers and in Depleted Natural Gas Reservoirs," *Energy Conversion and Management*, volume 33, nos. 5–8, pp. 619–626.

Kondratyev, K.Y.A. and Cracknell, A.P., 1998, *Observing Global Climate Change*, Taylor & Francis. ISBN- 0748401245, 544 pp.

Korbul, R. and Kaddour, A., 1995, "Sleipner Vest CO_2 disposal – injection of removed CO_2 into the Utsira Formation," *Energy Conversion and Management*, volume 36, nos. 6–9, pp. 509–512.

Kruglinski, S., 2006, "Whatever Happened to Cold Fusion?," *Discover*, March, 27 (03).

Krumbein, W.E., 1979, "Photolithotropic and Chemoorganotrophic Activity of Bacteria and Algae as Related to Beachrock Formation and Degradation Gulf of Aqaba Sinai," *Geomicrobiology*, vol. 1, No. 2, pp. 139–203.

Krumbein, W.K., 1974, "On the Precipitation of Aragronite on the Surface of Marine Bacteria," *Naturwissenschaften*, vol. 61, p. 167.

Krupa, I. and Luyt, A.S., 2001, "Physical properties of blends of LLDPE and an oxidized paraffin wax," *Polymer*, Vol. 42, pp. 7285–7289.

KRÜSS, 2006, "Instruments for Surface Chemistry, Measuring Principle of KRÜSS Tensio Meters," *KRÜSS GmbH*, Wissenschaftliche Laborgeräte, Hamburg, Germany.

Kulkarni, M.G. and Dalai, A.K., 2006, "Waste Cooking Oils an Economical Source for Biodiesel: A Review," *Ind. Eng. Chem. Res. 45*, pp. 2901–2913.

Kumar, A., Jain, S.K., and Bansal, N.K., 2003, "Disseminating energy-efficient technologies: a case study of compact fluorescent lamps (CFLs) in India," *Energy Policy* 31, pp. 259–272.

Kunisue, T., Masayoshi Muraoka, Masako Ohtake, Agus Sudaryanto, Nguyen Hung Minh, Daisuke Ueno, Yumi Higaki, Miyuki Ochi, Oyuna Tsydenova, Satoko Kamikawa et al., 2006, "Contamination status of persistent organochlorines in human breast milk from Japan: Recent levels and temporal trend," *Chemosphere*: in press.

Kurki, A., Hill, A., and Morris, M., 2006, "Biodiesel: The sustainability dimensions," *ATTRA*, pp. 1–12. http://attra.ncat.org/attra-pub/PDF/biodiesel_sustainable.pdf (accessed on October 27, 2007).

Kuroda, H., 2006, "Emerging Asia in the Global Economy: Prospects and Challenges," Remark by President, Asian Development Bank at the Council on Foreign Relations. February 17, Washington, D.C., USA.

Kutchko, B.G. and Kim, A.G., 2006, "Fly ash characterization by SEM–EDS," *Fuel*, Vol. 85(17–18), pp. 2537–2544.

Kuuskaraa, V.A., Charles, M., Boyer II, and Jonathan, A.K., 1992, "Coalbed Gas-1: Hunt for quality Basins goes abroad," *Oil & Gas Journal*, October, pp. 80–85.

Kyoto Protocol, 1997, Conference of the Parties Third Session Kyoto, 1–10 December 1997. Kyoto Protocol to the United Nations Framework Convention on Climate Change.

Labuschange, C., Brent, A.C., and Erck, R.P.G., 2005, "Assessing the sustainability performances of industries," *Journal of Cleaner Production*, 13, pp. 373–385.

Lacey, J., 1990, "Isolation of thermophilic microorganisms," in Labeda, D.P., ed., *Isolation of Biotechnological Organisms from Nature*, New York, MCGraw-Hill Publishing Co., pp. 141–181.

Lakhal, S. and H'Mida, S., 2003, "A gap analysis for green supply chain benchmarking," in *32th International Conference on Computers & Industrial Engineering*, Volume 1, August 11–13, Ireland, pp. 44–49.

Lakhal, S., H'mida. S., and Islam, R., 2005, "A Green supply chain for a petroleum company," *Proceedings of 35th International Conference on Computer and Industrial Engineering*, Istanbul, Turkey, June 19–22, 2005, Vol. 2, pp. 1273–1280.

Lakhal, S., Khan, M.I., and Islam, M.R., 2006a, "A framework for a green decommissioning of an offshore platform," in *Proceedings of 36th CIE Conference on Computer and Industrial Engineering*, July, Taiwan, pp. 4345–56.

Lang, X., Dalai, A.K., Bakhsi, N.N., Reaney, M.J., and Hertz, P.B., 2001, "Preparation and characterization of bio-diesels from various bio-oils," *Bioresource Technology* 80, pp. 53–62.

Lange, J-P., 2002, "Sustainable development: efficiency and recycling in chemicals manufacturing," *Green Chem*, 4, pp. 546–50.

Larrondo, L.E., Urness, C.M., and Milosz, G.M., 1985, "Laboratory Evaluation of Sodium Hydroxide, Sodium Orthosilicate, and Sodium Metasilicate as Alkaline Flooding," *Society of Petroleum Engineering*, Vol. 13577, pp. 307–315.

Lastella, G., Testa, C., Cornacchia, G., Notornicola, M., Voltasio, F., and Sharma, V.K., 2002, "Anaerobic digestion of semi-solid organic waste: Biogas production and its purification," *Energy Conversion and Management*, 43, pp. 63–75.

Le Roux, N., "Going for gold with microbes," *Chemical Eng.*, Jan., p. 432, 1987.

Leal Filho, W.L., 1999, "Sustainability and university life: some European perspectives," in W. Leal Filho, ed., *Sustainability and University Life: Environmental Education, Communication and Sustainability*, pp. 9–11. Berlin, Peter Lang.

Lean, G., 2007, "Oil and gas may run short by 2015," *The Independent*, UK. http://environment.independent.co.uk/climate_change/article2790960.ece, 22 July, 2007 (Accessed on 23 July 2007).

Leclercq, B., 2006, Beeswax, Accessed September 22, 2006, from Beekeeping website: http://www.beekeeping.com/leclercq/wax.htm.

Lecomte du Nou¨y, P., 1919, *J. Gen. Physiol.* Vol.1, 521.

Lee, I., Johnson, L.A., and Hammond, E.G., 1995, "Use of branched-chain esters to reduce the crystallization temperature of biodiesel," *JAOCS* 72 (10), pp. 1155–1160.

Lee, S.T., LO, H., and Dharmawardhana, B.T., 1988, "Analysis of Mass Transfer Mechanisms Occurring in Rich Gas Displacement Process," SPE paper 18062 presented at the Annual Technical Meeting, Houston Oct. 2–5.

Lee, S.C., Choi, B.Y., Lee, T.J., Ryu, C.K., Ahn, Y.S., and Kim, J.C., 2006, "CO_2 absorption and regeneration of alkali metal-based solid sorbents," *Catalysis Today*, 111, pp. 385–390.

Lee, T.R., 1996, "Environmental stress reactions following the Chernobyl accident, One Decade after Chernobyl accident, summing up the consequences of the accident," *Proceeding of an International Conference, Vienna*, STI/PUB/1001.IAEA, Vienna, 1996, pp. 238–310

Lehman-McKeeman, L.D. and Gamsky, E.A., 1999, "Diethanolamine inhibits choline uptake and phosphatidylcholine synthesis in Chinese hamster ovary cells," *Biochem. Biophys. Res. Commun.* 262(3), pp. 600–604.

Lems, S., van derKooi, H.J., deSwaan Arons, J., 2002, "The sustainability of resource utilization," *Green Chem*, Vol.4, pp. 308–13.

Leontieff, W., 1973, "Structure of the world economy: outline of a simple input-output formulation," Stockholm: Nobel Memorial Lecture, 11 December, 1973.

Letcher, T.M. and Williamson, A., 2004, "Forms and Measurement of Energy," *Encyclopedia of Energy*, 2, pp. 739–748.

Leung, D.Y.C. and Guo, Y., 2006, "Transesterification of neat and used frying oil: optimization for biodiesel production," *Fuel processing technology* 87, pp. 883–890.

Lewis, R.J., Sr., 2002, Hawley's Condensed Chemical Dictionary, 14th Edition, New York, John Wiley and Sons.

Li, D.H.W., and Lam, J.C., 2004, "Predicting solar irradiance on inclined surfaces using model for heat input and output of heat storing stoves," *Applied Thermal Engineering*, 25 (17–18), pp. 2878–2890.

Li, D.H.W., and Lam, J.C., 2004, "Predicting solar irradiance on inclined surfaces using sky radiance data," *Energy Conversion and Management* 45(11–12), pp. 1771–1783.

Li, T., Gao, J., Szoszkicz, R., Landman, U., and Riedo, E., 2007, "Structured and viscous water in subnanometer gaps," *Physical Review* B, vol. 75, 115415, March 15, pp. 115415-1-115415-6.

Liberman, J., 1991, *Light: Medicine of the future: How we can use it to heal ourselves now*, Bear & Company, Inc., Sata Fe, NM, USA.

Lindzen, R.S., 2002, "Global Warming: The Origin and Nature of the Alleged Scientific Consensus," *Regulation: The Cato Review of*

Business and Government. http://eaps.mit.edu/faculty/lindzen/153_Regulation.pdf.

Lindzen, R.S., 2006, "Climate Fear," *The Opinion Journal*, April, 12, 2006. www.opinionjournal.com/ extra/?id=110008220 (Accessed on June, 30, 2006).

Liodakis, S., Katsigiannis, G., and Kakali, G., 2005, "Ash Properties of Some Dominant Greek forest species," *Thermochimica Acta*, Vol. 437, pp. 158–167.

Liu, P., 1993, *Introduction to energy and the environment*, Van Nostrand Reinhold, New York.

Livingston, R.J., and Islam, M.R., 1999, "Laboratory modeling, field study and numerical simulation of bioremediation of petroleum contaminants," *Energy Sources*, vol. 21 (1/2), pp. 113–130.

Logan, R.K., 1986, The Alphabet Effect: The Impact of the Phonetic Alphabet on the Development of Western Civilization, St. Martin's Press, New York, 1986.

Low, N.M.P., Fazio, P., and Guite, P., 1984, ":Development of light-weight insulating clay products from the clay sawdust-glass system," *Ceramics International* 10(2), pp. 59–65.

Lowe, E.A, Warren, J.L., Moran, S.R., 1997 *Discovering industrial ecology—an executive briefing and sourcebook*, Columbus, Battelle Press.

Lowy, J., 2004, "Plastic left holding the bag as environmental plague. Nations around world look at a ban," <http://seattlepi.nwsource.com/national/182949_bags21.html>.

Lozada, D. and Farouq Ali, S.M., 1988, "Experimental Design for Non-Equilibrium Immiscible Carbon Dioxide Flood," Fourth UNITAR / UNDP International Conference on Heavy Crude and Tar Sands, Edmonton, August 7–12, paper no. 159.

Lu, Y., Zhang, Y., Zhang, G., Yang, M., Yan, S., and Shen, D., 2004, "Influence of thermal processing on the perfection of crystals in polyamide 66 and polyamide 66/clay nanocomposites," *Journal of Polymer*, Elsevier, V. 45, Issue 26, pp. 8999–9009.

Lubchenco, J.A., et al., 1991, "The sustainable biosphere initiative: an ecological research agenda," *Ecology* 72, pp. 371–412.

Lumley, S. and Armstrong, P., 2004, "Some of the Nineteenth Century Origins of the Sustainability Concept," *Env. Devlopment and Sustainability*, vol. 6, no. 3, Sept., pp. 367–378.

Lunder, S. and Sharp, R., 2003, "Mother's milk, record levels of toxic fire retardants found in American mother's breast milk," Environmental Working Group, Washington, USA.

Sugie, H., Sasaki, C., Hashimoto C., Takeshita, H., Nagai, T., Nakamura, S., Furukawa, M., Nishikawa, T., Kurihara., K., 2004, "Three cases of sudden death due to butane or propane gas inhalation: analysis of tissues for gas components," *Forensic Science International* 143 (2–3), pp. 211–214.

Sarwar, M. and Islam, M.R., "A Non-Fickian Surface Excess Model for Chemical Transport Through Fractured Porous Media," *Chem. Eng. Comm.*, vol. 160, pp. 1–34, 1996.

Ma, F. and Hanna, M.A., 1999, "Biodiesel production: a review," *Bioresource Technology*, 70, pp. 1–15.

Maclallum, M.F. and Guhathakurta, K., 1970, "The Precipitation of Calcium Carbonate from Seawater by Bacteria Isolated from Bahama Bank Sediments," *Journal of Applied Bacteria*, vol. 33, pp. 649–655.

MacLeod, F.A., Lappin-scott, H.M., and Costerton, J.W., 1988, "Plugging of a Model Rock System by Using Starved Bacteria," *Applied and Environment Microbiology*, vol. 51, pp. 1365–1372.

Malcolm, P., 1998, *Polymer chemistry: an introduction*, Oxford University Press, London, p.3.

Mallinson, R.G., 2004, "Natural Gas Processing and Products," *Encyclopedia of Energy*, Vol. IV, Elsevier Publication, Okhlahama, USA, Pp. 235–247.

Mancktelow, N.S., 1989, "The rheology of paraffin wax and its usefulness as an analogue for rocks," *Bulletin of the Geological Institutions of the University of Uppsala*, Vol. 14, pp. 181–193.

Mann, H., 2005, Personal communication, Professor, Civil Engineering Department, Dalhousie University, Halifax, Canada.

Manning, D.G., 1996, "Corrosion Performance of Epoxy-Coated Reinforcing Steel: North American Experience," *Construction and Building Materials* 10(5), p. 349.

Manser, C.E., 1996, "Effects of lighting on the welfare of domestic poultry: A review," *Animal Welfare*, Volume 5, Number 4, pp. 341–360.

Mansoori, G.A., 1997, "Modeling of Asphaltene and Other Heavy Organic Depositions," *J. Pet. Sci. Eng.*, vol. 17, pp. 101–111.

Mansour, E.M.E., Abdel-Gaber, A.M., Abd-El-Nabey, B.A., Khalil, N., Khamis, E., Tadros, A., Aglan, H., and Ludwick, A., 2003, *Corrosion* 59(3), p. 242.

Market Development Plan, 1996, Market status report: postconsumer plastics, business waste reduction, Integrated Waste Development Board, Public Affairs Office. California.

Markiewicz, G.S., Losin, M.A., and Campbell, K.M., 1988, "The Membrane Alternative for Natural Gas Treating: Two Case Studies," *SPE* 18230. *63rd Annual Technical Conference end Exhibition of the Society of Petroleum Engineers* held in Houston, Tx, October 2–5.

Marsden, J., *The Chemistry of Gold Extraction*, Ellis Horwood Ltd., pp. 221–235, 1992.

Martinot, E., 2005, Global Renewable Status Report. Paper prepared for the REN21 Network. The Worldwatch Institute.

Martinot, E., 2005, Renewable Energy Policy Network for the 21st Century. Global Renewables Status Report Prepared for the REN 21 Network by the Worldwatch institute.

Marx, K., 1883, *Capital: A critique of political economy Vol. II: The Process of Circulation of Capital*, London, Edited by Frederick Engels.

Maske, J., 2001, "Life in PLASTIC, it's fantastic," *GEMINI*, Gemini, NTNU and SINTEF Research News, N-7465 Trondheim, Norway.

Maskell, K., 2001, *Climate Change 2001: The Scientific Basis*, Technical Summary. Cambridge University Press, Cambridge, UK.

Matsuoka, K., Iriyama, Y., Abe, T., Matsuoka, M., and Ogumi, Z., 2005, "Electro-oxidation of methanol and ethylene glycol on platinum in alkaline solution: Poisoning effects and product analysis," *Electrochimica Acta*, Vol.51, pp. 1085–1090.

Matsuyama, H., Teramoto, M., and Sakakura, H., 1996, "Selective Permeation of CO_2 through poly{2-(*N,N*-dimethyl)aminoethyl methacrylate} Membrane Prepared by Plasma-Graft Polymerization Technique," *J. Membr. Sci.* 114, 1996, pp. 193–200.

Mayer, E.H., Berg, R.L., Carmichael, J.D., and Weinbrandt, R.M., 1983, "Alkaline Injection for Enhanced Oil Recovery – A Status Report," *Journal of Petroleum Technology*, pp. 209–221.

McCarthy, B.J. and Greaves, P.H., 1988, "Mildew-causes, detection methods and prevention," *Wool Science Review*, Vol. 85, pp. 27–48.

McHugh, S., Collins, G., and O'Flaherty, V., 2006, "Long-term, high-rate anaerobic biological treatment of whey wastewaters at psychrophilic temperatures," *Bioresource Technology* 97(14), pp. 1669–1678.

MEA (Millennium Ecosystem Assessment), 2005, The millennium ecosystem assessment, Commissioned by the United Nations, the work is a four-year effort by 1,300 scientists from 95 countries.

Meher, L.C., Vidya Sagar, D., and Naik, S.N., 2004, "Technical aspects of biodiesel production by transesterification-a review," *Renewable and sustainable energy review*, pp. 1–21.

Merriman, B. and Burchard, P., 1996, "An Attempted Replication of CETI Cold Fusion Experiment," 1996, http://www.lenr-canr.org/PDetail6.htm#2029.

Metivier, H., 2007, Update of Chernobyl: Ten Years On. Assessment of Radiological and Health Impacts. Nuclear Energy Agency, Organisation For Economic Co-Operation and Development, 2002. http://www.nea.fr/html/rp/ reports/2003/nea3508-chernobyl.pdf (Acessed on December 17, 2007).

Miao, X. and Wu., Q., 2006, "Biodiesel production from heterotrophic microalgal oil," *Bioresource Technology*, Vol. 97, (6), pp. 841–846.

Miller, G., 1994, *Living in the Environment: Principles, Connections and Solutions*, California, Wadsworth Publishing.

Miralai, S., Khan, M.M., and Islam, M.R., 2007, "Replacing artificial additives with natural alternatives," *J Nat. Sci. Sust. Tech.*, vol. 1, no. 2.

Mita, K., Ichimura, S. and James, T., 1994, "Highly repetitive structure and its organization of the fibroin gene," *Journal of Molecular Evolution*, V.38, pp.583–592.

Mittelstaedt, M., 2006, "Toxic Shock," 5-part series, *Globe and Mail*, starting from May 27-June 1.

Mittelstaedt, M., 2006a, "Chemical used in water bottles linked to prostate cancer," *The Globe and Mail*, Friday, 09 June 2006.

Mittelstaedt, M., 2007, "Vitamin D casts cancer prevention in new light," *The Globe and Mail* [Toronto], Sat 4 April 28, page A1.

Mittelstaedt, M., 2008, "Coalition of public health and environmental advocates says federal government hasn't gone far enough in regulating controversial chemical," *Globe and Mail*, Dec. 16.

Moffet, A. S., 1994, "Microbial mining boosts the environment, bottom line," *Science* 264, May 6.

Moire, L., Rezzonico, E., and Poirier, Y., 2003, "Synthesis of novel biomaterials in plants," Journal of Plant Physiol, V.160, pp. 831–839.

Molero, C., Lucas, A.D., and Rodrıguez, J.F., 2006, "Recovery of polyols from flexible polyurethane foam by 'split-phase' glycolysis: Glycol influence," *Polymer Degradation and Stability*, Vol. 91, pp. 221–228.

Mollet, C., Touhami, Y., and Hornof, V., 1996, "A Comparative Study of the Effect of Ready Made and in-Situ Formed Surfactants on IFT Measured by Drop Volume Tensiometry," *J. Colloid Interface Sci.*, Vol. 178, p. 523.

Morgan, J., Townley, S., Kemble, G., and Smith, R., 2002 "Measurement of physical and mechanical properties of beeswax," *Materials Science and Technology*, Vol. 18(4), pp. 463–467.

Moritis, G., 2004, "Point of View: EOR Continues to Unlock Oil Resources," *Oil and Gas Journal*. ABI/INFORM Global, Vol. 102 14, pp. 45–49.

Moritis, G., 1994, "EOR Dips in US but Remains a Significant Factor," *Oil and Gas J.*, pp. 51–79, Sept 26.

Moritis, G., 1998, "EOR Oil Production Up Slightly," *Oil and Gas Journal*, April 20, pp. 49–56.

Morrow, H., 2001, "Environmental and human health impact assessments of battery systems," *Industrial Chemistry Library* 10, pp. 1–34.

Mortimer, N., 1989, "Friends of Earth, Vol 9," in Thompson, B., *Nuclear Power and Global Warming*, 1997 (http://www.seaus.org.au/powertrip.html (acessed on November 10, 2007).

Mossop, G. and Shesten, I., 1994, Geological Atlas of Western Canada Sedimentary Basin, Cdn. Soc. Pet. Geol. and Alberta Research Council, Calgary, Alberta.

MSDS Material Safety Data Sheet, 2006, Canadian Centre for Occupational Health and Safety, 135 Hunter Street East, Hamilton ON Canada L8N 1M5.

MSDS, 2005, Ethylene Glycol Material Safety Data Sheet, www.sciencestuff.com/ msds/C1721.html.

MSDS, 2006, Material safety data sheet for ethylene glycol, www.jtbaker.com/msds/ englishhtml/o8764.htm (Accessed on January 28, 2008).

MTR, 2007, Membrane Technology and Research: Natural Gas Liquids Recovery/Dewpoint Control, http://www.mtrinc.com/natural_gas_liquids_recovery.html (accessed on 8th August, 2006).

Mudd, G.M., 2000, "Remediation of Uranium Mill Tailings Wastes," in *Australia: A Critical Review. Contaminated Site Remediation: From Source Zones to Ecosystems*, CSRC, Melbourne, Vic., 4–8 Dec.

Mundy, B., 1989, "Distant Action in Classical Electromagnetic Theory," *British Journal for the Philosophy of Science* 40 [1], pp. 39–68.

Munger, C., 1999, *Corrosion Prevention by Protective Coatings*, NACE International, Houston, USA.

Munger, C.G., 1990, "COC-Compliant Inorganic Zinc Coating," *Material Performance*, October, p. 27.

Munger, C.G., 1992, "Coating Requirements for Offshore Structures," *Material Performance*, June, p. 36.

Muntasser, Z., 2002, *The Use of Coatings to Prevent Corrosion in Harsh Environments*. M.Sc. Thesis, Dalhousie University, Halifax, NS, Canada.

Muntasser, Z.M., Al-Darbi, M.M., and Islam, M.R., 2001, "Prevention of Corrosion in a Harsh Environment using Zinc Coating," *SPE Production Conference*, Oklahoma, USA.

Murphy, M., 2003, "Technical Developments in 2002: Organic Coatings, Processes, and Equipment," *Metal Finishing* 101(2), p. 47.

Murrell, J.C., 1994, "Molecular genetics of methane oxidation," *Biodegradation* 5, pp. 145–149.

Mustafiz, S., 2002, *A Novel Method forHeavy Metal Removal from Aqueous Streams*, MASc Dissertation, Dalhousie University, Canada.

Nabi, M.N., Akhter, M.S., and Shahadat, M.M.Z., 2006, "Improvement of engine emissions with conventional diesel fuel and diesel–biodiesel blends," *Bioresource Technology*, Vol. 97, pp. 372–378

Nallinson, R.G., 2004, "Natural Gas Processing and Products," *Encyclopedia of Energy Vol. IV*, Elsevier Publication, Okhlahama, USA, pp. 235–247.

Narayan, R., "Rationale, Drivers, Standards, and Technology for Biobased Materials"; Ch 1, pg 3 in Renewable Resources and Renewable Energy, Ed. M. Graziani & P. Fornasiero; CRC Press, 2006..

NASA, 1999, "Biomass Burning and Global Change," http://asdwww.larc.nasa.gov/ biomass_burn/biomass_burn.html

NASA/ESA, 2004, "Sun's storms create spectacular light show on earth," *NOAA News*, National Oceanic and Atmospheric Administration (NOAA), United States Department of Commerce, accessed March 5, 2008, http://www.noaanews.noaa.gov/stories2004/s2337.htm.

Natural Gas Org., 2004, www.naturalgas.org/naturalgas/processing_ng.asp(accessed on August 08, 2007).

Natural Gas Org, 2004, "Overview of natural gas," http://www.naturalgas.org/overview/ background.asp (May 8, 2008).

Natural Resources Canada, 1998, "Alberta Post-Consumer Plastics Recycling Strategy Recycling," Texas, Society of Petroleum Engineers 1997. http://www.springerlink.com.ezproxy.library.dal.ca/content/x3t872344nx6/?p=be9a263b7ced4adaa1358862d23a1edd&pi=0Natural Resource Canada, 2004, Publications and Softwares. www.canren.gc.ca/prod serv/index.asp?CaId=196&PgId=1309 accessed on August 23, 2007).

Naylor, R.H., 1976, "Galileo: Real Experiment and Didactic Demonstration," *Isis*, Vol. 67, No. 3. (September), pp. 398–419.

Naylor, R. H., 1980, "Galileo's Theory of Projectile Motion," *Isis*, Vol. 71, No. 4. (December), pp. 550–570.

Naylor, R.H., 1990, "Galileo's Method of Analysis and Synthesis," *Isis*, Vol. 81, No. 4. (December), pp. 695–707.

NEA and IAEA, 2005, *Uranium 2005, Resources, Production and Demand*. OECD, International Atomic Energy Agency (IAEA). Published by OECD Publishing pp.388 ISBN: 9264024255, 2005.

NEA-OECD, 2003, *Nuclear Electricity Generation: What Are the External Costs?*, Nuclear Development, ISBN 92-64-02153-1.

Neep, J.P., 1995, "Robust Estimation of P-wave Attenuation from Full Waveform Array Sonic Data," *J. Seismic Exploration*, vol. 4, pp. 329–344.

Nichols, C., Anderson, S., and Saltzman, D., 2001, "A Guide to Greening Your Bottom Line Through a Resource-Efficient Office Environment," City of Portland, Office of Sustainable Development, Portland.

Nikiforuk, A., 1990, *Sustainable Rhetoric*, Harrowsmith, pp. 14–16.

Nivola, P.S., 2004, The Political Economy of Nuclear Energy in the United States. The Brooking Institution. Policy Brief, #138., www.brookings.edu/comm/policybriefs/pb138.htm (Accessed on January 15, 2007).

NOAA, 2005, Greenhouse Gases, Global Monitoring Division, Earth System Research Laboratory, National Oceanic and Atmospheric Administration, USA.

NOAA, 2005, Trends in Atmospheric Carbon Dioxide, NOAA-ESRL Global Monitoring Division, www.cmdl.noaa.gov/ccgg/trends/ (accessed on June 04, 2006).

Norris, P., Nixon, A., and Hart, A., 1989, "Acidophilic, mineral-oxidizing bacteria: The utilization of carbon dioxide with particular reference to autotrophy in sulfolobus," *Microbiology of Extreme Environments and its Potential for Biotechnology*, FEMS Symposium no. 49, Da Costa, M.S., Duarte, J.C., and Williams, R.A.D., pp. 24–39.

Novosad, Z. and Constain, T.G., 1988, "New Interpretation of Recovery Mechanisms in Enriched Gas Drives," *Journal of Canadian Pet. Tech.*, volume 27, pp. 54–60.

Novosad, Z. and Constain, T.G., 1989, "Mechanisms of Miscibility Development in hydrocarbons Gas Drives: New Interpretation," *SPE Res. Eng.*, pp. 341–347.

NTP (National Toxicology Program), 1993, NTP technical report on the toxicology and carcinogenesis studies of ethylene glycol (CAS Nos. 107-21-1) in B6C3F1 mice (feed studies), National Toxicology Program, US Department of Health and Human Services, NIH Publication 93–3144.

Obernberger, I., Biedermann, F., Widmann, W., and Riedl, R., 1997, "Concentrations of Inorganic Elements in Biomass Fuels and Recovery in the Different Ash Fractions," *Biomass and Bioenergy*, Vol. 12 3, pp. 211–224.

OCED, 1998, *Towards sustainable development: environmental indicators*, Paris, Organization for Economic Cooperation and Development, 132pp.

OECD, 1993, Organization for Economic Cooperation and development core set of indicators for environmental performance reviews. A synthesis report by the Group on State of the Environment, Paris.

Office of Energy Efficiency and Renewable Energy, 2004, Biodiesel Analytical Methods, Report NREL/SR-510–36240, 100 pp.

Oldenburg, C.M., Pruess, K., and Benson, S.M., 2001, "Process Modeling of CO_2 Injection into Natural Gas Reservoirs for Carbon Sequestration and Enhanced Gas Recovery," *Energy Fuels*, 15 (2), pp. 293–298.

Olsson, M. and Kjällstrand, J., 2004, "Emissions from burning of softwood pellets," *Biomass and Bioenergy*, Vol. 27, No. 6, pp. 607–611.

Omer, A.M. and Fadalla, Y., 2003, "Biogas energy technology in Sudan," *Renewable Energy* 28, pp. 499–507.

OSHA, 2003, Petroleum refining processes, OSHA Technical Manual, Section IV, Chapter 2, http://www.osha.gov/dts/osta/otm/ otm_iv/otm_iv_2.html (accessed on June 18, 2008).

Osujo, L.C. and Onoiake, M., 2004, "Trace heavy metals associated with crude oil: A case study of Ebocha-8 Oil-spill-polluted site in Niger Delta, Nigeria," *Chemistry and Biodiversity*, vol. 1, issue 11, pp. 1708–1715.

Ott, J.N., 2000, *Health and Light: The Effects of Natural and Artificial Light on Man and Other Living Things*, Pocket Books, New York, NY, USA.

Oyarzun, P., Arancibia, F., Canales, C., and Aroca, E.G., 2003, "Biofiltration of high concentration of hydrogen sulphide using *Thiobacillus thioparus*," *Process Biochemistry* 39, pp.165–170.

Patin, S., 1999, *Environmental Impact of the Offshore Oil and Gas Industry*, Eco Monitor Publishing, East Northport, New York.

Paul, D.B. and Spencer, H.G., 2008, "'It›s Ok, We›re Not Cousins by Blood': The Cousin Marriage Controversy in Historical Perspective," *PLoS Biol* 6(12): e320 doi:10.1371/journal.pbio.0060320.

Pearson, K.W., 1892, *The Grammar of Science*, London, Walter Scott.

Peart, J. and Kogler, R., 1994, "Environmental Exposure Testing of Low VOC Coatings for Steel Bridges," *Journal of Protective Coatings and Linings*, January, p. 60.

Pershing, J. and Cedric, P., 2002, "Promises and Limits of Financial Assistance and the Clean Development Mechanism," *Beyond Kyoto:*

Energy Dynamics and Climate Stabilization, Paris: International Energy Agency, pp. 94–98.

Peters, M.S. and Timmerhaus, K.D., 1991, *Plant design and economics for chemical engineers*, Fourth Edition, McGraw-Hill, Inc. New York, USA.

Peterson, B.E., 2008, Oregon State University, internet lecture, last accessed Nov. 12, 2008.

Peterson, C.L., Reece, D.L., Hammond, B.L., Thompson, J., and Beck, S.M., 1997, "Processing, characterization and performance of eight fuels from lipids," *Appl. Eng. Agric.* 13 (1), pp. 71–79.

Piipari, R., Tuppurainen, M., Tuomi, T., Mantyla, L., Henriks-Eckerman, M.L., Keskinen, H., and Nordman, H., 1998, "Diethanolamine-induced occupational asthma, a case report," *Clin. Exp. Allergy* 28(3), pp. 358–362.

Piro, G., Canonico, L.B., Galbariggi, G., Bertero, L., and Carniani, C., 1995, "Experimental Study on Asphaltene Adsorption onto Formation Rock: An Approach to Asphaltene Formation Damage prevention," SPE 30109, Proc. European Formation Damage Conf., The Hague.

Plastic Task Force, 1999, "Adverse health effects of plastics," <http://www.ecologycenter.org/erc /fact_sheets plastichealtheffects.html# plastichealthgrid>

Pokharel, G.R., Chhetri, A.B., Devkota, S., and Shrestha, P., 2003, "En route to strong sustainability: can decentralized community owned micro hydro energy systems in Nepal Realize the Paradigm? A case study of Thampalkot VDC in Sindhupalchowk District in Nepal," International Conference on Renewable Energy Technology for Rural Development. Kathmandu, Nepal.

Pokharel, G.R., Chhetri, A.B., Khan, M.I., and Islam, M.R., 2006, "Decentralized micro hydro energy systems in Nepal: en route to sustainable energy development," *Energy Sources*: in press.

Polsby, E., 1994, "Marketplace: what to do when the lights go out," *Home energy Magazine* online, November/December 1994: http://www.homeenergy. org/eehem/94/941115.html (accessed on March 24, 2008).

Pope, D.H. and Morris, E.A. III, 1995, "Mechanisms of microbiologically induced corrosion (MIC)," *Materials Performance*, vol. 34, May, pp. 24–28.

Postek, M.T., Howard, K.S., Johnson, A.H., and McMichael, K.L., 1980, *Scanning electron microscopy: A student's handbook*, Burlington, Ladd Research Industries.

Prescott, N.B., Kristensen, H.H., and Wathes, C.M., 2004, "Light," in Weeks, C. and Butterworth, A., eds., *Measuring and Auditing Broiler Welfare*, CABI Publishing, Wallingford, UK, pp. 101–116.

Puri, R. and Yee, D., 1990, "Enhanced Coalbed methane recovery," SPE paper 20732 presented at the 65th Annual Technical Conference and Exhibition, New Orleans, U.S.A., September 23–26.

Putin, S., 1999, *Environmental impact of the offshore oil and gas industry*, EcoMonitor Publishing, East Northport, New York, 425 pp.

Putthanarat, S., Eby, R.K., Rajesh, R.N., Shane, B.J., Walker, M.A., Peterman, E., Ristich, S., Magoshi, M., Tanaka, T., Stone, M.S., Farmer, B.L., Brewer, C., and Ott, D., 2004, "Nonlinear optical transmission of silk/green fluorescent Protein (GFP) films," *Journal of polymer*, Elsevier, V. 45, Issue 25, pp. 8451–8457.

Putthanarat, S., Zarkoob, S., Magoshi, J., Chen, J.A., Eby, R.K., Stone, M., and Adams, WW., 2002, "Effect of processing temperature on the morphology of silk membranes," *Journal of polymer*, V.43, Issue 12, pp. 3405–3413.

Quine, W.V., 1937, "Review of Harold Jeffreys, *Scientific Inference*," (Reissued with additions. pp. vii; 272. Cambridge UP 1937 [1st pub 1931]), in *Science* [New Series] 86 [2243], p. 590.

Radich, 2006, "Biodiesel performance, costs, and use," US Energy Information Administration website, /http://www.eia.doe.gov/oiaf/analysispaper/biodiesel/ index.htmlS.

Ragheb, M., 2007, "Chernobyl Accidents," https://netfiles.uiuc.edu/mragheb/www/NPRE%20402%20ME%20405%20Nuclear%20Power%20Engineering/Chernobyl%20 Accident.pdf

Rahbar S., Khan, M.M., Satish, M., Ma, F., and Islam, M.R., 2005, "Experimental & numerical studies on natural insulation materials," ASME Congress, 2005, Orlando, Florida, Nov 5–11, 2005, IMECE2005–82409.

Rahman, M.H., Wasiuddin, N.M., and Islam, M.R., 2004, "Experimental and Numerical Modeling Studies of Arsenic Removal With Wood Ash From Aqueous Streams," *Canadian Journal of Chemical Engineering* 82, pp. 968–977.

Rahman, M.S., 2006, "Effect of natural alkaline solution on EOR during chemical flooding," Project Report, Department of Civil and Resource Engineering, Dalhousie University, Canada, pp. 28.

Rahman, M.S., 2007, *The Prospect of Natural Additives in Enhanced Oil Recovery and Water Purification Operations*, M.A.Sc Thesis, Faculty of Engineering, Dalhousie University, Halifax, Canada.

Rahman, M.S., Hossain, M.E., and Islam, M.R., 2006, "An Environment-Friendly Alkaline Solution for Enhanced Oil Recovery," *JPST* in press, Ref# PET/06/076.

Rahman, M.S. and Islam, M.R., 2007, "Physico-Chemical Characterization of Ashes from *Acer nigrum* by SEM, XRD and NMR technique," *Int. J. Materials and Product Technology*, accepted.

Raily, K., 2007, "Hemptons," www.hemptons.co.za/Users/seeds/htm (Accessed on October 20, 2007).

Ramakrishnan, T.S. and Wasan, D.T., 1983, "A Model for Interfacial Activity of Acidic Crude Oil-Caustic Systems for Alkaline Flooding," *SPE Journal*, SPE-10716, pp. 602–618.

Ramesh, C. and Keller, A., 1994, "Eltink SJEA," *Journal of Polymer*, V.35, pp. 5293–9.

Randall, W., 1999, Technical handbook for marine biodiesel in recreational boats. Prepared for report, prepared by system lab services, a division of Williams pipe Lines Company.

Rangaswamy, N., ,Vedhalakshmi, R., and Balakrishnan, K., 1995, "Evaluation of Coated Rebar Validity of Short-Term Accelerated Corrosion Tests in Relation to Long-Term Field Evaluation," *ACM & M*, (vol. 42), p. 7.

Rao, M.B. and Sircar, S., 1996, "Performance and pore characterization of nanoporous carbon membranes for gas separation," *Journal of Membrane Science*, 3(7), pp. 109–18.

Rao, M.B. and Sirkar, S., 1993, "Liquid-phase adsorption of bulk ethanol-water mixtures by alumina," *Adsorption Science and Technology*, 10(1–4), pp. 93–104.

Rao, M.B., Sircar, S., and Golden, T.C., 1992, "Gas separation by adsorbent membranes," *US Patent*, 5, 104, 425.

Rao, P., Ankam, S., Ansari, M., Gavane, A.G., Kumar, A., Pandit, V.I., and Nema, P., 2005, "Monitoring of hydrocarbon emissions in a petroleum refinery," *Environmental Monitoring and Assessment*, 2005, 108, pp. 123–132.

Rassamdana, H., Mirzaee, N., Mehrabi, A.R., and Sahimi, M., 1997, "Field-Scale Asphalt Precipitation During Gas Injection into a Fractured Carbonate Reservoir," SPE 38313, SPE Western Regional Meeting, June, Long Beach, CA.

Ray, R., Little, B., Wagner, P., and Hart, K., 1997, *Scanning* 19, p. 98.

Rechtsteiner, G.A. and Ganske, J.A., 1998, "Using Natural and Artificial Light Sources to Illustrate Quantum Mechanical Concepts," *The Chemical Educator*, Volume 3, Number 4, pp. 1–12.

Rees, W., 1989, "Sustainable development: myths and realities," *Proceedings of the Conference on Sustainable Development*, Winnipeg, Manitoba: IISD.

Reis, J.C., 1996, *Environmental control in petroleum engineering*, Gulf Publishing Company, Houston, Texas.

Regert, M., Langlois, J., and Colinart, S., 2005, "Characterization of wax works of art by gas chromatographic procedures," *Journal of Chromatography A*, Vol. 1091, pp. 124–136.

Renton, J.J. and Brown, H.E., 1995, "An Evaluation of Fluidized Bed Combustor Ash as a Source of Alkalinity to Treat Toxic Rock Materials," *Engineering Geology*, Vol. 40, pp. 157–167.

Rice, D.D., Law, B.E., and Clayton, J.L., 1993, "Coal-bed Gas – an Undeveloped Resource. In the Future of Energy Gases," US Geological Survey professional paper no. 1570, United States Government Printing office, Washington, DC, 389–404.

Ridgeway, James, 2000, "Eco Spaniel Kennedy: Nipping at Nader's Heels," *Village Voice*, New York, 16–22 August.

Riemer, P., 1996, "Greenhouse Gas Mitigation Technologies, an Overview of the CO_2 Capture, Storage and Future Activities of the IEA Greenhouse Gas R&D Program," *Energy Conversion and Management*, volume 37, nos. 6–8, pp. 665–670.

Robinson, J.G., 1993, "The limits to caring: sustainable living and the loss of biodiversity," *Conservation Biology* 7, pp. 20–28.

Robinson, R.J., Bursell, C.G., and Restine, J.L., 1977, "A Caustic Steam flood Pilot-Kern River Field," paper SPE 6523 presented at SPE AIME 47th Annual California Regional Meeting, 13akerafield,California, April 13–15.

Roger, A.K. and Jaiduk, J.O., 1985, "A rapid engine test to measure injector fouling in diesel engines using vegetable oil fuels," *J. Am. Oil Chem. Soc.*62 (11), pp. 1563–4.

Rojas, G., Dyer, S., Thomas, S., and Farouq Ali, S.M., 1995, "Scaled Model Studies of CO_2 Floods," *SPE Res. Eng.* Vol. 10, no. 3, May, pp. 169–178.

Rojas, G. and Farouq Ali, S.M., 1985, "Dynamics of sub-critical CO_2 Brine Floods for Heavy Oil Recovery," SPE paper 13598 presented at the SPE California Regional Meeting, Bakersfield, March.

Rojey, A., Jaffret, C., Cornot-Gandolphe, S., Durand, B., Jullian, S., and Valais, M., 1997, "Natural Gas Production processing and Transport,"*Editions Technip*, pp. 252–276.

Roth, S.H., 1993, "Hydrogen sulfide," in *Handbook of Hazardous Materials*, New York, NY, Academic Press.

Rudner, R., 1961, "An Introduction to Simplicity," *Philosophy of Science* 28 [2], pp. 109–119.

Rydh, C.J. and Sande, B.A., 2005, "Energy analysis of batteries in photovoltaic systems Part II: Energy return factors and overall battery efficiencies," *Energy Conversion and Management* 46, pp. 1980–2000.

Rywotycki, R., 2002, "The effect of fat temperature on heat energy consumption during frying of food," *Journal of food engineering*, 54, pp. 257–261.

Saastamoinen, J., Tuomaala, P., Paloposki, T., and Klobut, K., 2005, "Simplified dynamic model for heat input and output of heat storing stoves," *Applied Thermal Engineering*, 25 (17–18), pp. 2878–2890.

Saeed, N.O., Ajijolaiya, L.O., Al-Darbi, M.M., and Islam, M.R., 2003, "Mechanical properties of mortar reinforced with hair fibre," *Proc. Oil and Gas Symposium, CSCE Annual Conference*, Moncton.

Saeed, N.O., Al-Darbi, M.M., and Islam, M.R., 2003, Canadian Society for Civil Engineering 31st Annual Conference. Moncton, NB, Canada, paper code: GCR-535.

Saeed, N.O., Al-Darbi, M.M., and Islam, M.R., 2003, "Antibacterial Effects of Natural Materials On *Shewanella Puterfaciens*," *Proc. Oil and Gas*

Symposium, CSCE Annual Conference, refereed proceeding, Moncton, June.

Saha, S. and Chakma, A., 1992, "An energy effi cient mixed solvent for the separation of CO_2," *Energy Conversion Management*, 33, p. 413.

Sahimi, M., Mehrabi, A.R., Mirzaee, N., and Rassamdana, H., 2000, "The effect of asphalt precipitation on flow behavior and production of a fractured carbonate oil reservoir during gas injection," *Transport in Porous Media*, vol. 41, no. 3, Dec., pp. 325–347.

Sahimi, M., Rasasmdana, H., and Dabir, B., 1997, "Asphalt Formation and Precipitation: Experimental studies and Theoretical Modeling," *SPE J*, vol. 2, June, pp. 157–169.

Saito, K., Ogawa, M., Takekuma, M., Ohmura, A., Migaku Kawaguchi, A., Rie Ito, A., Koichi Inoue, A., Yasuhiko Matsuki, C., and Hiroyuki Nakazawa, 2005, "Systematic analysis and overall toxicity evaluation of dioxins and hexachlorobenzene in human milk," *Chemosphere*, Vol. 61, pp. 1215–1220.

Saka, S. and Kudsiana, D., 2001, "Methyl esterification of free fatty acids of rapeseed oil as treated in supercritical methanol," *J Chem Eng Jpn* , 34(3), pp. 373–387.

SAL, 2006, "Soil Acidity and Liming: Internet Inservice Training, Best Management Practices for Wood Ash Used as an Agricultural Soil Amendment," <http://hubcap.clemson.edu/~blpprt/bestwoodash.html> accessed June 7, 2006.

Salbu, B., Janssens, K., Lind, O.C., Proost, K., Gijsels, L., and Danesi, P.R., 2005, "Oxidation States of Uranium in Depleted Uranium Particles from Kuwait," *Journal of Environmental Radioactivity* 2005, 78, pp. 125–135.

Santagata, D., Sere, P., Elsner, C., and Di Sarli, A., 1998, "Evaluation of the Surface Treatment Effect of the Corrosion Performance of Paint Coated Carbon Steel," *Progress in Organic Coatings* 33, p. 44.

Schewe, P.F. and Stein, B., 1999, "Light has been Slowed to a Speed of 17 m/s," American Institute of Physics, *Bulletin of Physics News*, No. 415, February 18.

Schlesinger, G., 1959, "The Principle of Simplicity and Verifiability," *Philosophy of Science* 26 [1], pp. 41–42.

Schlomach, J., Quarch, K., and Kind, M., 2006, "Investigation of Precipitation of Calcium Carbonate at High Supersaturations," *Chem. Eng. Technol*, vol. 29, No. 2, pp. 215–219.

Schroeder, D.V., 2003, "Radiant Energy," online chapter for the course, *Energy, Entropy, and Everything*, Physics Department, Weber State University, accessed March 5, 2008, http://physics.weber.edu/schroeder/eee/chapter6.pdf.

Schuchardt, U., Sercheli, R., and Vargas, R.M., 1998, "Transesterification of vegetable oils: a review," *J Braz Chem Sco*, 9(1), pp. 199–210.

Seo, K-S., Han, C., Wee, J-H., Park, J-K., and Ahn, J-W., 2005, "Synthesis of Calcium Carbonate in a Pure Ethanol and Aqueous Ethanol Solution as the Solvent," *Journal of Crystal Growth*, vol. 276, pp. 680–687.

Sercu, B., Nunez, D., Langenhove, V.H., Aroca, G., and Verstraete, W., 2005, "Operational and microbiological aspects of a bioaugmented two-stage biotrickling filter removing hydrogen sulfide and dimethyl sulfide," *Biotechnology and Bioengineering* 90(2), pp. 259–269.

Service, R.F., 2005, "Is it time to shoot for the sun?", *Science*, 309, pp. 549–551.

Sh, A.M.M., 2006, Murugappa Chettiar Research Centre, Photosynthesis and Energy Division, Tharamani, Madras- 600 113, India.

Shastri, C.M., Sangeetha, G., and Ravindranath, N.H., 2002, "Dissemination of Efficient ASTRA stove: case study of a successful entrepreneur in Sirsi, India," *Energy for Sustainable Development*. Volume VI., No. 2.

Shaw, J., 2002, "The Global Experiment," *Harvard Magazine*, Nov-Dec, 2002.

Sheehan, J., Dunahay, T., Benemann, J., and Roessler, P., 1998, "A Look Back at the U.S. Department of Energy's Aquatic Species Program—Biodiesel from Algae," NREL/TP-580–24190.

Shekhovtsov, G.A. and Shekhovtsov, B.A., 1970, "The influence of the reflecting powers of on the ranges of optical instruments," *Journal of Mining Science*, 6 (1), pp. 122–123.

Shimada, H., 1996, "Study on supported binary sulfide catalysts for secondary hydrogenation of coal-derived liquids," *Fuel and Energy* Abstracts, Volume 37, Number 2, March, p. 94.

Shumaker, G.A., McKissick, J., Ferland, C., and Doherty, B., 2003, "A Study on the Feasibility of Biodiesel Production in Georgia, February 2003," *FR-03–02*, Center of Agribusiness and Economic Development, 26 pp.

Siegel, D.M., 1975, "Completeness as a Goal in Maxwell's Electromagnetic Theory," *Isis* 66 [3], pp. 361–368.

Simpson, T.K., 1966, "Maxwell and the Direct Experimental Test of His Electromagnetic Theory," *Isis* 57 [4], pp. 411–32.

Singh, K.J. and Sooch, S.S., 2004, "Comparative study of economics of different models of family size biogas plants for State of Punjab, India," *Energy Conv. and Mant*. 45, pp. 1329–1341.

Sircar, S., Novosad, J., and Myers, A.L., 1972, "Adsorption from Liquid Mixtures on Solids. Thermodynamics of excess Properties and Their Temperature Coefficients," *I & EC Fund.*, vol. 11, p. 249.

Sky radiance data, *Energy Conversion and Management* 45(11–12), pp. 1771–1783.??

Smalley, R.E., 2005, "Materials Matters," *MRS Bulletin*, Vol.30, www.mrs.org/ publications/bulletin.

Smart, J.S., 1997, *Journal of Protective Coatings and Linings*, February, p. 56.

Smith, P., 2001, "How green is my process? A practical guide to green metrics," *Proceedings of the Conference Green Chemistry on Sustainable Products and Processes*, 2001.

Sokolov, Y., 2006, "Uranium Resources: Plenty to Sustain Growth of Nuclear Power," IAEA/NEA Press Conference on Uranium Resources, Vienna, Austria.www.iaea.org/NewsCenter/Statements/DDGs/2006/sokolov01062006,html, June 1.

Sondi, I. and Sondi, B.S., 2005, "Influence of the Primary Structure of Enzymes on the Formation of CaCO3 Polymorphs: A Comparison of Plant *Canavalia ensiformis)* and Bacterial *Bacillus pasteurii)* Ureases," *Langmuir*, vol. 21, pp. 8876–8882.

Sondi, I., and Matijevic, E., 2001, "Homogeneous Precipitation of Calcium Carbonates by Enzyme Catalyzed Reaction," *Journal of Colloid and Interface Science*, vol. 238, pp. 208–214.

Sorokhtin, O.G., Chilingar, G.V., and Khilyuk, L.F., 2007, *Global Warming and Global Cooling, Evolution of Climate on Earth*, Developments in Earth & Environmental Sciences series. ISBN:978-0-444-52815-5, ISSN:1571-9197, 313 pp.

Spangenberg, J.H. and Bonniot, O., 1998, "Sustainability indicators-a compass on the road towards sustainability," Wuppertal Paper No. 81, February 1998, ISSN No. 0949–5266.

Speight, J.C., 1991, *The chemistry and technology of petroleum*, New York, M. Dekker, 760.

Spies, P.H., 1998, "Millenium Megatrends: Forces Shaping the 21st Century," Key-Note Address to the *Annual Conference of the International Association of Technological University Libraries* (IATUL), at the University of Pretoria, South Africa.

Sraffa, P., 1960, *Production of Commodities by Means of Commodities*, Cambridge, Cambridge University Press.

SRI, 2003, *Chemical Economics Handbook (CEH) Product Review: Mono-,Di- and Triethylene Glycols*, SRI International, Menlo Park, CA. November, 2003.

Srivastava, R.K. and Huang, S.S., 1995, "Technical Feasibility of CO_2 Loading in Weyburn Reservoir – A Laboratory Investigation," CIM. paper no. 95–1119 presented at the 6th Saskatchewan Petroleum Conference, Regina, Oct. 16–18.

Srivastava, R.K. and Huang, S.S., 1997, "Laboratory Investigation of Weyburn CO_2 Miscible flooding," 7th Saskatchewan Petroleum Conference, Regina, Oct. 19–22, C. I. M., paper no. 97–154.

Srivastava, R.K., Huang, S.S., Dyer, S.B., and Mourits, F.M., 1993, "A Scaled Physical Model for Saskatchewan Heavy Oil Reservoirs – Design, Fabrication and Preliminary CO_2 Flood Studies," 5th Petroleum Conference of South Saskatchewan Section, the Petroleum Society of CIM, Regina, Oct. 18–20.

Srivastava, R.K., Huang, S.S., Dyer, S.B., and Mourits, F.M., 1994, "Heavy oil recovery by sub-critical Carbon Dioxide Flooding," SPE paper 27058 presented at the III LACPEC, Buenos Aires, Argentina, April 26–29.

Srivastava, R.K., Huang, S.S., Dyer, S.B., and Mourits, F.M., 1995, "Measurement and Prediction of PVT properties of Heavy and Medium Oil with Carbon Dioxide," 6th UNITAR, International Conference on Heavy Crude and Tar Sands, Houston, Feb. 12–17.

St. Clair, Jeffrey, 2004, *Been Brown So Long It Looked Like Green to Me: The Politics of Nature*. Monroe ME, Common Courage Press. 408 pp.

Statistics Canada, 2006, "Canada's Population Clock," Statistics Canada, Demography Division. Updated in October 27, 2006,

Steenari, B.M. and Lindqvist, O., 1997, "Stabilisation of Biofuel Ashes for Recycling to Forest soil," *Biomass and Bioenergy*, Vol. 13 1–2, pp. 39–50.

Stefest, H., 1970, "Algorithm 368 Numerical Inversion of Laplace Transforms," *Commun. ACM*, vol. 13 (1), pp. 47–49.

Stewart, D.J., Mikhael, N.Z., Nanji, A.A., Nair, R.C., Kacew, S., Howard, K., Hirte, W., and Maroun, J.A., 1985, "Renal and hepatic concentrations of platinum: relationship to cisplatin time, dose, and nephrotoxicity," *J Clin Oncol.*, 1985 Sep, 3(9), pp. 1251–6.

Stewart, J.T. and Klett, M.G., 1979, "Converting coal to liquid/gaseous fuels," *Mechanical Engineering*, vol. 101, June, pp. 34–41.

Stock, J., 1776, "An Account of the Life of George Berkeley, D.D. Late Bishop of Cloyne in Ireland," available at the website: http://www.maths.tcd.ie/~dwilkins/Berkeley/Stock/Life.pdf, also see Berkeley, Bishop George, 1735, ***A Defence of Free Thinking in Mathematics*** (Dublin) {Last accessed 24 March 2008}

Straube, J.F., 2000, Moisture Properties of Plaster and Stucco for Strawbale Buildings, Report for Canada Mortgage and Housing Corporation, June 2000.

Struik, D.J., 1967, A Concise History of Mathematics, 3rd ed., Dover Publications, New York, 1967.

Subramanian, A.K., Singal, S.K., Saxena, M., and Singhal, S., 2005, "Utilization of liquid biofuels in automotive diesel engines: An Indian perspective," *Biomass and Bioenergy*, 9, pp. 65–72.

Sudaryanto, A., Kunisue, T., Kajiwara, N., Iwata, H., Adibroto, T.A., Hartono, P., and Tanabe, S., 2006, "Specific accumulation of organochlorines in human breast milk from Indonesia: Levels, distribution, accumulation kinetics and infant health risk," *Environmental Pollution*, Vol. 139, No. 1, pp. 107–117.

Sugie, H., Sasaki, C., Hashimoto, C., Takeshita, H., Nagai, T., Nakamura, S., Furukawa, Supple, B., Howard, H.R., Gonzalez, G.E., and Leahy, J.J., 1999, "The effect of steam treating waste cooking oil on the yield of methyl ester," *J. Am. Oil Soc. Chem.*, 79 (2), pp. 175–178.

Suskind, R., 2004, "Without a doubt," [Sunday] New York Times Magazine (17 October).

Sustainability Institute, 2007, "Two Approaches to Sewage Treatment and to the World," [online]Available:(http://www.sustainabilityinstitute.org/dhm_archive/search.php?display_article=vn177todded) [February 15, 2007].

Sweis, F.K., 2004, "The effect of admixed material on the flaming and smouldering combustion of dust layers," *Journal of Loss Prevention in the Process Industries*, vol. 17 (6), pp. 505–508.

Syed, M., Soreanu, G., Falletta, P., and Beland, M., 2006, "Removal of hydrogen sulfide from gas streams using biological processes-A review," *Canadian Biosystem Engineering*, 48, pp. 2.1–2.14.

Szklo, A. and Schaeffer, R., 2007, "Fuel specification, energy consumption and CO_2 emission in oil refineries," *Energy* 2007, 32, pp. 1075–1092.

Szokolik, A., 1992, "Evaluating Single-Coat Inorganic Zinc Silicates for Oil and Gas Production Facilities in Marine Environment," *Journal of Protective Coatings and Linings*, March, p. 24.

Szostak-Kotowa, J., 2004, "Biodeterioration of textiles," *International Biodeterioration & Biodegradation*, Vol. 53, pp. 165 – 170.

Taber, J.J., 1988, "The Use of Flue Gas for Enhanced Recovery of Oil," EOR by Gas Injection, Symposium, International Energy Agency Collaborative Research Program on EOR, Copenhagen, Denmark, September 14.

Taber, J.J., 1990, "Environment, Improvements and Better Economics in EOR Operations," *In Situ*, vol. 14 no. 4, pp. 345–404.

Taber, J.J., 1994, "A Study of Technical Feasibility for the Utilization of CO_2 for Enhanced Oil Recovery," The Utilization of Carbon Dioxide from Fossil Fuel Fired Power Stations, IEA Greenhouse Gas R & D program, Cheltenham, England.

Tanaka, S., Koide, H., and Sasagawa, A., 1994, "Possibility of CO_2 underground sequestration in Japan," *Energy Conversion Management*, vol. 36, pp. 527–530.

Tang, D.E. and Peaceman, D.W., 1987, "New Analytical and Numerical Solutions for the Radial Convection Dispersion Problems," SPE 16001 presented at the Ninth SPE Symposium on Reservoir Simulation, San Antonio, TX, Feb.

Tanner, D., 1995, "Ocean Thermal energy Conversion: Current Overview and Future Outlook," *Renewable Energy* 6(3), pp. 367–373.

Tate, R.E., Watts, K.C., Allen, C.A.W., and Wilkie, K.I., 2006, "The densities of three biodiesel fuels at temperatures up to 300°C," *Fuel* 85, pp. 1004–1009.

Tator, K.B., 1977, "How Coatings Protect and Why they Fail," *Corrosion 77*, NACE, paper no. 4, p. 1.

Taylor, A.J.P., 1961, *The Origins of the Second World War*, London, Atheneum, 2nd Ed.

Teel, D., 1994, Liquid fuel solutions of methane and liquid hydrocarbons, US Patent 5315054.

Tester, J.W, Drake, E.M., Golay, M.W., Driscoll, M.J., and Peters, W.A., 2005, *Sustainable Energy, Choosing Among Options*, The MIT Press, Cambridge, Massachusetts, London, England, pp 864, 2005.

The Epoch Times, 2006, "Potato Farms a Hot Bed for Cancer," March 24–30, www.theepochtimes.ca.

The Globe and Mail, 2006, "Toxic shock: Canada's Chemical reaction," May 27, Saturday, 2006.

The New York Times, 2006, "Citing Security, Plants Use Safer Chemicals," April, 25, 2006.

Thibodeau, L., Sakanoko, M., and Neale, G. H., 2003, "Alkaline Flooding Processes in Porous Media in the Presence of Connate Water," *Powder Technology*, Vol. 32, pp. 101–111.

Thipse, S.S., Schoenitz, M., and Dreizin, E.L., 2002, "Morphology and composition of the fly ash particles produced in incineration of municipal solid waste," *Fuel Processing Technology*, Vol. 75(3), pp. 173–184.

Thomas, P. and Nowak, M.A., 2006, "Climate Change: All in the Game," *Nature* (441) June 1, 2006.

Thorpe, T.W., 1998, "Overview of Wave Energy Technologies," AEAT-3615 for the marine foresight panel, May 1998.

Tickell, J., 2003, "From the fryer to the fuel tank: The complete guide to using vegetable oil as an alternative fuel." , www.electricitybook.com/fryer

Tilley, J., 1997, "Technology Responses to Global Climate Change Concerns: The Benefits from International Collaboration," *Energy Conversion Management*, volume 38, S3-S12.

Tipton, T., Johnston, C.T., Trabue, S.L., Erickson, C., and Stone, D.A., 1993, "Gravimetric/FT-IR Apparatus for the Study of Vapor Sorption on Clay Films," *Rev.Sci.Instrum* 64 (4), pp. 1091–1092.

Tiwari, G.N., 2002, *Solar energy: fundamentals, design, modelling and application*, Narosa Publishing House, New Delhi, India.

Toninello, A., Pietrangeli, P., De Marchi, U., Salvi, M., and Mondov, B., 2006, "Amine Oxidases in Apoptosis and Cancer," *Biochimica et Biophysica Acta* 1765, pp. 1–13.

Tornquist, C., 1997, "Nuclear Fusion Still No Dependable Energy Source," *CNN News*, April 5, www.cnn.com/US/9704/05/fusion.confusion/ (accessed on Jan 16, 07).

Trujillo, E.M., 1983, "The Static and Dynamic Interracial Tensions between Crude Oils and Caustic Solutions," *SPEJ*, 645.

Tschulakow, A.V., Yan, Y., and Klimek, W., 2005, "A New Approach to the Memory of Water," *Homeopathy* 94 (4), pp. 241–247.

Tsoutsos, T., Frantzeskaki, N., and Gekas, V., 2005, "Environmental impacts from the solar energy technologies," *Energy Policy* 33, pp. 289–296.

Tulloch, A.P., 1970, "The composition of beeswax and other waxes secreted by insects," *Lipids*, volume 5, no. 2, pp. 247–258.

Turkenberg, WIM C., 1997, "Sustainable Development, Climate Change, and Carbon Dioxide Removal," *Energy Conversion Management*, vol. 38, S3-S12.

Twu, C.H, Tassone, V., Sim, W.D., and Watanasiric, S., 2005, "Advanced Equation of State Method for Modeling TEG–Water for Glycol Gas Dehydration," *Fluid Phase Equilibria* 228–229, pp. 213–221.

UNCSD (United Nations Commission on Sustainable Development), 2001, Indicators of Sustainable Development: Guidelines and Methodologies, United Nations, New York.

United Kingdom Offshore Operations Association, Atmospheric Emission (UK OOA), 2003, Available http://www.ukooa.co.uk/issues/1999report/enviro99_atmospheric.htm, March 11, 2003.

United States Coast Guard, 1990, *Update of inputs of petroleum hydrocarbons into the oceans due to marine transportation activities*, National Research Council, National Academy Press, Washington, D.C., 1990.

Uranium Enrichment, 2006, Nuclear Issues Briefing Paper 33, March 2006, Uranium Information Centre Ltd, GPO Box 1649N, Melbourne 3001, Australia, 2006.

Uranium Information Center, 2006, The Economics of Nuclear Power, Briefing Paper no 8, Australia, http://www.uic.com.au/nip08.htm <accessed on June 05, 2006>

U.S. BoLS (Bureau of Labor Statistics), 2006, *Consumer Price Index*, Washington DC, January 18.

U.S. DoE, 2004, *Annual Energy Review 2004*, p 98.

USDoE, 2006, <http://www.fossil.energy.gov/programs/oilgas/eor/index.html>, US Department of Energy, accessed April 06, 2006.

USHR, 1999, Oil refineries fail to report millions of pounds of harmful emissions. A report prepared for Rep. Henry A. Waxman, by minority staff, special investigation division, committee on government reforms, U.S.House of Representative, 10th November, pp.19.

Vallejo, F., Tomas-Barberan, F.A., and Garcıa-Viguera, C., 2003, "Phenolic compound contents in edible parts of broccoli inflorescences after domestic cooking," *J. of the Science of Food and Agriculture* 83, pp. 1511–1516.

van der Meer, L.G.H., 1992, "Investigation Regarding the Storage of Carbon Dioxide in Aquifers in the Netherlands," *Energy Conversion Management*, volume 33, no. 5–8, pp. 611–618.

van der Meer, L.G.H., 1995, "The CO_2 storage efficiency of Aquifers," *Energy Conversion and Management*, volume 36, nos. 6–9, pp. 513–518.

Van Niel, C.B., 1931, "On the morphology and physiology of the purple and green sulfur bacteria," *Archiv fur Mikrobiologie* 3, pp. 1–112.

Vassilev, S.V. and Vassileva, C.G., 2005, "Methods for characterization of composition of fly ashes from coal-fired power stations: A critical overview," *Energy Fuels*, Vol. 19, pp. 1084–98.

Veil, J.A., 2002, "Drilling Waste Management: past, present and future," Annual Technical Conference and Exhibition, San Antonio, Texas, 29 Septermber-2 October, 2002, SPE paper no. 77388.

Venkataraman, C., Joshi, P., Sethi, V., Kohli, S., and Ravi, M.R., 2004, *Aerosol Science and Technology*, vol. 38, no. 1, pp. 50–61.

Vikram, V.B., Ramesh, M.N., and Prapulla, S.G., 2005, "Thermal degradation kinetics of nutrients in orange juice heated by electromagnetic and conventional methods," *Journal of Food Engineering* 69(1), pp. 31–40.

Volckova, E., Evanics, F., Yang, W.W., and Bose, R.N., 2003, "Unwinding of DNA polymerases by the antitumor drug,*cis*-diamminedichloroplatinum(II)," *Chem. Commun.*, 2003, pp. 1128–1129.

Voss, A., 1979, "Waves Currents, Tides-Problems and Prospects," *Energy* 4 (5), pp. 823–831.

Wackernagel, M. and Rees, W., 1996, *Our ecological footprint*, Gabriola Island, New Society Publishers.

Wangnick, K., 2002, IDA Worldwide Desalting Plants Inventory Report No.17, Wangnick Consulting GmbH and the International Desalination Association (IDA), Vienna, July.

Wareham, S., 2006, The Health Impacts of Nuclear Power, Nuclear Power Forum, UNSW, October 18, 2006, Medical Association for Prevention of War. www.mapw.org.au.

Wasiuddin, N.M., Ali, N., and Islam, M.R., 2002b, "Use of Offshore Drilling Waste in Hot Mix Asphalt (HMA) Concrete as Aggregate Replacement," paper no. EE 29168, ETCE '02, Feb. 4–6, 2002, Houston, Texas.

Wasiuddin, N.M., Tango, M., and Islam, M.R., 2002, "A Novel Method for Arsenic Removal at Low Concentrations," *Energy Sources* 24, pp. 1031–1041.

Waste Online, 2005, "Plastic recycling information sheet," <http://www.wasteonline.org.uk/ resources/InformationSheets/Plastics.htm> [Accessed: February 20, 2006].

WCED (World Commission on Environment and Development) 1987, "Our common future," *World Conference on Environment and Development*, Oxford, Oxford University Press; 1987, 400pp.

Website 1: http://linguisticmystic.com/2007/07/09/what-do-assassins-and-sofas-have-in-common-english-words-with-arabic-origins/

Website 1: http://en.wikipedia.org/wiki/Two_New_Sciences (last accessed 1 February 2008)

Website 2: http://en.wikipedia.org/wiki/Galileo_Galilei

Website 3a: Newton, Isaac, 1687, ***Philosophiae naturalis principia mathematica*** [1729 translation by Andrew Motte], Ch 1. "Of the method of first and last ratios of quantities, by the help whereof we demonstrate the propositions that follow" {Last accessed 24 March 2008}

Website 3b: Newton, ***op.cit.***, Ch. 2 "Of the invention of centripetal forces"{Last accessed 24 March 2008}
Website 3c: Newton, ***op.cit.***, Ch. 3 "Of the motion of bodies in eccentric conic sections" {Last accessed 24 March 2008}
Website 3d: Newton, ***op.cit.***, Ch. 4 "Of the finding of elliptic, parabolic, and hyperbolic orbits, from the focus given" {Last accessed 24 March 2008}
Website 3e: Newton, ***op.cit.***, "Axioms, or Laws of Motion" (precedes Ch. 1 of Book I) {Last accessed 24 March 2008}
Website 4: http://www.emachineshop.com/engine/
Website 5: British plastic federation, "The history of plastic," Available online at: http://www.bpf.co.uk/bpfindustry/History_of_Plastics.cfm Accessed November 15th, 2005.
Website 6: www.wasteonline.org.uk/resources/InformationSheets/Plastics.htm [Accessed: May 12, 2006].
Website 7: http://www.solcomhouse.com/recycling.html (Last accessed, Feb. 20, 2010)
Website8:http://www.americanchemistry.com/s_plastics/doc.asp?CID=1102&DID=4664 Accessed on November 20th, 2005.
Website 9: available at: http://www.plastics.ca/news/default.php?id=197 Accessed on February 2nd, 2007
Website 10: American chemistry council, Available online at: http://www.americanchemistry.com/plastics/ Accessed November 10th, 2005.
Website11:http://invsee.eas.asu.edu/nmodules/engmod/manipulation.html, Accessed on September 4th, 2006.
Website 12: http://www.plasticsresource.com/s_plasticsresource/ Accessed on November 10th, 2005.
Website 13a: http://www.pslc.ws/mactest/natupoly.htm Accessed on July 12th , 2005.
Website 14: http://www.psigate.ac.uk/roads/cgibin/
Website 15: (http://www.fpl.fs.fed.us/documnts/techline/fuel-value-calculator.pdf, http://www.epa.gov/ ttn /chief/ ap42/ch01/final/c01s04.pdf).
Website 16: (http://www.hrt.msu.edu/Energy/Notebook/pdf/Sec4/Approximate_Heating_Values_by_ %20Bartok.pdf)
Website 17: (http://www.etc-cte.ec.gc.ca/databases/OilProperties/oil_prop_e.html)
Website 18: http://www.mindfully.org/Plastic/Ethylene-Gas.htm
Website 25: http://unfcc.int/resource/docs/2009/cop15/eng/l07.pdf
Website 1a: www.cameco.com/sustainable_development/ cleanenvironment/act.php.
Website 1b: www.oakdenehollins.co.uk/ [Accessed: May 03, 2006].
Website 10a: http://encarta.msn.com/media_461531189/Oil_Refining_and_Fractional_Distillation.html (accessed on May 20, 2008).

Website 11a: http://www.tribecaradio.net/blog/categories/stealThis Radio/

Website 13a: Balanced Solution.com. Moisture Properties of Plaster and Stucco for Strawbale Buildings, www.ecobuildnetwork.org/pdfs/ Straube_Moisture_Tests.pdf (accessed on 8th Aug, 20).

Website 14:a www.hemptons.co.za/Users/seeds/htm (accessed on December 15, 2006)

Website 15a: http://hyperphysics.phy-astr.gsu.edu/hbase/ems3.html (accessed on November 5, 2006)

WEC, 2006, "The World Energy Council: How to Avoid a Billion Tones of CO_2 Emission," http://www.worldenergy.org/wec-geis/default.asp <accessed on May 30, 2006>.

Welford, R., 1995, *Environmental strategy and sustainable development: the corporate challenge for the 21st Century*, London, Routledge.

Wenger, L.M., Davis, C.L., Evensen, J.M., Gormly, J.R., and Mankiewicz, P.J., 2004, "Impact of modern deepwater drilling and testing fluids on geochemical evaluations," *Organic Geochemistry*, Vol. 35, pp. 1527–1536.

Weyl, H., 1944, "How Far Can One Get With a Linear Field Theory of Gravitation in Flat Space-Time?," *American Journal of Mathematics* 66 [4], pp. 591–604.

Wiener, P.P., 1943, "A Critical Note on Koyré's Version of Galileo," *Isis*, Vol. 34, No. 4. (Spring, 1943), pp. 301–302.

Wikipedia, 2008, http://en.wikipedia.org/wiki/Lemon_juice, accessed Nov., 2008.

Williams, L.P., 1965, *Michael Faraday: A Biography*, London, Chapman & Hall. xvi, 531 pp.

Wills, J., Shemaria, M., and Mitariten, M.J., 2004, "Production of Pipeline Quality Natural Gas," SPE 87644, *SPE/EPA/DOE Exploration and Production Environmental Conference*, San-Antonio-Texas, 10–12 March 2003.

Winter, E.M. and Bergman, P.D., 1996, "Potential for Terrestrial Disposal of Carbon Dioxide in the US," US/Japan Joint Technical Workshop, US Dept. of Energy, State College, PA, Sept. 30–Oct. 2.

Winterton, N., 2001, "Twelve more green chemistry principles," *Green Chem* Vol. 3, G73–5.

Wise Uranium Project, 2005, "Uranium Radiation Properties," www.wise-uranium.org/ rup.html, (accessed on March 19, 2006).

Wittwer, R.F. and Immel, M.J., 1980, "Chemical composition of five deciduous tree species in four-year-old closely spaced plantations," *Plant and Soil*, vol. 54, no. 3, Oct., pp. 461–467.

WNA (World Nuclear Association), 2010, http://www.world-nuclear.org/info/inf63.html, last viewed Feb. 22, 2010.

Woodruff, A.E., 1968, "The Contributions of Hermann von Helmholtz to Electrodynamics," *Isis* 59 [3], pp. 300–311.

World Health Organization (WHO), 1994, "Brominated diphenyl ethers," *Environmental Health Criteria*, Vol.162, International Program on Chemical Safety.

Wright, T., 2002, "Definitions and frameworks for environmental sustainability in Higher education," *International Journal of Sustainability In Higher Education Policy*, Vol. 15, (2).

Wu, H., Zong, M.H., Luo, Q., and Wu, H.C., 2003, "Enzymatic conversion of waste oil to biodiesel in a solvent free medium," Prepr. Pap.-Am. Chem. Soc., Div. Fuel Chem. 48 (2) 533.

Xiaoling, M. and Qingyu, W., 2006, "Biodiesel production from heterotrophic microalgal oil," *Bioresource Technology*, vol.97, (6), pp. 841–846

Yang, H-H., Chien, S-M., Lo, M-Y., Lan, J.C.-W., Lu, W-C., and Ku, Y-Y., 2007, "Effects of biodiesel on emissions of regulated air pollutants and polycyclic aromatic hydrocarbons under engine durability testing," *Atmospheric Environment* 41, pp. 7232–7240.

Yen, T.F., Preface, *True Sustainability in Technological Development and Natural Resource Management*, Nova Science Publishers, NY, 381 pp.

York, M., 2003, "One Spoonful of Bee Pollen Each Day, and You, Too, Might Make It to 113," *The New York Times*, December, 2003 <accessed on June 06, 2006>.

Yu, J., Lei, M., Cheng, B., and Zhao, X., 2004, "Facile Preparation of Calcium Carbonate Particles with Unusual Morphologies by Precipitation Reaction," *Journal of Crystal Growth*, vol. 261, pp. 566–570.

Zatzman, G., 2007, "The Honey $\rightarrow$ Sugar $\rightarrow$ Saccharin™ $\rightarrow$ Aspartame™ Syndrome: A Note," *Journal of Nature Science and Sustainable Technology*, vol. 1, no. 3, pp. 397–401.

Zatzman, G.M., 2008, "Some Inconvenient Truths About Al Gore's Inconvenient Truth," *J. Nat.Sci. and Sust.Tech.*, vol. 1, no. 4, pp. 699–707.

Zatzman, G.M., Chhetri, A.B., Khan, M.M., Al-Maamari, R., and Islam, M.R., 2007, "Colony collapse disorder: the case for a science of intangibles," *J. Charact. Dev. Novel Mat.*, vol. 1, no. 2.

Zatzman, G., Chhetri, A.B., Khan, M.M., Maamari, R., and Islam, M.R., 2008, "Colony Collapse Disorder- The Case for a Science of Intangibles," *Journal of Nature Science and Sustainable Technology*, vol. 2, no. 3.

Zatzman, G.M. and Islam, M.R., 2006, "Natural Gas Energy Pricing," Chapter 2 in Mokhatab, S., Speight, J.G., and Poe, W.A., eds., *Handbook of Natural Gas Transmission and Processing*, Gulf Professional Publishing, Elsevier.

Zatzman, G.M. and Islam, M.R., 2007, "Truth, Consequences and Intentions: The Study of Natural and Anti-Natural Starting Points and Their Implications," *J. Nature Science and Sustainable Technology*, vol. 1, no. 2, pp. 169–174.

Zatzman, G.M. and Islam, M.R., 2007, *The Economics of Intangibles*, Nova Science Publishers, New York, 407 pp.

Zatzman, G.M., Khan, M.M., Chhetri, A.B., and Islam, M.R., 2008, "A Delinearized History Of Time And Its Roles In Establishing And Unfolding Knowledge Of The Truth," *Journal of Information, Intelligence and Knowledge*, vol. 1, no. 1, pp. 1–38.

Zero-Waste, 2005, "The Case for Zero Waste," <http://www.zerowaste.org/> [Accessed on August 12, 2006].

Zevenhoven, R. and Kohlmann, J., 2001, "CO_2 sequestration by magnesium silicate mineral carbonation in Finland," *Second Nordic Minisymposium on Carbon Dioxide Capture and Storage*, Goteborg, October 26, page 13–18.

Zhang, Y., Dube, M.A., McLean, D.D., and Kates, M., 2003, "Biodiesel Production from Waste Cooking Oil: 1. Process Design and Technological Assessment," *Bioresour. Technol.*, 89, pp. 1–16.

Zheng, S., Kates, M., Dube, M.A., and McLean, D.D., 2006, "Acid-catalyzed production of biodiesel from waste frying oil," *Biomass and bioenergy* 30, pp. 267–272.

Zick, A.A., 1986, "A Combined Condensing/Vapourizing Mechanism in the Displacement of Oil By Enriched Gases," SPE paper 15493 presented at the 61st SPE Technical Meeting, New Orleans, LA, October 5–8.

Zucchetti, M., 2005, "The zero-waste option for nuclear fusion reactors: Advanced fuel cycles and clearance of radioactive materials," Technical note, *Annals of Nuclear Energy*, 32, pp. 1584–1593.

Index

Also of Interest

Check out these other related titles from Scrivener Publishing

Zero-Waste Engineering, by Rafiqul Islam, ISBN 9780470626047. In this controversial new volume, the author explores the question of zero-waste engineering and how it can be done, efficiently and profitably. *NOW AVAILABLE!*

Sustainable Energy Pricing, by Gary Zatzman, ISBN 9780470901632. In this controversial new volume, the author explores a new science of energy pricing and how it can be done in a way that is sustainable for the world's economy and environment. *NOW AVAILABLE!*

Flow Assurance, by Boyun Guo and Rafiqul Islam, January 2013, ISBN 9780470626085. Comprehensive and state-of-the-art guide to flow assurance in the petroleum industry.

An Introduction to Petroleum Technology, Economics, and Politics, by James Speight, ISBN 9781118012994. The perfect primer for anyone wishing to learn about the petroleum industry, for the layperson or the engineer. *NOW AVAILABLE!*

Ethics in Engineering, by James Speight and Russell Foote, ISBN 9780470626023. Covers the most thought-provoking ethical questions in engineering. *NOW AVAILABLE!*

Formulas and Calculations for Drilling Engineers, by Robello Samuel, ISBN 9780470625996. The most comprehensive coverage of solutions for daily drilling problems ever published. *NOW AVAILABLE!*

Emergency Response Management for Offshore Oil Spills, by Nicholas P. Cheremisinoff, PhD, and Anton Davletshin,

ISBN 9780470927120. The first book to examine the Deepwater Horizon disaster and offer processes for safety and environmental protection. *NOW AVAILABLE!*

Advanced Petroleum Reservoir Simulation, by M.R. Islam, S.H. Mousavizadegan, Shabbir Mustafiz, and Jamal H. Abou-Kassem, ISBN 9780470625811. The state of the art in petroleum reservoir simulation. *NOW AVAILABLE!*

Energy Storage: A New Approach, by Ralph Zito, ISBN 9780470625910. Exploring the potential of reversible concentrations cells, the author of this groundbreaking volume reveals new technologies to solve the global crisis of energy storage. *NOW AVAILABLE!*